Exclusive Papers of the Editorial Board Members (EBMs) of the Materials Chemistry Section of Molecules

Exclusive Papers of the Editorial Board Members (EBMs) of the Materials Chemistry Section of Molecules

Editor

Giuseppe Cirillo

MDPI • Basel • Beijing • Wuhan • Barcelona • Belgrade • Manchester • Tokyo • Cluj • Tianjin

Editor
Giuseppe Cirillo
Department of Pharmacy, Health
and Nutritional Sciences
University of Calabria
Rende
Italy

Editorial Office
MDPI
St. Alban-Anlage 66
4052 Basel, Switzerland

This is a reprint of articles from the Special Issue published online in the open access journal *Molecules* (ISSN 1420-3049) (available at: www.mdpi.com/journal/molecules/special_issues/ Materials_Chemistry_papers_fromEBMs).

For citation purposes, cite each article independently as indicated on the article page online and as indicated below:

LastName, A.A.; LastName, B.B.; LastName, C.C. Article Title. *Journal Name* **Year**, *Volume Number*, Page Range.

ISBN 978-3-0365-2230-2 (Hbk)
ISBN 978-3-0365-2229-6 (PDF)

© 2021 by the authors. Articles in this book are Open Access and distributed under the Creative Commons Attribution (CC BY) license, which allows users to download, copy and build upon published articles, as long as the author and publisher are properly credited, which ensures maximum dissemination and a wider impact of our publications.
The book as a whole is distributed by MDPI under the terms and conditions of the Creative Commons license CC BY-NC-ND.

Contents

About the Editor vii

Preface to "Exclusive Papers of the Editorial Board Members (EBMs) of the Materials Chemistry Section of Molecules" ix

Marta A. Andrade and Luísa M. D. R. S. Martins
New Trends in C–C Cross-Coupling Reactions: The Use of Unconventional Conditions
Reprinted from: *Molecules* **2020**, *25*, 5506, doi:10.3390/molecules25235506 1

Daniel García Velázquez, Rafael Luque and Ángel Gutiérrez Ravelo
Microwave-Assisted Synthesis and Properties of Novel Hexaazatrinaphthylene Dendritic Scaffolds
Reprinted from: *Molecules* **2020**, *25*, 5038, doi:10.3390/molecules25215038 33

Nikolaos Chalmpes, Konstantinos Spyrou, Konstantinos C. Vasilopoulos, Athanasios B. Bourlinos, Dimitrios Moschovas, Apostolos Avgeropoulos, Christina Gioti, Michael A. Karakassides and Dimitrios Gournis
Hypergolics in Carbon Nanomaterials Synthesis: New Paradigms and Perspectives
Reprinted from: *Molecules* **2020**, *25*, 2207, doi:10.3390/molecules25092207 45

Maxime Balestrat, Abhijeet Lale, André Vinícius Andrade Bezerra, Vanessa Proust, Eranezhuth Wasan Awin, Ricardo Antonio Francisco Machado, Pierre Carles, Ravi Kumar, Christel Gervais and Samuel Bernard
In-Situ Synthesis and Characterization of Nanocomposites in the Si-Ti-N and Si-Ti-C Systems
Reprinted from: *Molecules* **2020**, *25*, 5236, doi:10.3390/molecules25225236 57

Carlo Nazareno Dibenedetto, Teresa Sibillano, Rosaria Brescia, Mirko Prato, Leonardo Triggiani, Cinzia Giannini, Annamaria Panniello, Michela Corricelli, Roberto Comparelli, Chiara Ingrosso, Nicoletta Depalo, Angela Agostiano, Maria Lucia Curri, Marinella Striccoli and Elisabetta Fanizza
PbS Quantum Dots Decorating TiO_2 Nanocrystals: Synthesis, Topology, and Optical Properties of the Colloidal Hybrid Architecture
Reprinted from: *Molecules* **2020**, *25*, 2939, doi:10.3390/molecules25122939 81

Piersandro Pallavicini, Lorenzo De Vita, Francesca Merlin, Chiara Milanese, Mykola Borzenkov, Angelo Taglietti and Giuseppe Chirico
Suitable Polymeric Coatings to Avoid Localized Surface Plasmon Resonance Hybridization in Printed Patterns of Photothermally Responsive Gold Nanoinks
Reprinted from: *Molecules* **2020**, *25*, 2499, doi:10.3390/molecules25112499 99

Jelena Bijelić, Dalibor Tatar, Sugato Hajra, Manisha Sahu, Sang Jae Kim, Zvonko Jagličić and Igor Djerdj
Nanocrystalline Antiferromagnetic High- Dielectric Sr_2NiMO_6 (M = Te, W) with Double Perovskite Structure Type
Reprinted from: *Molecules* **2020**, *25*, 3996, doi:10.3390/molecules25173996 117

Maria Luisa De Giorgi, Stefania Milanese, Argyro Klini and Marco Anni
Environment-Induced Reversible Modulation of Optical and Electronic Properties of Lead Halide Perovskites and Possible Applications to Sensor Development: A Review
Reprinted from: *Molecules* **2021**, *26*, 705, doi:10.3390/molecules26030705 135

Hossein Arshid, Mohammad Khorasani, Zeinab Soleimani-Javid, Rossana Dimitri and Francesco Tornabene
Quasi-3D Hyperbolic Shear Deformation Theory for the Free Vibration Study of Honeycomb Microplates with Graphene Nanoplatelets-Reinforced Epoxy Skins
Reprinted from: *Molecules* **2020**, *25*, 5085, doi:10.3390/molecules25215085 . . . . **177**

Sergey Vyazovkin
Kissinger Method in Kinetics of Materials: Things to Beware and Be Aware of
Reprinted from: *Molecules* **2020**, *25*, 2813, doi:10.3390/molecules25122813 . . . . **199**

Antonella Curulli
Nanomaterials in Electrochemical Sensing Area: Applications and Challenges in Food Analysis
Reprinted from: *Molecules* **2020**, *25*, 5759, doi:10.3390/molecules25235759 . . . . **217**

Inbar Schlachet, Hen Moshe Halamish and Alejandro Sosnik
Mixed Amphiphilic Polymeric Nanoparticles of Chitosan, Poly(vinyl alcohol) and Poly(methyl methacrylate) for Intranasal Drug Delivery: A Preliminary In Vivo Study
Reprinted from: *Molecules* **2020**, *25*, 4496, doi:10.3390/molecules25194496 . . . . **253**

Giuseppe Cirillo, Elvira Pantuso, Manuela Curcio, Orazio Vittorio, Antonella Leggio, Francesca Iemma, Giovanni De Filpo and Fiore Pasquale Nicoletta
Alginate Bioconjugate and Graphene Oxide in Multifunctional Hydrogels for Versatile Biomedical Applications
Reprinted from: *Molecules* **2021**, *26*, 1355, doi:10.3390/molecules26051355 . . . . **265**

About the Editor

Giuseppe Cirillo

Dr. Giuseppe Cirillo received his PhD in Methodologies for the Development of Molecules of Pharmaceutical Interest at University of Calabria, Italy, in 2008, with subsequent 6-year postdoctoral fellows experiences between University of Calabria, Italy, and Leibniz Institute for Solid State and Materials Research Dresden, Germany. From 2010 to 2013 he was co-founder and CEO of University of Calabria spin-off company. He got the Italian National Scientific Qualification for associate professorship in Drug Technology, Socioeconomics and Regulations and Principles of Chemistry for Applied Technologies. He is currently researcher at the Department of Pharmacy, Health and Nutritional Sciences, University of Calabria, Italy. He has co-authored 118 scientific journal publication.

Preface to "Exclusive Papers of the Editorial Board Members (EBMs) of the Materials Chemistry Section of Molecules"

This book covers the recent contributions to the development of the "Materials Chemistry" research fields by the Editorial Board Members of the "Materials Chemistry" Section of Molecules in 2020.

The "Materials Chemistry" Section of Molecules, taking advantages from knowledge in chemistry, biotechnology, chemical engineering, physics, and materials science, aims to be an open access place for the dissemination of theoretical and experimental studies related to the chemical approaches to materials-based problems.

The Materials Section benefits from the cooperation of 137 Editorial Board Members, whose scientific contributions highlight the importance of an interdisciplinary environment where the different expertise act synergistically to focus the progress in the field of creating and manipulating new materials.

In this book, 13 papers (either research and review article) dealing with some of facet of Materials Chemistry are collected. The aim is to provide readers with a vision of the relevance of papers covering studies related to both the synthesis and characterization of organic and inorganic materials.

Some of the collected papers focused on synthetic approach to materials science.

Andrade et al. provided an overview of the 2015-2020 literature works covering the implementation of unconventional methodologies in carbon–carbon (C–C) cross-coupling reactions, including the use of alternative energy sources to solvent- free and green media protocols.

García Velázquez et al. described a microwave-assisted synthetic approach for the fabrication of water-soluble π-conjugated hexaazatrinaphthylenes-based dendritic architectures constructed by hexaketocyclohexane and 1,2,4,5-benzenetetramine units.

Chalmpes et al. presented the preparation of carbon nanosheets and fullerols through spontaneous ignition processes where coffee and fullerenes played the role of the combustible fuel, whereas sodium peroxide acted as strong oxidizer.

Balestrat et al. investigated the pyrolysis of a liquid poly(vinylmethyl-co-methyl)silazane modified by tetrakis(dimethylamido)titanium in flowing ammonia, nitrogen and argon followed by the annealing of as-pyrolyzed ceramic powders for the synthesis of Si-Ti-N and Si-Ti-C nanocomposite systems.

Dibenedetto et al. focused the attention of their research on TiO_2/PbS heterostructures, interesting materials for applications in different fields including catalysis, biomedicine, and energy conversion. Authors investigated the synthesis as a function of the experimental conditions and the heterostructures in terms of topology, structural properties, and optical properties

Pallavicini et al. examined different polymeric coatings for spherical gold nanoparticles, with the aim of maintaining their spectral stability both in liquid inks and in dry prints, key feature for their use in secure writing of photothermally readable information.

Bijelić et al. synthesized perovskites with Sr_2NiMO_6 (M = Te, W) structure using aqueous citrate sol-gel route. The work aimed to address the great interest on double perovskites properties, which is related to the coexistence of ferro/ferri/antiferro-magnetic ground state and semiconductor band gap within the same material.

Other contributions are related on the theoretical and experimental characterization of materials properties.

To this regard, De Giorgi et al. focused their research on the comprehension of the environmental effects on the optical and electronic properties of lead halide perovskites, with the ultimate aim to describe the state of the art for development of perovskite-based sensors.

Arshid et al. employed a quasi-3D hyperbolic shear deformation theory, together with the Hamilton's principle and the modified couple stress theory, to analyze the vibrational behavior of rectangular micro-scale sandwich plates resting on a visco-Pasternak foundation. Authors aimed to provide new insights many modern engineering applications and their optimization design.

Vyazovkin's paper is focused on the applications of the Kissinger method to estimate the activation energy of different processes, including chemical reactions, crystallization and glass transition. Author gave a theoretical discussion to explain the origins of the complex temperature dependence of the respective rate, and then applied the Kissinger method to some available experimental data.

Finally, biomedical application of materials is the topic of the last three papers of the book.

In the first of them, Curulli made an overview on the application of different nanomaterials and nanocomposites with tailored morphological properties as sensing platforms for food analysis, with particular attention to the sensors based on carbon-based nanomaterials, metallic nanomaterials, and related nanocomposites

Schlachet et al. aimed to evaluate the biodistribution of mixed amphiphilic mucoadhesive nanoparticles made of chitosan-g-poly(methyl methacrylate) and poly(vinyl alcohol)-g-poly(methyl methacrylate) and ionotropically crosslinked with sodium tripolyphosphate in the brain after intravenous and intranasal administration.

The biomedical part of the book is closed by Cirillo et al., who provided evidence that a multifunctional hybrid hydrogel made of a caffeic-acid-sodium alginate bioconjugate and graphene oxidecan act as a platform to either provide the electro-responsive release of biologically active molecules such as Lysozyme or facilitate its crystallization under oxidative stress.

Giuseppe Cirillo
Editor

Review

New Trends in C–C Cross-Coupling Reactions: The Use of Unconventional Conditions

Marta A. Andrade and Luísa M. D. R. S. Martins *

Centro de Química Estrutural and Departamento de Engenharia Química, Instituto Superior Técnico, Universidade de Lisboa, 1049-001 Lisboa, Portugal; marta.andrade@tecnico.ulisboa.pt
* Correspondence: luisammartins@tecnico.ulisboa.pt; Tel.: +35-121-841-9389

Academic Editor: Giuseppe Cirillo
Received: 30 October 2020; Accepted: 20 November 2020; Published: 24 November 2020

Abstract: The ever-growing interest in the cross-coupling reaction and its applications has increased exponentially in the last decade, owing to its efficiency and effectiveness. Transition metal-mediated cross-couplings reactions, such as Suzuki–Miyaura, Sonogashira, Heck, and others, are powerful tools for carbon–carbon bond formations and have become truly fundamental routes in catalysis, among other fields. Various greener strategies have emerged in recent years, given the widespread popularity of these important reactions. The present review comprises literature from 2015 onward covering the implementation of unconventional methodologies in carbon–carbon (C–C) cross-coupling reactions that embodies a variety of strategies, from the use of alternative energy sources to solvent-free and green media protocols.

Keywords: cross-coupling reactions; microwave irradiation; ultrasounds; mechanochemistry; solvent-free; water; ionic liquids; deep eutectic solvents; sustainability

1. Introduction

Cross-coupling reactions have attracted and inspired researchers in the academia and industry for decades, given its significance as a synthetic tool in modern organic synthesis. The continuous interest in cross-coupling reactions, with more than 40 years of history, has been mostly driven by its valuable contributions and applications in the medicinal and pharmaceutical industries. Especially noteworthy are reactions furnishing carbon-carbon bonds, regarded as one of the most challenging tasks in organic synthesis. C–C Cross-coupling protocols developed by Suzuki [1], Heck [2], and Sonogashira [3], among others, changed the way in which organic synthesis is conceptualized, endowing a practical synthetic route for the direct formation of C–C bonds [4–6].

The continuous challenge of the design of reactions from a sustainable viewpoint (e.g., more efficient energy consumption, easy catalyst recovery, avoidance of high amounts of organic solvents) has led to the development of greener reaction protocols [7–10]. Tremendous efforts have been paid to develop alternative energy inputs that can dramatically reduce the reaction time, resulting in higher yields of the desired product. On the other hand, the use of eco-friendly media or solvent-free protocols is preferred over common organic solvents to obtain the maximum conversion in an eco-friendly manner [11–13]. Other issues such as efficient separation and subsequent recycling of the catalyst have been effectively solved, mostly by heterogeneous catalytic systems, many of them involving metallic nanoparticles [14,15], and, also, by the development of magnetic-supported catalysts [16–18].

The implementation of green methodologies in C–C cross-coupling reactions relies on a variety of strategies, such as the use of microwave irradiation, ultrasounds, and mechanochemistry as energy input under green and sustainable reaction media [11,19]. In this review, we wish to cover recent advances reported in the literature from 2015 to date that have demonstrated a special focus on unconventional methodologies for the generation of C–C bonds using cross-coupling reactions. Out of the scope of this review are photo-induced cross-coupling reactions. The synthesis of novel heterogeneous catalysts is not analyzed separately.

2. Alternative Activation Methods for Cross-Coupling Reactions

The use of alternative means of activation can improve many established protocols, providing superior results when compared to reactions performed under conventional conditions. Microwave (MW) irradiation, ultrasounds, and mechanochemistry can be used to enhance conventional experimental techniques for cross-coupling reactions.

2.1. Microwave Irradiation

Microwave (MW)-assisted synthesis has developed into a popular branch of synthetic organic chemistry, being currently employed for a variety of chemical applications. MW irradiation produces efficient internal heating by direct coupling of MW energy with the molecules (solvents, reagents, and catalysts) that are present in a reaction mixture, in contrast to conventional heating methods [14,20,21]. Reaction vessels employed in MW reactors, typically made out of microwave transparent materials (borosilicate glass, quartz, or Teflon), allow the radiation to pass through the walls of the vessel, creating an inverted temperature gradient as compared to conventional thermal heating [22]. This unconventional MW energy results in the acceleration of chemical reactions due to the selective absorption of MW radiation by polar molecules [20,21,23]. The rate enhancements are essentially a result of a thermal effect, a change in temperature compared to heating by standard convection methods [24,25]. The use of MW irradiation fulfils the requirement of green chemistry, providing rapid heating, reduced reaction times, and, in many cases, increased yields and selectivity with significant energy savings [23,26]. MW-assisted reactions have been studied in detail by many researchers in numerous scientific areas, highlighting the advantages of this methodology [22]. High-yielding C–C cross-coupling reactions under the influence of MWs have been described in the literature for the past decade. There are extensive studies about Suzuki–Miyaura and, to a lesser extent, Mizoroki–Heck and Sonogashira-type C–C cross-coupling reactions incorporating MW irradiation.

In 2015, Martínez et al. described the MW-assisted Mizoroki–Heck reaction, under solvent-free conditions, using a new family of solid catalysts, consisting in supported Pd nanoparticles (NP) on a synthetic clay (laponite) (Scheme 1) [27].

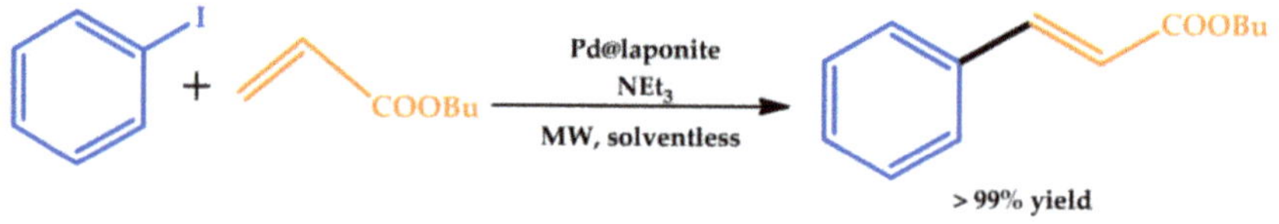

Scheme 1. Heck reaction of iodobenzene with butyl acrylate using laponite-supported Pd nanoparticles (NPs) [27]. MW: microwave.

The influence of reaction time and MW power employed on the catalytic results was thoroughly studied by the authors. Complete conversions and excellent product yields were achieved in a few minutes, representing a remarkable increase of around 24 to 60 times in comparison to the use of conventional heating. It was observed that the time necessary to reach complete conversion was dependent on the power of the MW source. The catalyst was efficiently recovered and reused several times, achieving total conversion, outperforming most of the cases described in the literature. Nevertheless, the use of high MW power was found to be detrimental for the catalyst lifetime, leading to catalyst poisoning by coke deposition. In this sense, the authors highlighted that the best performances could be obtained using MW power up to 25 W.

Shah et al. provided an example of synergism between PdNP and MW heating in ligand-free C–C cross-coupling reactions—namely, Suzuki, Hiyama, Heck, and Sonogashira [28]. PdNPs < 10 nm were impregnated onto a commercially available microporous polystyrene resin and successfully used to synthesize asymmetric terphenyls by the sequential addition of aryl boronic acids. Highly pure coupled products were obtained in 6–35 min under MW heating, using benign low-boiling solvents that otherwise would involve 4–30 h of inert atmosphere under conventional heating (Figure 1). No side-products were observed, and, in most cases, the products were purified by simple crystallization. The catalyst was successfully recycled up to six times without significant leaching or a decrease in efficiency. A TEM analysis showed that the size of the NP remained unchanged after the reaction.

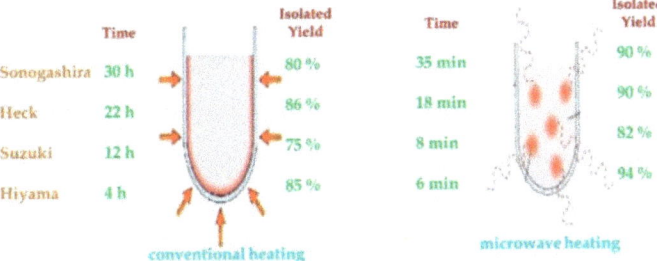

Figure 1. A comparison of time and yield for standard protocols of various coupling reactions. Reproduced with the permission of reference [28], Copyright 2016 Elsevier B.V.

Massaro and colleagues prepared a covalently linked thermo-responsive polymer on halloysite nanotubes (HNTs) by means of MW irradiation, used as a support and stabilizer for PdNP [29]. The prepared HNT-poly(N-isopropylacrylamide) (PNIPAAM) nanomaterial showed a good catalytic activity towards the Suzuki cross-coupling reaction of phenylboronic acid and several aryl halides under MW irradiation in aqueous media and low catalyst loading (0.016 mol%), reaching turnover numbers (TONs) and turnover frequencies (TOFs) up to 6250 and 37,500 h^{-1}, respectively (Table 1). The catalyst was easily separated from the reaction mixture and reused for five cycles, with a minor drop in its catalytic activity and negligible Pd leaching.

Table 1. Suzuki cross-coupling reaction of phenylboronic acid with various aryl halides under optimized reaction conditions under microwave (MW) irradiations [29].

Ar-X	Conversion (%)	TON	TOF (h^{-1})
4-bromoacetophenone	>95	6250	37,500
4-iodoanisole	85	5310	31,870
4-bromobenzaldheyde	94	5880	35,250
4-bromoanisole	85	5310	31,880
3-bromobenzaldheyde	82	5130	30,750
4-iodoacetophenone	94	5880	35,250
2-iodotoluene	73	4560	27,380

Turnover number (TON) calculated as moles of substrate converted/moles of Pd. Turnover frequency (TOF) calculated as TON/hours. HNT: halloysite nanotube; PNIPAAM: HNT-poly(N-isopropylacrylamide); PDNPs: Pd nanoparticles, TBAB: tetrabutylammonium bromide.

The beneficial use of MWs was also observed in a Sonogashira-type coupling reaction and reported by Lei et al. [30]. After the study of the effect of different catalysts, bases, solvents, temperatures, and reaction times on the coupling of 4-bromobenzonitrile to 4-((trimethylsilyl)ethynyl)benzonitrile, chosen as the model reaction (Scheme 2), the authors extended the scope of the MW-assisted reaction to a variety of substrates. Efficient conversion of the substrates into the desired 1-aryl-2-(trimethylsilyl)acetylene products with excellent yields was observed within a short reaction time under MWs. Compared to the 4-bromobenzonitrile substrate, the 4-iodobenzonitrile substrate also afforded quantitative yields. Moreover, several functional groups were well-tolerated under optimized conditions, with excellent yields. The use of MWs was found to significantly improve the efficiency of the method based on Cu and Pd-catalyzed Sonogashira-type coupling processes in a short reaction time, making it a useful approach for the synthesis of structurally diverse 1-aryl-2-(trimethylsilyl)acetylene.

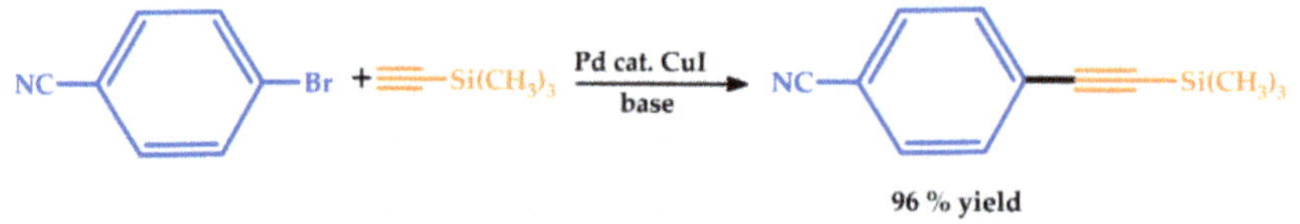

Scheme 2. Microwave-assisted Sonogashira-type coupling of 4-bromobenzonitrile to 4-((trimethylsilyl) ethynyl)benzonitrile [30].

Savitha et al. explored the MW-assisted Suzuki–Miyaura cross-coupling reaction of the chloro uracil analog with a wide array of (hetero) aryl potassium organotrifluoroborates, in water [31]. Aryl and heteroaryl trifluoroborates were found to be excellent reagents for Pd-catalyzed cross-coupling reactions with 6-chloro-3-methyluracil (Scheme 3). A set of electron-rich phosphines was employed, along with Pd(OAc)$_2$, as the catalyst. Although none of the catalytic systems provided exceptional results, the catalytic activity was significantly enhanced when a Pd-XPhos precatalyst, a coordinatively unsaturated complex, was employed, with excellent conversions (72–90%). The screening of solvents showed that the reaction afforded exceptional conversion in plain water, which, in combination with MW irradiation, increased its green potential.

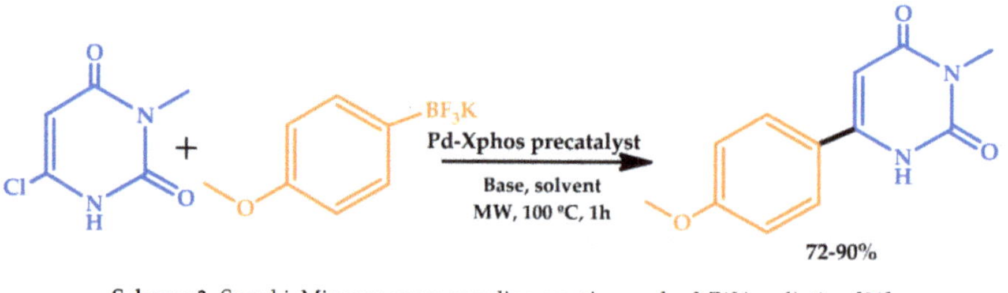

Scheme 3. Suzuki–Miyaura cross-coupling reaction under MW irradiation [31].

In 2017, Baran and coworkers designed a heterogeneous Pd(II) catalyst containing a O-carboxymethyl chitosan Schiff base (CS-NNSB) as the support material. The catalytic activity of CS-NNSB-Pd(II) was examined in Suzuki cross-coupling reactions under mild conditions using a simple MW heating technique and solvent-free reaction media [32]. CSNNSB-Pd(II) presented a high catalytic performance and excellent selectivity for the coupling reaction of phenyl boronic acid with 4-bromoanisol, with very low catalyst loading (6 mmol%) in a very short reaction time (5 min), offering high TON and TOF values (up to 16,167 and 202,087), confirming the suitability of the proposed catalytic system. The substrate scope was extended under the optimum conditions with a range of aryl halides, containing withdrawing groups or electron-donating functional groups. While the coupling reactions of aryl bromides and iodides provided the desired biaryls products with excellent yields, relatively lower reaction yields were observed for the aryl chlorides owing to the poor reaction activity. In addition, the reusability and catalytic behavior of the catalyst was tested, and it could be reused up to seven runs. Verbitskiy and colleagues proposed a convenient method for the synthesis of a new fluorophore on the basis of 4,5,6-tri(het)-arylpyrimidine containing three electron-donating carbazole moieties by consisting of a series of MW-assisted Suzuki cross-coupling reactions [33]. The synthesis of the target structure of fluorophore was accomplished by using the sequence of MW-assisted Suzuki cross-coupling and bromination with N-bromosuccinimide, using as a starting material 5-bromo-4,6-di(thiophen-2-yl)pyrimidine. The reaction time never exceeded 25 min, and the yield of the target fluorophore 3,3'-(5,5'-(5-[5-(9-ethyl-9H-carbazol-3-yl)-3-hexylthiophen-2-yl]pyrimidine-4,6-diyl)bis(thiophene-5,2-diyl))bis(9-ethyl-9H-carbazole) reached 56%, with the 29% combined yield over three steps. The optical properties of the obtained compound were explored, envisioning possible applications in sensors for nitroaromatic compounds.

MW irradiation was employed in another study for both the preparation of the catalyst and its application for the Suzuki cross-coupling reaction of bromobenzene and phenylboronic acid [34]. Elazab et al. developed a simple and efficient synthetic protocol to produce highly active PdNP catalysts supported on a copper oxide matrix using MW irradiation through a simple and fast approach under mild temperature and pressure (Scheme 4). The synthesized Pd/CuO bimetallic catalyst with an average size of 20–40 nm was investigated towards the Suzuki cross-coupling reactions in 50% aqueous ethanol solvent using MW irradiation.

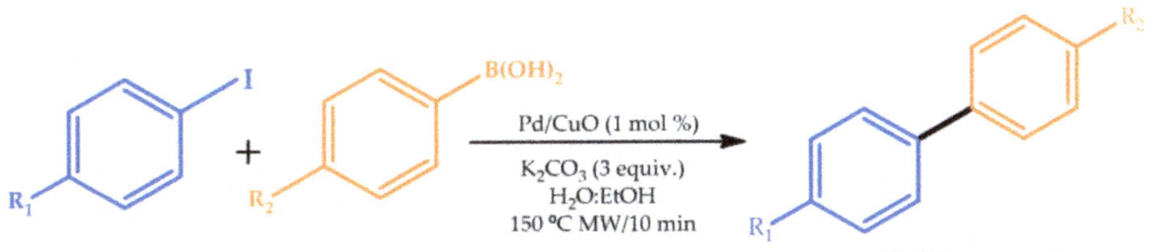

Scheme 4. Suzuki cross-coupling reactions using a Pd/CuO catalyst [34].

The prepared Pd/CuO catalyst was found to be stable, showing excellent conversion within 15 min at 150 °C, highlighting the crucial role played by the CuO solid support in preventing NP agglomeration. Moreover, it could be simply recovered and recycled up to five times with a negligible loss in performance or catalytic activity under the batch reaction. The implemented MW irradiation method was recognized as simple, reliable, and rapid, allowing the synthesis of controlled-size NPs. The results obtained for the Suzuki cross-coupling reaction are in line with those observed by Massaro et al. [29] and discussed above in this section, even though the comparison is not straightforward, as the experimental conditions are distinct—namely, catalyst support and loading and the microwave power employed.

Zakharchenko et al. described the efficient MW-assisted C–C cross Suzuki–Miyaura coupling reaction with a series of catalysts, including Pd(II) complexes. The synthesized materials, consisting of 3-(2-pyridyl)-5R-1,2,4-triazoles Pd(L^R)$_2$ (R = ethyl, n-propyl, i-propyl, and t-butyl) catalyzed the MW-assisted Suzuki–Miyaura cross-coupling reaction of bromoanisole with phenyl boronic acid in the presence of a base. Furthermore, these heterogeneous catalysts were recovered and reused without losing activity for quite a few consecutive cycles [35]. The low power of the MW irradiation (10 W) used provided a much more efficient synthetic method than conventional heating, allowing for the attainment of significantly better yields in much shorter times.

2.2. Sonochemistry

Ultrasound-promoted organic synthesis is a powerful and important green methodology. Sonochemistry is the use of ultrasound irradiation in chemical processes, leading to the generation of intense mechanical, thermal, and chemical effects in a liquid medium as the consequence of the cavitation phenomenon throughout the generation, growth, and sudden collapse of gaseous microbubbles [14,36]. This feature creates extremely high local pressures (up to 1000 bar) and high temperatures (up to 5000 K) that can trigger high-energy radical mechanisms in mild conditions [24]. The physical properties of the irradiated mixture are crucial for the effectiveness of cavitation, as well as for the proper transfer of acoustic energy to the reactants.

Ultrasounds can therefore be used as an important tool to perform a number of chemical reactions, enhancing rates and yields at shorter reaction times, with simpler and easier workup conditions compared to conventional procedures [26,36]. Ultrasound-promoted procedures that combine sonochemistry with green, nonconventional solvents or even solvent-free have been increasingly reported in the literature, including, most frequently, the use of water, ionic liquids, poly(ethylene glycol) (PEG), and others. Despite the large and diverse number of reports on this topic, very few data are available on comparative results obtained for direct and indirect sonochemistry procedures.

In 2017, Rezaei developed a highly efficient procedure involving PEDOT nanofibers/Pd(0) composites (PEDOT-NFs@Pd) as recyclable catalysts for the C–C cross-coupling reactions of aryl halides with olefinic compounds under ultrasonic irradiation in water (Scheme 5) [37]. The catalytic performance of the catalyst toward Heck cross-coupling reactions under ultrasound irradiation was examined, being clear that the coupling reaction could be finished in a shorter time with better yields. The author further explored the catalytic applicability of PEDOT-NFs@Pd towards Heck coupling reactions of styrene and n-butyl acrylate with different aryl halides under optimized conditions, showing that various aryl bromide or iodide with different substituent groups could be efficiently converted to the corresponding products in high yields. Moreover, the catalyst was recovered by simple filtration and reused quite a few times without a significant loss of activity.

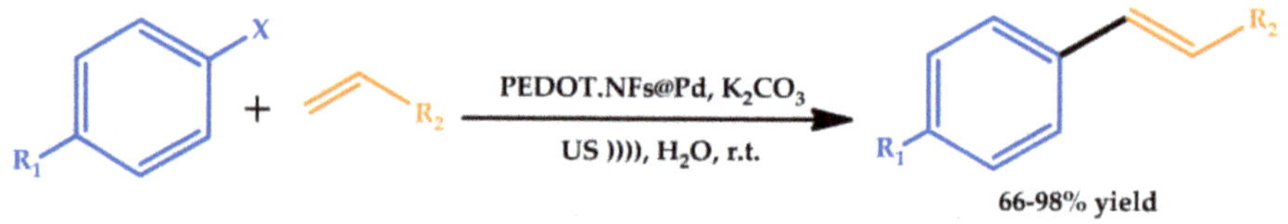

Scheme 5. US-assisted Mizoroki–Heck cross-coupling reactions [37]. NF: nanofibers.

Panahi and colleagues reported the synthesis of a Pd nanomaterial based on a Cu-metal-organic framework (MOF), Pd@Cu-MOF, as an effective catalyst for a Suzuki–Myaura C-C cross-coupling under ultrasonic irradiation [38]. Operational parameters, such as temperature, time, solvent, and base, were the first objects of optimization in a model reaction of phenylboronic acid and p-bromobenzene. Subsequently, different derivatives of biaryl compounds were synthesized by the US-assisted reaction of different aryl halide derivatives and phenylboronic acid derivatives in shorter reaction times and considerably higher yields (75–99%) in the presence of a base. The catalyst was recoverable, being reused for at least four consecutive reaction cycles.

In 2018, there was a US-assisted and Pd-catalyzed reaction of 4-bromoanisole and phenylboronic acid in a 1:1 mixture of ethanol and water [39]. The authors explored the effects of several operating parameters, such as US power, temperature, catalyst loading, and molar ratio on the catalytic performance (Figure 2).

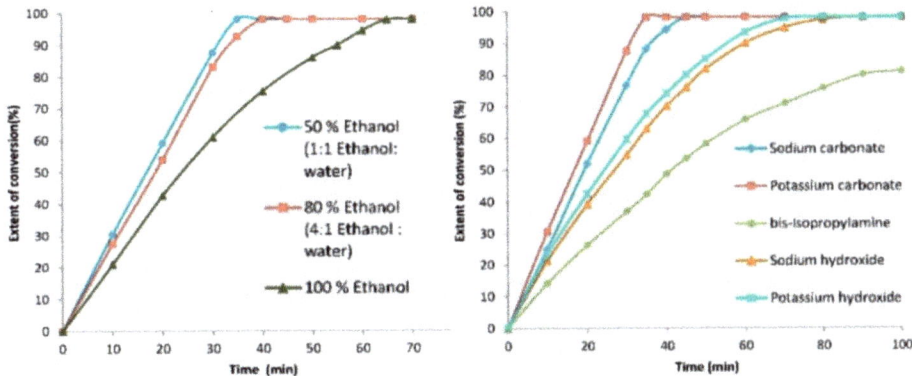

Figure 2. (**Left**): Effect of solvent combination on the US-assisted Suzuki cross-coupling reaction. (**right**): Effect of different bases on the US-assisted Suzuki cross-coupling reaction. Reaction Conditions: 4-bromoanisole (1 mmol) and phenylboronic acid (1.5 mmol) using 1.5 mol% of catalyst Pd/C, base (2.0 mmol), 30 °C, ultrasonic bath, 22 kHz, and the irradiation power of 40 W. Adapted with the permission of reference [39], Copyright 2017 Elsevier B.V.

Overall optimum conditions established were the catalyst loading of 1.5 mol%, molar ratio of (phenylboronic acid:4-bromoanisole) 1.5, US power of 40 W, and duty cycle of 90% at a frequency of 22 kHz for a maximum conversion of 98%, achieved in 35 min. This study also demonstrated a very good catalytic performance using a volume ten times higher as compared to the normally used volumes in the case of a simple ultrasonic horn. It was also observed that the reaction rate increased with an increase in temperature over the range of 30–60 °C, decreasing beyond this temperature.

The development of a simple US-assisted protocol for the Suzuki–Miyaura reaction yielding biphenyl compounds was reported by Baran et al. [40]. The catalytic performance of PdNPs@CMC/AG, consisting of Pd(0)NP on a natural composite composed of carboxymethyl cellulose/agar polysaccharides (CMC/AG), was evaluated in the synthesis of various biphenyl compounds by using ultrasounds.

The prepared nanocomposite exhibited an excellent catalytic performance for cross-coupling reactions that contained aryl iodides, achieving high reaction yields. In addition, PdNPs@CMC/AG was successfully reused up to six reaction cycles without a loss of its catalytic activity, indicating a high reproducibility of PdNPs@CMC/AG. Compared to conventional methods, the US-assisted synthesis technique reported in this study exhibited some advantages, such as a shorter reaction time, greener reaction conditions, higher yields, and easier workup.

In 2018, Naeimi et al. reported a novel and efficient catalytic system consisting of Cu(I)-immobilized on functionalized graphene oxide for the Sonogashira coupling reaction under ultrasonic irradiation [41]. The prepared nanocatalyst GO@SiO$_2$-HMTA-Cu(I) (GO: graphene oxide; HMTA: hexamethylenetetramine) afforded excellent yields of diarylethene compounds and short reaction times under mild reaction conditions (Table 2). The catalytic activity of GO@SiO$_2$-HMTA-Cu(I) was found to be dependent on the type of leaving group: all aryl halides (chloride, bromide, and iodide). The products can be easily separated from the reaction mixture by filtration. Furthermore, the catalyst was simply recycled and reused eight times without a significant loss of activity.

Table 2. Study of the effect of ultrasonic radiation on the Sonogashira reaction [41].

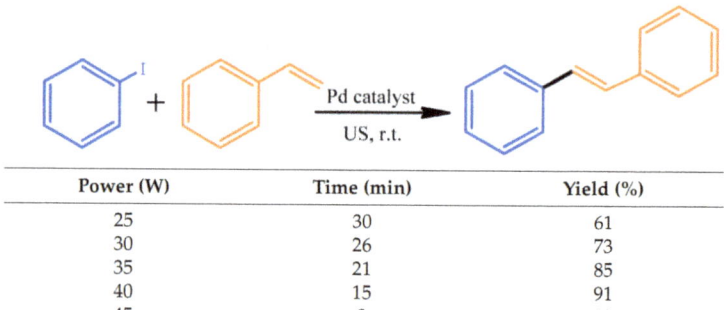

Power (W)	Time (min)	Yield (%)
Silent	120	-
35	60	81
40	50	87
45	35	93
50	25	98
55	25	98

Reaction conditions: aryl halide (1 mmol), phenyl acetylene (1.2 mmol) GO@SiO$_2$-HMTA-Cu(I) and NEt$_3$ (3 mmol) in water. GO: graphene oxide; HMTA: hexamethylenetetramine.

In 2019, the same authors reported the synthesis of a magnetic thiamine Pd complex nanocomposite, a novel, highly active, and reusable catalyst for the Mizoroki–Heck coupling reaction of several types of iodo, bromo, and even aryl chlorides under US irradiation [42]. The prepared Pd catalyst demonstrated an excellent performance for the efficient, mild, fast, clean, and safe sonochemical synthesis of trans-stilbene derivatives at room temperature, producing the corresponding trans-coupled products in excellent yields and moderate reaction times. The authors first examined the efficiency of the solvent in the Mizoroki–Heck reaction, DMF being the most efficient for this transformation. The gathered data demonstrated that, in the presence of ultrasonic irradiation and 0.02 g of the Pd(0) catalyst, the reaction time was shortened to 8 min, with a sharp increase of the reaction yield to 99% (Table 3). The catalyst was separated by the use of an external magnet and reused several times without a marked loss of activity.

Table 3. Study of the effects of ultrasonic irradiation on the Mizoroki-Heck cross-coupling [42].

Power (W)	Time (min)	Yield (%)
25	30	61
30	26	73
35	21	85
40	15	91
45	8	99
50	8	99

Reaction conditions: Iodobenzene (1 mmol), styrene (1.2 mmol) catalyst (20 mg), and NEt$_3$ (2 mmol) in DMF.

2.3. Mechanochemistry

A mechanochemical reaction is defined as a chemical reaction that is induced by the direct absorption of mechanical energy [43]. Mechanochemical processes are completely different from thermal processes, being usually performed using grinding in ball mills. The high speed achieved under ball milling conditions results in a homogeneous mixture of reactants, which successively facilitates chemical reactions. Particle refinement leads to an increase of surface area, surface energy, and number of defects. Mechanochemistry can be used for the mechanical activation of solids, due to the increased surface energy, alterations in structure, chemical composition, or chemical reactivity occurring throughout the milling process. The interest in mechanochemical synthesis is growing at a fast pace, as it provides convenient, feasible, and greener methods for the synthesis of valuable chemical compounds.

This method has been successively applied to several organic chemical reactions, involving many methodologies under solvent-free conditions and, also, finding enhanced selectivities and reactivities compared to reactions performed in a solution. Examples of mechanochemical-assisted cross-coupling reactions are now becoming notorious in the literature.

In 2016, Jiang et al. studied the effects of the liquid-assisted grinding in the mechanochemical Suzuki–Miyaura reaction of aryl chlorides [44]. The catalytic systems, using Davephos or PCy$_3$, were tested, showing strong influences from different liquids, with an unexpected improvement of the yield over 55% when using alcohols as additives, which is most likely explained by in-situ formed alkoxides and their participation in the oxidative addition. Further expansion of the substrate scope using the Pd(OAc)$_2$/PCy$_3$/MeOH system afforded the desired products in good-to-high yields, proving that it can be applied to both activated and unactivated aryl chlorides with good-to-high yields (Figure 3). The result implicates that the liquid-assisted grinding effect does not only provide a sub-stoichiometric solvent environment but may also participate in the formation of mechanically induced transition species. Thus, this technique may have the potential to induce higher catalytic activity for existing systems under high-speed ball milling conditions.

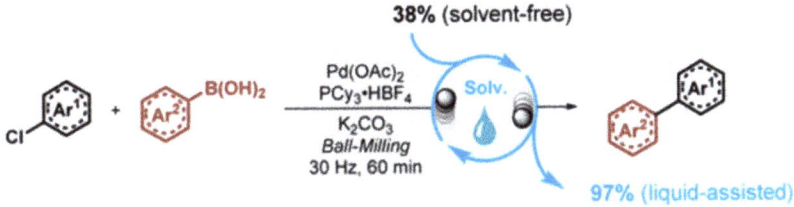

Figure 3. Mechanical Suzuki–Miyaura reaction of aryl chlorides using liquid-assisted grinding. Reproduced with the permission of reference. [44], Copyright 2016 American Chemical Society.

The result implicates that the liquid-assisted grinding effect does not only provide a sub-stoichiometric solvent environment but may also participate in the formation of mechanically induced transition species. Thus, this technique may have the potential to induce a higher catalytic activity for existing systems under high-speed ball milling conditions.

In 2017, Shi et al. described the synthesis of a cheap and reusable Pd catalyst supported on MgAl-layered double hydroxides (Pd/MgAl-LDHs) to be used in Heck reactions under high-speed ball milling conditions at room temperature [45]. The authors explored the effects of the ball milling size, ball milling filling degree, reaction time, rotation speed, and grinding auxiliary category in the yields of mechanochemical Heck reactions. The obtained results point out the suitability of this catalyst for high-speed ball milling (HSBM) systems, affording a wide assortment of Heck coupling products in satisfactory yields. Moreover, the catalyst could be easily recovered and reused for at least five times without a significant loss of catalytic activity.

Cao et al. reported in 2018 a novel mechanochemical method for the synthesis and subsequent reaction of organozinc species in a Negishi cross-coupling reaction [46]. Organozincs were generated from an alkyl halide with the addition to the reaction mixture of a coupling partner along with a Pd catalyst and tetrabutylammonium bromide (TBAB) to perform the Negishi reaction in a one-pot, two-step process (Scheme 6).

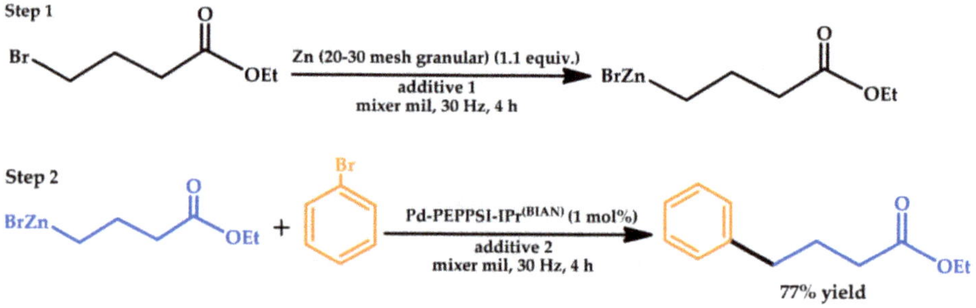

Scheme 6. Optimization of the one-jar, two-step reaction for mechanochemical Negishi cross-coupling reactions [46]. PEPPSI: pyridine-enhanced pre-catalyst preparation stabilization and initiation.

The approach showed a broad substrate scope, offering broad opportunities for the in-situ generation of organometallic compounds from base metals and their simultaneous engagement in mechanochemical synthetic reactions without the need to use inert atmosphere techniques or dry solvents.

Seo et al. reported a very complete study on the development of a broadly applicable mechanochemical protocol for the solid-state Pd-catalyzed organoboron Suzuki–Miyaura cross-coupling reaction [47]. The authors studied the influence of olefin additives that might act as molecular dispersants and the effect of the phosphine ligand on the solid-state organoboron cross-coupling reaction. The influence of the mechanochemical reaction parameters on the solid-state reaction of 4-bromo-1,1′-biphenyl with 4-dimethylaminophenylboronic acid was also investigated (Table 4). Aiming to explore the scope of the present solid-state coupling reaction, a variety of solid aryl bromides was used, affording good-to-excellent yields (%).

Table 4. Investigation on the effects of varying the mechanochemical parameters [47].

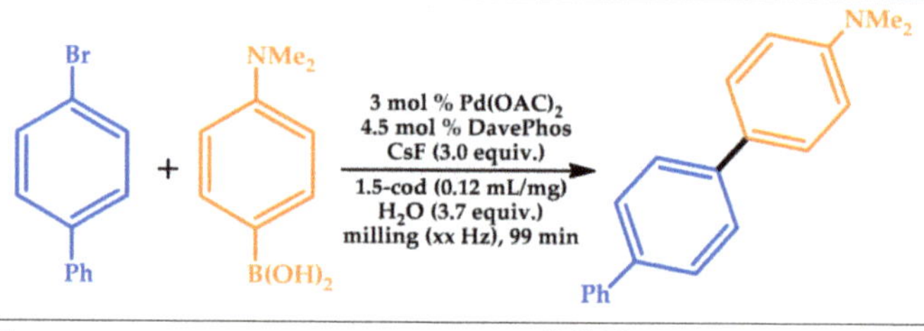

Frequency (Hz)	Number of Balls	Ball Size' (mm)	Yield (%)
10	1	5	70
15	1	5	75
20	1	5	70
25	1	5	97
30	1	5	99
25	1	3	90
25	2	5	97

Reaction conditions: Pd(OAc)$_2$ (0.009 mmol), DavePhos (0.0135 mol), CsF (0.9 mmol), H$_2$O (20 μL), and 1,5-cod (0.12 μL/mg) in a stainless-steel ball milling jar (1.5 mL).

Additionally, the use of solid aryl chlorides, boronic acids, aryl dibromides and dichlorides, and polyaromatic hydrocarbon substrates were also studied for this reaction. Furthermore, the authors explored the synthetic utility of the proposed protocol, conducting the solid-state cross-coupling reaction on the gram scale under mechanochemical conditions (Figure 4) of 9-bromoanthracene with

4-dimethylaminophenylboronic acid at the 8.0-mmol scale in a stainless-steel ball milling jar using four stainless-steel balls, affording 87% yield in 3 h.

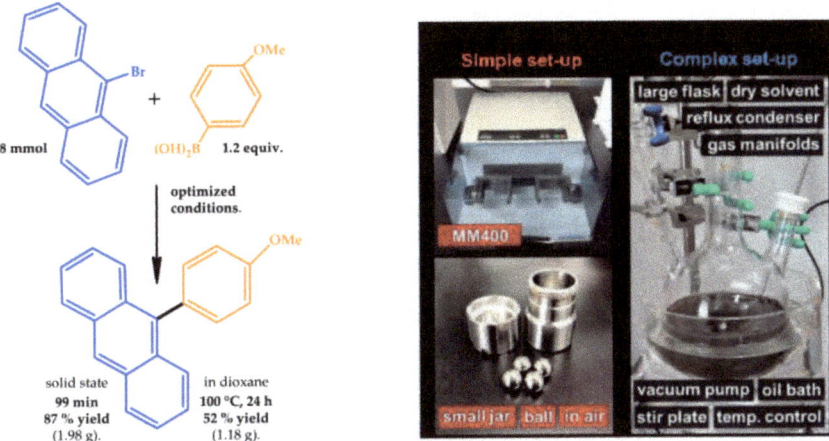

Figure 4. (**Left**): Reaction scheme for the solid-state cross-coupling of 9-bromoanthracene with *p*-methoxyboronic acid. (**Right**): Solid-state coupling reaction on the gram scale. Reproduced with the permission of reference [47], Copyright 2019 Royal Society of Chemistry.

For comparison, a solution-based Suzuki–Miyaura cross-coupling reaction was also conducted in a large amount of solvent (dioxane) at a high temperature (100 °C) and extended reaction time (24 h), obtaining, in this case, a moderate yield (52%). The mechanistic data obtained suggested that olefin additives could act as dispersants for the Pd-based catalyst suppressing the higher aggregation of the NPs, which could lead to the catalyst deactivation, acting also as stabilizer for the active Pd(0) species upon coordination (Figure 5) [47].

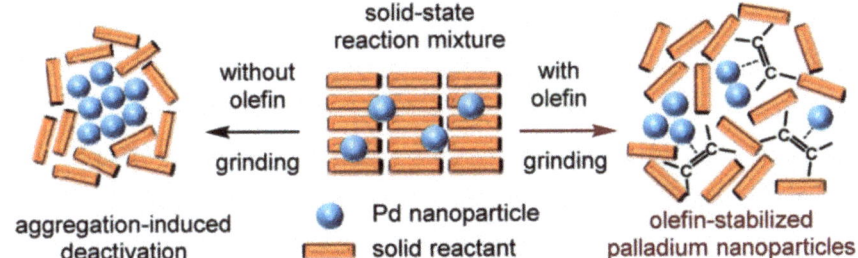

Figure 5. Possible roles for the olefin additives in solid-state cross-coupling reactions. Reproduced with the permission of reference [47], Copyright 2019 Royal Society of Chemistry.

Vogt et al. introduced a facile and highly sustainable synthesis concept for a Pd-catalyzed mechanochemical reaction, turning catalyst powders, ligands, and solvents obsolete (Figure 6). The authors performed the Pd-catalyzed C-C Suzuki polymerization of 4-bromo or 4-iodophenylboronic acid, providing poly(para-phenylene) [48], and were surprised to observe one of the highest degrees of polymerization (199) reported in the literature. The solvent-free environment of a ball mill enabled the direct use of Pd milling equipment or Pd black catalyst, instead of conventional Pd(II) salts or Pd complexes. With 4-iodophenylboronic acid as the monomer, a good yield and high degree of polymerization were achieved in planetary ball milling using Si_3N_4 milling material, while full conversion to long-chain polymers was obtained in a mixer ball mill using a softer poly(methyl methacrylate) PMMA vessel. In addition, the degree of polymerization achievable by this method surpassed those obtained by solution or electrochemical processes. The obtained results indicate a most likely heterogeneous reaction that was not improved by using established ligands from solution-based homogeneous procedures.

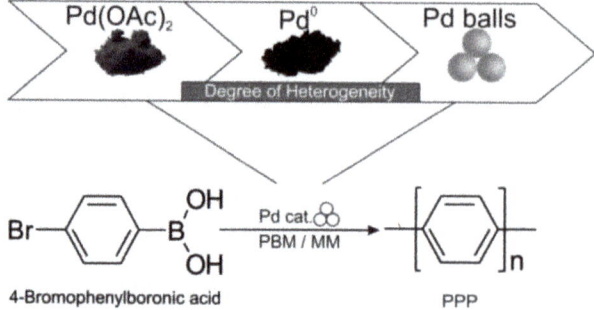

Figure 6. Mechanochemical Suzuki polymerization reaction of 4-bromophenylboronic acid to poly (para-phenylene) (PPP). The catalyst was simplified from a Pd salt to Pd black and to simply Pd milling balls. PMB: planetary ball mill; MM: mixer ball mill. Reproduced with the permission of reference [48], Copyright 2019 Wiley-VCH Verlag GmbH & Co. KGaA, Weinheim, Germany.

Pentsak et al. investigated the solid-state reaction of aryl halides with arylboronic acids in the absence of a solvent and without any liquid additives [49]. A few important conditions for performing the Suzuki–Miyaura reaction were analyzed in detail, pointing out the prominent role of water, formed as a by-product in the side reaction of arylboronic acid trimerization. Minimal amounts of water allow the initiation of the reaction with potassium tetraphenylborate and potassium trifluoroborate. Electron microscopy studies revealed surprising changes that occurred within the reaction mixture, such as the formation of spherical nano-sized particles containing the reaction product. Catalyst recycling was easily performed, involving the product isolation by sublimation and, thus, providing the possibility to completely avoid the use of solvents at all stages (Table 5). Under optimized conditions, the quantitative yields could be achieved by conventional heating of the solid-phase reaction mixture, without the need for additional mixing. The PdNPs/multi-walled carbon nanotubes (MWCNT) catalyst can be reused multiple times without a loss of efficiency. The absence of a liquid phase is an important factor to avoid metal leaching.

Table 5. Reuse of the PdNPs/MWCNT catalyst in the solid-phase Suzuki–Miyaura process [49].

Conversion%	Cycle						
	1	2	3	4	5	6	7
0.1 mol% of the catalyst	95	75	94	84	100	98	100
0.5 mol% of the catalyst	79	80	88	86	96	97	93

Reaction conditions: 1 mmol of aryl halide, 1.2 mmol of phenylboronic acid, 1.2 mmol of K2CO3, and 95 °C for 6 h. MWCNT: multi-walled carbon nanotubes.

Soliman and co-workers explored the preparation of Pd(II) and Pt(II) composites with activated carbon (AC), graphene oxide, and multiwalled carbon nanotubes by ball milling and their use as catalysts for the mechanical-assisted Suzuki–Miyaura reaction [50]. The Pd-AC composites exhibited high catalytic activity towards the mechanochemical cross-coupling of bromobenzene and phenylboronic acid, with yields up to 75%, achieving a TON and TOF of 222 and 444 h^{-1}, respectively, using the Pd(4.7 wt%)-AC catalyst (Table 6). The effect of small amounts of olefins on the reaction yield was investigated for the most promising catalytic system (using cyclooctene as the additive), obtaining relevant yields of ca. 80%. The authors highlighted the importance of ball milling as a promising synthetic tool under eco-friendly conditions, in comparison to other approaches for the energy input, such as MW and conventional heating.

Table 6. Mechanochemical Suzuki—Miyaura cross-coupling reactions of bromobenzene with phenylboronic acid [50].

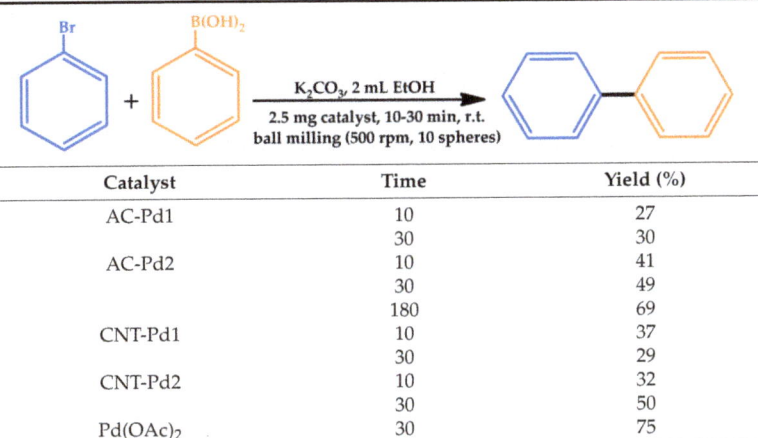

Catalyst	Time	Yield (%)
AC-Pd1	10	27
	30	30
AC-Pd2	10	41
	30	49
	180	69
CNT-Pd1	10	37
	30	29
CNT-Pd2	10	32
	30	50
Pd(OAc)$_2$	30	75

3. Unconventional Media for Cross-Coupling Reactions

The use of organic solvents as the traditional medium for organic synthesis continues to be extensively applied. From a sustainable point of view, unconventional and more environmentally benign reaction media should be explored [9,51]. Widespread effort has been made to develop efficient catalytic systems under alternative media for cross-coupling reactions.

3.1. Solvent-Free

According to the principles of green chemistry, solvent-free protocols for organic reactions are of great interest, reducing or eliminating the utilization of hazardous and expensive organic solvents. Solvent-free methodologies have been commonly used under MW irradiation or mechanochemical conditions often using supported reagents, achieving maximum conversion and minimizing wastes.

Dumonteil et al. described a Pd-catalyzed Mizoroki–Heck cross-coupling performed in solvent and ligand-free conditions for the synthesis of abscisic acid (ABA), a well-known phytohormone of great biological interest [52]. Upon optimization of the reaction parameters, various dienes and trienes were obtained in moderate-to-good yields without isomerization. The optimized solvent-free Mizoroki–Heck reaction was, furthermore, successfully applied to the synthesis of ABA, offering a short new pathway using environmentally friendlier solvents and reagents as an alternative to the synthesis already described in the literature.

Baran et al. devised a highly efficient heterogeneous Pd catalyst prepared using guar gum (GG), a natural biopolymer [53]. The catalytic studies showed that GG-Pd is a very active catalyst for Suzuki cross-coupling reactions under solvent-free media without any additives. Aryl halides bearing electron-attracting or donating groups were converted to the desired coupled product in excellent yields within 5 min under green reaction conditions, showing significant tolerance against different functional groups. Furthermore, the catalytic results indicated the yield order of biaryl yields as follows: Ar-I > Ar-Br > Ar-Cl (Table 7).

Table 7. Catalytic activity of the guar gum (GG)-Pd catalyst in Suzuki–Miyaura coupling reactions [53].

R	X		
	I	Br	Cl
4-OCH$_3$	98	88	56
3-NO$_2$	87	85	70
3-CH$_3$	62	59	33
4-CH$_3$	71	69	42

Additional stability tests showed that the catalyst could be reused at least ten times, with a minor decrease of its catalytic performance and negligible Pd leaching. The authors foresaw the general applicability of the designed GG-Pd catalyst and in industrial applications due to its economic and catalytic advantages.

In 2019, Gribanov and colleagues reported the environmentally friendly and efficient synthesis of fully substituted 1,2,3-triazoles comprising a solvent-free Pd-catalyzed Suzuki cross-coupling reaction [54]. The efficiencies of the Stille and Suzuki reactions of halo-1,2,3-triazoles with pinacol arylboronates were compared, revealing the preference of the Suzuki method. The elaborated protocol proceeded without the use of solvents in aerobic conditions, low catalyst loadings, and KOH as a base. The authors demonstrated the wide scope of this methodology, which can be extended to heteroaromatic halides—particularly, challenging 4- and 5-halo-1,2,3-triazoles.

Bharamanagowda and colleagues synthesized a new hybrid core shell catalyst, Fe$_3$O$_4$-lignin@PdNPs, through the support of PdNPs in a Fe$_3$O$_4$-lignin nanocomposite, obtained by sonication (Figure 7) [18]. The performance of this novel catalyst was evaluated towards the Mizoroki–Heck C-C cross-coupling reaction. The reaction between iodobenzene and n-butyl acrylate afforded a yield as high as 99% under solvent-free conditions.

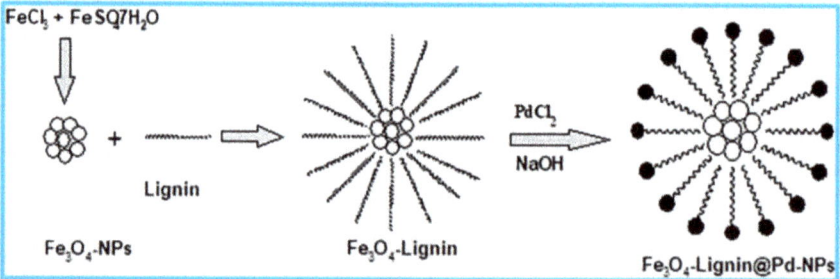

Figure 7. Synthesis of Fe$_3$O$_4$, Fe$_3$O$_4$-lignin, and Fe$_3$O$_4$-lignin@Pd-NPs. Reproduced with the permission of reference [18], Copyright 2020 John Wiley & Sons, Ltd.

The scope of the catalyst was also explored for the same reaction of various aryl/heterocyclic halides and *n*-butyl acrylate/styrene under optimized conditions, attaining the respective products in high yields (73–99%). The catalyst was magnetically recovered and reused for five cycles of the Mizoroki–Heck reaction of iodobenzene and *n*-butyl acrylate, achieving yields of the desired products of 90–95% without a significant activity loss and without any contamination. No visible change was observed in the shapes and sizes of the PdNPs after recycling, as confirmed by TEM images.

3.2. Water as Solvent

Water is an abundant, safe, and green medium and has been applied to an array of organic reactions in the last two decades [10,55]. Water plays a dual role as a medium and as a cocatalyst in quite a few synthetic routes [9], playing an important role in the rate acceleration of reactions through the establishment of extensive hydrogen bonds with the functional groups of reactants, thus activating them in the process. In recent years, water has been used as a greener sustainable reaction medium and applied to cross-coupling reactions in different supported catalytic systems for a wide variety of nanoparticles as catalysts. The good yields might be due to a combination of the hydrophobic effect of water and the nature of the support and the stabilization of leached metallic reactive species in the water.

In 2016, Handa and co-workers proposed a new Cu/ppm Pd technology for selective Suzuki–Myaura reactions of aryl iodides carried out under the mildest conditions based on the remarkable synergistic effects between copper and ppm levels of Pd as the cocatalyst. Several biaryl couplings were run to investigate the scope of this process using Na_2CO_3 as the base, at 45 °C, the mildest conditions yet described for such Cu-catalyzed couplings, achieving, in general, moderate-to-good yields. Control experiments were performed to confirm the synergistic effects by Cu/ppm Pd (Scheme 7). [56]. The entire aqueous system, containing nanomicelles that function as the reaction medium, the base, copper, and Pd can be recycled. The yields for biaryl were all comparable throughout five cycles. The associated E_{Factor}, an environmental factor that accounts for the ratio of the mass of waste per mass of product for these couplings was of about 5.8, ca. ten times lower than those reported and typically observed in the industry.

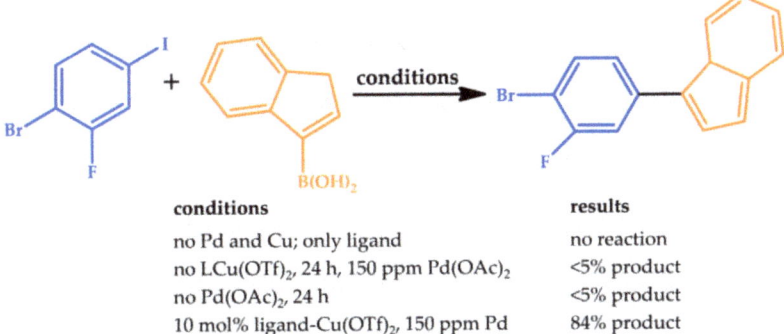

Scheme 7. Control experiments confirming a synergistic effect in the Suzuki–Miyaura reaction of aryl iodides [56].

Guarnizo and co-workers studied the formation of Pd/magnetite NPs through the anchorage of partially water-soluble 4-(diphenylphosphino)benzoic acid (dpa) on its surface [57]. The immobilized dpa enables the easy capture of Pd ions deposited on the surface of the magnetite nanoparticles after reduction with $NaBH_4$. The catalytic efficiency for the Suzuki C–C coupling reaction was studied for Fe_3O_4dpa@Pdx, with different Pd loadings. The highest TOF was obtained for Fe_3O_4dpa@Pd0.5, affording the highest value reported to date for the reaction of bromobenzene with phenylboronic acid in a mixture of ethanol/water. Remarkably, the same reaction carried out in water also returned excellent yields. The small size of the PdNPs supported on magnetite in Fe_3O_4-dba@Pd0.3 and Fe_3O_4dba@Pd0.5, along with the presence of Pd single-atom catalysts, explains the excellent results for the Suzuki C–C coupling reaction both in neat water and in a mixture of ethanol/water.

In 2016, Ghazali-Esfahani et al. prepared a highly robust and active heterogeneous catalyst for Suzuki cross-coupling reactions that operated in water in the presence of an inexpensive base. For this matter, PdNPs immobilized on a cross-linked imidazolium-containing polymer were evaluated as a catalyst for Suzuki carbon-carbon cross-coupling reactions in aqueous media [58]. The prepared PdNP-polymeric ionic liquid (PIL) nanocatalysts showed good catalytic activities for aryl iodides and aryl bromides and moderate activity with aryl chloride substrates at low catalyst loadings (10 mg, 1.7 mol%). The coupling of sterically hindered substrates was also accomplished in reasonable yields (Table 8). Aryl iodides and aryl bromides were efficiently converted to biaryl products in the presence of phenylboronic acid in a high yield using short reaction times. The PdNP-PIL catalyst in water tends to be superior to various catalysts in ionic liquids media. Moreover, the PdNP-PIL catalyst is also able to couple aryl chloride substrates in a moderate yield (16–23%). The ability to reuse the PdNP-PIL catalyst was explored using two different aryl iodide substrates. For both substrates, good recyclability was observed over five catalytic runs with > 90% of the original activity retained with negligible leaching, confirming the high stability of the catalyst, which is rather unusual. The catalyst is simple to prepare, stable for prolonged periods, and easy to use and recycle.

Table 8. Suzuki cross-coupling reactions between phenylboronic acid and various aryl halides catalyzed by PdNP-polumeric ionic liquid (PIL) [58].

R	X	Yield (%)
H	I	98
Me	I	93
MeO	I	99
NO_2	I	99
CN	I	97
Me	Br	70
MeO	Br	99
NO_2	Br	92
CN	Br	98
Me	Cl	23
MeO	Cl	17
CN	Cl	16

Ahadi et al. reported the synthesis of novel Pd-supported periodic mesoporous organosilica based on the bipyridinium IL catalyst (Pd@Bipy–PMO) (periodic mesoporous organosilica) for the Suzuki–Miyaura coupling reaction in water [59]. The reaction of 4–bromoacetophenone with phenylboronic acid was chosen to perform the optimization of the overall process, studying the effects of the solvent, temperature, catalyst amount, and base. Furthermore, the stabilized Pd species inside the mesochannels provided high-to-excellent catalytic efficiency for the Suzuki–Miyaura coupling of various iodo– and bromo–aryl derivatives and different boronic acids, presenting excellent yields in a short reaction time. The activity of the heterogeneous catalyst was retained for six consecutive recycling runs.

The role of the base in the ligand-free Pd-catalyzed Suzuki–Miyaura reaction in water under mild conditions was explored by Wang et al. [60] through the establishment of a simple and efficient system. The tested bases were shown to stabilize active Pd species, preventing aggregation and deactivation of the species. For this matter, long-chain quaternary ammonium hydroxides were found to be the most suitable candidates. The effect of Pd loading on the reactivity of the reaction was studied by the authors (Table 9). The limited solubility of biaryl compounds in water allowed the development of an efficient method of filtration for product purification from the aqueous catalytic system. The ligand-free Pd-catalyzed Suzuki–Miyaura reaction showed improved durability in water with ppm levels of Pd loading. The entire catalytic system could be recycled after product separation. This water-compatible and air-stable effective catalytic protocol represented an attractive green synthetic improvement in Suzuki–Miyaura couplings.

Table 9. Reduction of Pd loading for the ligand-free Suzuki–Miyaura reaction between 4-methoxyphenylboronic acids and bromobenzene [60].

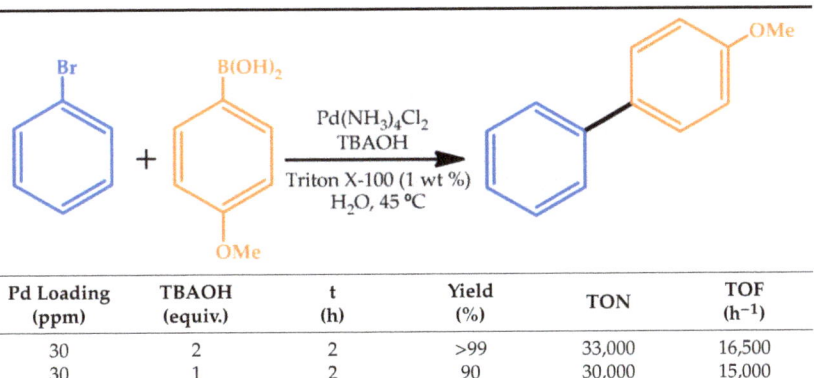

Pd Loading (ppm)	TBAOH (equiv.)	t (h)	Yield (%)	TON	TOF (h^{-1})
30	2	2	>99	33,000	16,500
30	1	2	90	30,000	15,000
30	4	6	75	25,000	4100

Reaction conditions: bromobenzene (1 mmol), 4-methoxyphenylboronic acid (1.2 mmol), TBAOH (1–4 mmol), Triton X-100 (0.1 mL), and H_2O (7 mL), 45 °C, air.

Lambert et al. reported the highly efficient micellar catalysis in pure water provided by robust polymeric micelles consisting of Pd(II)–NHC (NHC: N-heterocyclic carbene) units [61]. The authors highlighted the several benefits conferred by the proposed approach to both the Suzuki–Miyaura and Heck cross-coupling reactions, such as a very broad substrate scope, outstanding catalytic activity in water, low catalyst loadings (0.1 mol%), easy recycling, and absence of metal leaching. For comparison purposes, a molecular compound and a copolymer both exhibiting an analogous structure were prepared separately and also tested (Figure 8). The three catalytic systems were compared for the Suzuki–Miyaura reaction of 4-(hydroxymethyl)phenylboronic acid and iodotoluene. Remarkably, the micellar catalyst led to full conversion, while the molecular and linear species provided 40% and 10% conversions, respectively. The substantial increase of the catalytic activity of the micellar catalyst in water was ascribed to the increase in the local concentration around the Pd(II)–NHC sites favored by the hydrophobic sequestrating effect.

All the features for the polymer are significantly distinct from the molecular and/or nonmicellar catalytic versions, affording promising nanostructured catalytic supports from a green chemistry point of view.

A sustainable and robust protocol for the copper-free Sonogashira cross-coupling under micellar aqueous reaction conditions using the commercially available catalyst CataCXium A Pd G3 was proposed by Jakobi et al. [62]. Several alkyne substrates were efficiently cross-coupled with a wide range of aryl halides, affording improved yields and low catalyst loadings. The Pd-catalyzed Sonogashira reaction (0.30 mol% catalyst loading) provided yields up to 98%, with the surfactant TPGS-750-M in H_2O and THF as the cosolvent. Overall, the optimization of the reaction parameters rendered a simple and scalable operationally process able to provide a higher selectivity for heterocyclic compounds, alkynylated arenes, and monofunctionalized products achieved by the micellar aqueous reaction conditions.

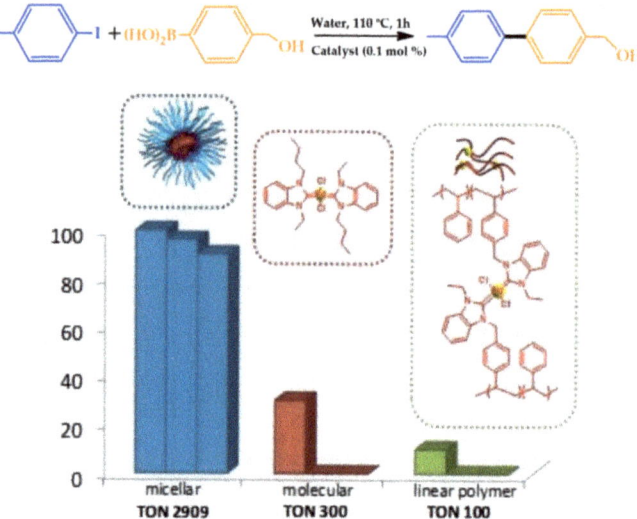

Figure 8. (**Top**): Reaction scheme for Suzuki coupling of 4-(hydroxymethyl)phenylboronic acid and iodotoluene. (**Bottom**): The three bars refer to the 1st, the 2nd, and the 3rd catalytic cycles, respectively. Reproduced with the permission of reference [61], Copyright 2018 The Royal Society of Chemistry. TON: turnover numbers.

3.3. Poly(Ethylene Glycol)

Poly(ethylene glycol) (PEG) oligomers are efficient, eco-friendly, and low-volatility liquids with a broad spectrum of chemical, industrial, and medical applications [63]. PEGs are readily miscible with water and polar organic solvents but immiscible with nonpolar organic solvents, aliphatic hydrocarbons, and supercritical CO_2. The use of relatively inexpensive PEGs in recent years as a greener media in several organic reactions has gained considerable attention [19,64]. PEGs are suitable media for oxidation/reduction transformations, as they are nearly harmless, with a very low vapor pressure, high catalytic capacity, high thermal stability, and stability in both acidic and basic media. Furthermore, PEGs have been used as efficient additives in aqueous-phase cross-coupling transformations to improve the interaction between water-soluble catalysts and organic reactants.

In 2015, PEG-400 was used as a solvent catalytic system for the Ni-catalyzed Sonogashira reaction by Wei et al. [65]. The coupling reaction of aryl iodides with terminal alkynes was carried out in a mixture of poly(ethylene glycol) (PEG-400) and water at 100 °C with K_2CO_3 as the base in the presence of $NiCl_2(PPh_3)_2$ and CuI, affording a range of arylacetylenes in good-to-excellent yields (Scheme 8). The recyclability of the $NiCl_2(PPh_3)_2$/CuI/PEG-400/H_2O system was assessed, allowing the recovery and recycling of the catalyst up to six times without a noticeable loss of catalytic activity. Apart from catalyst reuse, this system also avoided the use of easily volatile organic solvents.

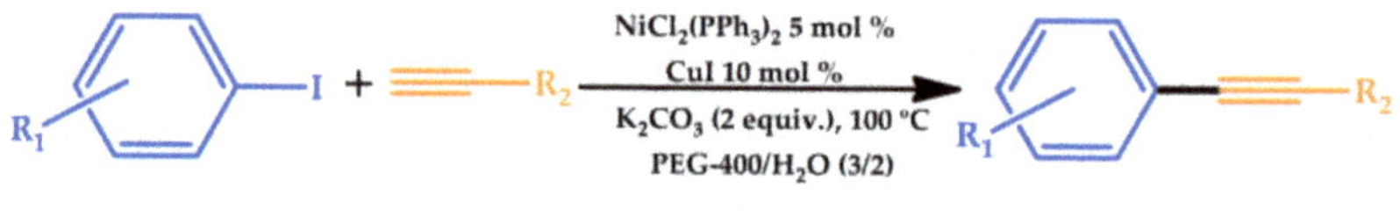

Scheme 8. Nickel-catalyzed Sonogashira coupling in poly(ethylene glycol) (PEG-400)/H_2O [65].

PEG-400 was used a green solvent in the first Mn-catalyzed Sonogashira coupling reaction of aryl iodides under mild reaction conditions [66]. First, the reaction was performed with iodobenzene and phenyl acetylene as the model substrates in PEG-400 at 70 °C, using Mn(OAc)$_2$2H$_2$O as the catalyst and Et$_3$N as the base. Different bases and other commonly applied Mn catalysts were also tested, as well as the effects of organic solvents, with no better results obtained compared with PEG-400. A variety of diarylacetylenes was the obtained in moderate-to-good yields under standard conditions using the proposed catalytic system.

The use of PEG was also the choice of Gautam et al. as an environmentally benign solvent for the first palladacycle-catalyzed carbonylative Sonogashira cross-coupling of aryl iodides (Scheme 9) [67]. The oxime palladacycle (Figure 9) provided a phosphine-free approach for the synthesis of ynones at low Pd loadings, therefore resulting in high catalytic TONs and TOFs extremely higher than the best Pd catalyst reported in the literature for the model reaction of carbonylative cross-coupling between 4-iodoanisole and phenylacetylene. The oxime palladacycle used in this protocol behaved as an efficient and robust catalyst, thereby providing a phosphine-free protocol for the synthesis of ynones. The palladacycle catalyst showed a good recyclability, being reused up to four times with a slight decrease in activity.

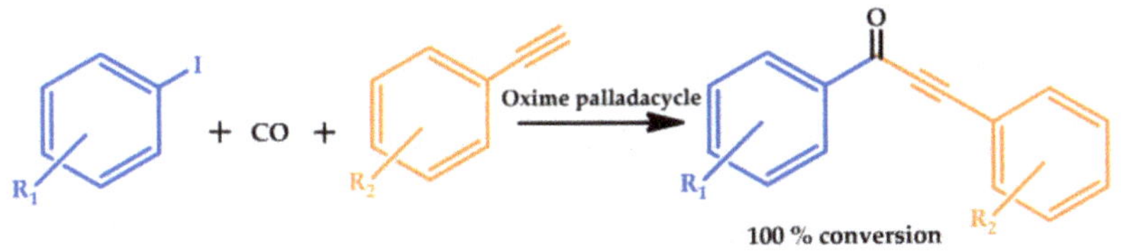

Scheme 9. Oxime palladacycle-catalyzed carbonylative Sonogashira cross-coupling [67].

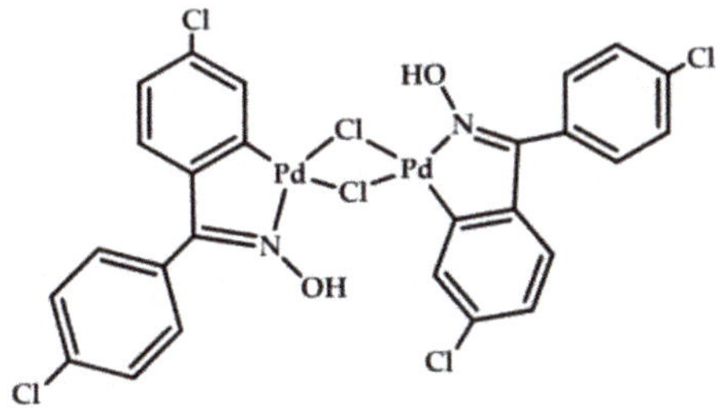

Figure 9. Oxime palladacycle used in the carbonylative Sonogashira cross-coupling of aryl iodides [67].

Khanmoradi et al. described the use of a Pd-vanillin Schiff base complex immobilized on mesoporous MCM-41 (Figure 10) as an efficient catalyst for the Suzuki–Miyaura, Stille, and Mizoroki–Heck reactions carried out in green solvents (H$_2$O and PEG-400) [68].

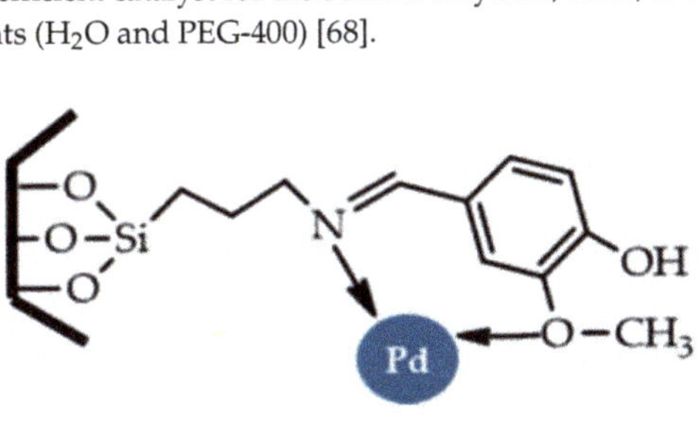

Figure 10. Structure of the Pd-vanillin-MCM-41 catalyst [68].

The results indicated the Pd-vanillin-MCM-41 catalyst was a highly efficient catalyst for various aryl halides under mild conditions (Table 10). Phenyl iodides and phenyl bromides demonstrated a good reactivity, providing the corresponding products in good-to-very good yields for all reactions. The catalyst was reused for five consecutive cycles without a significant loss of its catalytic activity or metal leaching.

Table 10. Catalytic results for the Heck reaction [68].

Solvent	Base (eq.)	Catalyst (mol%)	Yield (%)
PEG-400	K_2CO_3	0.53	72
PEG-400	K_2CO_3	0.71	91
PEG-400	K_2CO_3	0.88	91
EtOH	K_2CO_3	0.71	-
H_2O	K_2CO_3	0.71	75
DMF	K_2CO_3	0.71	83
PEG-400	Et_3N	0.71	90
PEG-400	$NaHCO_3$	0.71	75
PEG-400	$NaHCO_3$	0.71	88
PEG-400	$NaHCO_3$	0.71	35

Reaction conditions: aryl halide (1 mmol), *n*-butyl acrylate (1.2 mmol), and base (3 mmol), 120 °C, 25 min. PEG-400: poly(ethylene glycol).

Zeynizadeh and co-workers recently reported the use of PEG-400 as a green solvent for a Pd-free catalytic system in the Suzuki–Miyaura cross-coupling reaction [69]. A robust magnetic and reusable catalyst, Fe_3O_4@APTMS@$Cp_2ZrCl_{x(x\,=\,0,1,2)}$ magnetic nanoparticles (MNPs) (APTMS: (3-aminopropyl)trimethoxysilane), presenting a core shell structure and previously prepared by the authors, was used in this system. After optimization of the reaction conditions, the scope of the presented procedure was examined for the reaction between various aryl(naphthyl) halides and phenylboronic acid. The current strategy afforded relatively short and acceptable reaction times and good-to-excellent yields of the pure biaryl products (Table 11). The reaction rate was found to be strongly influenced by the presence of electron-donating and electron-withdrawing groups, the latter presenting a higher rate rather comparatively. Furthermore, the TON and TOF values were obtained up to 481 and 2884, respectively, for the mentioned Suzuki–Miyaura cross-coupling reaction.

3.4. Ionic Liquids

Ionic liquids (ILs) are understood as liquids composed of poorly coordinated ions with a melting point below 100 °C that have recently attracted great interest as a "greener" alternative to conventional organic solvents due to their notable physicochemical properties [51]. These include thermal stability; nonflammability; recyclability; low vapor pressure; and catalytic properties, such as nonflammability, nonvolatility, polarity, and stability [70]. The remarkable properties of ILs can be substantially modified by changing the cationic and/or the anionic components, which can give rise to ILs with specific properties via numerous combinations of cations and anions [71]. One of the most extensively studied class of ILs is based on imidazolium cations with an appropriate counter anion, which are known to support many organic transformations. In view of these attributes, ILs are increasingly being used as popular reaction media in synthesis, analysis, catalysis, and separation, as well as medicine and pharmaceuticals, as efficient and sustainable media and, not surprisingly, have been used in cross-coupling reactions as efficient and sustainable media [10].

Table 11. Suzuki–Miyaura cross-coupling reaction in the presence of Fe_3O_4@APTMS@$Cp_2ZrCl_{x(x\ =\ 0,\ 1,\ 2)}$ as a Pd-free nanocatalyst in PEG-400 [69]. APTMS: (3-aminopropyl)trimethoxysilane.

$$R-X + (HO)_2B-C_6H_5 \xrightarrow[\text{PEG-400, }K_2CO_3,\ 70\ °C,\ 10\text{-}90\ min]{Fe_3O_4@APTMS@Cp_2ZrCl_{x(x=0,1,2)}} R-C_6H_5$$

R	Yield (%)	TON	TOF (h^{-1})
C$_6$H$_5$–I	98	481	1424
H$_3$C–C$_6$H$_4$–I	96	471	628
H$_3$CO–C$_6$H$_4$–I	83	407	444
C$_6$H$_5$–Br	95	466	1398
2-CH$_3$-C$_6$H$_4$–Br	80	392	336
H$_3$C–C$_6$H$_4$–Br	83	407	543
CN–C$_6$H$_4$–Br	97	476	1903

In 2015, Patil and coworkers described the use of dual-functionalized task-specific hydroxyl functionalized IL as an excellent promoter in the Pd-catalyzed Suzuki–Miyaura cross-coupling of aryl halides with arylboronic acids in water [72]. The IL cation with hydroxyl functionality induced a reaction in water, acting as a reducing agent, whereas the anion with prolinate functionality served as the ligand and stabilized and/or activated the in-situ generation of PdNPs. The generated PdNPs exhibited excellent catalytic activity towards the Suzuki cross-coupling of aryl halide with aryl boronic acid in water, without the need of a phosphine ligand (Table 12).

Strikingly, the coupling of less activated aryl chlorides also proceeded smoothly with aryl boronic acid. The aqueous system containing ionic liquid along with PdNPs presented a good recyclability, up to seven times, without a noticeable loss of catalytic activity, showing potential for industrial applications.

Hejazifar and coworkers reported the design and properties of surface-active ILs able to form stable microemulsions with heptane and water and their application as a reaction media for Pd-catalyzed cross-coupling reactions [73]. The application of the microemulsions as a reaction media resulted in high reactivity even at a low catalyst loading, with excellent yields > 90% for most products resulting from the coupling of aryl halides or boronic acids with electron-deficient substituents. The gathered data demonstrated the dual role of IL of a surfactant and ligand not limited to the formation of a suitable reaction media, providing the outstanding reactivity reported. The catalytic system behavior allowed the simple and successful product separation and catalyst recycling.

The effects of the IL concentration on the catalytically active species towards Pd-catalyzed cross-coupling reactions was studied by Taskin et al. [74]. A series of surface-active IL were applied as additives in the Heck reaction of ethyl acrylate and iodobenzene, obtaining high yields (> 90%) in water, pointing out the IL concentration as the key factor affecting the formation of the catalytically active species in the reaction, the morphology, and chemical state of the Pd species (Figure 11).

Table 12. Screening of ionic liquids for the Suzuki–Miyaura coupling [72].

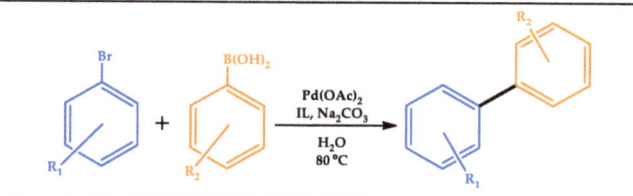

IL	Loading (mol%)	Time (h)	Isolated Yield (%)
[BMIM][Cl]	20	2	65
[BMIM][BF4]	20	1	76
[BMIM][Pro]	20	1	80
[HEMPy][Cl]	20	2	80
[HEMPy][BF4]	20	1	84
[HEMPy][Pro]	20	1	98
1-proline	20	2	50
—	20	2	40
[HEMPy][Pro]	10	1	84
[HEMPy][Pro]	15	1	90
[HEMPy][Pro]	25	1	98
[HEMPy][Pro]	30	1	98

(BMIM) = 1-butyl-3-methylimidazolium, (HEMPy) = 1-(2-hydroxyethyl)-1-methylpyrrolidinium, (Cl) = chloride, (BF4) = tetrafluoroborate, (PF6) = hexafluorophosphate, and (Pro) = prolinate.

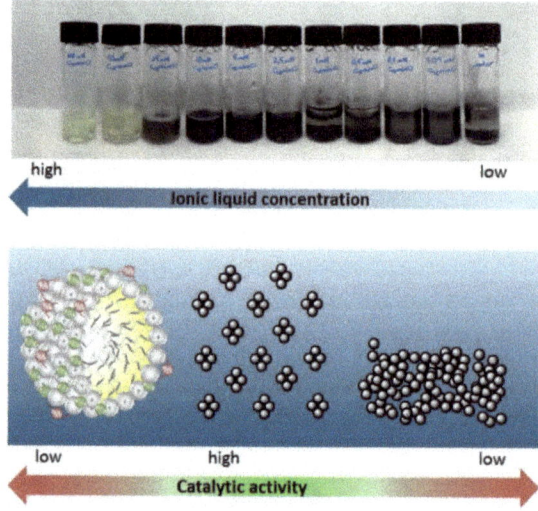

Figure 11. Concentration dependence of Pd-metallomicelles and PdNPs in an aqueous-ionic liquid (IL) micellar solution. Conditions: 2 mL (C$_{12}$mimCl) 1 solution, 0.02 mmol Pd$_2$(allyl)$_2$Cl$_2$, and 0.02 mmol K$_2$CO$_3$ 60 min at 80 °C under air. Reproduced with the permission of reference [74], Copyright 2017 The Royal Society of Chemistry.

At higher concentrations, the formation of a N-heterocyclic carbene complex was observed, while, at lower surfactant concentrations, the rapid decomposition of metallomicelles into catalytically active Pd(0) nanoclusters occurred, a unique behavior only observed for amphiphilic IL that differs significantly from dimethylimidazolium-based IL or from conventional surfactants. The dual role of imidazolium-based surfactants as templates for NP formation through NHC carbene complex formation, as well as stabilizers and surfactants for a consecutive reaction, was postulated by the authors.

The Hayouni research group used their previous expertise in the use of bio-sourced ILs as solvents in hydrogenation processes to explore the potentialities of these ILs in the Heck reaction [75]. Bio-sourced ammonium and phosphonium carboxylate ILs obtained from l-lactic acid or l-malic acid were easily synthesized and used as the solvent for the Heck coupling reaction between iodoarene or halogenoaromatics with *tert*-butylacrylate used $PdCl_2$ as catalyst.

With the use of $NaHCO_3$, good conversion and selectivity were obtained using tetrabutylammonium (TBA) ILs, presenting better results than those for a typical coupling media as the usual organic solvents or commercial ILs, although the catalytic system could not be recycled. The replacement of the base by triethylamine NEt_3, a relative strong homogeneous base, afforded good efficiency, with the advantageous possibility of recycling of the catalytic system without a loss of reactivity up to five times (Table 13).

Table 13. Catalytic results for the Heck reaction with organic solvents and ionic liquids (ILs) with NEt_3 as the base [75].

Solvent or ILs	Cycle	Conversion (%)	Selectivity (%)
toluene	1	1	100
water	1	30	100
DMSO	1	76	80
TBP Br	1	47	100
TBA Cl	1	99	100
	2	100	100
	3	61	100

Reactions conditions: 1.2 mmol of iodobenzene and 600 mg of solvent. TBP: tetrabutylphosphonium, TBA: tetrabutylammonium.

In 2018, solid salt precipitation issues in cross-coupling reactions were addressed by combining basic anions with the trihexyl(tetradecyl)phosphonium ((P_{66614})$^+$) cation, ensuring an IL by-product [76]. Among all synthesized basic salts, the novel IL base (P_{66614})(OH)$_4$MeOH showed the highest basic strength, being the most efficient in the Pd-catalyzed Suzuki–Miyaura cross-coupling reaction (Figure 12). No precipitate was observed after the catalytic run due to the generation of the room temperature IL (P_{66614})Cl. In addition, an easy recycling of the generated IL was predicted simply by elution with MeOH on an ion exchange column, since (P_{66614})Cl was used in the synthesis of the active base ((P_{66614})(OH)$_4$MeOH) in an ion-exchange process.

The enhanced catalytic reaction rates for the Suzuki–Miyaura cross-coupling reaction was highlighted by the authors as another advantage of the basic salt being an IL itself, when (P_{66614})(OH)$_4$MeOH employed as both the base and solvent was preferred over the use of 1,4-dioxane as the solvent.

In 2019, Matias et al. used, for the first time, a C-scorpionate Ni(II) complex (NiCl(κ^3-HC[pz]$_3$))Cl (pz = pyrazol-1-yl) as a catalyst for the MW-assisted Heck–Mizoroki C–C cross-coupling in IL medium, obtaining good yields of the corresponding products [77]. The replacement of common organic solvents by suitable room temperature ILs as reaction medium addressed both the recyclability of the catalyst and sustainability using greener solvents (Scheme 10). Furthermore, the use of the IL allowed an easy separation of the product from the catalyst, since, at high temperatures, the IL dissolves all components of the reaction mixture, but, at room temperature, the substrates and products form a second phase. The stability of the catalyst in the IL allowed its recycling and reuse for several consecutive catalytic cycles with preserved activity.

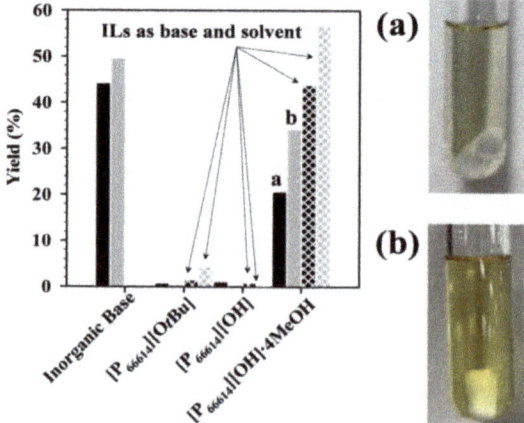

Figure 12. (**Left**): Effect of $(P_{66614})^+$ bases vs. inorganic bases on Pd-catalyzed Buchwald–Hartwig amination (BHA) (black) and Suzuki–Miyaura coupling (SMC) (grey). Solid bars: 1,4-dioxane as the solvent. Crosshatched bars: ILs as both the base and solvent. Inorganic base: KO*t*Bu for BHA and K_2CO_3 for SMC. (**Right**): Final reaction mixtures (with magnetic stir bar) for BHA (**a**) and SMC (**b**) using $(P_{66614})(OH)\cdot 4MeOH$ as the base and dioxane as the solvent. Reproduced with the permission of reference [76], Copyright 2018 The Royal Society of Chemistry.

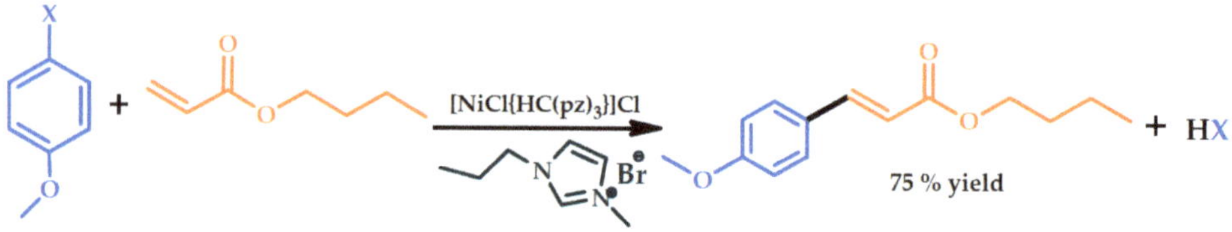

Scheme 10. Heck reaction of 4-bromoanisole and butyl acrylate in 1-butyl-3-methylimidazolium bromide, catalyzed by $[NiCl\{\kappa^3\text{-}HC(pz)_3\}]Cl$ (pz = pyrazol-1-yl) [77].

3.5. Deep Eutectic Solvents

The concept of deep eutectic solvents (DES) was introduced by Abbott and coworkers in 2003 [78] with the deployment of $ChCl:ZnCl_2$ acidic solvent in several reactions as a green alternative for ILs. The concept of DES is, however, quite different from that of traditional ILs, since they are not entirely composed of ionic species. DES are nowadays usually understood as combinations of two or three safe and inexpensive components that are able to establish hydrogen bond interactions with each other to form an eutectic mixture that presents a lower melting point than either of the individual components [12,79]. DES constitute important alternatives to conventional organic solvents due to their interesting properties and benefits, such as an attractive low price, no flammability, ease synthetic accessibility, and benign and safe nature, along with recyclability and biodegradability and a low ecological footprint. The most widely deployed DES are prepared by mixing a hydrogen bond acceptor, exemplified by an inexpensive quaternary ammonium salt such as choline chloride (ChCl, included in the so-called vitamin B4), with a hydrogen bond donor such as sugars, organic and amino acids, urea (most of them are from renewable resources), or glycerol (a biowaste). The use of DES allows to overcome drawbacks of ILs, such as a complicated synthesis procedure, difficult high purity, and sensitivity to air and moisture. For example, the melting points of the DES are lower than the melting points of the ILs. In addition, DES consists of relatively cheaper compounds when compared to IL components. DES allows the solvation of various solutes in high concentrations much higher than their solubility in water. When compared to the conventional solvents, DES are recognized as highly soluble, relatively less toxic, thermally stable, biodegradable, nonflammable, and nonvolatile [12,80].

DES are now a matter of growing interest both at the academic and industrial levels. Extensive and valuable reports have been published highlighting the beneficial impacts of DES in organic reactions, playing a dual role of DES as a solvent and catalyst for many chemical processes [10,12]. Although, the application of neoteric media in metal-catalyzed reactions has witnessed a remarkable growth in recent years, fewer reports of catalytic systems regarding cross-coupling reactions carry out in DES media are available in the literature.

In 2016, the synthesis of different tetrahydroisoquinolines using a DES (choline chloride:ethylene glycol) and a catalyst based on magnetite-supported copper(II) oxide was explored by Marset and coworkers [81]. The presence of a DES medium proved to be essential for the reaction to occur after the testing of other VOC solvents as the reaction medium, being also important for minimizing the lactam formation (Figure 13, left). Excellent results were obtained for the aryl group substituent, bearing both electron-withdrawing or electron-donating groups), whereas, for phenyl derivatives, moderate yields were obtained. A direct proportional relationship was also found between the conductivity of the DES medium and the yield obtained. Moreover, the amount of copper in the catalyst was the lowest loading ever reported [82–85].

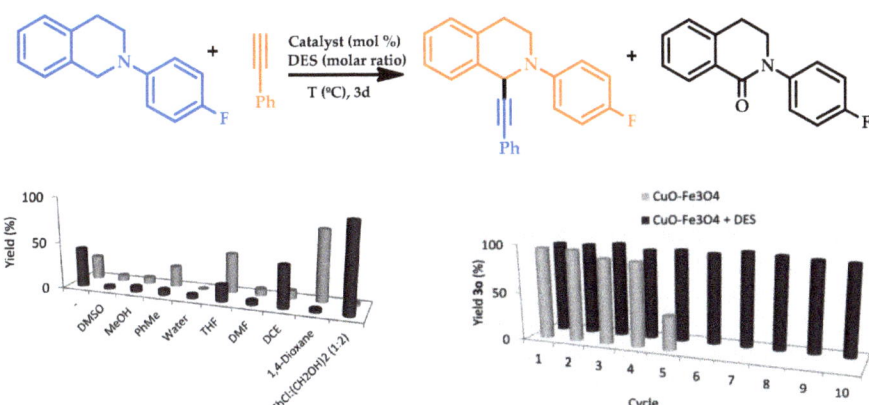

Figure 13. (**Top**) Reaction scheme of 2-(4-fluorophenyl)-1,2,3,4-tetrahydroisoquinoline and phenylacetylene. (**Bottom left**) Obtained yield in different solvents for the products of the reaction shown on top. (**Bottom right**) Recycling of the CuO–Fe$_3$O$_4$ catalyst and CuO–Fe$_3$O$_4$ and deep eutectic solvent (DES). Reproduced with the permission of reference [81], Copyright 2016 The Royal Society of Chemistry.

The mixture (solvent and catalyst) was reused for ten consecutive cycles without any significant loss of activity with this highly sustainable protocol, with water as the only stoichiometric waste, demonstrating the high recyclability of the system (Figure 13, right). Upon catalyst recovery by magnetic decantation, the obtained yield demonstrated a sharp decrease after four reaction cycles.

In 2017, the same research group reported the synthesis of cationic pyridiniophosphine ligands to develop DES-compatible catalytic systems. The ligands were successfully employed to different Pd-catalyzed cross-coupling reactions—namely, Suzuki–Miyaura, Sonogashira, or Heck couplings. Optimization studies were performed to determine the best catalyst for the different reaction. Furthermore, the scope of the cross-coupling reactions was explored, obtaining moderate-to-good yields in general. The cationic phosphines enhanced the catalytic activity of Pd in this polar medium. Moreover, the recyclability of these processes was also investigated, allowing the reuse of both catalyst and DES up to five times without a significant loss of catalytic activity. The DES choline chloride:glycerol (1:2) was used as sustainable medium for the Pd-catalyzed Suzuki–Miyaura cross-couplings between (hetero)aryl halides (Cl, Br, and I) and mono- and bifunctional potassium aryltrifluoroborates. The reaction proceeded efficiently and chemoselectively in the air and under mild conditions, with a low catalyst loading (1 mol%) and employing Na_2CO_3 as the base, without further additional ligands [86]. Overall, valuable biaryls and terphenyl derivatives were obtained in excellent yields (>98%). The DES, catalyst, and base were easily recovered and reused up to six cycles, with an E_{factor} as low as 8.74. The proposed methodology was also applied by the authors to the synthesis of two nonsteroidal anti-inflammatory drugs—namely, Felbinac and Diflunisal.

The above-mentioned DES, ChCl:Gly (1:2), was found to promote Sonogashira cross-couplings using the commercially available Pd/C in the absence of external ligands. Heteroaryl iodides were effectively coupled with both aromatic and aliphatic alkynes under heterogeneous conditions, affording yields in the range 50–99% for 3 h at 60 °C by Messa et al. [87]. This catalytic system was also active towards the Pd-catalyzed coupling electron-poor and electron-rich (hetero)aryl iodides, known to be poorly reactive, with several aromatic alkynes, presenting a broad substrate scope and affording yields ranging from 54% to 99%. The catalyst and the DES used were successfully recovered and reused up to four times (Figure 14), presenting an E_{factor} as low as 24.4. A small change in the ligand structures, from phosphines to cationic pyridiniophosphines, was enough to keep the excellent results in this DES medium, such as those obtained in classical organic solvents. Moreover, the catalyst load increased up to 5 mol% with the use of CuI (20 mol%) as the cocatalyst, demonstrating the feasibility of the difficult SH Pd-catalyzed cross-couplings of alkyl-, trimethylsilyl-, and hydroxybenzyl-substituted alkynes, obtaining yields in the range of 63–93%.

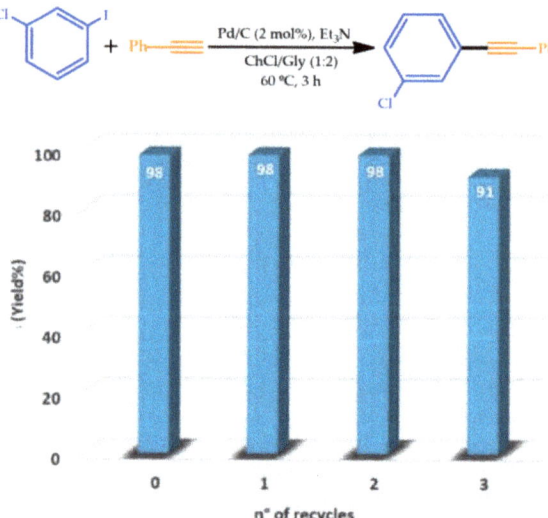

Figure 14. Recycling of Pd/C (2 mol%) and DES in the coupling reaction between 1-chloro-3-iodobenzene and phenylacetylene. Adapted with the permission of reference [87], Copyright 2020 Wiley VCH.

In 2019, Saavedra et al. developed a novel, versatile, and DES-compatible bipyridine Pd complex as a general precatalyst for several cross-coupling reactions [88]. The application of the prepared Pd precatalyst to the Hiyama, Suzuki–Miyaura, Heck–Mizoroki, and Sonogashira cross-coupling reactions in DES under aerobic conditions demonstrated a high catalytic activity. The same Pd precatalyst was able to catalyze a broad number of cross-coupling reactions, keeping the excellent results obtained in classical organic solvents given the hydrogen bond capacity of the ligand.

The catalyst and DES system were easily and successfully recycled up to three or five consecutive cycles, depending on the cross-coupling reaction, without a significant loss of activity. Overall, the authors highlighted the great applicability, versatility, and sustainability of this system.

The first application of DES in the bioamination of ketones was devised by Paris and colleagues [89]. The stability of the amine transaminases (ATAs) was confirmed in DES-buffer mixtures up to 75% (w/w) neoteric solvent, affording a good catalytic performance with high conversions (Figure 15). The solubility of DES allowed the metal-catalyzed step at 200-mM loading of the substrate and the following biotransformation at 25 mM. A chemoenzymatic cascade toward enantiopure biaryl amines was therefore efficiently established owing to the unique properties of DESs. The authors presented an excellent proof of concept for the unparalleled enzymatic activity of ATAs, highlighting the practical value of biorenewable solvents in organic synthesis.

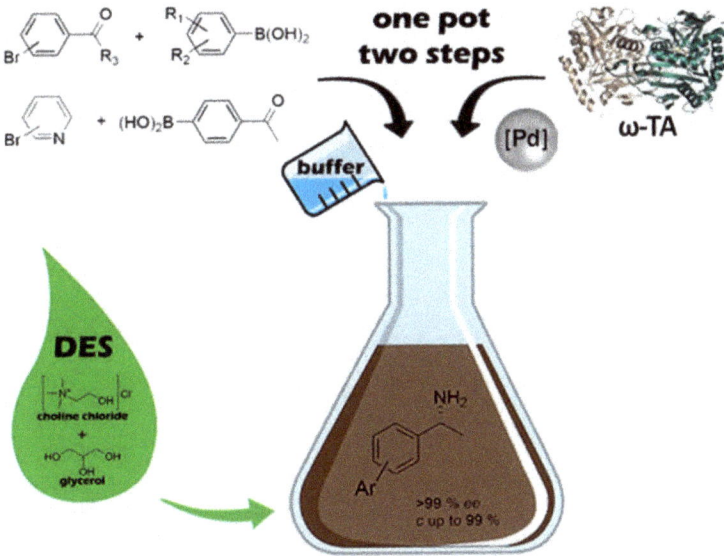

Figure 15. Strategy proposed for the one-pot synthesis of biaryl-substituted amines by combining palladium and enzyme catalysis in DES. Reproduced with the permission of reference [89], Copyright 2019 American Chemical Society.

4. Conclusions

The deployment of unconventional conditions for cross-coupling reactions, whether alternative energy inputs, as well as new green unconventional media individually or combined, has brought tremendous benefits and innovative approaches to synthetic chemistry. The use of these groundbreaking approaches is in complete agreement with the principles of green chemistry through their numerous advantages: change of reactivity, improvement of yields and selectivity, reduction of reaction time, energy saving, and waste minimization.

Regarding alternative activation energy for cross-coupling reactions, important aspects must be taken into consideration. The suggested approaches should be simple, without being too specific, with an underlying mechanism and a systematic characterization of operational parameters and a description of the experimental conditions of the MW, US, and mechanochemistry-assisted reactions to facilitate comparisons between studies. The coupling of different technologies with other unconventional media or methods of activation and the possible synergetic effects arising thereof, as well as new fields of applications, should also be highlighted. The possibility of scaling-up the lab-scale methodologies that provide excellent results would be the true realization of the sustainability goals aimed for by the authors. Considering the novelty witnessed in the scrutinized publications, a bright future for the development of sustainable cross-coupling reactions is envisioned in the years to come.

Author Contributions: Writing—original draft preparation, M.A.A. and writing—review and editing, M.A.A. and L.M.D.R.S.M. All authors have read and agreed to the published version of the manuscript.

Funding: This research was partially funded by FCT through UIDB/00100/2020 of the Centro de Química Estrutural. M.A.A. acknowledges financial support from an UID/QUI/00100/2019-BL/CQE-2017-022 FCT grant.

Conflicts of Interest: The authors declare no conflict of interest.

References

1. Miyaura, N.; Suzuki, A. Palladium-Catalyzed Cross-Coupling Reactions of Organoboron Compounds. *Chem. Rev.* **1995**, *95*, 2457–2483. [CrossRef]
2. Dieck, H.A.; Heck, F.R. Palladium catalyzed synthesis of aryl, heterocyclic and vinylic acetylene derivatives. *J. Organomet. Chem.* **1975**, *93*, 259–263. [CrossRef]
3. Sonogashira, K.; Tohda, Y.; Hagihara, N. A convenient synthesis of acetylenes: Catalytic substitutions of acetylenic hydrogen with bromoalkenes. *Tetrahedron Lett.* **1975**, *50*, 4467–4470. [CrossRef]
4. Chen, X.; Engle, K.M.; Wang, D.H.; Jin-Quan, Y. Palladium(II)-catalyzed C-H aetivation/C-C cross-coupling reactions: Versatility and practicality. *Angew. Chem. Int. Ed.* **2009**, *48*, 5094–5115. [CrossRef] [PubMed]
5. de Barros, S.D.T.; Senra, J.D.; Lachter, E.R.; Malta, L.F.B. Metal-catalyzed cross-coupling reactions with supported nanoparticles: Recent developments and future directions. *Catal. Rev. Sci. Eng.* **2016**, *58*, 439–496. [CrossRef]
6. Molnár, Á. Efficient, selective, and recyclable palladium catalysts in carbon-carbon coupling reactions. *Chem. Rev.* **2011**, *111*, 2251–2320. [CrossRef]
7. Buskes, M.J. Impact of Cross-Coupling Reactions in Drug Discovery and Development. *Molecules* **2020**, *25*, 3493. [CrossRef]
8. Floris, B.; Sabuzi, F.; Galloni, P.; Conte, V. The beneficial sinergy of MW irradiation and ionic liquids in catalysis of organic reactions. *Catalysts* **2017**, *7*. [CrossRef]
9. Sherwood, J.; Clark, J.H.; Fairlamb, I.J.S.; Slattery, J.M. Solvent effects in palladium catalysed cross-coupling reactions. *Green Chem.* **2019**, *21*, 2164–2213. [CrossRef]
10. Hooshmand, S.E.; Heidari, B.; Sedghi, R.; Varma, R.S. Recent advances in the Suzuki-Miyaura cross-coupling reaction using efficient catalysts in eco-friendly media. *Green Chem.* **2019**, *21*, 381–405. [CrossRef]
11. Yousaf, M.; Zahoor, A.F.; Akhtar, R.; Ahmad, M.; Naheed, S. Development of green methodologies for Heck, Chan–Lam, Stille and Suzuki cross-coupling reactions. *Mol. Divers.* **2020**, *24*, 821–839. [CrossRef] [PubMed]
12. Hooshmand, S.E.; Afshari, R.; Ramón, D.J.; Varma, R.S. Deep eutectic solvents: Cutting-edge applications in cross-coupling reactions. *Green Chem.* **2020**, *22*, 3668–3692. [CrossRef]
13. Santoro, S.; Ferlin, F.; Luciani, L.; Ackermann, L.; Vaccaro, L. Biomass-derived solvents as effective media for cross-coupling reactions and C-H functionalization processes. *Green Chem.* **2017**, *19*, 1601–1612. [CrossRef]
14. Baig, R.B.N.; Varma, R.S. Alternative energy input: Mechanochemical, microwave and ultrasound-assisted organic synthesis. *Chem. Soc. Rev.* **2012**, *41*, 1559–1584. [CrossRef] [PubMed]
15. Philippot, K.; Serp, P. Concepts in Nanocatalysis. In *Nanomaterials in Catalysis*, 1st ed.; Wiley-VCH: Weinheim, Germany, 2012; pp. 1–54. ISBN 9783527331246.
16. Kandathil, V.; Koley, T.S.; Manjunatha, K.; Dateer, R.B.; Keri, R.S.; Sasidhar, B.S.; Patil, S.A.; Patil, S.A. A new magnetically recyclable heterogeneous palladium(II) as a green catalyst for Suzuki-Miyaura cross-coupling and reduction of nitroarenes in aqueous medium at room temperature. *Inorg. Chim. Acta* **2018**, *478*, 195–210. [CrossRef]
17. Hajipour, A.R.; Sadeghi, A.R.; Khorsandi, Z. Pd nanoparticles immobilized on magnetic chitosan as a novel reusable catalyst for green Heck and Suzuki cross-coupling reaction: In water at room temperature. *Appl. Organomet. Chem.* **2018**, *32*, 1–11. [CrossRef]
18. Madrahalli Bharamanagowda, M.; Panchangam, R.K. Fe_3O^4-Lignin@Pd-NPs: A highly efficient, magnetically recoverable and recyclable catalyst for Mizoroki-Heck reaction under solvent-free conditions. *Appl. Organomet. Chem.* **2020**, *34*, 1–18. [CrossRef]

19. García-Álvarez, J.; Hevia, E.; Capriati, V. Reactivity of Polar Organometallic Compounds in Unconventional Reaction Media: Challenges and Opportunities. *Eur. J. Org. Chem.* **2015**, *2015*, 6779–6799. [CrossRef]
20. Kappe, C.O.; Dallinger, D.; Murphree, S.S.; Warren, P. *Practical Microwave Synthesis for Organic Chemists*; KGaA: Weinheim, Germany, 2009; ISBN 9783527314522.
21. Kappe, C.O. Controlled microwave heating in modern organic synthesis. *Angew. Chem. Int. Ed.* **2004**, *43*, 6250–6284. [CrossRef]
22. Palma, V.; Barba, D.; Cortese, M.; Martino, M.; Renda, S.; Meloni, E. Microwaves and heterogeneous catalysis: A review on selected catalytic processes. *Catalysts* **2020**, *10*. [CrossRef]
23. Kappe, C.O. Microwave-assisted chemistry. In *Comprehensive Medicinal Chemistry II*; Elsevier: Amsterdam, The Netherlands, 2006; Volume 3, pp. 837–860. [CrossRef]
24. Chatel, G.; Varma, R.S. Ultrasound and microwave irradiation: Contributions of alternative physicochemical activation methods to Green Chemistry. *Green Chem.* **2019**, *21*, 6043–6050. [CrossRef]
25. De La Hoz, A.; Díaz-Ortiz, Á.; Moreno, A. Microwaves in organic synthesis. Thermal and non-thermal microwave effects. *Chem. Soc. Rev.* **2005**, *34*, 164–178. [CrossRef] [PubMed]
26. Gawande, M.B.; Bonifácio, V.D.B.; Luque, R.; Branco, P.S.; Varma, R.S. Solvent-free and catalysts-free chemistry: A benign pathway to sustainability. *ChemSusChem* **2014**, *7*, 24–44. [CrossRef] [PubMed]
27. Martínez, A.V.; Invernizzi, F.; Leal-Duaso, A.; Mayoral, J.A.; García, J.I. Microwave-promoted solventless Mizoroki-Heck reactions catalysed by Pd nanoparticles supported on laponite clay. *RSC Adv.* **2015**, *5*, 10102–10109. [CrossRef]
28. Shah, D.; Kaur, H. Supported palladium nanoparticles: A general sustainable catalyst for microwave enhanced carbon-carbon coupling reactions. *J. Mol. Catal. A Chem.* **2016**, *424*, 171–180. [CrossRef]
29. Massaro, M.; Schembri, V.; Campisciano, V.; Cavallaro, G.; Lazzara, G.; Milioto, S.; Noto, R.; Parisi, F.; Riela, S. Design of PNIPAAM covalently grafted on halloysite nanotubes as a support for metal-based catalysts. *RSC Adv.* **2016**, *6*, 55312–55318. [CrossRef]
30. Lei, Y.; Hu, T.; Wu, X.; Wu, Y.; Xiang, H.; Sun, H.; You, Q.; Zhang, X. Microwave-assisted copper- and palladium-catalyzed sonogashira-type coupling of aryl bromides and iodides with trimethylsilylacetylene. *Tetrahedron Lett.* **2016**, *57*, 1100–1103. [CrossRef]
31. Savitha, B.; Sajith, A.M.; Reddy, E.K.; Kumar, C.S.A.; Padusha, M.S.A. Suzuki-Miyaura cross–coupling reaction in water: Facile synthesis of (hetero) aryl uracil bases using potassiumorganotrifluoroborates under microwave irradiation. *ChemistrySelect* **2016**, *1*, 4721–4725. [CrossRef]
32. Baran, T. A new chitosan Schiff base supported Pd(II) complex for microwave-assisted synthesis of biaryls compounds. *J. Mol. Struct.* **2017**, *1141*, 535–541. [CrossRef]
33. Verbitskiy, E.V.; Baranova, A.A.; Khokhlov, K.O.; Yakovleva, Y.A.; Chuvashov, R.D.; Kim, G.A.; Moiseykin, E.V.; Dinastiya, E.M.; Rusinov, G.L.; Chupakhin, O.N.; et al. New push–pull system based on 4,5,6-tri(het)arylpyrimidine containing carbazole substituents: Synthesis and sensitivity toward nitroaromatic compounds. *Chem. Heterocycl. Compd.* **2018**, *54*, 604–611. [CrossRef]
34. Elazab, H.A.; Sadek, M.A.; El-Idreesy, T.T. Microwave-assisted synthesis of palladium nanoparticles supported on copper oxide in aqueous medium as an efficient catalyst for Suzuki cross-coupling reaction. *Adsorpt. Sci. Technol.* **2018**, *36*, 1352–1365. [CrossRef]
35. Zakharchenko, B.V.; Khomenko, D.M.; Doroshchuk, R.O.; Raspertova, I.V.; Starova, V.S.; Trachevsky, V.V.; Shova, S.; Severynovska, O.V.; Martins, L.M.D.R.S.; Pombeiro, A.J.L.; et al. New palladium(ii) complexes with 3-(2-pyridyl)-5-alkyl-1,2,4-triazole ligands as recyclable C-C coupling catalysts. *New J. Chem.* **2019**, *43*, 10973–10984. [CrossRef]
36. Lupacchini, M.; Mascitti, A.; Giachi, G.; Tonucci, L.; D'Alessandro, N.; Martinez, J.; Colacino, E. Sonochemistry in non-conventional, green solvents or solvent-free reactions. *Tetrahedron* **2017**, *73*, 609–653. [CrossRef]
37. Tabatabaei Rezaei, S.J. PEDOT nanofiber/Pd(0) composite-mediated aqueous Mizoroki–Heck reactions under ultrasonic irradiation: An efficient and green method for the C–C cross-coupling reactions. *J. Iran. Chem. Soc.* **2017**, *14*, 585–594. [CrossRef]
38. Panahi, L.; Naimi-Jamal, M.R.; Mokhtari, J. Ultrasound-assisted Suzuki-Miyaura reaction catalyzed by Pd@Cu2(NH2-BDC)2(DABCO). *J. Organomet. Chem.* **2018**, *868*, 36–46. [CrossRef]
39. Sancheti, S.V.; Gogate, P.R. Intensification of heterogeneously catalyzed Suzuki-Miyaura cross-coupling reaction using ultrasound: Understanding effect of operating parameters. *Ultrason. Sonochem.* **2018**, *40*, 30–39. [CrossRef] [PubMed]

40. Baran, T. Ultrasound-accelerated synthesis of biphenyl compounds using novel Pd(0) nanoparticles immobilized on bio-composite. *Ultrason. Sonochem.* **2018**, *45*, 231–237. [CrossRef]
41. Naeimi, H.; Kiani, F. Hexamethylenetetramine Copper Diiodide Immobilized on Graphene Oxide Nanocomposite as Recyclable Catalyst for Sonochemical Green Synthesis of Diarylethynes. *ChemistrySelect* **2018**, *3*, 13311–13318. [CrossRef]
42. Naeimi, H.; Kiani, F. Magnetically thiamine palladium complex nanocomposites as an effective recyclable catalyst for facile sonochemical cross coupling reaction. *Appl. Organomet. Chem.* **2019**, *33*, 1–12. [CrossRef]
43. Stolle, A.; Szuppa, T.; Leonhardt, S.E.S.; Ondruschka, B. Ball milling in organic synthesis: Solutions and challenges. *Chem. Soc. Rev.* **2011**, *40*, 2317–2329. [CrossRef]
44. Jiang, Z.J.; Li, Z.H.; Yu, J.B.; Su, W.K. Liquid-Assisted Grinding Accelerating: Suzuki-Miyaura Reaction of Aryl Chlorides under High-Speed Ball-Milling Conditions. *J. Org. Chem.* **2016**, *81*, 10049–10055. [CrossRef] [PubMed]
45. Shi, W.; Yu, J.; Jiang, Z.; Shao, Q.; Su, W. Encaging palladium(0) in layered double hydroxide: A sustainable catalyst for solvent-free and ligand-free Heck reaction in a ball mill. *Beilstein J. Org. Chem.* **2017**, *13*, 1661–1668. [CrossRef] [PubMed]
46. Cao, Q.; Howard, J.L.; Wheatley, E.; Browne, D.L. Mechanochemical Activation of Zinc and Application to Negishi Cross-Coupling. *Angew. Chem. Int. Ed.* **2018**, *57*, 11339–11343. [CrossRef] [PubMed]
47. Seo, T.; Ishiyama, T.; Kubota, K.; Ito, H. Solid-state Suzuki-Miyaura cross-coupling reactions: Olefin-accelerated C-C coupling using mechanochemistry. *Chem. Sci.* **2019**, *10*, 8202–8210. [CrossRef]
48. Vogt, C.G.; Grätz, S.; Lukin, S.; Halasz, I.; Etter, M.; Evans, J.D.; Borchardt, L. Direct Mechanocatalysis: Palladium as Milling Media and Catalyst in the Mechanochemical Suzuki Polymerization. *Angew. Chem. Int. Ed.* **2019**, *58*, 18942–18947. [CrossRef]
49. Pentsak, E.O.; Ananikov, V.P. Pseudo-Solid-State Suzuki–Miyaura Reaction and the Role of Water Formed by Dehydration of Arylboronic Acids. *Eur. J. Org. Chem.* **2019**, *2019*, 4239–4247. [CrossRef]
50. Soliman, M.M.A.; Peixoto, A.F.; Ribeiro, A.P.C.; Kopylovich, M.N.; Alegria, E.C.B.A.; Pombeiro, A.J.L. Mechanochemical preparation of Pd(II) and Pt(II) composites with carbonaceous materials and their application in the Suzuki-Miyaura reaction at several energy inputs. *Molecules* **2020**, *25*. [CrossRef]
51. Lipshutz, B.H.; Gallou, F.; Handa, S. Evolution of solvents in organic chemistry. *ACS Sustain. Chem. Eng.* **2016**, *4*, 5838–5849. [CrossRef]
52. Dumonteil, G.; Hiebel, M.A.; Berteina-Raboin, S. Solvent-freesolvent-free mizoroki-heckmizoroki-heck reaction applied to the synthesis of abscisic acid and some derivatives. *Catalysts* **2018**, *8*. [CrossRef]
53. Baran, T.; Yılmaz Baran, N.; Menteş, A. Highly active and recyclable heterogeneous palladium catalyst derived from guar gum for fabrication of biaryl compounds. *Int. J. Biol. Macromol.* **2019**, *132*, 1147–1154. [CrossRef]
54. Gribanov, P.S.; Chesnokov, G.A.; Dzhevakov, P.B.; Kirilenko, N.Y.; Rzhevskiy, S.A.; Ageshina, A.A.; Topchiy, M.A.; Bermeshev, M.V.; Asachenko, A.F.; Nechaev, M.S. Solvent-free Suzuki and Stille cross-coupling reactions of 4- and 5-halo-1,2,3-triazoles. *Mendeleev Commun.* **2019**, *29*, 147–149. [CrossRef]
55. Chatterjee, A.; Ward, T.R. Recent Advances in the Palladium Catalyzed Suzuki-Miyaura Cross-Coupling Reaction in Water. *Catal. Lett.* **2016**, *146*, 820–840. [CrossRef]
56. Handa, S.; Smith, J.D.; Hageman, M.S.; Gonzalez, M.; Lipshutz, B.H.; Handa, S.; Smith, J.D.; Hageman, M.S.; Gonzalez, M.; Lipshutz, B.H. Synergistic and Selective Copper/ppm Pd-Catalyzed Suzuki-Miyaura Couplings: In Water, Mild Conditions, with Recycling. *ACS Catal.* **2016**, *6*, 8179–8183. [CrossRef]
57. Guarnizo, A.; Angurell, I.; Muller, G.; Llorca, J.; Seco, M.; Rossell, O.; Rossell, M.D. Highly water-dispersible magnetite-supported Pd nanoparticles and single atoms as excellent catalysts for Suzuki and hydrogenation reactions. *RSC Adv.* **2016**, *6*, 68675–68684. [CrossRef]
58. Ghazali-Esfahani, S.; PĂunescu, E.; Bagherzadeh, M.; Fei, Z.; Laurenczy, G.; Dyson, P.J. A simple catalyst for aqueous phase Suzuki reactions based on palladium nanoparticles immobilized on an ionic polymer. *Sci. China Chem.* **2016**, *59*, 482–486. [CrossRef]
59. Ahadi, A.; Rostamnia, S.; Panahi, P.; Wilson, L.D.; Kong, Q.; An, Z.; Shokouhimehr, M. Palladium comprising dicationic bipyridinium supported periodic mesoporous organosilica (PMO): Pd@Bipy-PMO as an efficient hybrid catalyst for suzuki-miyaura cross-coupling reaction in water. *Catalysts* **2019**, *9*. [CrossRef]

60. Wang, Y.; Liu, Y.; Zhang, W.Q.; Sun, H.; Zhang, K.; Jian, Y.; Gu, Q.; Zhang, G.; Li, J.; Gao, Z. Sustainable Ligand-Free, Palladium-Catalyzed Suzuki–Miyaura Reactions in Water: Insights into the Role of Base. *ChemSusChem* **2019**, *12*, 5265–5273. [CrossRef]
61. Lambert, R.; Wirotius, A.L.; Vignolle, J.; Taton, D. C-C couplings in water by micellar catalysis at low loadings from a recyclable polymer-supported Pd(ii)-NHC nanocatalyst. *Polym. Chem.* **2019**, *10*, 460–466. [CrossRef]
62. Jakobi, M.; Gallou, F.; Sparr, C.; Parmentier, M. A General Protocol for Robust Sonogashira Reactions in Micellar Medium. *Helv. Chim. Acta* **2019**, *102*. [CrossRef]
63. Pires, M.J.D.; Purificação, S.I.; Santos, A.S.; Marques, M.M.B. The Role of PEG on Pd- and Cu-Catalyzed Cross-Coupling Reactions. *Synthesis* **2017**, *49*, 2337–2350. [CrossRef]
64. Soni, J.; Sahiba, N.; Sethiya, A.; Agarwal, S. Polyethylene glycol: A promising approach for sustainable organic synthesis. *J. Mol. Liq.* **2020**, *315*, 113766. [CrossRef]
65. Wei, T.; Zhang, T.; Huang, B.; Tuo, Y.; Cai, M. Recyclable and reusable NiCl2(PPh3)2/CuI/PEG-400/H2O system for the sonogashira coupling reaction of aryl iodides with alkynes. *Appl. Organomet. Chem.* **2015**, *29*, 846–849. [CrossRef]
66. Qi, X.; Jiang, L.B.; Wu, X.F. Manganese-catalyzed Sonogashira coupling of aryl iodides. *Tetrahedron Lett.* **2016**, *57*, 1706–1710. [CrossRef]
67. Gautam, P.; Bhanage, B.M. Oxime Palladacycle Catalyzed Carbonylative Sonogashira Cross-Coupling with High Turnovers in PEG as a Benign and Recyclable Solvent System. *ChemistrySelect* **2016**, *1*, 5463–5470. [CrossRef]
68. Khanmoradi, M.; Nikoorazm, M.; Ghorbani-Choghamarani, A. Synthesis and Characterization of Pd Schiff Base Complex Immobilized onto Functionalized Nanoporous MCM-41 and its Catalytic Efficacy in the Suzuki, Heck and Stille Coupling Reactions. *Catal. Lett.* **2017**, *147*, 1114–1126. [CrossRef]
69. Zeynizadeh, B.; Mousavi, H.; Sepehraddin, F. A green and efficient Pd-free protocol for the Suzuki–Miyaura cross-coupling reaction using Fe3O4@APTMS@Cp2ZrClx(x = 0, 1, 2) MNPs in PEG-400. *Res. Chem. Intermed.* **2020**, *46*, 3361–3382. [CrossRef]
70. Hallett, J.P.; Welton, T. Room-temperature ionic liquids: Solvents for synthesis and catalysis. 2. *Chem. Rev.* **2011**, *111*, 3508–3576. [CrossRef]
71. Hayes, R.; Warr, G.G.; Atkin, R. Structure and Nanostructure in Ionic Liquids. *Chem. Rev.* **2015**, *115*, 6357–6426. [CrossRef]
72. Patil, J.D.; Korade, S.N.; Patil, S.A.; Gaikwad, D.S.; Pore, D.M. Dual functionalized task specific ionic liquid promoted in situ generation of palladium nanoparticles in water: Synergic catalytic system for Suzuki-Miyaura cross coupling. *RSC Adv.* **2015**, *5*, 79061–79069. [CrossRef]
73. Hejazifar, M.; Earle, M.; Seddon, K.R.; Weber, S.; Zirbs, R.; Bica, K. Ionic Liquid-Based Microemulsions in Catalysis. *J. Org. Chem.* **2016**, *81*, 12332–12339. [CrossRef]
74. Taskin, M.; Cognigni, A.; Zirbs, R.; Reimhult, E.; Bica, K. Surface-active ionic liquids for palladium-catalysed cross coupling in water: Effect of ionic liquid concentration on the catalytically active species. *RSC Adv.* **2017**, *7*, 41144–41151. [CrossRef] [PubMed]
75. Hayouni, S.; Ferlin, N.; Bouquillon, S. High catalytic and recyclable systems for heck reactions in biosourced ionic liquids. *Mol. Catal.* **2017**, *437*, 121–129. [CrossRef]
76. Choudhary, H.; Berton, P.; Gurau, G.; Myerson, A.S.; Rogers, R.D. Ionic liquids in cross-coupling reactions: "liquid" solutions to a "solid" precipitation problem. *Chem. Commun.* **2018**, *54*, 2056–2059. [CrossRef] [PubMed]
77. Matias, I.A.S.; Ribeiro, A.P.C.; Martins, L.M.D.R.S. New C-scorpionate nickel(II) catalyst for Heck C–C coupling under unconventional conditions. *J. Organomet. Chem.* **2019**, *896*, 32–37. [CrossRef]
78. Abbott, A.P.; Capper, G.; Davies, D.L.; Rasheed, R.K.; Tambyrajah, V. Novel solvent properties of choline chloride/urea mixtures. *Chem. Commun.* **2003**, 70–71. [CrossRef] [PubMed]
79. Porcheddu, A.; Colacino, E.; De Luca, L.; Delogu, F. Metal-Mediated and Metal-Catalyzed Reactions under Mechanochemical Conditions. *ACS Catal.* **2020**, *10*, 8344–8394. [CrossRef]
80. Paiva, A.; Craveiro, R.; Aroso, I.; Martins, M.; Reis, R.L.; Duarte, A.R.C. Natural deep eutectic solvents—Solvents for the 21st century. *ACS Sustain. Chem. Eng.* **2014**, *2*, 1063–1071. [CrossRef]
81. Marset, X.; Pérez, J.M.; Ramón, D.J. Cross-dehydrogenative coupling reaction using copper oxide impregnated on magnetite in deep eutectic solvents. *Green Chem.* **2016**, *18*, 826–833. [CrossRef]

82. Zhao, H.Y.; Wu, F.S.; Yang, L.; Liang, Y.; Cao, X.L.; Wang, H.S.; Pan, Y.M. Catalyst- and solvent-free approach to 2-arylated quinolines via [5 + 1] annulation of 2- methylquinolines with diynones. *RSC Adv.* **2018**, *8*, 4584–4587. [CrossRef]
83. Kumar, R.; Kumar, I.; Sharma, R.; Sharma, U. Catalyst and solvent-free alkylation of quinoline N-oxides with olefins: A direct access to quinoline-substituted α-hydroxy carboxylic derivatives. *Org. Biomol. Chem.* **2016**, *14*, 2613–2617. [CrossRef]
84. Boess, E.; Schmitz, C.; Klussmann, M. A comparative mechanistic study of Cu-catalyzed oxidative coupling reactions with N-phenyltetrahydroisoquinoline. *J. Am. Chem. Soc.* **2012**, *134*, 5317–5325. [CrossRef] [PubMed]
85. Baslé, O.; Li, C.J. Copper catalyzed oxidative alkylation of sp3 C-H bond adjacent to a nitrogen atom using molecular oxygen in water. *Green Chem.* **2007**, *9*, 1047–1050. [CrossRef]
86. Dilauro, G.; García, S.M.; Tagarelli, D.; Vitale, P.; Perna, F.M.; Capriati, V. Ligand-Free Bioinspired Suzuki–Miyaura Coupling Reactions using Aryltrifluoroborates as Effective Partners in Deep Eutectic Solvents. *ChemSusChem* **2018**, *11*, 3495–3501. [CrossRef] [PubMed]
87. Messa, F.; Dilauro, G.; Perna, F.M.; Vitale, P.; Capriati, V.; Salomone, A. Sustainable Ligand-Free Heterogeneous Palladium-Catalyzed Sonogashira Cross-Coupling Reaction in Deep Eutectic Solvents. *ChemCatChem* **2020**, *12*, 1979–1984. [CrossRef]
88. Saavedra, B.; González-Gallardo, N.; Meli, A.; Ramón, D.J. A Bipyridine-Palladium Derivative as General Pre-Catalyst for Cross-Coupling Reactions in Deep Eutectic Solvents. *Adv. Synth. Catal.* **2019**, *361*, 3868–3879. [CrossRef]
89. Paris, J.; Telzerow, A.; Ríos-Lombardía, N.; Steiner, K.; Schwab, H.; Morís, F.; Gröger, H.; González-Sabín, J. Enantioselective One-Pot Synthesis of Biaryl-Substituted Amines by Combining Palladium and Enzyme Catalysis in Deep Eutectic Solvents. *ACS Sustain. Chem. Eng.* **2019**, *7*, 5486–5493. [CrossRef]

Publisher's Note: MDPI stays neutral with regard to jurisdictional claims in published maps and institutional affiliations.

© 2020 by the authors. Licensee MDPI, Basel, Switzerland. This article is an open access article distributed under the terms and conditions of the Creative Commons Attribution (CC BY) license (http://creativecommons.org/licenses/by/4.0/).

Article

Microwave-Assisted Synthesis and Properties of Novel Hexaazatrinaphthylene Dendritic Scaffolds

Daniel García Velázquez [1,*], Rafael Luque [2] and Ángel Gutiérrez Ravelo [3,*]

1. Departamento de Ciencias, Colegio Hispano Inglés, S.A. Rambla de Santa Cruz, 94. 38004 S/C Tenerife, Spain
2. Departamento de Química Orgánica, Universidad de Cordoba, Campus de Rabanales, Edificio Marie Curie (C-3) Ctra Nnal IV-A, Km. 396 E-14014 Cordoba, Spain; q62alsor@uco.es
3. Instituto Universitario de Bio-Orgánica "Antonio González", Universidad de La Laguna, C/Astrofísico Francisco Sánchez, 2, 38206 La Laguna, Tenerife, Spain
* Correspondence: dgvelazq@ull.es (D.G.V.); agravelo@ull.es (Á.G.R.)

Academic Editor: Giuseppe Cirillo
Received: 31 August 2020; Accepted: 16 October 2020; Published: 30 October 2020

Abstract: A novel family of water-soluble π-conjugated hexaazatrinaphthylenes-based dendritic architectures constructed by hexaketocyclohexane and 1,2,4,5-benzenetetramine units is developed in a microwave-assisted organic synthesis (MAOS) approach. The structures and purity of these compounds are verified by ^{1}H and ^{13}C-NMR, MALDI-TOF MS, UV-vis, elemental analysis, DSC, AFM, STM and cyclic voltammetry.

Keywords: dendrimers; microwaves; hexaazatrinaphthylenes

1. Introduction

π-conjugated dendritic architectures have attracted a great deal of attention in recent years as their design and synthesis was shown to render unusual molecular structures and interesting assemblies [1,2]. These dendrimers also possess relevant applications as active chemical components in electronic and optoelectronic devices [2], in biological and material sciences [3], and as photocrosslinkable [4] and photoswitchable devices [5]. Water-compatibility is one of the key properties of such dendritic scaffolds, particularly interesting in view of their utilisation in biological fluids and potentially anti-cancer treatment.

Hexaazatrinaphthylene (HATNA) derivatives are interesting compounds that have a variety of properties [6], including liquid-crystal (discotic) [7], *n*-type semiconduction [8], magnetism [9] and even fluorescence [10], depending on the type of substituent within the structure. Due to this range of properties, an efficient, simple and tuneable preparation of such compounds to make them water-soluble will be highly desirable with regards to their applications and compatibility in aqueous chemistry.

With this important concept in mind, herein we report the design and simple preparation of a novel series of water-soluble π-conjugated HATNAs (**G1**, **G2** and **G3**, see Figure 1).

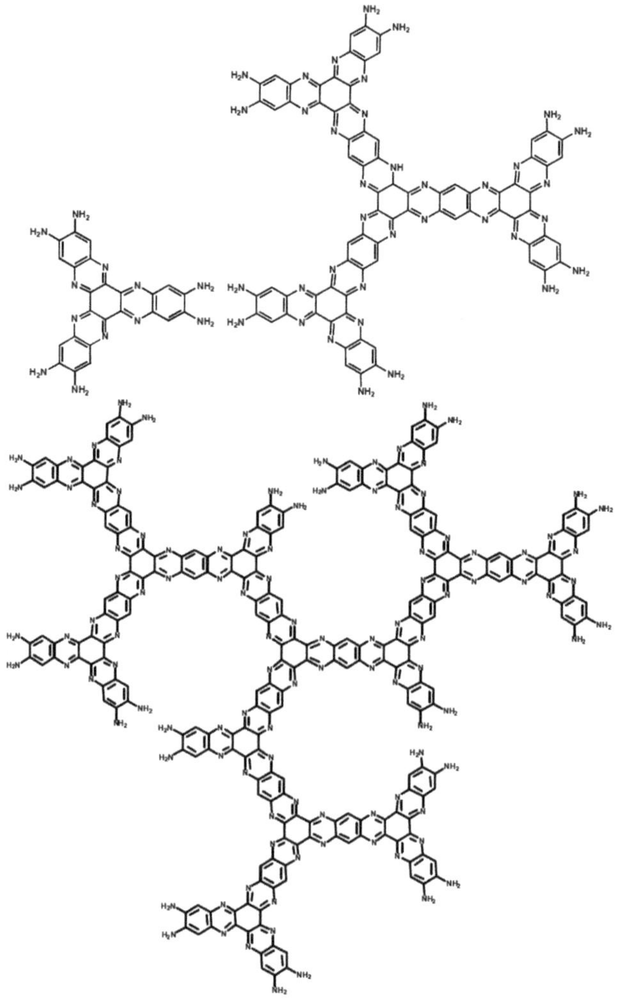

Figure 1. New π-conjugated dendritic architectures of Hexaazatrinaphthylenes (HATNAs) (**G1**, **G2** and **G3**).

These molecules were synthesized using an efficient microwave-assisted approach from hexaketocyclohexane octahydrate (**1**) and 1,2,4,5-benzenetetramine tetrahydrochloride (**2**) as building blocks. To the best of our knowledge, these compounds are the first examples of dendritic scaffolds based on HATNAs units.

2. Results and Discussion

The optimized conditions for the synthesis of **G1** were achieved [11] when **1** was heated with 3.75 equiv. of **2** in a mixture of EtOH-HOAc glacial 8:2 under microwave irradiation for 30 min at 160 °C (87% yield) (Figure 2). **G2** and **G3** could be respectively obtained in 82% and 85% yields, under similar reaction conditions (1 equiv. **1** and 3 equiv. **G1** and **G2**, respectively; see ESI). Condensing **G1** and an excess of corresponding acyl chlorides, five derivatives (**3a–e**) were synthesized (Figure 3). These types of materials (**3a–e**) have six amide groups in the aromatic π-electron system that contribute to the electron-withdrawing effect. Compound **5** was synthesized by condensation of **G1** and orthoquinone **4** [12] as shown in Figure 3.

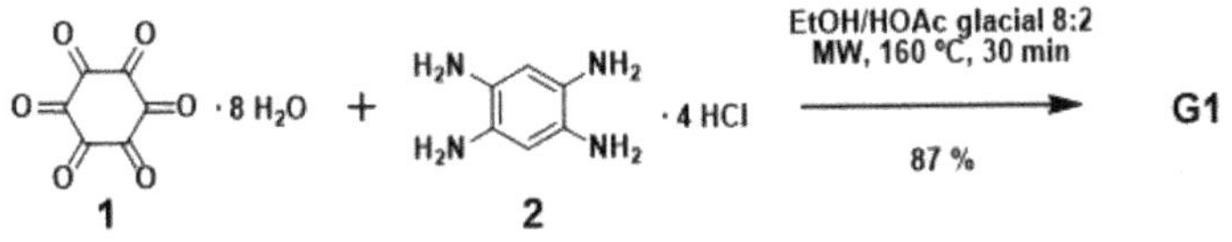

Figure 2. Synthesis of compound **G1**.

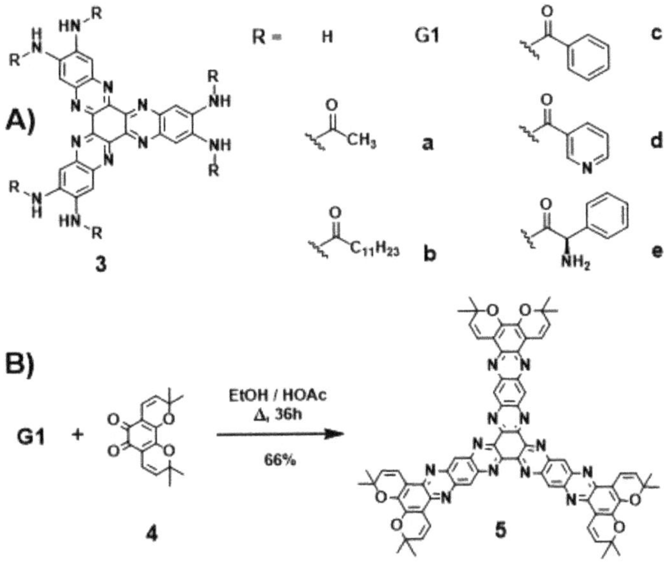

Figure 3. (**A**) Chemical structure of hexaamides **3a–e** derived from **G1**. (**B**) Synthesis of **5** for condensation of **4** and **G1**.

All compounds were purified by chromatography and crystallization. ^{1}H-NMR and ^{13}C-NMR spectra, MALDI-TOF MS, UV-vis, FT-IR and elemental analysis, unambiguously proved the structures (see ESI). The self-organization of **G1** into supramolecular nanostructures resulted from the interplay balancing of intramolecular, intermolecular and interfacial interactions. This self-assembly phenomenon was further investigated by ^{1}H-NMR, DSC, STM and AFM (see below).

^{1}H-NMR spectra showed that chemical shifts and line widths of **G1** are strongly dependent on the concentration (Figure 4) due to aggregation effects, in good agreement with previous reports.[13] Molecular interactions are indeed stronger at dilute concentration (ca. 10^{-5} M) [13]. ^{1}H-NMR chemical shifts (DMSO-d_6) of the aromatic protons for **3a–e** and **5** are around δ 6.76–7.83 ppm, moving to higher/lower field as compared with unsubstituted derivate **G1**. The dendritic structures **G1**, **G2** and **G3** present a low solubility in chloroform, dichloromethane and acetone, but are readily soluble in DMF, DMSO, ethanol and water.

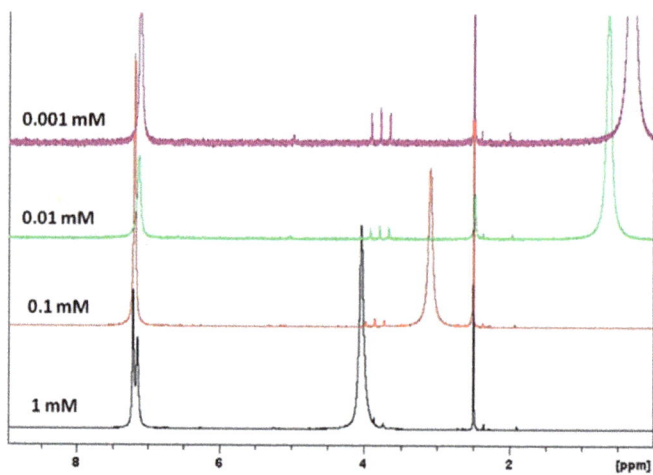

Figure 4. ^{1}H-NMR spectra of compound **G1** in DMSO-d^6 at 0.001, 0.01, 0.1 and 1 mM at 20 °C.

Therefore, the formation of hydrogen bonds causes the insolubility due to structural defects in columnar ordering that might crosslink neighboring columns via H-bonding, enforcing the intra-columnar stacking order [14]. Thus, neighboring columns crosslinking via hydrogen bonding promote intra-columnar stacking order. Nevertheless, the distortion from the planarity of the aromatic frameworks of **3a–e** due to the bulky groups brought high solubilities, presumably through the suppression of aggregation of the aromatic π-systems. Several attempts to crystallize all compounds in different solvent mixtures were unsuccessful, until now. According to molecular modeling, the diameters of **G1** and **G2** are about ca. 16.6 and 29.1 nm (see ESI) with a molecular weight of 474 and 1488 u.m.a., respectively (Figure 5).

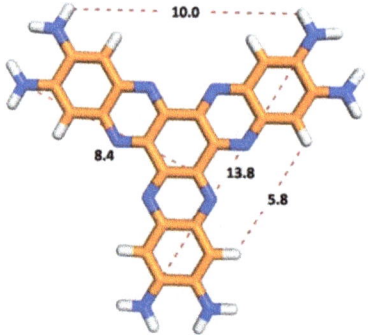

Figure 5. Optimized geometry of **G1** (B3LYP/6-31g*, vacuum) and distances (Å) between atoms.

When the aggregate of **G1** was formed in a homogeneous aqueous solution at moderate or dilute concentration, the aggregation behavior was analyzed conveniently by spectroscopic methods such as ^{1}H-NMR, UV-vis and MALDI-TOF MS (see ESI). The amine groups can maintain a subtle balance between HATNA-HATNA interaction and HATNA-solvent interaction to provide the one-dimensional aggregate, which was confirmed by means of UV-vis spectroscopy (Table 1). In the ethanolic solution, **G1** provides two absorption bands around 209 and 338 nm (Figure 6). The position of the emission maximum peaks undergoes a pronounced bathochromic and hyperchromic effect [15] with an increasing number of days from its preparation (Table 1), which indicates the formation of aggregates. The former two bands can be assigned to the transition from the highest ground state to the ν = 0 level of the lowest excited state (0–0 transition) and to the ν = 1 level (0–1 transition), respectively. The concentration-dependent spectral change was observable in aqueous solutions, which is attributed to dynamic exchange between monomer and aggregate species. Similar photophysical behaviors of three dendritic systems, **G1**, **G2** and **G3**, implied that the effective conjugation length did not improve as the dendritic generation increased.

Table 1. UV-vis spectral data for **G1** 10^{-5} M in ethanol at 20 °C.

	Days			
Max. Peak 1	1	3	5	7
Absorbance	0.1831	0.2611	0.3744	0.6018
λ (nm)	209	244	248	260
ε (M^{-1} cm^{-1})	18,310	26,110	37,440	50,180
	Days			
Max. Peak 2	1	3	5	7
Absorbance	0.1373	0.2225	0.3326	0.4463
λ (nm)	338	349	355	367
ε (M^{-1} cm^{-1})	13,730	22,250	33,260	44,630

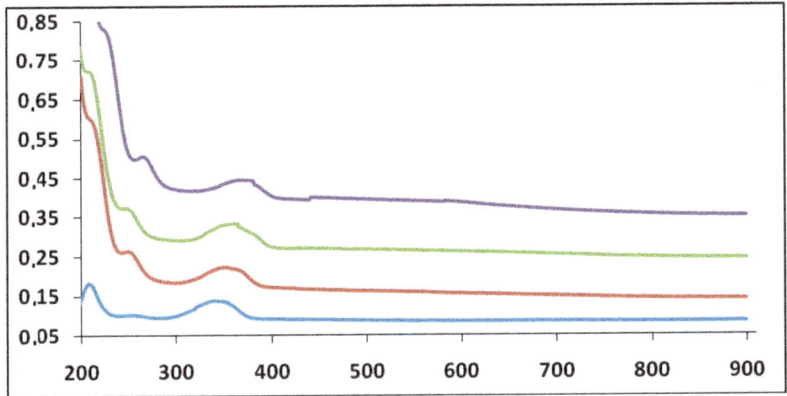

Figure 6. UV-vis spectra (A vs. λ nm) for compound **G1** 10^{-5} M on different days at 20 °C (ethanol as solvent). Day 1 (blue line). Day 3 (red line). Day 5 (green line). Day 7 (violet line) from the preparation of the solution.

FT-IR data in the solution state confirm the presence of amino groups and the 1,2,4,5-tetrasubstituted aromatic ring of compounds **G1**, **G2** and **G3**. Theoretically, up to six hydrogen bonds can be formed between successive coplanar disks within the same column. However, the fractions of intra- and inter-molecular hydrogen bonds were not quantified in the present study. FTIR data for **3a–e** in the solid-state provides evidence for the existence of hydrogen bonds. The two NH stretching vibrations in IR spectra located at 2910 and 3100 cm^{-1} are shifted to lower energy as compared to that of free NH groups [16]. The presence of only one signal around 1650–1690 cm^{-1} corresponding to the carbonyl group is indicative of the participation of all CO groups in the hydrogen bonds [17].

Table 2 shows the thermal behaviour of **G1**, **G2** and **G3** dendritic assemblies. All compounds possessed high thermal stability and decomposed above 250 °C. Thermal gravimetric analysis (TGA) showed no weight loss up to 275 °C. Glass transition temperatures (T_g) ranged from 142 to 163 °C, while the crystallization transition temperature (T_c) range was 165–238 °C.

Table 2. Mesophase assignment and transition temperatures, °C (onset)[a] of dendritic architectures. Glass-transition (T_g), Crystallization (T_c), and Melting (T_m) temperatures of **G1**, **G2** and **G3** compounds (transition enthalpies between parenthesis; J g^{-1}).

Compound	T_g [a] [°C]	T_{c1} [a] [°C]	T_{c2} [a] [°C]	T_m [a] [°C]
G1	163	192 (69)	238 (44)	300
G2	142	165 (90)	220 (26)	275
G3	160	196 (72)	—	254

[a] Measured by DSC at a heating and cooling rate of 10 °C min^{-1}. The data from second heating scan and first cooling scan are given and were found to be fully reproducible.

DSC results showed that **G1**-derivatives **3a–e** (HATNA-NHCOR) and **5** did not form columnar liquid crystalline phases as a consequence of the repulsion between adjacent cores (due to the large negatively charged nitrogen atoms) [18]. DSC curves of **3a–e** and **5** displayed a broad endothermic peak increasing in intensity from 120 to 270 °C (maximum intensity peak) upon heating from RT to 350 °C (Figure 7).

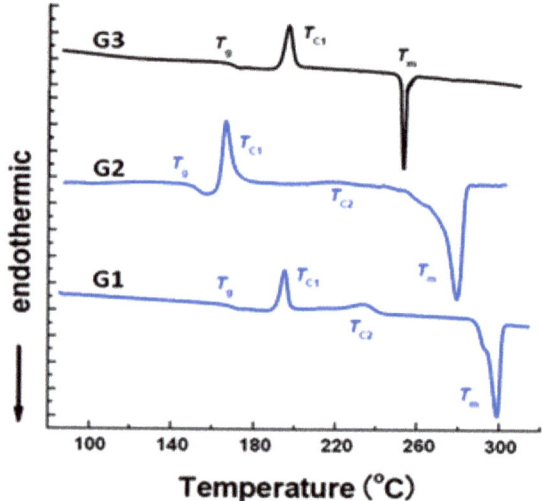

Figure 7. DSC traces of the second heating for compounds **G1**, **G2** and **G3**. All measurements were carried out with heating rate of 10 °C/min.

The associated enthalpy variation (25–76 J·g^{-1}) suggests that phase transitions have a strong first-order character. The non-mesogenic behaviour could be related to stabilizing forces induced by van der Waals interactions linked to the aromatic cores charge distribution (see the Mulliken population analysis performed using DFT calculation).

X-ray scattering experiments of **G1** were performed with unoriented powder samples at room temperature (see ESI, Figure S42) and confirmed the columnar mesophase. The X-ray patterns revealed two main features: a series of reflections at relatively small angles and a reflection at large angles corresponding to Bragg spacing of 0.37 nm (core–core separation), indicating a two-dimensional arrangement of the columnar cross-sections in a hexagonal lattice. These data point to the self-organisation of compound **G1** into a columnar π–π stacking phase.

STM and AFM experiments were subsequently conducted using different supports, namely Au(111) and mica, in order to confirm the aggregation behaviour of **G1** in aqueous solutions. Isolated discrete particles (less than 200 particles μm^{-2}) could be found on the surface of Au(111) as shown in Figure 8a. Although the smallest spots in Figure 8a correspond to particles with sizes in the range of

1.5–3 nm, the majority of them, statistically speaking, are around 2.4 nm (Figure 8b) and range from 2–3 Å width.

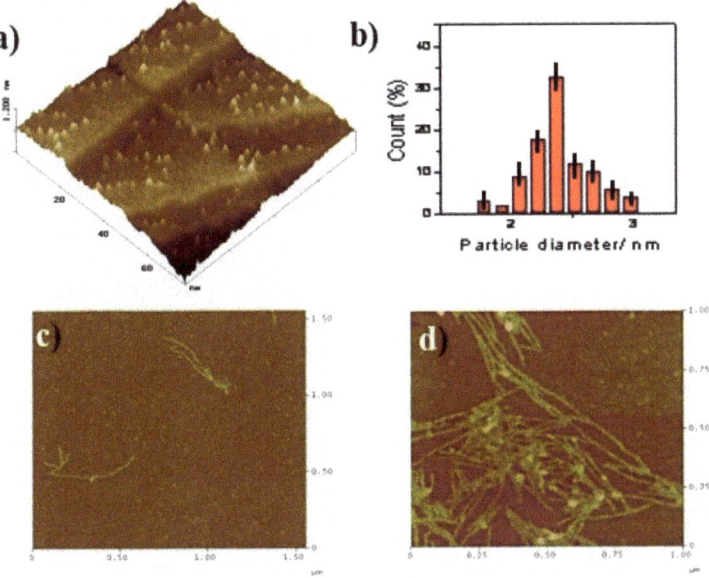

Figure 8. (a) 75 nm × 75 nm STM 3D image of the Au(111) surface after 1 min immersion into a **G1** 10^{-9} M water solution. (b) Particle size histogram of **G1**. AFM image of mica surface after different times of immersion into a **G1** 10^{-5} M water solution. (c) 1.75 µm × 1.75 µm, $t = 1$ min. (d) 1.00 µm × 1.00 µm, $t = 5$ min (additional images, see ESI).

Taking into account the molecular dimensions of **G1** (*ca.* 16 Å in size × 2.4 Å high), the simpler units found in aqueous solution should correspond at least to the **G1**-dimer. Comparable results have indeed been observed in related molecules [19].

Comparatively, results obtained for **G1** adsorbed on a mica surface were remarkably different (Figure 8c–d). Two different types of structures grow very fast. Firstly, particles around 12–25 Å in size and 3 Å width appeared randomly distributed on the surface. Secondly, fibers [20] (30–40 nm in size and 4–6 Å width) developed in the material. The number and length of these fibers were increased at longer times of immersion. Therefore, **G1** molecules self-assemble promoting a network of cross-linked fibers in mica. The fact that the width of the fiber is slightly larger than the molecular width would be in good agreement with an "*edge-on*" packing of **G1** molecules giving rise to 1D fiber growth, as previously reported in similar disc-like moieties [21].

Interestingly, **G1** molecules seemed to be tilted with respect to a normal surface packing as we can conclude by comparing the diameter of the **G1** molecule (*ca.* 16 Å) with the averaged width of the fibers, (4–6 Å). This is likely to be due to the repulsive interactions between the hydrophobic HATNA cores and the strong hydrophilic mica surface which would in principle restrict a conventional "*lying flat*" position of the molecules. Considering the width of the fibers (30–40 nm), the fibrilar structures most probably comprise of several single stacks in an "*edge-on*" arrangement and parallel assembled. These hypotheses may point to a compromise between two main driving forces in the self-assembly of the compounds, namely the π-stacking interfacial interactions (involved in the aromatic cores of **G1** within a single column) and the hydrogen bonding of amine side groups (which promote the intercolumnar packing) [20,21].

In order to ascertain the role played by the π-stacking interactions in **G1** self-assembly, the microscopy study was subsequently extended to the use of highly ordered pyrolytic graphite (HOPG) as a substrate. HOPG has a comparatively larger hydrophobic surface than those of Au(111) and mica.

Figure 9 shows a monolayer can be clearly seen growing near the HOPG terraces (Figure 9a–b, black arrows). Some big particles can also be found randomly distributed on the clean HOPG terraces. The size of this monolayer (2.6–3.2 Å) is in close agreement with the width of the **G1** molecule lying flat on the HOPG surface, i.e., in a *"face-on"* arrangement [21]. Increasing the time of immersion and/or the **G1** concentration leads to an almost complete covering of the HOPG surface by multiple layers resulting from self-assembled molecules (only some void areas left, Figure 9c–g). The majority of the aforementioned voids mostly comprise of the HOPG free surface, a partial **G1** monolayer and a second superposed monolayer (Figure 9c,e,g).

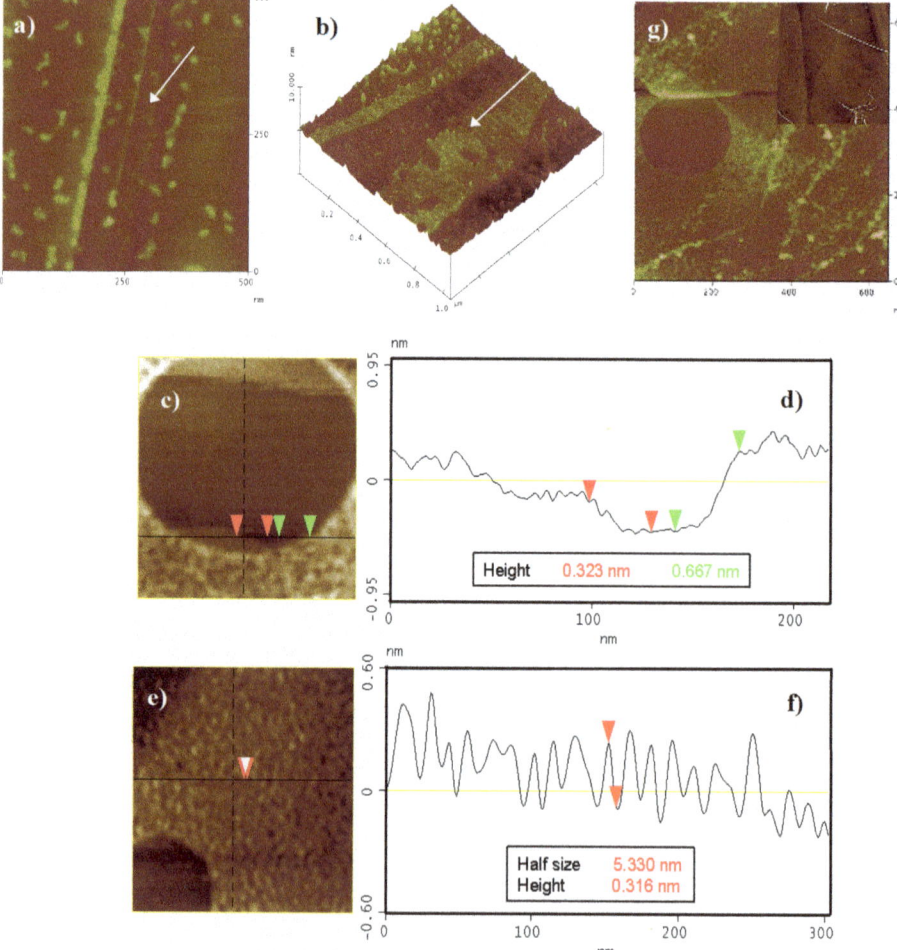

Figure 9. AFM images of pyrolytic graphite (HOPG) after different times of immersion into a **G1** 10^{-5} M water solution. (**a**) 420 nm × 420 nm, t = 5 min, (**b**) 1.1 μm × 1.1 μm 3D, t = 5 min, (**c**) 220 nm × 220 nm, t = 10 min and (**d**) corresponding cross-section shows the first layer (red arrows) and the second layer (green arrows). (**e**) 300 nm × 300 nm, t = 10 min and (**f**) cross-section showing the overlayer morphology. (**g**) 620 nm × 620 nm, t = 10 min. Inset: 2.8 μm × 2.8 μm, t = 5 min, **G1** 10^{-4} M.

The size of the second layer (*ca.* 6–7 Å, Figure 9d) was twice as great as that observed for the first monolayer, pointing to a π-stacking assembly [22]. No noticeable differences in AFM measurements under phase contrast mode could be observed (not even in the thickness between the first and the second monolayer) [23]. Nevertheless, a different type of packing (molecules in an "*edge-on*" arrangement) cannot be ruled out, especially considering the fiber-like structures shown in Figure 9e,f [21]. Last and most interestingly, new assemblies appear at greater **G1** concentration and/or time of immersion (i.e., long fibers 5–10 nm width and more than one micron long are observed as depicted in Figure 9g and inset). These fibers could only be found on HOPG when the surface was fully covered by several layers of **G1**. Under the investigated conditions, a maximum of three layers was observed. AFM studies on mica and HOPG consequently prove that these layers and fibers grow selectively on the appropriate substrate. Such motifs, which constitute a direct consequence of the π-stacking interactions, were not observed on Au(111) (only discrete particles were obtained).

3. Experimental Section

Preparation of 5,6,11,12,17,18-hexaazatrinaphthylene-2,3,8,9,14,15-hexaamine (**G1**): To a 10 mL reaction vial was added hexaketocyclohexane octahydrate (20 mg, 0,06 mmol, 20 mM) and 1,2,4,5-benzenetetramine tetrahydrochloride (3.75 equiv., 64 mg, 0.22 mmol, 73 mM) followed by 3 mL of 8:2 EtOH-HOAc glacial. The closed vessel was heated and stirred in CEM Discover© reaction cavity for 30 min at 180 °C. Then the reaction vessel was rapidly cooled at 60 °C. Upon cooling, solvents were removed and the black residue was washed with hot glacial acetic acid (3 × 10 mL) and ice water (2 × 10 mL). Drying for 48 h (under vacuum, 5–10 mmHg, 60–80 °C) afforded a violet-black solid as pure product (25 mg, 87%). A sample for analysis was recrystallized from a dichloromethane-ethanol mixture.

STM and *AFM* imaging were performed in air with a Nanoscope IIIa microscope from Digital Instruments (Veeco). *Preparation of samples:* Ultrathin dry films of **G1** were prepared from MilliQ water solutions on atomically-flat substrates at room temperature. Samples were prepared by drop casting from diluted water solutions during different times, and then subsequently were thoroughly rinsed with MilliQ water and finally dried during several hours under N_2 current flow before imaging.

4. Conclusions

A simple and efficient synthetic route towards the preparation of HATNA systems was prepared. These peculiar π-conjugated compounds can offer the opportunity to synthesize hierarchically high ordered self-assemblies (e.g., disk-like dendritic supramolecular systems) via π-stacking and the formation columnar anisotropic architectures. The compound **G1** can successfully self-assemble into nanofibers on HOPG and mica surfaces, while only discrete particles were observed on Au(111) surfaces. Optical and electrochemical properties of HATNA compounds as electron-transport materials are currently under investigation in our laboratories.

Supplementary Materials: The following are available on ESI (Electronic Supporting Information).

Author Contributions: Conceptualization, methodology, analysis, investigation, writing and supervision, R.L., Á.G.R. and D.G.V. All authors have read and agreed to the published version of the manuscript.

Funding: This research received no external funding.

Acknowledgments: D.G.V. acknowledges Gobierno de Canarias for pre-doctoral fellowship. Authors acknowledge to Alejandro G. Orive and Alberto Hernández by STM and AFM studies.

Conflicts of Interest: The authors declare no conflict of interest.

References and Note

1. Lo, S.C.; Burn, P.L. Development of Dendrimers: Macromolecules for Use in Organic Light-Emitting Diodes and Solar Cells. *Chem. Rev.* **2007**, *107*, 1097–1116. [PubMed]
2. Gillies, E.R.; Jonsson, T.B.; Fréchet, J.M.J. Stimuli-Responsive Supramolecular Assemblies of Linear-Dendritic Copolymers. *J. Am. Chem. Soc.* **2004**, *126*, 11936–11942. [CrossRef] [PubMed]
3. Kwak, G.; Choi, J.U.; Seo, K.H.; Park, L.S.; Hyun, S.H.; Kim, W.S. Three-Dimensional PtRu Nanostructures. *Chem. Mater.* **2007**, *19*, 2898–2904.
4. Kay, K.Y.; Han, K.J.; Yu, Y.J.; Park, Y.D. Dendritic fullerenes (C60) with photoresponsive azobenzene groups. *Tetrahedron Han, Lett.* **2002**, *43*, 5053–5057.
5. Liao, L.X.; Junge, D.M.; McGrath, D.V. Functional Polymers from Novel Carboxyl-Terminated Trithiocarbonates as Highly Efficient RAFT Agents. *Macromolecules* **2002**, *35*, 319–329.
6. Crispin, X.; Cornil, J.; Friedlein, R.; Okudaira, K.K.; Lemaur, V.; Crispin, A.; Kestemont, G.; Lehmann, M.; Fahlman, M.; Lazzaroni, R.; et al. Electronic Delocalization in Discotic Liquid Crystals: A Joint Experimental and Theoretical Study. *J. Am. Chem. Soc.* **2004**, *126*, 11889–11893. [CrossRef] [PubMed]
7. Goodby, J.W.; Saez, I.M.; Cowling, S.J.; Görtz, V.; Draper, M.; Hall, A.W.; Sia, S.; Cosquer, G.; Lee, S.E.; Raynes, E.P. Transmission and amplification of information and properties in nanostructured liquid crystals. *Angew. Chem. Int. Ed.* **2008**, *47*, 2754–2787. [CrossRef]
8. Ishi-I, T.; Yaguna, K.; Kuwahara, R.; Taguri, Y.; Mataka, S. Self-Assembling of n-Type Semiconductor Tri(phenanthrolino)hexaazatriphenylenes with a Large Aromatic Core. *Org. Lett.* **2006**, *8*, 585–588.
9. Marshall, S.R.; Rheingold, A.L.; Dawe, L.N.; Shum, W.W.; Kitamura, C.; Miller, J.S. Corner Sharing Tetrahedral Network in Co3(HAT)[N(CN)2]6(OH2)2 (HAT = 1,4,5,8,9,12-Hexaazatriphenylene). *Inorg. Chem.* **2002**, *41*, 3599–3601.
10. Hirayama, T.; Yamasaki, S.; Ameku, H.; Ishi-i, T.; Thiemann, T.; Mataka, S. Fluorescent solvatochromism of bi-polar N, N-diphenylaminoaryl-substituted hexaazatriphenylenes, tetraazaphenanthrene, and quinoxalines. *Dyes Pigmemts* **2005**, *67*, 105–110. [CrossRef]
11. García-Velázquez, D.; González-Orive, A.; Hernández-Creus, A.; Luque, R.; Ravelo, A.G. Novel organogelators based on amine-derived hexaazatrinaphthylene. *Org. Biomol. Chem.* **2011**, *9*, 6524–6527. [CrossRef]
12. Alonso, S.J.; Estévez-Braun, A.; Ravelo, A.G.; Zárate, R.; López, M. Double domino Knoevenagel hetero Diels–Alder strategy towards bis-pyrano-1, 4-benzoquinones. *Tetrahedron* **2007**, *63*, 3066–3074. [CrossRef]
13. Kaafarani, B.R.; Kondo, T.; Yu, J.; Zhang, Q.; Dattilo, D.; Risko, C.; Jones, S.C.; Barlow, S.; Domercq, B.; Amy, F.; et al. High charge-carrier mobility in an amorphous hexaazatrinaphthylene derivative. *J. Am. Chem. Soc.* **2005**, *127*, 16358–16359. [CrossRef]
14. Albouy, P.A.; Guillon, D.; Heinrich, B.; Levelut, A.-M.; Malthête, J. Structural study of the nematic and hexagonal columnar phases of wire shaped self assemblies of thermotropic mesogens. *J. Phys. II* **1995**, *5*, 1617–1634. [CrossRef]
15. Kestemont, G.; De Halleux, V.; Lehmann, M.; Ivanov, D.A.; Watson, M.; Geerts, Y.H. Discotic mesogens with potential electron carrier properties. *Chem. Commun.* **2001**, *20*, 2074–2076. [CrossRef] [PubMed]
16. Brunsfeld, L.; Schenning, A.P.H.J.; Broeren, M.A.C.; Janssen, H.M.; Vekemans, J.A.J.M.; Meijer, E.W. Chiral Amplification in Columns of Self-Assembled N,N',N"-Tris((S)-3,7-dimethyloctyl)benzene-1,3,5-tricarboxamide in Dilute Solution. *Chem. Lett.* **2000**, *29*, 292–293. [CrossRef]
17. Hanabusa, K.; Kawakami, A.; Kimura, M.; Shirai, H. Remarkable viscoelasticity of organic solvents containing trialkyl-1, 3, 5-benzenetricarboxamides and their intermolecular hydrogen bonding. *Chem. Lett.* **1997**, *26*, 429–430. [CrossRef]
18. Roussel, O.; Kestemont, G.; Tant, J.; De Halleux, V.; Gomez Aspe, R.; Levin, J.; Remacle, A.; Ivanov, D.; Gearba, R.I.; Lehmann, M.; et al. Discotic liquid crystals as electron carrier materials. *Mol. Cryst. Liq. Cryst.* **2003**, *396*, 35–39. [CrossRef]
19. Hierlemann, A.; Campbell, J.K.; Baker, L.A.; Crooks, R.M.; Ricco, A.J. Structural distortion of dendrimers on gold surfaces: A tapping-mode AFM investigation. *J. Am. Chem. Soc.* **1998**, *120*, 5323–5324. [CrossRef]
20. Palma, M.; Levin, J.; Debever, O.; Geerts, Y.; Lehmann, M.; Samorí, P. Self-assembly of hydrogen-bond assisted supramolecular azatriphenylene architectures. *Soft Matter* **2008**, *4*, 303–310. [CrossRef]

21. Scmaltz, B.; Weil, T.; Müllen, K. Polyphenylene-Based Materials: Control of the Electronic Function by Molecular and Supramolecular Complexity. *Adv. Mater.* **2009**, *21*, 1067–1078. [CrossRef]
22. Palermo, V.; Palma, M.; Samorí, P. Electronic characterization of organic thin films by Kelvin probe force microscopy. *Adv. Mater.* **2006**, *18*, 145–164. [CrossRef]
23. Wu, J.; Watson, M.D.; Zhang, L.; Wang, Z.; Müllen, K. Hexakis(4-iodophenyl)-peri-hexabenzocoronene- A Versatile Building Block for Highly Ordered Discotic Liquid Crystalline Materials. *J. Am. Chem. Soc.* **2004**, *126*, 177–186. [CrossRef] [PubMed]

Sample Availability: Samples of the compounds are not available from the author.

Publisher's Note: MDPI stays neutral with regard to jurisdictional claims in published maps and institutional affiliations.

© 2020 by the authors. Licensee MDPI, Basel, Switzerland. This article is an open access article distributed under the terms and conditions of the Creative Commons Attribution (CC BY) license (http://creativecommons.org/licenses/by/4.0/).

Article

Hypergolics in Carbon Nanomaterials Synthesis: New Paradigms and Perspectives

Nikolaos Chalmpes [1], Konstantinos Spyrou [1], Konstantinos C. Vasilopoulos [1], Athanasios B. Bourlinos [2,*], Dimitrios Moschovas [1], Apostolos Avgeropoulos [1], Christina Gioti [1], Michael A. Karakassides [1] and Dimitrios Gournis [1,*]

1 Department of Materials Science and Engineering, University of Ioannina, 45110 Ioannina, Greece; chalmpesnikos@gmail.com (N.C.); konstantinos.spyrou1@gmail.com (K.S.); kovasil@auth.gr (K.C.V.); dmoschov@cc.uoi.gr (D.M.); aavger@uoi.gr (A.A.); chgioti@cc.uoi.gr (C.G.); mkarakas@uoi.gr (M.A.K.)
2 Physics Department, University of Ioannina, 45110 Ioannina, Greece
* Correspondence: bourlino@cc.uoi.gr (A.B.B.); dgourni@uoi.gr (D.G.); Tel.: +30-26510-07141 (D.G.)

Academic Editor: Giuseppe Cirillo
Received: 12 April 2020; Accepted: 6 May 2020; Published: 8 May 2020

Abstract: Recently we have highlighted the importance of hypergolic reactions in carbon materials synthesis. In an effort to expand this topic with additional new paradigms, herein we present novel preparations of carbon nanomaterials, such-like carbon nanosheets and fullerols (hydroxylated fullerenes), through spontaneous ignition of coffee-sodium peroxide (Na_2O_2) and C_{60}-Na_2O_2 hypergolic mixtures, respectively. In these cases, coffee and fullerenes played the role of the combustible fuel, whereas sodium peroxide the role of the strong oxidizer (e.g., source of highly concentrated H_2O_2). The involved reactions are both thermodynamically and kinetically favoured, thus allowing rapid product formation at ambient conditions. In addition, we provide tips on how to exploit the released energy of such highly exothermic reactions in the generation of useful work.

Keywords: hypergolic reactions; sodium peroxide; carbon nanosheets; fullerols; useful energy

1. Introduction

Carbon plays a central role in material science due to its variety of forms and enchanted properties [1]. Traditionally, carbon synthesis is an energy-consuming process that requires heating of an organic precursor in an oven at elevated temperature for certain periods of time. In this respect, the development of fast, spontaneous and energy-liberating (e.g., exothermic) preparative methods at ambient conditions would be of great value in carbon materials synthesis. Recently, we have introduced hypergolic reactions [2–4] as a useful tool in the rapid and spontaneous synthesis of a wide range of carbon nanomaterials at ambient conditions [5–7]. It is worth noting, the released energy from such highly exothermic reactions could be further exploited in the generation of useful work (chemical, mechanical, electrical, etc.). In one case, we have shown the formation of carbon nanosheets by the self-ignition of pyrophoric lithium dialkylamides salts in air, with the released energy being utilized for the generation of thermoelectric power [5]. In another case, the spontaneous ignition of an acetylene-chlorine mixture produced highly crystalline graphite at ambient conditions [6]. Lastly, hypergolic mixtures based on nitrile rubber or Girard's reagent T and fuming HNO_3 as a strong oxidizer, led to the formation of carbon nanosheets or photoluminescent carbon dot respectively, with the released heat being utilized for the thermal transformation of a triazine precursor into graphitic carbon nitride or of coffee grains into a lightweight carbon absorbent [7].

In an effort to further build upon these results from our group, in the present work we provide additional new paradigms of hypergolic reactions in the service of carbon nanomaterials synthesis. Sodium peroxide Na_2O_2 was used as a strong oxidizer [8], the latter being a source of highly concentrated H_2O_2 (a reactive oxygen species) upon contact with water. On the other hand, coffee [9,10] and fullerenes C_{60} acted as the carbon source in the corresponding hypergolic mixtures. As far as the carbon materials of interest is concerned here, these included carbon nanosheets [11] and fullerols (hydroxylated fullerenes) [12].

In a first paradigm, the ignition of instant coffee grains by Na_2O_2 resulted in carbon nanosheets, the latter shown here to be an effective solar energy absorbent. Interestingly, the energy released from the reaction could be drain off to photovoltaics or to the preparation of important magnetic materials, such as magnetic iron oxides from the thermal decomposition of ferric acetate. In a second paradigm, simply crushing C_{60} in the presence of Na_2O_2 caused ignition of the mixture with simultaneous formation of fullerols, a well-established fullerene derivative with interesting physico-chemical properties and numerous applications.

2. Results and Discussion

2.1. Carbon Nanosheets

The XRD pattern of the coffee-derived carbon nanosheets exhibited a very broad reflection centred at $d_{002} = 3.9$ Å (Figure 1, top), signalling the formation of amorphous carbon [13]. Likewise, Raman spectroscopy gave broad D (1355 cm^{-1}) and G (1590 cm^{-1}) bands with a relatively intensity ratio of $I_D/I_G = 0.9$ (Figure 1, bottom), both being typical features of non-crystalline carbon [13]. On the other hand, the XPS spectrum of the nanosheets was identical to those reported in references 5 and 7 for oxidized carbon nanosheets. AFM study of the nanosheets showed the presence of large plates with thickness 1.5–2.5 nm (Figure 2, top). The morphology and size of the sheets were additionally confirmed by TEM microscopy (Figure 2, bottom).

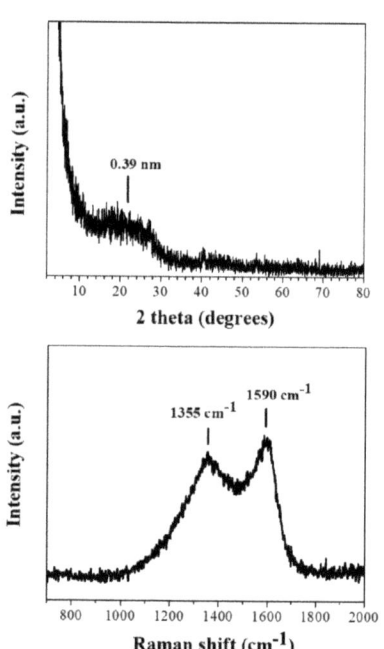

Figure 1. XRD pattern (top) and Raman spectrum (bottom) of the coffee-derived carbon nanosheets.

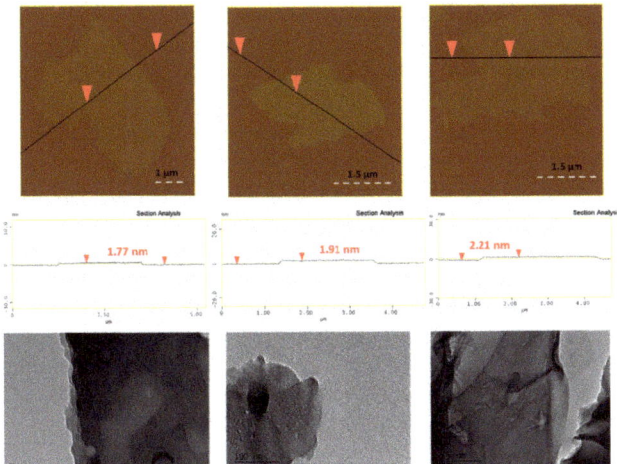

Figure 2. Top: AFM images of cross sectional analysis of selected carbon nanosheets. Bottom: representative TEM images of the sheets.

The N_2 adsorption-desorption isotherms for the carbon nanosheets is shown in Figure 3. The sample exhibited isotherm of type I according to IUPAC classification, with a H4 type hysteresis loop which is characteristic of slit-shaped pores. The surface area of the sample and the total pore volume were calculated using BET equation (S_{BET}) or alternatively, by the "t-plot" and Quenched Solid Density Functional Theory (QSDFT) methods. According to these methods the specific surface area was calculated to be 130 m^2/g (S_{BET}), 164 m^2/g (S_t, t-plot) and 179 m^2/g (QSDFT) respectively, whereas the total pore volume found to be 0.06 cm^3/g (cumulative at P/P_o = 0.97), 0.06 cm^3/g (t-plot) and 0.085 cm^3/g (QSDFT). From the pore size distribution PSD (inset) according to the QSDFT model (Figure S1), carbon nanosheets seemed to exhibit micropores with two average sizes 1.2 nm and 1.7 nm and mesoporous 3.7 nm. However, the volume analysis from "t-plot" (Figure S2) showed only the presence of micropores in agreement with classification type of adsorption isotherm. It is obvious that the observed step down in the desorption branch between relative pressures 0.4 to 0.6 (Figure 3), is responsible for the observed mesoporous peak in the calculated PSD. That peak is probably an artefact, caused by the spontaneous evaporation of metastable pore liquid (i.e., cavitation). Besides, H3 or H4 hysteresis is often attributed to the occurrence of pore blocking and percolation phenomena and it is not only associated with the pore condensation [14]. From a practical point of view, the surface area of the nanosheets in combination with the oxygen functionalities present on their surface could make the material useful in adsorption processes (e.g., removal of heavy metals or dyes from water).

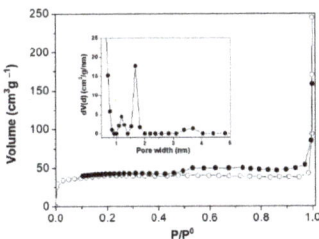

Figure 3. N_2 adsorption-desorption isotherms of carbon nanosheets. Inset: Quenched Solid Density Functional Theory (QSDFT) pore size distribution.

The black solid could be used as a pigment in water glass paints for solar energy harvesting. Water glass refers to an aqueous solution of the inorganic polymer sodium silicate (40%, Aldrich) that can wrap and electrostatically stabilize any dispersed solid. To this aim, carbon nanosheets were simply mechanically mixed with water glass to create the paint. The paint was then spread over a piece of paper with a brush and left to dry at room temperature (see the corresponding black square drawing in Figure 4, top). A similar drawing using an analogous CuO paint was also sketched in a separate piece of paper (Figure 4, top); cupric oxide is a reference black pigment that is often used in solar water heaters [15]. Following, both coatings were illuminated under an infrared lamp at the same distance and for the same time (20 s) prior to scanning with a thermal camera. As it can be seen from the thermal camera images in Figure 4, bottom, carbon nanosheets and cupric oxide developed comparable temperatures (100–120 °C) under identical conditions. Hence, thanks to their flat surface and blackness, the coffee-derived carbon nanosheets could be promising solar energy absorbents [16]. It should be noted that although the CuO film seems homogeneously darker, the nanosheets yield higher temperature rise in places that appear even darker due to a higher mass of deposited material. Thus, the temperature distribution as presented is valid.

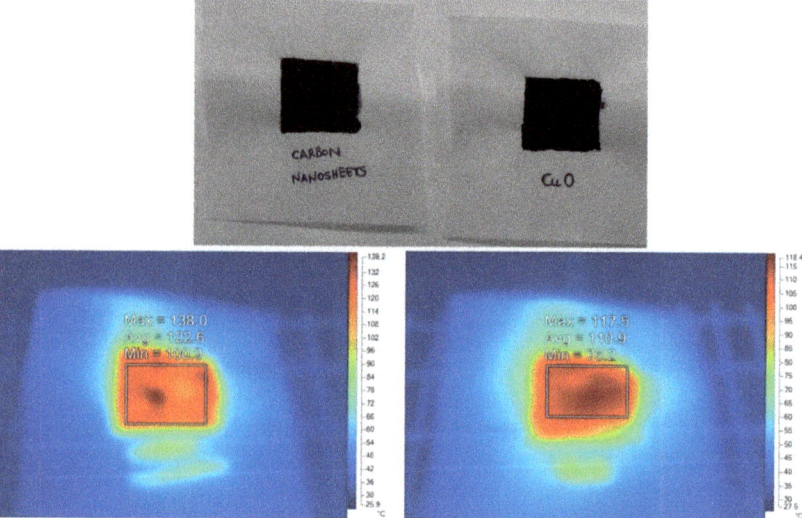

Figure 4. Top: dry carbon nanosheet and CuO paints on paper sheet (black squares). Bottom: the corresponding thermal camera images after 20 s illumination with infrared lamp (left: carbon nanosheets; right: CuO). Both samples developed comparable temperature after illumination under identical conditions, thus demonstrating that the coffee-derived carbon nanosheets are effective solar energy absorbents.

The heat and light produced from the reaction of the coffee grains with Na_2O_2 was utilized in the generation of useful work. In one example, a miniature silicon photovoltaic panel connected with a green LED light was placed above the ignition mixture with the front side facing down the mixture at certain distance (Figure 5, top). The band-gap energy of silicon is 1.1 eV, the latter corresponding to the infrared part of the electromagnetic radiation. Upon ignition, the thermal radiation and light produced from the flame turned on the LED light (Figure 5, top), thus acting as a sort of thermophotovoltaic [17]. In another example, the heat produced from the hypergolic mixture was exploited in the thermal decomposition of ferric acetate into magnetic iron oxide [18,19]. For this purpose, a quartz tube charged with ferric acetate was dipped into an alumina crucible containing the hypergolic mixture (Figure 5, bottom). Ignition of the mixture provided the necessary heat for the thermal decomposition of the precursor inside the tube into magnetic iron oxide (Figure 5, bottom).

Figure 5. Top: ignition of the coffee-Na_2O_2 mixture generates enough thermal radiation and light to turn on a miniature silicon photovoltaic device. Bottom: the ignitable mixture served as a heat source for the in-situ thermal transformation of ferric acetate (orange powder at far-left) into magnetic iron oxide (black solid at far-right). For this purpose, a small quartz tube was charged with the precursor and then merely dipped into the ignition mixture. Notice the release of smoke from the top rim of the tube due to thermal decomposition of ferric acetate.

2.2. Fullerols

The ignition of coffee by sodium peroxide presented above is considered a classic demonstration experiment in the area of hypergolic reactions that, as shown here, results in carbon nanosheets. Another interesting hypergolic mixture reported for first time is the C_{60}-Na_2O_2 system (Figure 10). Although several carbon allotropes and their derivatives have been used as booster additives in hypergolic fuels [20], however, it is very seldom to observe spontaneous ignition between elemental carbon and a strong oxidizer upon direct contact. Perhaps a sole example in the literature refers to the activated charcoal-Na_2O_2 pair [8] (see also *Bretherick's Handbook of Reactive Chemical Hazards*). Hence, the C_{60}-Na_2O_2 system adds another example in the list, thus paving the way for the advancement of novel hypergolics strictly based on carbon. It should be mentioned that similar treatment of other carbon allotropes (e.g., nanodiamonds, carbon nanotubes) with Na_2O_2 gave no ignition. Apparently, the small size and strain of fullerenes make them prone to ignition.

The C_{60}-Na_2O_2 system is of synthetic value as well, since it can afford fullerols, an important hydroxylated fullerene derivative. In general, the formation of fullerols from C_{60} requires the simultaneous presence of H_2O_2 and NaOH [21]. In our case, both chemicals were released in-situ by the peroxide hydrolysis in humid air (e.g., $Na_2O_2 + 2\ H_2O \rightarrow H_2O_2 + 2\ NaOH$). The ATR-IR spectrum of the dark brown solid derived from the reaction displayed several broad bands of moderate intensity at 3450 cm^{-1} (O-H), <3000 cm^{-1} (aliphatic C-H), 1717 cm^{-1} (C=O), 1618 cm^{-1} (enol C=C) and 1045 cm^{-1} (C-O) (Figure 6, top). The sharp peaks at 1183 cm^{-1} and 1427 cm^{-1} are ascribed to unreacted C_{60}, which cannot be completely removed by further washing (e.g., with toluene). This issue should be cautiously taken into consideration upon scaling-up synthesis. These observations are similar to those reported by Afreen et al. elsewhere for fullerol formation from C_{60} [22]. Especially the O-H, enol C=C and C-O stretchings at 3450 cm^{-1}, 1618 cm^{-1} and 1045 cm^{-1} respectively, are indicative of the attachment of -OH groups onto the C_{60} cage. Of particular interest is the appearance of aliphatic C-H and carbonyl C=O bands in the spectrum. The carbonyl group is often ascribed in the literature to the pinacol-pinacolone rearrangement in fullerene 1,2-diols [23]. However, such rearrangement seems rather unlikely for fullerenes since it would require the movement of a fixed skeletal carbon within the cage. In addition, it cannot explain the appearance of C-H aliphatics in the spectrum. Hence, these bands should be better ascribed to the keto-enol tautomerism, which simpler involves the movement of a hydrogen atom from the hydroxyl group of C=C-OH (enol form) to an alpha carbon (e.g., H-C-C=O, keto from). In this way, both C-H and C=O bands can be rationalized in the spectrum. According to TGA under N_2 (Figure 6, bottom), fullerol exhibited a first weight loss below 150 °C due to the release of hydration water (ca. 8% w/w) and another one between 150 and 600 °C due to dehydroxylation of the cluster (ca. 13% w/w) [24]. These weight losses agree with the formula $C_{60}(OH)_7 \cdot 4H_2O$, i.e., low-hydroxylated fullerol [25]. It is worth noting, the latter formula is close to that of $C_{60}(OH)_8 \cdot 2H_2O$ previously reported by Afreen et al. [22].

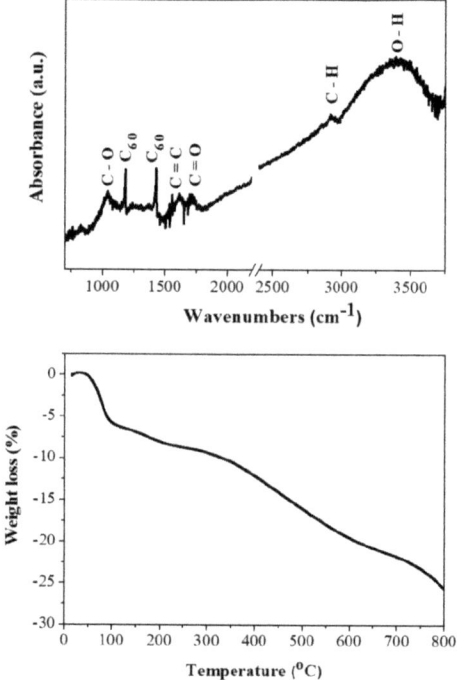

Figure 6. Top: ATR-IR spectrum of the fullerols. Bottom: the corresponding thermogravimetric analysis (TGA) trace under N_2.

AFM study of the aqueous-dispersed solid revealed the presence of large globular nanoparticles with size in the range of 15–25 nm (Figure 7). Fullerol typically has a size of 1.5 nm; however, it is well-known to form larger aggregates and eventually fullerol nanoparticles (10–60 nm) as a result of strong hydrogen bonding and π–π interactions between adjacent clusters [26].

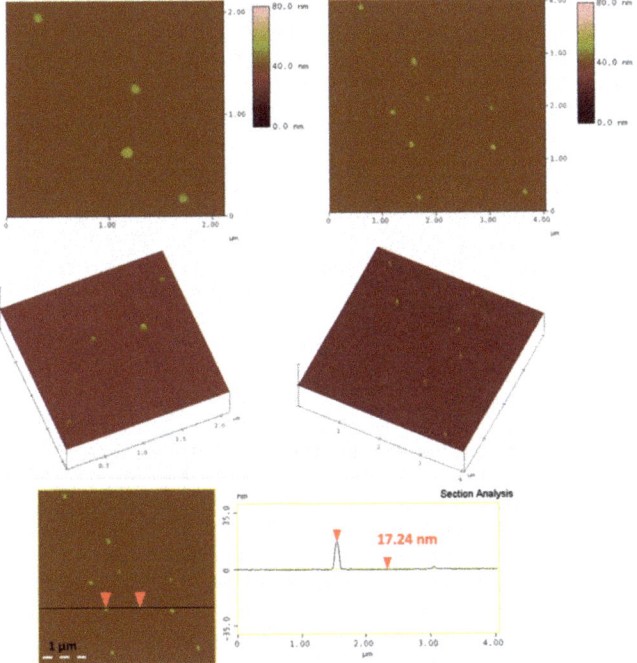

Figure 7. Representative AFM height images (top), 3D morphology (middle) and cross section analysis profile (bottom) of the globular fullerol nanoparticles.

High resolution C1s XPS gave useful information about the structure of the synthesized fullerols (Figure 8). The dominant peak at 285.2 eV is due to the carbon cage of the fullerols [27]. Additionally, the shake-up features associated with the C_{60}-cage atoms were identified at 288.5 eV and 291.2 eV [28,29]. On the other hand, the small fraction of sp^3 carbons observed at 284.2 eV is consistent with a low degree of functionalization (e.g., low-hydroxylated fullerols). The peak at 286.0 eV is typical of hydroxyl C-OH groups, whereas those at 287.2 and 289.9 eV of C=O/C-O-C/C(O)O species [30]. Lastly, the feature at 283.0 eV is ascribed to iron carbides, the latter resulting from iron contamination of the sample by the metallic container used for reaction.

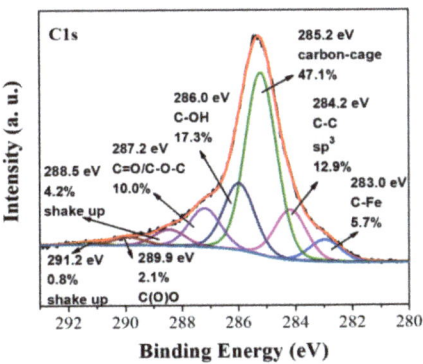

Figure 8. C1s photoelectron spectrum of fullerols.

To sum-up, the ATR-IR, TGA, AFM and XPS data are well-consistent with the formation of low-hydroxylated fullerol, a useful precursor to higher fullerols.

3. Materials and Methods

For safety reasons all experiments were conducted in a fume hood with ceramic tiles bench.

3.1. Carbon Nanosheets

2 g of instant coffee grains (Nescafe®) were mixed with 2 g Na_2O_2 beads (93%, Sigma-Aldrich, St. Louis, MO, USA) in a porcelain dish followed by the quick addition of 1 mL water. Ignition started with a short delay (30–45 s after water addition) giving off an intense yellow flame due to the sodium ions present. The reaction led to a crude carbon residue after cooling of the product. Thorough washings with water, dimethylformamide (DMF) and acetone afforded a fine black powder containing carbon nanosheets (yield: 2%). The whole process is visualized in Figure 9. Any minor inorganic contaminants due to the porcelain dish were removed by treating the sample with 48% HF aqueous solution.

Figure 9. Left-to-right: instant coffee grains were mixed with Na_2O_2 beads in a porcelain dish. Subsequent addition of water caused ignition of the mixture towards the formation of a crude carbon residue. After thorough washings, a fine black powder was obtained containing carbon nanosheets.

3.2. Fullerols

150 mg of C_{60} powder (98%, Sigma-Aldrich, St. Louis, MO, USA) were crushed in the presence of 1.7 g Na_2O_2 beads (93%, Sigma-Aldrich, St. Louis, MO, USA) using a stainless steel mortar and pestle. Due to rapid oxidation of the carbon clusters by the peroxide, the mixture ignited spontaneously within few seconds. The product washed successively with concentrated HCl aqueous solution (37%) in order to remove excess of Na_2O_2 (slow addition), water, and finally acetone. A fine dark brown powder was obtained made up of fullerol clusters (yield: 10%). The clusters were soluble in water by alkaline treatment with NaOH and sonication, due to neutralization of the acidic phenolic protons into phenolate ionic groups [12]. The whole process is visualized in Figure 10. It should be mentioned that the stainless steel mortar and pestle was used in order to crush the very hard Na_2O_2 beads. Also, no ignition was observed upon crushing the Na_2O_2 beads inside the metallic container in the absence of fullerenes, suggesting that no reaction is taking place between the peroxide and the walls of the metallic container.

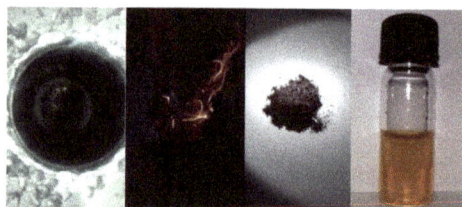

Figure 10. Crushing fullerenes in the presence of Na_2O_2 in a stainless steel mortar and pestle caused ignition of the mixture as evidenced by the generation of bright flash. Washing of the product led to a dark brown powder, namely fullerols. Fullerols were soluble in water by mild alkaline treatment and sonication.

At this point we would like to first emphasize that the yields of the above reactions, although low at first glance, are generally considered fair enough in materials science. Second, the selection of a combustible material is rather a trial and error effort at this stage. Several other organic compounds were tested in our lab but failed ignition with sodium peroxide. In many cases though, the reagents reacted exothermically without giving off flame. We speculate that the quantity of mixed reactants play a key role for ignition. Third, the reported hypergolic pairs are unique in terms of matching. For instance, the reaction of coffee grains or C_{60} with fuming HNO_3 (100%), though exothermic, gave no ignition.

Powder X-ray diffraction (XRD) was performed using background-free Si wafers and Cu Kα radiation from a Bruker Advance D8 diffractometer (Bruker, Billerica, MA, USA). Raman spectra were recorded with a RM 1000 Renishaw micro-Raman system using a laser excitation line at 532 nm (Renishaw, Wotton-under-Edge, England). Attenuated total reflection infrared spectroscopy (ATR-IR) measurements were performed using a Jasco IRT-5000 microscope coupled with a FT/IR-4100 spectrometer (Jasco, Easton, MD, USA). The ZnSe prism of the ATR objective had a 250 μm area in contact with the sample. Background was subtracted and the baseline was corrected for all spectra. Thermogravimetric analysis (TGA) was performed using a Perkin Elmer Pyris Diamond TG/DTA (PerkinElmer, Inc., Waltham, MA, USA). The fullerol sample was heated under N_2 flow at a rate of 5 °C min^{-1}. X-ray photoelectron spectroscopy (XPS) measurements were performed in an ultra-high vacuum at a base pressure of 4×10^{-10} mbar with a SPECS GmbH spectrometer equipped with a monochromatic Mg K$_\alpha$ source (hv = 1253.6 eV) and a Phoibos-100 hemispherical analyser (Berlin, Germany). Spectral analysis included a Shirley background subtraction and peak separation using Gaussian-Lorentzian functions in a least square fitting program (Winspec) developed at the LISE laboratory, University of Namur, Belgium. The N_2 adsorption-desorption isotherms were measured at 77 K on a Sorptomatic 1990, ThermoFinnigan porosimeter (Thermo Finnigan LLC, San Jose, CA, USA). Carbon nanosheets were outgassed at 150 °C for 20 h under vacuum before measurements. Specific surface areas were determined with the Brunauer–Emmett–Teller (BET) method. Atomic force microscopy (AFM) images were collected in tapping mode with a Bruker Multimode 3D Nanoscope (Ted Pella Inc., Redding, CA, USA) using a microfabricated silicon cantilever type TAP-300G, with a tip radius of <10 nm and a force constant of approximately 20–75 N m^{-1}. The transmission electron microscopy (TEM) study of carbon nanosheets deposited on carbon coated copper grids (CF300-CU-UL, carbon square mesh, CU, 300 mesh from Electron Microscopy Science) was performed using the instrument JEM HR-2100, JEOL Ltd., Tokyo, Japan operated at 200 kV in bright-field mode.

4. Conclusions

We have presented new hypergolic reactions towards the synthesis of functional carbon nanomaterials, such-like carbon nanosheets and fullerols. The new systems composed of coffee or fullerene as the combustible fuel and sodium peroxide as the strong oxidizer. In all instances, synthesis was fast, spontaneous and exothermic at ambient conditions, thus enabling not only facile carbon formation but also the generation of useful work (herein photovoltaic or chemical). The coffee-derived carbon nanosheets, due to their flat surface and blackness, seemed to serve as an effective solar energy absorbent, whereas fullerols itself is a well-known fullerene derivative with important properties and uses. Overall, the present work builds upon previous results from our group, showing altogether the wider applicability and generic character of hypergolic reactions in carbon materials synthesis.

Supplementary Materials: The following are available online at http://www.mdpi.com/1420-3049/25/9/2207/s1, Figure S1: Fitting comparison between experimental data of carbon nanosheets and QSDFT model (N_2, carbon equilibrium transition kernel at 77 K based on a slit pore model), Figure S2: t-plot for nitrogen adsorbed at 77 K in carbon nanosheets.

Author Contributions: Conceptualization, experiments and writing A.B.B. and D.G.; characterization, formal analysis, and writing N.C., K.S., K.C.V., D.M., A.A., C.G. and M.A.K. All authors have read and agreed to the published version of the manuscript.

Funding: We acknowledge support of this work by the project "National Infrastructure in Nanotechnology, Advanced Materials and Micro-/Nanoelectronics" (MIS 5002772) which is implemented under the Action "Reinforcement of the Research and Innovation Infrastructure", funded by the Operational Programme "Competitiveness, Entrepreneurship and Innovation" (NSRF 2014-2020) and co-financed by Greece and the European Union (European Regional Development Fund). This research is also co-financed by Greece and the European Union (European Social Fund- ESF) through the Operational Programme "Human Resources Development, Education and Lifelong Learning" in the context of the project "Strengthening Human Resources Research Potential via Doctorate Research" (MIS-5000432), implemented by the State Scholarships Foundation (IKY).

Acknowledgments: The authors greatly acknowledge Ch. Papachristodoulou for the XRD measurements and lab technician P. Triantafyllou for the construction of the silicon photovoltaic.

Conflicts of Interest: The authors declare no conflict of interest.

References

1. Georgakilas, V.; Perman, J.A.; Tucek, J.; Zboril, R. Broad Family of Carbon Nanoallotropes: Classification, Chemistry, and Applications of Fullerenes, Carbon Dots, Nanotubes, Graphene, Nanodiamonds, and Combined Superstructures. *Chem. Rev.* **2015**, *115*, 4744–4822. [CrossRef] [PubMed]
2. Hypergolic Propellant. Available online: https://en.wikipedia.org/wiki/Hypergolic_propellant (accessed on 28 December 2019).
3. Silva, G.D.; Iha, K. Hypergolic Systems: A Review in Patents. *J. Aerosp. Technol. Manag.* **2012**, *4*, 407–412. [CrossRef]
4. Schneider, S.; Hawkins, T.; Rosander, M.; Vaghjiani, G.; Chambreau, S.; Drake, G. Ionic Liquids as Hypergolic Fuels. *Energy Fuels* **2008**, *22*, 2871–2872. [CrossRef]
5. Baikousi, M.; Chalmpes, N.; Spyrou, K.; Bourlinos, A.B.; Avgeropoulos, A.; Gournis, D.; Karakassides, M.A. Direct production of carbon nanosheets by self-ignition of pyrophoric lithium dialkylamides in air. *Mater. Lett.* **2019**, *254*, 58–61. [CrossRef]
6. Chalmpes, N.; Spyrou, K.; Bourlinos, A.B.; Moschovas, D.; Avgeropoulos, A.; Karakassides, M.A.; Gournis, D. Synthesis of Highly Crystalline Graphite from Spontaneous Ignition of In Situ Derived Acetylene and Chlorine at Ambient Conditions. *Molecules* **2020**, *25*, 297. [CrossRef] [PubMed]
7. Chalmpes, N.; Asimakopoulos, G.; Spyrou, K.; Vasilopoulos, K.C.; Bourlinos, A.B.; Moschovas, D.; Avgeropoulos, A.; Karakassides, M.A.; Gournis, D. Functional Carbon Materials Derived through Hypergolic Reactions at Ambient Conditions. *Nanomaterials* **2020**, *10*, 566. [CrossRef]
8. Sodium Peroxide + Cornflakes + Milk = Explosion! Available online: https://www.youtube.com/watch?v=M57TOMvsNsI (accessed on 1 February 2018).
9. Wen, X.; Liu, H.; Zhang, L.; Zhang, J.; Fu, C.; Shi, X.; Chen, X.; Mijowska, E.; Chen, M.-J.; Wang, D.-Y. Large-scale converting waste coffee grounds into functional carbon materials as high-efficient adsorbent for organic dyes. *Bioresour. Technol.* **2019**, *272*, 92–98. [CrossRef]
10. Gao, G.; Cheong, L.-Z.; Wang, D.; Shen, C. Pyrolytic carbon derived from spent coffee grounds as anode for sodium-ion batteries. *Carbon Resour. Convers.* **2018**, *1*, 104–108. [CrossRef]
11. Fan, H.; Shen, W. Carbon Nanosheets: Synthesis and Application. *ChemSusChem* **2015**, *8*, 2004–2027. [CrossRef]
12. Vileno, B.; Marcoux, P.R.; Lekka, M.; Sienkiewicz, A.; Fehér, T.; Forró, L. Spectroscopic and Photophysical Properties of a Highly Derivatized C60 Fullerol. *Adv. Funct. Mater.* **2006**, *16*, 120–128. [CrossRef]
13. Roh, J.-S. Structural Study of the Activated Carbon Fiber Using Laser Raman Spectroscopy. *Carbon Lett.* **2008**, *9*, 127–130. [CrossRef]
14. Thommes, M. Physical Adsorption Characterization of Nanoporous Materials. *Chem. Ing. Tech.* **2010**, *82*, 1059–1073. [CrossRef]
15. Karthick Kumar, S.; Suresh, S.; Murugesan, S.; Raj, S.P. CuO thin films made of nanofibers for solar selective absorber applications. *Sol. Energy* **2013**, *94*, 299–304. [CrossRef]
16. Jönsson, G.; Fredriksson, H.; Sellappan, R.; Chakarov, D. Nanostructures for Enhanced Light Absorption in Solar Energy Devices. *Int. J. Photoenergy* **2011**, *2011*, 939807.

17. Coutts, T.J. A review of progress in thermophotovoltaic generation of electricity. *Renew. Sustain. Energy Rev.* **1999**, *3*, 77–184. [CrossRef]
18. Jewur, S.S.; Kuriacose, J.C. Studies on the thermal decomposition of ferric acetate. *Thermochim. Acta* **1977**, *19*, 195–200. [CrossRef]
19. Pinheiro, E.A.; Pereira de Abreu Filho, P.; Galembeck, F.; Correa da Silva, E.; Vargas, H. Magnetite crystal formation from iron(III) hydride acetate. An ESR study. *Langmuir* **1987**, *3*, 445–448. [CrossRef]
20. Yan, Q.-L.; Gozin, M.; Zhao, F.-Q.; Cohen, A.; Pang, S.-P. Highly energetic compositions based on functionalized carbon nanomaterials. *Nanoscale* **2016**, *8*, 4799–4851. [CrossRef]
21. Wang, S.; He, P.; Zhang, J.M.; Jiang, H.; Zhu, S.Z. Novel and Efficient Synthesis of Water-Soluble [60]Fullerenol by Solvent-Free Reaction. *Synth. Commun.* **2005**, *35*, 1803–1808. [CrossRef]
22. Afreen, S.; Kokubo, K.; Muthoosamy, K.; Manickam, S. Hydration or hydroxylation: Direct synthesis of fullerenol from pristine fullerene [C60] via acoustic cavitation in the presence of hydrogen peroxide. *RSC Adv.* **2017**, *7*, 31930–31939. [CrossRef]
23. Meier, M.S.; Kiegiel, J. Preparation and Characterization of the Fullerene Diols 1,2-C60(OH)2, 1,2-C70(OH)2, and 5,6-C70(OH)2. *Org. Lett.* **2001**, *3*, 1717–1719. [CrossRef] [PubMed]
24. Goswami, T.H.; Singh, R.; Alam, S.; Mathur, G.N. One-Pot Synthesis of a Novel Water-Soluble Fullerene-Core Starlike Macromolecule via Successive Michael and Nucleophilic Addition Reaction. *Chem. Mater.* **2004**, *16*, 2442–2448. [CrossRef]
25. Kokubo, K. Water-Soluble Single-Nano Carbon Particles: Fullerenol and Its Derivatives. In *The Delivery of Nanoparticles*; InTech: London, UK, 2012.
26. Djordjevic, A.; Srdjenovic, B.; Seke, M.; Petrovic, D.; Injac, R.; Mrđanović, J. Review of Synthesis and Antioxidant Potential of Fullerenol Nanoparticles. *J. Nanomater.* **2015**, *2015*, 567073. [CrossRef]
27. Maxwell, A.J.; Brühwiler, P.A.; Nilsson, A.; Mårtensson, N.; Rudolf, P. Photoemission, autoionization, and x-ray-absorption spectroscopy of ultrathin-film C60 on Au(110). *Phys. Rev. B* **1994**, *49*, 10717–10725. [CrossRef]
28. Enkvist, C.; Lunell, S.; Sjögren, B.; Brühwiler, P.A.; Svensson, S. The C1s shakeup spectra of Buckminsterfullerene, acenaphthylene, and naphthalene, studied by high resolution x-ray photoelectron spectroscopy and quantum mechanical calculations. *J. Chem. Phys.* **1995**, *103*, 6333–6342. [CrossRef]
29. Leiro, J.A.; Heinonen, M.H.; Laiho, T.; Batirev, I.G. Core-level XPS spectra of fullerene, highly oriented pyrolitic graphite, and glassy carbon. *J. Electron Spectrosc. Relat. Phenom.* **2003**, *128*, 205–213. [CrossRef]
30. Felicissimo, M.; Jarzab, D.; Gorgoi, M.; Forster, M.; Scherf, U.; Scharber, M.; Svensson, S.; Rudolf, P.; Loi, M. Determination of vertical phase separation in a polyfluorene copolymer: Fullerene derivative solar cell blend by X-ray photoelectron spectroscopy. *J. Mater. Chem.* **2009**, *19*, 4899–4901. [CrossRef]

Sample Availability: Samples of the compounds are not available from the authors.

 © 2020 by the authors. Licensee MDPI, Basel, Switzerland. This article is an open access article distributed under the terms and conditions of the Creative Commons Attribution (CC BY) license (http://creativecommons.org/licenses/by/4.0/).

Article

In-Situ Synthesis and Characterization of Nanocomposites in the Si-Ti-N and Si-Ti-C Systems

Maxime Balestrat [1], Abhijeet Lale [1,2], André Vinícius Andrade Bezerra [1,3], Vanessa Proust [2], Eranezhuth Wasan Awin [4], Ricardo Antonio Francisco Machado [3], Pierre Carles [1], Ravi Kumar [4], Christel Gervais [5] and Samuel Bernard [1,*]

1. CNRS, IRCER, UMR 7315, University of Limoges, F-87000 Limoges, France; maxime.balestrat@unilim.fr (M.B.); abhijeet.lale@unilim.fr (A.L.); andrevbezerra@gmail.com (A.V.A.B.); pierre.carles@unilim.fr (P.C.)
2. Institut Europeen des Membranes (IEM), UMR 5635 (CNRS-ENSCM-UM2), Universite Montpellier 2, Place E. Bataillon, F-34095 Montpellier, France; vanessa.proust6@gmail.com
3. Chemical Engineering, Federal University of Santa Catarina, Florianópolis 88010-970, Brazil; ricardo.machado@ufsc.br
4. Laboratory for High Performance Ceramics, Department of Metallurgical and Materials Engineering, Indian Institute of Technology-Madras (IIT Madras), Chennai 600036, India; eranezhuth@gmail.com (E.W.A.); nvrk@iitm.ac.in (R.K.)
5. Laboratoire Chimie de la Matière Condensée de Paris, Collège de France, CNRS, Sorbonne Université, 4 Place de Jussieu, 75005 Paris, France; christel.gervais_stary@sorbonne-universite.fr
* Correspondence: samuel.bernard@unilim.fr; Tel.: +33-(0)587-502-444

Academic Editor: Giuseppe Cirillo
Received: 29 September 2020; Accepted: 3 November 2020; Published: 10 November 2020

Abstract: The pyrolysis (1000 °C) of a liquid poly(vinylmethyl-*co*-methyl)silazane modified by tetrakis(dimethylamido)titanium in flowing ammonia, nitrogen and argon followed by the annealing (1000–1800 °C) of as-pyrolyzed ceramic powders have been investigated in detail. We first provide a comprehensive mechanistic study of the polymer-to-ceramic conversion based on TG experiments coupled with in-situ mass spectrometry and ex-situ solid-state NMR and FTIR spectroscopies of both the chemically modified polymer and the pyrolysis intermediates. The pyrolysis leads to X-ray amorphous materials with chemical bonding and ceramic yields controlled by the nature of the atmosphere. Then, the structural evolution of the amorphous network of ammonia-, nitrogen- and argon-treated ceramics has been studied above 1000 °C under nitrogen and argon by X-ray diffraction and electron microscopy. HRTEM images coupled with XRD confirm the formation of nanocomposites after annealing at 1400 °C. Their unique nanostructural feature appears to be the result of both the molecular origin of the materials and the nature of the atmosphere used during pyrolysis. Samples are composed of an amorphous Si-based ceramic matrix in which TiN_xC_y nanocrystals ($x + y = 1$) are homogeneously formed "in situ" in the matrix during the process and evolve toward fully crystallized compounds as TiN/Si_3N_4, TiN_xC_y ($x + y = 1$)/SiC and TiC/SiC nanocomposites after annealing to 1800 °C as a function of the atmosphere.

Keywords: polymer-derived ceramics; nanocomposites; TiN; TiC; Si_3N_4; SiC; structural properties

1. Introduction

Silicon carbide (SiC) and silicon nitride (Si_3N_4) are technologically relevant high-performance ceramics which attract nowadays strong interest for driving the development of energy, environment and health sectors [1–4]. By the synergistic cooperation between SiC or Si_3N_4—as a matrix—and a dispersed ceramic nanophase of different chemical composition—as a nano-precipitate—nanocomposites with peculiar physical and chemical properties (e.g., mechanical, electrical, optical, catalytic, etc.) that are of high scientific and technological importance can be generated [5–8].

As developed functional properties closely depend on the synthesis route of these materials. For this purpose; ceramic processing methods based on molecular engineering and precursor chemistry are well appropriate approaches to design nanocomposites that can reach performances far beyond those developed by more conventional synthesis routes. In general, such nanocomposites exhibit improved properties when compared to those prepared via classical high temperature metallurgical techniques because of the homogeneous distribution of the nanophase within the matrix (i.e., absence of agglomeration of the nano-precipitates), the small size of the nano-precipitates (no sintering because of the generally low temperature of preparation) and the lack of undesirable elements. A very convenient precursor route to produce these materials in non-oxide ceramic systems such as those based on SiC and Si_3N_4 is the polymer derived ceramics (PDCs) route [9–19]. Such a method uses preformed preceramic polymers to be modified at molecular scale to form the nanocomposite precursors. The latter offers the advantages for the in-situ synthesis of the ceramic nano-precipitates in a ceramic (possibility of different nature) matrix during the pyrolysis and annealing experiments. In addition, it allows processing materials in particular shapes and morphologies (dense or porous) that are difficult, or even impossible to obtain from conventional routes [20–25].

In the present paper, we investigate the PDCs route to design titanium nitride/silicon nitride (TiN/Si_3N_4) and titanium carbide/silicon carbide (TiC/SiC) nanocomposites using a commercial liquid polysilazane. The preparation of such nanocomposites usually involves the use of polysilazanes (TiN/Si_3N_4) [26–28] and polycarbosilanes (TiC/SiC) [5,29,30] as preformed preceramic polymers. Herein, our approach involves first the synthesis of the nanocomposite precursor containing Si, Ti, C, N and H elements from a poly(vinylmethyl-co-methyl)silazane which has been modified upon reaction with tetrakis(dimethylamido)titanium. The thermo-chemical behavior of this precursor under various atmospheres is directed to tailor the structure of the final materials: the polymer is converted by heat-treatments (=pyrolysis) at 1000 °C under ammonia, nitrogen or argon into single-phase amorphous ceramics with adjusted compositions; in particular their C and N contents. The latter are subsequently annealed at higher temperatures in flowing nitrogen or argon to initiate the crystallization of the nanophase, i.e., TiN, TiC_xN_{1-x} or TiC, and then of the matrix, i.e., Si_3N_4 or SiC. Thus, this approach provides the material with tuned phase composition and nano-/microstructure organization according to the temperature of annealing and the nature of the atmosphere applied during the pyrolysis. To build further knowledge toward a more rational approach to the preparation of ceramic nanocomposites in nitride, carbonitride and carbide systems—Si-Ti-N (TiN/Si_3N_4), Si-Ti-N-C (TiN_xC_y (x + y = 1)/SiC), Si-Ti-C (TiC/SiC)—as reported in the schematic representation of the synthetic process in Figure 1, the present work aims at (i) investigating the structure of a novel titanium-modified polysilazane, (ii) providing a comprehensive mechanistic study of the polymer-to-amorphous ceramic conversion during the pyrolysis of the precursor at 1000 °C in flowing ammonia, nitrogen and argon and (iii) characterizing the high temperature behavior of single-phase amorphous ceramics during their annealing and transformation into the titled nanocomposites. Thus, the results allow for a knowledge-based preparative path toward nanostructured polymer-derived ceramics with adjusted compositions and microstructures.

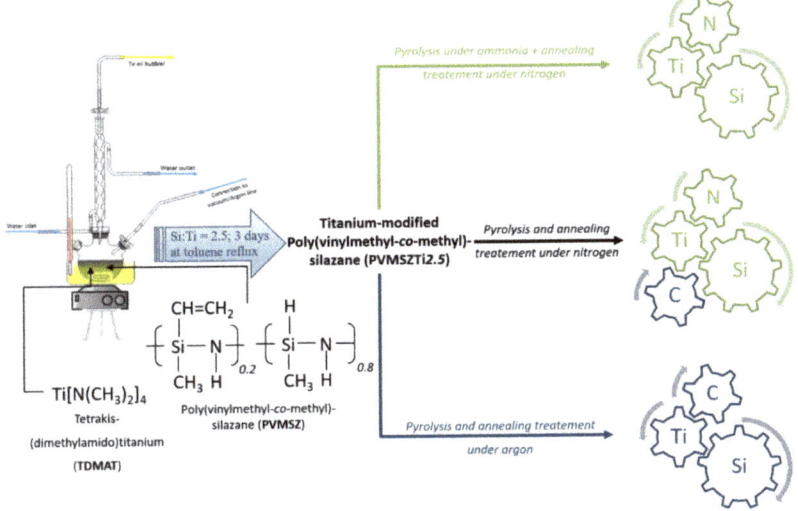

Figure 1. Schematic diagram of the general process of design of nanocomposites in the Si-Ti-N and Si-Ti-C systems from poly(vinylmethyl-*co*-methyl)silazane.

2. Results and Discussion

2.1. Polymer Synthesis and Characterization

The titanium-modified polysilazane labeled PVMSZTi2.5 has been synthesized via the reaction between tetrakis(dimethylamido)titanium Ti[N(CH$_3$)$_2$]$_4$ (TDMAT as titanium source) and a commercially available poly(vinylmethyl-*co*-methyl)silazane labeled PVMSZ by fixing a PMVSZ (monomeric unit):TDMAT ratio of 2.5.

The modification of polysilazanes with TDMAT relies in general on the reaction between the methyl-/dimethylamino groups of TDMAT and the silicon centers of PVMSZ, causing the decrease in the number of Si–NH–Si (Equation (1)) and Si–H (Equation (2)) groups in the obtained precursors and the concomitant evolution of dimethylamine and methane during the synthesis [26,27]. It should be noted that TDMAT can also react with NH groups from the monomeric unit (Si(CH$_3$)(CH=CH$_2$)NH)$_{0.2}$. Moreover, as-formed -N-Ti[N(CH$_3$)$_2$]$_3$ groups and -Si-N(CH$_3$)-Ti[N(CH$_3$)$_2$]$_3$ units can subsequently condense with free SiH and/or NH units or stay as they stand. To support this discussion, we have characterized the precursor by elemental analysis, FTIR and NMR spectroscopies.

$$\text{-Si-H} + \text{H}_3\text{C-N(CH}_3\text{)-Ti[N(CH}_3\text{)}_2\text{]}_3 \rightarrow \text{-Si-N(CH}_3\text{)-Ti[N(CH}_3\text{)}_2\text{]}_3 + \text{CH}_4 \quad (1)$$

$$\text{-N-H} + \text{N(CH}_3\text{)}_2\text{-Ti[N(CH}_3\text{)}_2\text{]}_3 \rightarrow \text{-N-Ti[N(CH}_3\text{)}_2\text{]}_3 + \text{HN(CH}_3\text{)}_2 \quad (2)$$

Compared to the chemical formula of PVMSZ ([Si$_{1.0}$C$_{1.5}$N$_{1.1}$H$_{5.5}$]$_n$ (oxygen content in the sample is 0.4 wt% and can be therefore neglected), the elemental analysis data of the air- and moisture-sensitive solid PVMSZTi2.5 sample—which allowed to determine a chemical formula of [Si$_{1.0}$Ti$_{0.3}$C$_{3.0}$N$_{1.6}$H$_{8.5}$]$_n$ (oxygen content in the sample is 0.6 wt% and can be therefore neglected)—confirms that (i) Ti is incorporated at molecular scale in the structure of PVMSZ and (ii) the carbon, nitrogen and hydrogen contents significantly increase. The measured formula shows that the Si:Ti ratio is higher (i.e., 3.3) than the targeted value fixed before the synthesis, i.e., 2.5, indicating that the Ti content is lower than expected. Most probably, the TDMAT introduced during the polymer synthesis has not completely reacted with PVMSZ. Thus, a certain amount of unreacted TDMAT could be recovered during the extraction of the solvent. Therefore, it is suggested that the reaction cannot be completed when too much TDMAT is added due to a high steric hindrance imposed by the evolutive PVMSZ structure, i.e., the Si:Ti ratio becomes too small and the reactive functions of the neat PVMSZ are no longer accessible. Based on Equations (1) and (2), the significant increase of the carbon, nitrogen and hydrogen contents in the PVMSZTi2.5 sample can be caused by the formation of Si-N(CH$_3$)-Ti bridges. However, this significant increase is most probably related to the high portion of -Ti[N(CH$_3$)$_2$]$_3$ units which are present in as-formed units. They can be considered as side groups that do not further react during the synthesis progress. Thus, this means that the PVMSZTi2.5 sample displays a higher degree of crosslinking than PVMSZ because of the formation of Ni-Ti bonds and -Si-N(CH$_3$)-Ti- bridges and a relative high portion of surrounding-[N(CH$_3$)$_2$]$_3$ groups. In order to obtain a complete view of the titanium-modified polysilazane structure, the PVMSZTi2.5 sample has been characterized by infrared sand solid-state NMR spectroscopies.

The FTIR spectrum of PVMSZTi2.5 (Figure S1 in Supplementary Materials) displays some of the characteristic absorption bands of neat PVMSZ [31,32] attributed to the stretching of N-H bonds at 3376 cm^{-1} coupled to N-H deformations at 1174 cm^{-1}, the stretching of C-H bonds in CH$_x$ groups at 2960 cm^{-1} and the stretching of Si-H bonds and deformations of Si-CH$_3$ units at 2123 cm^{-1} and at 1250 cm^{-1}, respectively. Below 1056 cm^{-1}, absorption bands attributed to the stretching and deformation vibrations involving Si-C, Si-N, C-H, C-C and N-Ti bonds overlap and cannot be assigned unambiguously. The main changes occur in the intensity of some of the characteristic bands quoted above. Thus, a decrease in the intensity of the absorption bands assigned to Si-NH-Si groups at 3376 and 1174 cm^{-1} and Si-H bonds at 2123 cm^{-1} is observed in the spectrum of PVMSZTi2.5—more significantly for the bands assigned to N-H bonds. In parallel, there is appearance of a set of broad bands in the range of 2750–2900 cm^{-1} attributed to the vibration of C-H bonds from the methyl/dimethylamino groups present in TDMAT. The band at 1294 cm^{-1} can be assigned to the N-C bonds in -N(CH$_3$) groups. Another set of bands that can be attributed to deformation of C-H bands appears at around 1417 and 1457 cm^{-1}. Thus, FTIR spectroscopy confirms the incorporation of -N(CH$_3$) groups in PVMSZ possibly via Equations (1) and (2):

(i) those involving -NH units in PVMSZ and -N(CH$_3$)$_2$ groups in TDMAT to form -N-Ti[N(CH$_3$)$_2$]$_3$ units releasing dimethylamine according to Equation (1). This reaction occurs probably majoritarly;
(ii) those involving the silicon centers of PVMSZ and -N(CH$_3$)$_2$ groups in TDMAT causing the decrease of Si-H groups while forming -Si-N(CH$_3$)-Ti- bridges in the obtained precursor and the concomitant evolution of methane according to Equation (2).

Moreover, the band at 3046 cm^{-1}, which is assigned to the vibration of C–H bond in the vinyl groups and the typical absorption band arising from the stretching of C=C double bonds in vinyl groups at 1591 cm^{-1} present in PVMSZ are almost vanished in the FTIR spectrum of PVMSZTi2.5. It is rather suggested that TDMAT acts as a catalyst for the polymerization of the vinyl groups (Equation (3)) and/or the hydrosilylation reaction between -Si–H and -Si-CH=CH$_2$ units leading to the formation of carbosilane bonds (-Si-C-C-Si-) (Equation (4)). Indeed, inorganic catalysts such as transition metals or metal complexes could remarkably increase the hydrosilylation rate as well as lower the temperature required for hydrosilylation [33–36].

$$n \text{ -Si-CH=CH}_2 \rightarrow \text{-(CH(-Si)-CH}_2)_n \tag{3}$$

$$\text{-Si-CH=CH}_2 + \text{-Si-H} \rightarrow \text{-Si-CH}_2\text{-CH}_2\text{-Si- and/or -Si-CH(CH}_3)\text{-Si-} \tag{4}$$

To support this discussion, the FTIR spectra of PVMSZTi2.5 and a PVMSZ having undergone the same thermal procedure than PVMSZTi2.5 without TDMAT—it is labeled PVMSZ_T—have been compared (Figure S1 in Supplementary Materials). In the FTIR of PVMSZ_T, the bands attributed to vinyl groups are still present although their intensities decrease. Thus, this confirms the catalytic activity of TDMAT towards reactions involving vinyl groups. These results point out at three or four main effects that took place during the reaction of PVMSZ with TDMAT as depicted in Equations (1) to (4). To achieve an in-depth understanding of the local carbon, silicon and nitrogen environments in the polymer, we investigated ^{13}C, ^{29}Si and ^{15}N solid-state NMR spectroscopy of the PVMSZTi2.5 sample (Figure 2). The cross-polarization (CP) technique has been used for ^{13}C and ^{15}N-NMR experiments to obtain spectra with reasonable acquisition times and signal-to-noise ratios.

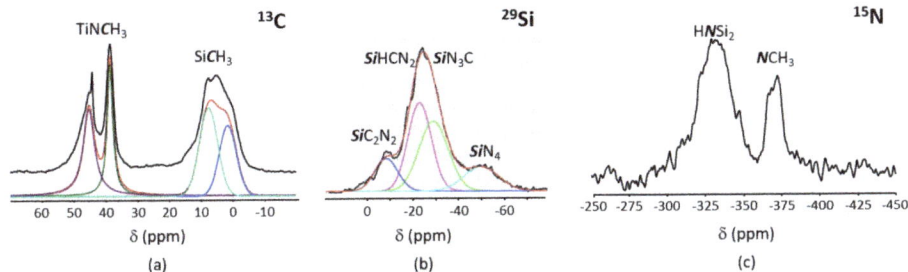

Figure 2. Experimental ^{13}C CP MAS NMR (**a**), ^{29}Si MAS NMR (**b**) and ^{15}N CP MAS NMR (**c**) spectra recorded for the PVMSZTi2.5 sample at 7 T.

The solid-state ^{13}C CP MAS NMR spectrum of PVMSZTi2.5 (Figure 2a) exhibits signals which can be simulated with four components as already performed with titanium-modified polymethylsilazane [27]. The signal emerging around 6 ppm is very broad with shoulders present around 0 and 14 ppm. It is deconvoluted into two main components at 1 and 8 ppm to reproduce the shape of the signal although the shoulder at 14 ppm could not be included in this deconvolution. ^{13}C-NMR signals at around 0 ppm are typical of carbon atoms of aliphatic groups bonded to a silicon atom [31,32,37–41], i.e., in this case, the silylmethyl group -SiCH$_3$ as identified in neat PVMSZ. The presence of two signals can be due to the two types of SiCH$_3$ unit-containing environments proposed in Equations (1) and (2):

(i) in SiCN$_3$ environments (i.e., Si environment after reaction of the SiH groups in the monomeric unit [Si(CH$_3$)(H)NH]$_{0.8}$ with TDMAT,

(ii) in SiCRN$_2$ (R = H (i.e., Si environment in the monomeric unit [Si(CH$_3$)(H)NH]$_{0.8}$) or R = C (i.e., Si environment in the monomeric unit [Si(CH$_3$)(CH=CH$_2$)NH]$_{0.2}$)); thus units present in PVMSZ that did not react upon chemical modification with TDMAT.

Moreover, this broad signal probably also contains minor signals corresponding to aliphatic carbon atoms directly linked to a silicon as Si-(CH-CH$_2$)$_n$- groups formed as proposed through Equation (3) or Si-CH$_2$-CH$_2$-Si or Si-CH(CH$_3$)-Si units formed by hydrosilylation reactions as proposed through Equation (4). The weak broad signal at around 14 ppm is in the aliphatic region and can be tentatively assigned to aliphatic carbons not directly linked to a silicon such as Si-(CH-CH$_2$)$_n$ or Si-CH(CH$_3$)-Si groups. The resonances at 39 and 45 ppm are assigned to -TiNCH$_3$ groups [27]. In particular, the peak at 45 ppm is attributed to NCH$_3$ groups linked to titanium, i.e., -Si-N(Si-)-Ti[NCH$_3$)$_2$]$_3$ units, resulting from Equation (1). Indeed, its position is similar to the position of the signal identified in the liquid-state ^{13}C-NMR spectrum of TDMAT (Figure S2 in Supplementary Materials). Consequently, the signal at 39 ppm is assigned to -N(CH$_3$) groups in N$_2$Si(CH$_3$)-N(CH$_3$)-Ti[NCH$_3$)$_2$]$_3$ units resulting from Equation (2). It should be mentioned that signals at 124 and 138 ppm—attributed to the carbon of the vinyl groups present in PVMSZ—are identified the PVMSZTi2.5 sample (Figure S3 in Supplementary Materials), but the weak intensity of the signals confirms that hydrosilylation is almost complete during the synthesis of the PVMSZTi2.5 sample. The ^{29}Si MAS spectrum of the PVMSZTi2.5 sample (Figure 2b) is composed of a main broad resonance at around −24 ppm that can be fitted with two components at −21 and −29 ppm similarly to titanium-modified polymethylsilazane [27]. The signal at −21 ppm is related to HSiCN$_2$ environments, i.e., CH$_3$-Si(H)N$_2$ units as present in PVMSZ [31,32]. The other signal at −29 ppm could correspond to SiN$_3$C environments and more precisely to N$_2$Si(CH$_3$)-N(CH$_3$)-Ti[NCH$_3$)$_2$]$_3$ units resulting from Equation (2). In addition, there is a signal at around −9 ppm that is attributed to SiC_2N_2 environments. The ^{29}Si chemical shift of SiC_2N_2 environments depends on the conformation of the silazane [41]. Thus, the PVMSZTi2.5 sample is preferentially composed of six- and eight-membered Si-N rings. The signal around −50 ppm is typically for SiN_4 environment although it is difficult to imagine such an environment in our polymeric system. One reason could be the presence of a small portion of ending groups in PVMSZ that react with TDMAT for the existence of such an environment. The experimental ^{15}N CP MAS NMR spectrum of the PVMSZTi2.5 sample (Figure 2c) shows two broad signals centered at −330 ppm and −370 ppm and confirms the previous discussion. The signal at −330 ppm is attributed to $HNSi_2$ environments since these groups in the silazane backbone are expected between −335 ppm and −325 ppm [31]. The additional signal centered at −370 ppm is assigned to NCH_3 environments based on our data collected for titanium-modified polymethylsilazane [27].

The combination of multinuclear solid-state NMR data with results derived from elemental analyses and FTIR allows to have a complete understanding of the chemistry behind the reaction between PVMSZ and TDMAT and a full view of the structure of the PVMSZTi2.5 sample. At least three reactions depicted in Equations (1) to (4) occur during the synthesis of the PVMSZTi2.5 sample. They allow building the polymer network and extending the degree of cross-linking of the polymeric backbone. Titanium atoms are homogeneously distributed within the PVMSZ structure as bridges i.e., those involving -(Si-N)$_n$-Ti- units (Equation (1)) and those involving $N_2Si(CH_3)$-$N(CH_3)$-Ti- units (Equation (2)). In addition, hydrosilylation occurs during the synthesis to form carbosilane bridges (Equation (4)). This leads to a relatively high crosslinked polyme containing a certain portion of side grops of the type -Ti[$N(CH_3)_2$]$_x$. The modification of the structure of PVMSZ—because of its reaction with TDMAT—will affect its thermo-chemical behavior. The latter is investigated in the following section through pyrolysis procedures up to 1000 °C in various atmospheres that allowed delivering compounds with controlled phase composition.

2.2. Ceramic Conversion

Here, we first discuss the thermo-chemical transformation of the PVMSZTi2.5 sample into ceramic materials up to 1000 °C in flowing ammonia, nitrogen and argon as monitored by TG experiments. The collection of TG experiments allows having a good overview of the reactivity of the PVMSZTi2.5 sample with the different types of atmospheres. Then, the polymer-to-ceramic conversion has been investigated in more details using tools complementary to TG experiments: (i) MS to identify the gases evolving from the polymer during TG experiments under argon and nitrogen-MS was not possible to be used in flowing ammonia, (ii) ex-situ spectroscopic analyses (FTIR and/or solid-state NMR) of intermediates isolated during pyrolysis in such atmospheres to follow the evolution of the chemical bonding and environments. All samples produced at 1000 °C have been characterized by solid-state NMR.

2.2.1. TG Experiments

The thermo-chemical conversion of preceramic polymers into ceramics occurs through the evolution of gaseous by-products involving a weight loss upon the heat treatment as reported in Figure 3. Results are compared to data recorded from PMVSZ (Figure S4 in Supplementary Materials). In contrast to TG curves recorded in nitrogen and argon atmospheres that exhibit one single-step weight loss which is almost achieved at 700 °C, the TG curve of the PVMSZTi2.5 sample recorded in an ammonia atmosphere is more complex and displays a three-step weight loss similarly to PVMSZ (Figure S4 in Supplementary Materials) in the temperature range 30–1000 °C, i.e., from 30 to 250 °C (1st weight loss), from 250 to 550 °C (2nd weight loss) and from 550 to 1000 °C (3rd weight loss).

The PVMSZTi2.5 sample is decomposed with weight losses measured at 1000 °C higher than those measured for PMVSZ (Figure S4 in Supplementary Materials) in the same atmospheres whereas the degree of crosslinking in PVMSZTi2.5 is supposed to be higher than in PVMSZ. As-formed units are indeed expected to reduce the segment mobility of PVMSZ hindering depolymerization reactions in the polymeric network in the low temperature regime of the polymer-to-ceramic conversion; thereby reducing the total weight loss of the precursor. However, in the present case, we observe that a high cross-linking is not the only prerequisite to design preceramic polymers with high and optimized ceramic yield. Another important issue is a sufficient latent reactivity of the precursors, i.e., the ability to undergo further cross-linking reactions during the heat treatment. This is particularly the case when -Si-H, -N-H and vinyl groups are present in the polymer to occur dehydrocoupling and hydrosilylation reactions. Such units are supposed to be in a significant lower portion in PVMSZTi2.5 compared to PVMSZ because they reacted with TDMAT during its synthesis. Furthermore, TDMAT catalyzed hydrosilylation reactions during the PVMSZTi2.5 synthesis. Therefore, dehydrocoupling and hydrosilylation reactions are limited during the heat-treatment of the PVMSZTi2.5 sample. Finally, the highest weight losses recorded for PVMSZTi2.5 is also a consequence of the presence of -N(CH$_3$)$_3$ as side groups in its structure. They are not able to undergo cross-linking reactions and are decomposed during the polymer-to-ceramic conversion. This is particularly understandable in flowing ammonia because -N[(CH$_3$)$_x$]$_y$ units (x = 1 or 2 and y = 1 (for x = 1) and 1 → 3 (for x = 2)) are highly reactive with ammonia to release amines as gaseous by-products via transamination reactions as illustrated in Equation (5).

$$-Ti[N(CH_3)_x]_y + yNH_3\ (g) \rightarrow -Ti[NH_2]_y + y(CH_3)_xNH_{3-x} \tag{5}$$

Since the polymer-to-ceramic conversion of preceramic polymers into ceramic materials is a complex process, spectroscopic analyses (MS, FTIR and/or solid-state NMR) must be very extensive. Thus, detailed discussions of these efforts are reported in detail hereafter.

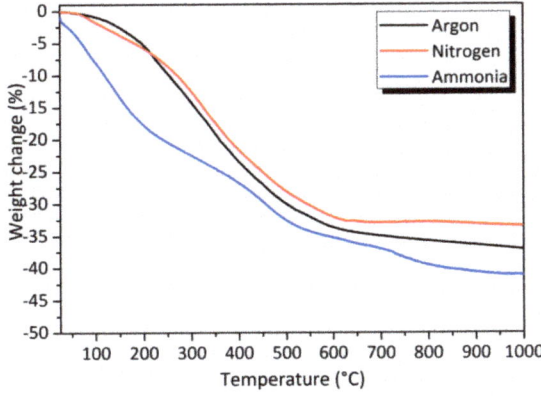

Figure 3. TG curves recorded during decomposition of the PVMSZTi2.5 sample in flowing ammonia, nitrogen and argon.

2.2.2. Pyrolysis in Flowing Ammonia

To elucidate the mechanisms governing the polymer-to-ceramic conversion in flowing ammonia, intermediates have been isolated during the pyrolysis of PVMSZTi2.5 at 200, 450, 700 and 1000 °C with a dwelling time of 2 h at each temperature. Pyrolysis intermediates are labeled **P-TNH** (T being the temperature at which the sample was exposed; NH corresponds to ammonia atmosphere). Representative FTIR has been first carried out. The corresponding spectra are shown in Figure 4.

The **P-200NH** sample spectrum displays the characteristic absorption bands of the PVMSZTi2.5 sample with a main decrease in intensity of the C-H (from methylamino groups) vibration and deformation bands from 2750 to 2900 cm^{-1} and from 1375 to 1500 cm^{-1}. These changes are confirmed in the spectrum of the **P-450NH** sample. Thus, these observations prove the occurrence of the reaction depicted in Equation (5) in the low temperature regime of the pyrolysis (T < 450 °C) as suggested based on TG experiments. To support our observations, the N-H vibration band around 3380 cm^{-1} becomes much broader toward lower wavenumbers and more intense suggesting the formation of both -NH and -NH$_2$ units as expected through Equation (5). This can be explained by the fact that -Ti[NH$_2$]$_y$ groups formed in Equation (5) can further condense with formation of NH-containing bridges such as Ti-NH-Ti bridges by release of ammonia according to Equation (6).

$$\text{-Ti-NH}_2 + \text{-Ti-NH}_2 \rightarrow \text{-Ti-NH-Ti-} + \text{NH}_3 \text{ (g)} \tag{6}$$

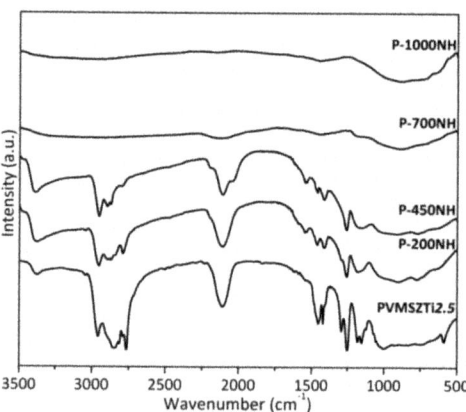

Figure 4. FTIR spectra of pyrolysis intermediates isolated in flowing ammonia and derived from PVMSZTi2.5.

In contrast, the intensities of the Si-H vibration band around 2120 cm^{-1} and SiCH$_3$ deformation band around 1250 cm^{-1} are relatively stable in the corresponding spectra indicating their poor reactivity under ammonia below 450 °C. Changes occur in the spectrum of the **P-700NH** sample. The intensity of the bands attributed to these units significantly decrease demonstrating that Si-H and Si-CH$_3$ groups (as well as Si-CH$_2$-CH$_2$-Si/Si-CH(CH$_3$)-Si units) react with ammonia at intermediate temperatures. It is—in general—reported in the temperature range 400–700 °C for polysilazanes [42,43] according to a nucleophilic substitution by ammonia that releases dihydrogen and methane (Equation (7)).

$$\text{-Si-R} + \text{NH}_3 \rightarrow \text{-Si-NH}_2 + \text{RH} \quad (\text{R = H, CH}_3\text{, C}_2\text{H}_4\text{Si-}) \tag{7}$$

As proposed by Choong Kwet Yive et al. for polysilazanes [42,43], homolytic cleavages are the most probable reactions occurring in a parallel way according to Equations (8)–(10).

$$\text{-Si-R} \rightarrow \text{-Si}^\bullet + \text{R}^\bullet \quad (\text{R = H, CH}_3\text{, C}_2\text{H}_4\text{Si-}) \tag{8}$$

$$\text{R}^\bullet + \text{NH}_3 \text{ (g)} \rightarrow \text{RH} + {}^\bullet\text{NH}_2 \tag{9}$$

$$\text{-Si}^\bullet + {}^\bullet\text{NH}_2 \rightarrow \text{-SiNH}_2 \tag{10}$$

As-formed -SiNH$_2$ groups could self-condense at intermediate temperature (Equation (11)) and in the highest temperature regime of the pyrolysis (T > 700 °C, Equation (12)) during the conversion process with formation of Si-NH-Si and -N(Si)$_3$ units by release of ammonia.

$$\text{-Si-NH}_2 + \text{-Si-NH}_2 \rightarrow \text{-Si-NH-Si-} + \text{NH}_3 \tag{11}$$

$$\text{-Si-NH-Si-} + \text{-Si-NH}_2 \rightarrow \text{-N(Si)}_3 + \text{NH}_3 \tag{12}$$

Similarly, -Si-NH$_2$ groups (Equation (11)) as well as -TiNH-Ti- units (Equation (6)) can condense with -TiNH$_2$ (Equation (5)) to form -Si-NH-Ti- and -N(Ti)$_3$ units, respectively. In parallel, we cannot exclude that unreacted SiH and NH units can react together at high temperature to release hydrogen. Thus, the decomposition of the majority of functional groups is observed in the **P-700NH** sample to form -Si-N- and -Ti-N- bonds. In addition to this observation, the alteration in the shape of the absorption bands compared to the earlier state is due to the amorphous character of the material. This is confirmed with the broadness of the main signal in the spectrum recorded after pyrolysis at 1000 °C (**P-1000NH**). There are no more organic groups and the spectrum is expected to exhibit a mixture of 'SiN' and 'TiN' units as discussed earlier.

Multinuclear solid-state NMR spectroscopy (Figure 5) has been used to identify the most important structural rearrangements occurring during the thermo-chemical conversion of PVMSZTi2.5 in flowing ammonia. Within this context, the pyrolysis intermediates isolated at 450 °C (**P-450NH**) and 1000 °C (**P-1000NH**) have been analyzed by probing the local environment around silicon (^{29}Si, Figure 5a) and carbon (^{13}C, Figure 5b).

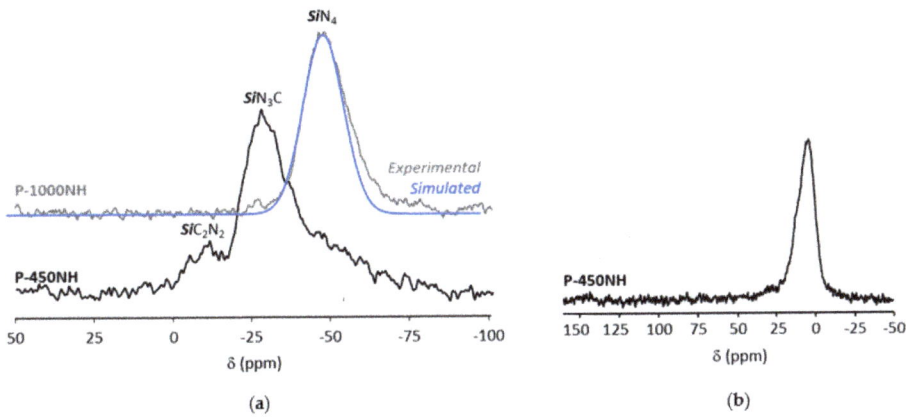

Figure 5. Experimental ^{29}Si MAS NMR (**a**) and ^{13}C CP MAS NMR (**b**) spectra recorded for the pyrolysis intermediates derived from PVMSZTi2.5 at 7 T.

Compared to the ^{29}Si MAS spectrum of the PVMSZTi2.5 sample (Figure 2b), the ^{29}Si MAS spectrum of the **P-450NH** sample (Figure 5a) results in a small shift of the main signal toward the resonance at −28 ppm attributed to *Si*N$_3$C units while the signal assigned to *Si*C$_2$N$_2$ units is retained. This suggests that H*Si*CN$_2$ units (and *Si*N$_4$ units) are the main environment affected by the pyrolysis under ammonia in the low temperature regime of the pyrolysis. In parallel, the ^{13}C CP MAS signals at 39 and 45 ppm in the spectrum of the **P-450NH** sample attributed to -TiNCH$_3$ units disappear (Figure 5b) compared to the ^{13}C CP MAS spectrum of the **PVMSZTi2.5** sample (Figure 2a). This confirms that such units are decomposed during the first step of the pyrolysis under ammonia as suggested by FTIR. Thus, the spectrum of the **P-450NH** sample shows one main signal around 10 ppm assigned to carbon atoms of aliphatic groups bonded to a silicon atom and/or a carbon atom linked to another aliphatic carbon. The presence of Si-CH(Si-)-CH$_3$ units or Si-(CH-CH$_2$)$_n$ environments confirms that such units persist upon further heat-treatment at intermediate temperature in flowing ammonia. More changes take place in the **P-1000NH** sample. Obviously, nucleophilic substitutions and rearrangements occur in the temperature range 450–1000 °C that cause a modification of the chemical environment around the silicon atoms. Solid-state NMR suggests a predominantly covalent material in which silicon is mainly present in *Si*N$_4$ environments (the corresponding signal can be simulated with one component) confirming the occurrence of reactions depicted from Equations (7) and (12). Thus, based on FTIR and NMR spectroscopies and deduced Equations, the **P-1000NH** sample is made of -N(Si)$_3$ and -N(Ti)$_3$

units. As a consequence, the thermo-chemical conversion of the PVMSZTi2.5 sample in flowing ammonia can be seen as represented in Figure 6.

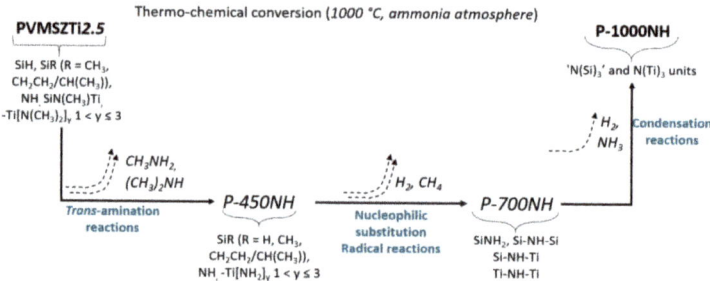

Figure 6. Schematic diagram of the thermo-chemical conversion of titanium-modified poly(vinylmethyl-co-methyl)silazanes into Si-Ti-N ceramics in flowing ammonia.

2.2.3. Pyrolysis in Flowing Nitrogen and Argon

- Pyrolysis in flowing nitrogen

Pyrolysis intermediates have been also isolated under nitrogen during the pyrolysis of PVMSZTi2.5 at 200, 450, 700 and 1000 °C with a dwelling time of 2 h at each temperature. Pyrolysis intermediates are labeled **P-TN** (T being the temperature at which the sample was exposed; N corresponds to nitrogen atmosphere). Representative FTIR has been first carried out. The corresponding spectra are shown in Figure 7.

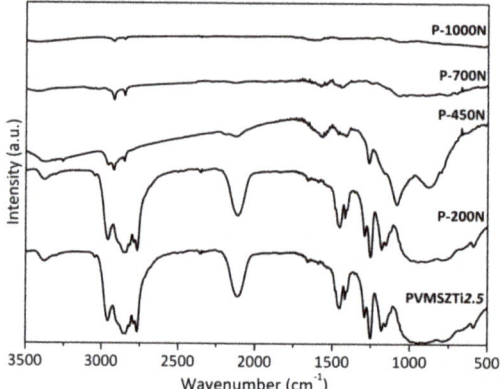

Figure 7. FTIR spectra of pyrolysis intermediates isolated in flowing nitrogen and derived from PVMSZTi2.5.

In good agreement with the relative stability of the PVMSZTi2.5 sample during TG experiments up to 200 °C (around 6 wt% of weight loss to be compared to around 17 wt% in flowing ammonia at the same temperature), the **P-200N** sample spectrum displays the characteristic absorption bands of the PVMSZTi2.5 sample with a relative stability in terms of intensity of bands attributed to the bonds characterizing the functional groups such as C-H, N-H and Si-H. To corroborate this observation, we investigated MS during the TG experiment in flowing nitrogen (Figure 8).

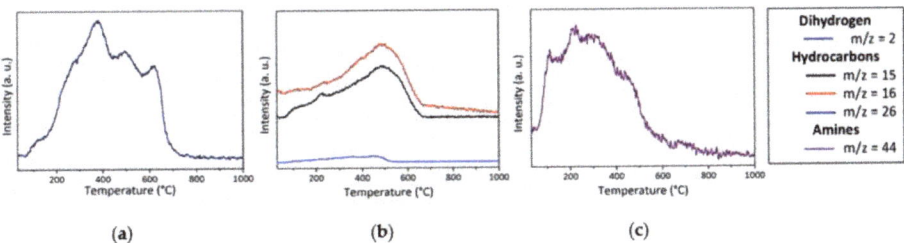

Figure 8. MS curves recorded during TG experiments of the PVMSZTi2.5 sample in flowing nitrogen: dihydrogen (**a**), hydrocarbons (**b**) and amines (**c**).

The MS curves show that the temperature range R.T. 200 °C is only associated with a release of dimethylamine ($m/z = 44$) fragments indicating the occurrence of condensation reactions involving side groups of the type -N((CH$_3$)$_2$)$_y$ ($y = 1 \rightarrow 3$) and NH units which already took place during the synthesis of the PVMSZTi2.5 sample (Equation (1)). Although the methylamine fragments could not be detected, the release of this gas, however, cannot be excluded in the low temperature regime of the thermo-chemical conversion through the cleavage of the -Si-N(CH$_3$)-Ti units formed by Equation (2), then abstracting hydrogen atoms. In this temperature range, MS also identifies the release of dihydrogen ($m/z = 2$), most probably because of dehydrocoupling reactions involving NH and SiH groups that did not react during the synthesis of the polymer (-Si-H + -Si-H → -Si-Si- + H$_2$ and -N-H + -Si-H → -N-Si- + H$_2$). These reactions are well known to occur at relatively low temperature [43] and methane ($m/z = 15, 16$) as a continuity of Equation (2). It should be pointed out that these signals can be also attributed to ammonia because of the occurrence of transamination reactions as depicted in Equations (13)–(15) [44].

$$2 \text{ -Si-NH-Si-} \rightarrow \text{-Si-NH}_2 + \text{-N(Si)}_3 \tag{13}$$

$$\text{-Si-NH}_2 + \text{-Si-NH-Si-} \rightarrow \text{-N(Si)}_3 + \text{NH}_3 \text{ (g)} \tag{14}$$

$$3\text{-Si-NH-Si-} \rightarrow 2\text{-N(Si)}_3 + \text{NH}_3 \text{ (g)} \tag{15}$$

The release of dihydrogen and methane—in addition to acetylene, ethylene or ethane fragments ($m/z = 26$)—becomes much more intense above 200 °C in excellent agreement with the significant band intensity changes occurring in the spectrum of the **P-450N** sample: the band intensity of the bonds characterizing the functional groups significantly decreases indicating that the polymer-to-ceramic conversion mainly takes place between 200 and 450 °C as suggested through the TG experiment that identified a weight loss of around 25 wt% at 450 °C. Associated with the main release of dihydrogen and methane, this band intensity decrease continues between 450 and 700 °C (**P-700N**); temperature at which the weight loss (32.5 wt%) is almost complete as indicated through the corresponding TG curve. Only the vibration bands attributed to C-H bond are still present as functional groups. Thus, from 200 to 750 °C, the decomposition reactions involve—in part—radical mechanisms [45–47]. As observed with FTIR spectra recorder in flowing ammonia, the absorption bands shape is characteristic of the amorphous character of the material after pyrolysis at 1000 °C (**P-1000N**). They are representative of amorphous ceramics showing no more organic groups and exhibiting a mixture of 'SiN', 'SiC' units probably also with 'TiC$_x$N$_y$' units ($x + y = 1$). Hence, based on these results, it seems that the decomposition of the PVMSZTi2.5 sample under nitrogen first mainly occurs via the polymerization mechanisms depicted in Equations (1) and (2) to crosslink the polymer backbone. Then, radical reactions (Equation (16)) followed by hydrogen abstraction (Equation (17)) take place at intermediate temperature to consolidate the ceramic structure already at 700 °C through the formation of dihydrogen and hydrocarbons.

$$\text{-Si-R} \rightarrow \text{-Si}^{\bullet} + \text{R}^{\bullet} \quad (\text{R = H, -CH}_3, \text{-CH}_2\text{-CH}_2\text{-Si-/-CH(CH}_3\text{)Si-}) \tag{16}$$

$$R^{\bullet} + H^{\bullet} \to RH \ (R = H, -CH_3, -CH_2-CH_2-Si-/-CH(CH_3)Si-) \tag{17}$$

Hydrogen radicals are formed by the decomposition of C-H bonds in the high temperature regime of the thermo-chemical conversion. It should be pointed out that the identification of methane can be due to the cleavage of C-C bonds as suggested through Equation (18). Then, the $CH_3^{\bullet}$ radical can abstract hydrogen from Si-H bonds (Equation (19)).

$$-Si-CH(CH_3)-Si \to -Si-CH^{\bullet}-Si + CH_3^{\bullet} \tag{18}$$

$$CH_3^{\bullet} + -SiH \to -CH_4 + Si^{\bullet} \tag{19}$$

The large escape of ethylene $CH_2=CH_2$ may arise from the degradation of the $Si-CH_2-CH_2-Si$ units formed by hydrosilation during the synthesis of the PVMSZTi2.5 sample according to Equations (20) and (21).

$$Si-CH_2-CH_2-Si \to -Si^{\bullet} + {}^{\bullet}CH_2-CH_2-Si \tag{20}$$

$${}^{\bullet}CH_2-CH_2-Si \to Si^{\bullet} + CH_2 = CH_2 \tag{21}$$

The temperature range 700–1000 °C is probably associated with dihydrogen release due to homolytic cleavage of residual C-H bonds followed by H abstraction (Equations (22) and (23)).

$$-C-H \to C^{\bullet} + H^{\bullet} \tag{22}$$

$$H^{\bullet} + H^{\bullet} \to H_2 \tag{23}$$

- Pyrolysis in flowing argon

The thermo-chemical conversion of the PVMSZTi2.5 sample in an argon atmosphere is similar to that one observed in flowing nitrogen based on the profile of the weight loss curve recorded during the TG experiments. The main difference takes place in the high temperature regime of the TG experiment since the weight loss is not stabilized after pyrolysis at 1000 °C in flowing argon. To understand the behavior of the PVMSZTi2.5 sample in an argon atmosphere, in-situ MS experiments were performed and the corresponding curves are represented in Figure 9.

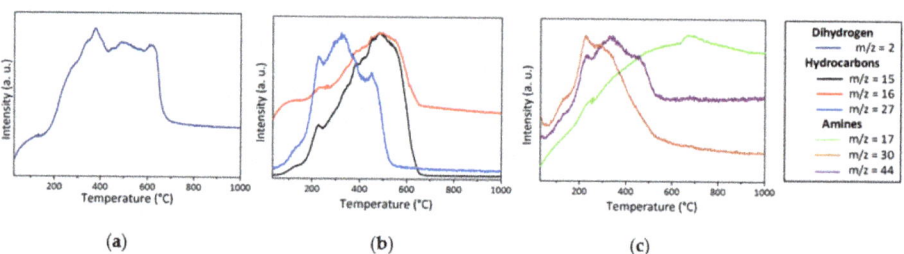

(a)   (b)   (c)

Figure 9. MS curves recorded during TG experiments of the PVMSZTi2.5 sample in flowing argon: dihydrogen (**a**), hydrocarbons (**b**) and amines (**c**).

The nature of the gases that are released from the PVMSZTi2.5 sample in flowing argon and the temperature ranges in which their release significantly occurs matches with the data recorded in nitrogen atmosphere. Only the fragments at m/z = 26, 27 and 30 are identified in either nitrogen (m/z = 26) or argon (m/z = 27 and 30). The ion signals at m/z = 26 and 27 most probably correspond to the $C_2H_2^+$ and $C_2H_3^+$ fragments; thereby arising due to the same molecules, i.e., ethane and ethylene evolving in both argon and nitrogen. The ion signal at m/z = 30 corresponds to the release of methylamine that takes place under argon at low temperature as dimethylamine release occurs. The release of methylamine is suggested in flowing nitrogen despite the fact that it is not detected by

MS. Thus, the discussion on the thermo-chemical conversion of the PVMSZTi2.5 sample in flowing argon fits the one given previously under nitrogen and the same mechanisms occur in both atmospheres in similar temperature range. Based on the results from FTIR and MS obtained in both nitrogen and argon atmospheres, it is suggested that the thermo-chemical conversion involves three main steps:

(1) Below 300 °C, the precursor undergoes thermal cross-linking via condensation reactions involving both side groups of the type -N((CH$_3$)$_2$)$_x$ (x = 0, 1, 2, 3) and NH units (Equation (1)) and -Si-N(CH$_3$)-Ti- that cleaves upon heating to abstract hydrogen releasing methylamine.
(2) From 300 to 700 °C, mainly hydrocarbons such as methane and ethane/ethylene and hydrogen are released from the evolving system via radical reactions then H abstraction.
(3) Above 700 °C, the remaining hydrogen atoms are gradually removed through homolytic cleavage of C-H bonds and then H abstraction to lead to a single-phase amorphous covalently bonded ceramic.

As reported in Figure 10, these reactions lead to **P-1000N** and **P-1000A** samples both displaying in their ^{29}Si MAS spectra a relatively broad signal at around −48 ppm ascribed to SiN$_4$ environments. However, its broadness—especially on the left part of the main signal—tends to indicate an enrichment in carbon around silicon to form SiC$_x$N$_{4-x}$ (0 ≤ x ≤ 4) type units compared to the sharper signal identified in the spectrum of the **P-1000NH** sample that can be simulated with one component as indicated in Figure 5.

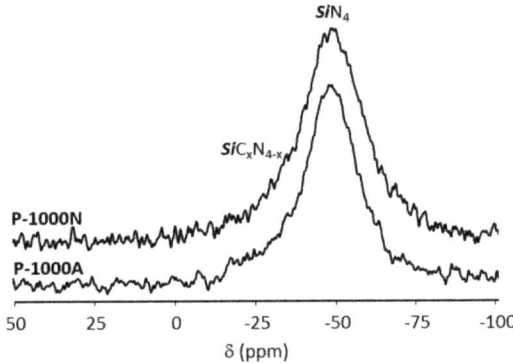

Figure 10. Experimental ^{29}Si MAS NMR for the pyrolysis samples isolated at 1000 °C in flowing argon and nitrogen at 7 T.

Thus, the thermo-chemical conversion of the PVMSZTi2.5 sample in flowing nitrogen and argon can be seen as represented in Figure 11 based on the TG data coupled with in-situ MS and ex-situ FTIR and solid-state spectroscopies of pyrolysis intermediates.

2.3. High-Temperature Phase and Microstructure Evolution of Single-Phase Amorphous Ceramics

After pyrolysis experiments, the X-ray diffractograms reveal that as-pyrolyzed samples are poorly crystallized as shown by the diffuse peaks identified in the patterns of the **P-1000NH, P-1000N, P-1000A** samples (Figure 12a). In a first approximation, they tend to reveal the slow nucleation of face-centered cubic (*fcc*) TiN crystals (powder diffraction file (ICDD PDF) number: 00-038-1420) according to the presence of peaks at 2θ around 36.5°, 42.6° and 62.3 which can be attributed to the (111), (200) and (220) TiN reflections, respectively. To support this observation, the very broad peak in the range 72.8–75.2° (peak positions changes according to the nature of the atmosphere) could be attributed to the (311) TiN reflection. The crystallite sizes are low and range from 1.4 (**P-1000NH**) to 1.7 nm (**P-1000A**).

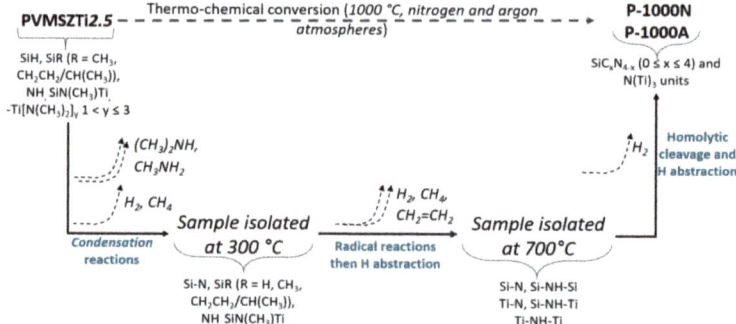

Figure 11. Schematic diagram of the thermo-chemical conversion of titanium-modified poly(vinylmethyl-*co*-methyl)silazanes into Si-Ti-C-N ceramics in flowing nitrogen and argon.

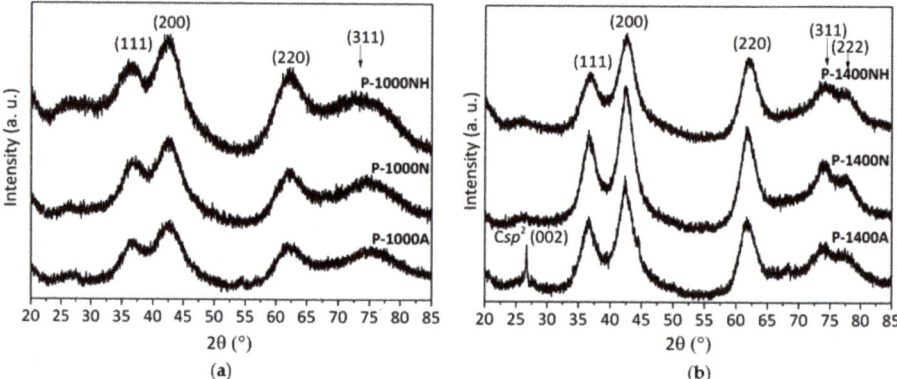

Figure 12. XRD patterns of **P-1000NH**, **P-1000N** and **P-1000A** (a) and P-1400NH, P-1400N and the P-1400A (b).

For all samples prepared at 1000 °C, EDS analyses demonstrate that:

(i) the initial Si:Ti ratio (2.5) fixed in the polymer is retained in the derived ceramics,
(ii) a strong tendency to decrease the carbon content from the sample treated under argon (**P-1000A**, carbon content of 16.4 wt%) to samples treated under ammonia (**P-1000NH**, carbon content of 1.5 wt%).

Thus, we can suggest that the **P-1000NH** sample is only made of Si, Ti and N whereas the **P-1000N** and **P-1000A** samples are composed of Si, Ti, C, N. The increase of the annealing temperature to 1400 °C under nitrogen (**P-1400NH** and **P-1400N** samples) and argon (**P-1400A** sample) with a dwelling time of 2 h at 1400 °C does not change the solid-state ^{29}Si NMR response: all spectra report the broad peak characteristic of SiN_4 environments with—for the **P-1400N** and **P-1400A** samples—emerging peaks in the range −10 to −35 ppm due to SiC_4 and SiC_xN_{4-x} ($0 \leq x \leq 4$) type units (see Figure S5 in Supplementary Materials). SiC_4 environments can be identified in such spectra by comparing them with the spectrum of the **P-1800A** sample which exhibits only this signal. Although solid-state NMR does not show any changes in the chemical environment of Si, XRD investigations show that annealing to 1400 °C involves further crystallization of the materials (Figure 12b). XRD patterns of the **P-1400NH**, **P-1400N**, **P-1400A** samples reveal better defined and shaper peaks at 2θ = 36.5° (**P-1400N** and **P-1400A**) and 36.7° (**P-1400NH**), from 42.3° (**P-1400A**) to 42.6° (**P-1400NH**), from 61.7° (**P-1400A**) to 62.0° (**P-1400NH**), 74.1° and 77.5° (all samples). Since the positions tend to slightly change between samples according to the used atmosphere, we tentatively suggested that the XRD patterns exhibit the (111), (200), (220), (311) and (222) peaks from a TiC_xN_y ($x + y = 1$) crystalline phase with different stoichiometries. For instance, the peaks (111), (200) and (220) can be indexed according to the $TiC_{0.3}N_{0.7}$ phase (ICDD PDF 00-042-1488) at 36.45°, 42.35° and 61.44°; the TiC phase (ICDD PDF 00-032-1383) at 35.90°, 41.70° and 60.44°; and the TiN phase (ICDD PDF 00-038-1420) at 36.66°, 42.59° and 61.81°, diffraction patterns, respectively.

As previously reported using the lattice parameter values for pure TiN and TiC and the lattice parameters of various TiC_xN_y phases, the lattice constant of TiC_xN_y ($x + y = 1$) is a linear function of its chemical composition which obeys Vegard's law [48,49]. Thus, the observed shift of the XRD (200) reflection in Figure 12b was used to estimate the chemical composition of the TiC_xN_y ($x + y = 1$) phase in the ceramics prepared at 1400 °C (Table 1) [50].

Table 1. Structural parameters measured based on XRD pattern of nanocomposite samples.

Samples	(200) Peak Position (2θ, °)	d_{200} (nm)	Lattice Parameter a (nm)	Composition of the TiC_xN_y from XRD
P-1400NH	42.6	0.212	0.4245	TiN
P-1400N	42.45	0.213	0.4264	$TiC_{0.2}N_{0.8}$
P-1400A	42.3	0.214	0.4273	$TiC_{0.3}N_{0.7}$
P-1500NH	42.6	0.212	0.4245	TiN
P-1500N	42.3	0.214	0.4273	$TiC_{0.3}N_{0.7}$
P-1500A	42.25	0.214	0.4273	$TiC_{0.4}N_{0.6}$
P-1600NH	42.6	0.212	0.4245	TiN
P-1600N	42.2	0.214	0.4283	$TiC_{0.4}N_{0.6}$
P-1600A	42.1	0.215	0.4293	$TiC_{0.6}N_{0.4}$
P-1800NH	42.6	0.212	0.4245	TiN
P-1800N	42.15	0.214	0.4288	$TiC_{0.5}N_{0.5}$
P-1800A	41.9	0.216	0.4312	$TiC_{0.8}N_{0.2}$

No further XRD peaks are observed except the (002) peak attributable to sp^2 carbon identified in the **P-1400A** sample. Based on the crystallite size (from 1.9 nm (**P-1400NH**) to 3.1 nm (**P-1400N/P-1400A**)), the **P-1400NH** sample can be considered as a TiN-based ceramic displaying the lowest crystallization degree.

To assess the phase micro-/nanostructure of these materials, the **P-1400NH**, **P-1400N**, **P-1400A** samples have been further studied by means of TEM (Figure 13). Figure 13a depicts a featureless low magnification TEM image of **P-1400NH** which is significantly different from images recorded for **P-1400N** and **P-1400A** samples. In the latter, specimens consist of homogeneously dispersed small nuclei (dark contrast) embedded in an amorphous network. The Selected Area Electron Diffraction (SAED) patterns of the **P-1400N** and **P-1400A** samples identify more distinct rings which can be indexed to the (111), (200) and (220) plans of the TiC_xN_y phase indicating an extent of the crystallization in those samples. In the high-resolution micrographs (Figure 13d–f), the analysis of the **P-1400NH** sample (Figure 13d) reveals the presence of nanocrystals with a lattice spacing of 0.24 nm corresponding to the d-spacing of the lattice plane of the TiN structure, i.e., the (111) direction of the *fcc* cubic rocksalt TiN structure which agrees with the phase identified in the X-ray diffractogram of the same sample (Figure 12b). The HRTEM micrographs in Figure 13e,f demonstrate that crystallization occurs in-situ in the **P-1400N** and **P-1400A** samples. Lattice fringes of the TiC_xN_y nanocrystals—with diameter less than 5 nm as better calculated in Figure 13e,f—are observed which confirms that the samples represent nanocomposites with a fringe spacing of 0.25 nm. As a conclusion, heat-treatment above 1400 °C provides access to nanocomposites composed of nano-precipitate homogeneously distributed in an amorphous matrix. Thus, in the following section, we investigated separately ammonia-treated samples and nitrogen/argon-treated samples.

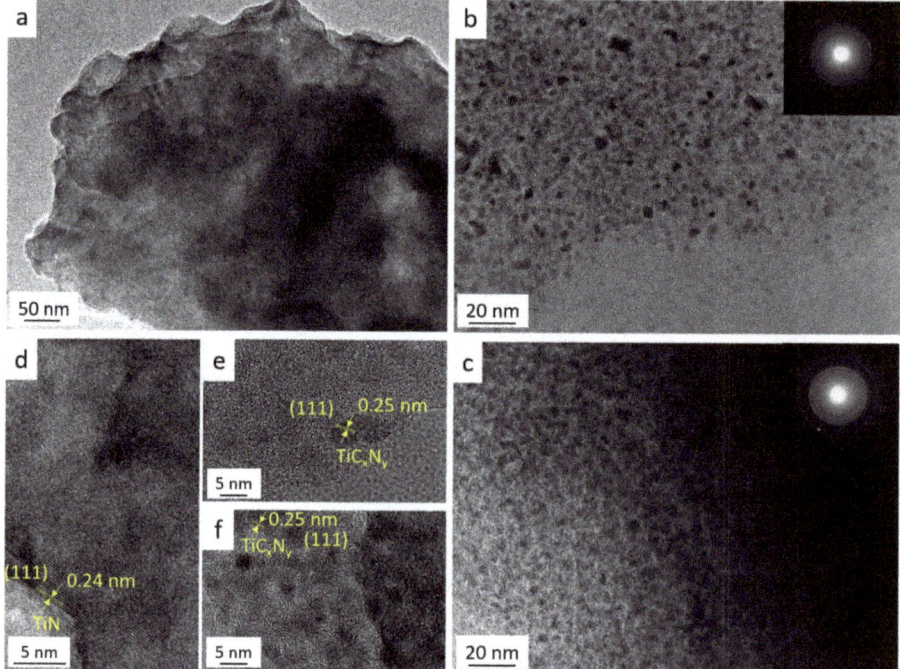

Figure 13. TEM images of P-1400NH (**a,d**), P-1400N (**b,e**) and the P-1400A samples (**c,f**) with inserted selected area electron diffraction patterns in (**a–c**).

2.3.1. Ammonia-Treated Samples Heat-Treated in the Temperature Range 1500–1800 °C

After heat-treatment at 1500 °C, the **P-1500NH** sample shows sharper and more intense TiN XRD peaks (crystallite size = 16 nm) as well as emerging XRD peaks corresponding to nanosized α-Si$_3$N$_4$ (ICDD PDF number: 04-005-5074) crystals (Figure 14a). HRTEM (Figure 15a) confirms clear evidence for the crystallization process of the TiN nanophase (**P-1500NH**) appearing as small nuclei (around 10 nm). We can clearly distinguish the fcc structure of TiN nanocrystals with a fringe spacing of 0.24 nm (Figure S7a in Supplementary Materials). This is confirmed through the SAED pattern (inset in Figure 15a).

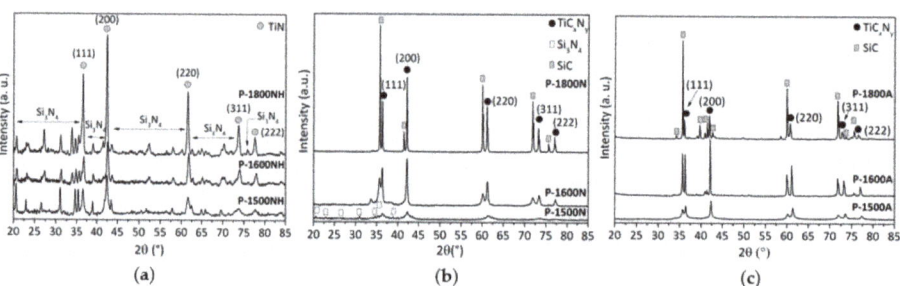

Figure 14. Evolution of the XRD patterns of P-1000NH (**a**), P-1000N (**b**) and P-1000A (**c**) samples in the temperature range 1500–1800 °C.

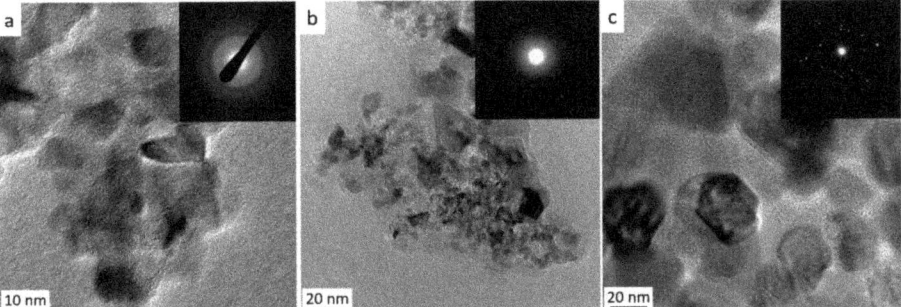

Figure 15. TEM micrographs of **P-1500NH** (a), **P-1500N** (b) and **P-1500A** (c) samples with selected area electron diffraction pattern as insets.

The increase of the annealing temperature confirms TiN crystal growth as shown through the XRD patterns of samples annealed at 1600 °C (**P-1600NH**, crystallite size = 22.4 nm) and 1800 °C (**P-1800NH**, crystallite size = 27.4 nm). Samples represent nanocomposites made of crystallized TiN and Si_3N_4 (α and β (ICDD PDF number: 04-033-1160)) phases (Figure 14a). In the following section, we discussed the crystallization behavior of the samples pyrolyzed then annealed in flowing nitrogen and argon together.

2.3.2. Nitrogen and Argon-Treated Samples Heat-Treated in the Temperature Range 1500–1800 °C

The increase of the temperature to 1500 °C (**P-1500N** (Figure 14b) and **P-1500A** (Figure 14c)) involves further crystallization of both $TiC_{0.3}N_{0.7}$ (crystalline size = 10.8 nm) and α and β-Si_3N_4 phases in **P-1500N** (see Figure S6 in Supplementary Materials) and both $TiC_{0.4}N_{0.6}$ (crystalline size = 28.4 nm) and β-SiC (ICDD PDF number: 00-029-1129) phases in **P-1500A**. In the latter, we cannot exclude the presence of α/β-Si_3N_4 phases because of the relative broadness of the XRD peaks. Thus, compared to the (200) XRD peak position in the XRD pattern of the **P-1400N** and **P-1400A** samples (Figure 12b, Table 1), the reflections of the TiC_xN_y ($x + y = 1$) phase slightly shift towards lower diffraction angles indicating that the TiC_xN_y ($x + y = 1$) phase becomes enriched in carbon. This is illustrated in Figure 16 for the **P-1500A** sample and its derivatives at 1600 °C (**P-1600A**) and 1800 °C (**P-1800A**).

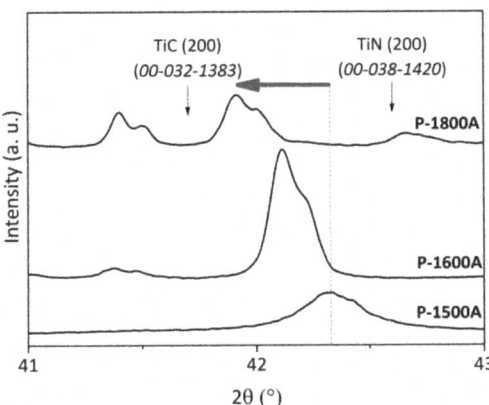

Figure 16. Evolution of the (200) peak position in the X-ray patterns of **P-1500A**, **P-1600A** and **P-1800A** samples.

Figure 15b,c displays the low magnification images of the **P-1500N** and **P-1500A** samples. They show that nanocrystals with a size that can exceed 20 nm—in particular for the sample formed under argon—co-exist with an amorphous phase appearing as a bright-appearing network. The SAED patterns (insets in Figure 15b,c) are composed of distinct spots that indicate the high degree of crystallinity of TiC$_x$N$_y$ (x + y = 1) nanocrystals. The high magnification of HRTEM images of both samples exhibit lattice fringe of the TiC$_x$N$_y$ (x + y = 1) nanophase with an equal fringe spacing of 0.25 nm. (Figure S7 in Supplementary Materials). Pyrolysis temperatures of 1600 °C identified compositions of TiC$_{0.4}$N$_{0.6}$ (**P-1600N**; crystalline size = 25.8 nm) and TiC$_{0.6}$N$_{0.4}$ (**P-1600A**; crystalline size = 43.3 nm) (Table 1) along with SiC. The heat-treatment to 1800 °C involves further crystallization of the phases and we noticed that the diffraction (200) peaks shift toward lower diffraction angles of TiC indicating an increase in the lattice parameter and the formation of the TiC$_{0.5}$N$_{0.5}$ (**P-1800N**; crystalline size = 109.6 nm) and TiC$_{0.8}$N$_{0.2}$ (**P-1800A**; crystalline size = 109.8 nm) solid solutions whereas β-/α-SiC could be assigned to other peaks (Figure 14b,c). From line broadening of the (220) reflection, the β-SiC crystallite size is estimated to be 99.8 nm.

EDS analyses (See EDS mapping in Figures S8 and S9 in Supplementary Materials) of the **P-1600A** and **P-1800A**, samples indicated the gradual formation of TiC/SiC nanocomposites with the increase of the temperature to 1800 °C, especially using argon as an atmosphere. The identification of *SiC*$_4$ environments has been confirmed in the **P-1800A** sample via ^{29}Si solid-state NMR (See Figure S5 in Supplementary Materials). To be more accurate in the C, N and O contents, we investigated C/S and N/O/H analysis for these samples and, by considering the measurement of Si and Ti contents by EDS, we found that they exhibited chemical formulas of Si$_{1.0}$Ti$_{0.2}$C$_{0.8}$N$_{0.2}$O$_{0.02}$ (**P-1600A**) and Si$_{1.0}$Ti$_{0.2}$C$_{1.1}$N$_{0.1}$O$_{0.01}$ (**P-1800A**) indicating the formation of a nanocomposite composed of 20 wt% of TiC in a SiC matrix free of carbon whereas nitrogen could be neglected according to the very low content (1.5 wt%). Thus, through the modulation of the atmosphere of pyrolysis, we succeeded in the design of nitride, carbonitride and carbide nanocomposites using an unique titanium-modified polysilazane.

3. Materials and Methods

3.1. General Comments

The synthesis of the precursor is carried out in a purified argon atmosphere passing through a column of phosphorus pentoxide and then a vacuum/argon line by means of standard Schlenk techniques. The cleaned glassware is stored in an oven at 90 °C overnight before being connected to the vacuum/argon line, assembled and pumped under vacuum for 30 min and then filled with argon. All chemical products are handled in an argon-filled glove box (Jacomex, Campus-type; O$_2$ and H$_2$O concentrations kept at ≤0.1 ppm and ≤0.8 ppm, respectively). Toluene (99.85%, Extra Dry over Molecular Sieve, AcroSeal(R)) was purchased from Acros Organics™. The poly(vinylmethyl-*co*-methyl)silazane (commercial name: Durazane® 1800) was provided by Merck company, Germany, stored in a fridge and used as-received. It is labeled PVMSZ. Anal. Found (wt%): Si, 41.3; C, 27.3; N, 22.7; H, 8.3; O, 0.4. [Si$_{1.0}$C$_{1.5}$N$_{1.1}$H$_{5.5}$]$_n$ (Referenced to Si$_{1.0}$ and oxygen content was omitted in the empirical formulae). FTIR (ATR/cm^{-1}): ν (N–H) = 3388 (m), ν (C–H) = 3046 (s), 3010 (s), 2954 (s), 2895 (s), 2852 (m), ν (Si–H) = 2121 (vs), δ (vinyl) = 1591 (m), δ (CH$_2$) = 1405 (m), δ (Si–CH$_3$) = 1251 (s), δ (N-H): 1166 (m), δ (C-H + C-C + N-Si-N + Si-C) = 1005-630 (vs); ^{1}H NMR (300 MHz, CDCl$_3$, δ/ppm): 0.4–0.1 (br, SiCH$_3$), 1.1–0.5 (br, NH), 4.9–4.4 (br, SiH), 6.3–5.7 (br, vinyl). Tetrakis(dimethylamido)titanium (Ti[N(CH$_3$)$_2$]$_4$, 99.99%) was obtained from Acros Organics™, stored in a fridge and used without further purification. It is labeled TDMAT.

3.2. Polymer Synthesis

The reaction between PVMSZ and TDMAT is done in toluene at temperature of reflux (115 °C) in a three-neck round-bottom flask. In a typical experiment, 5.7 g of TDMAT (16.5 mmol) is quickly added with a syringe under flowing argon to a solution of 2.7 g of PVMSZ (41.2 mmol referred to the theoretical monomeric unit of the polymer) in 80 mL of toluene at RT under vigorous stirring. Then, the temperature is increased up to 115 °C under static argon and kept at this temperature under vigorous stirring for three days. After cooling down, the solvent is extracted via an ether bridge (100 °C/$1.5.10^{-1}$ mbar) to release an air- and moisture-sensitive titanium-modified PVMSZ labeled PVMSZTi2.5 (2.5 being the Si:Ti ratio) that appears as a brownish-black powder. Anal. Found (wt %): Si 25.8, Ti 11.9, C 33.1, N 20.7, H 7.9, O 0.6. $[Si_{1.0}Ti_{0.3}C_{3.0}N_{1.6}H_{8.5}]_n$ (Referenced to $Si_{1.0}$ and oxygen content was omitted in the empirical formulae). FTIR (KBr/cm^{-1}): ν(N–H) = 3376 (w), ν (C–H) = 2960 (s), 2848 (s), 2761 (s), ν(Si–H) = 2123 (m), δ(CH_3) = 1457 (s), δ(CH_2) = 1417 (m), δ(C-N) = 1294 (m), δ(Si–CH_3) = 1250 (s), δ (N-H): 1174 (m), δ (C–H + C-C + N–Si–N + Si-C + Si–N–Ti) = 1054-655 (vs).

3.3. Synthesis of the Ceramic Materials

Polymeric powders are placed in alumina boats in the glove-box, then introduced in a sealed tube under argon atmosphere to prevent any oxygen contamination of the samples during the transfer to the furnace. Powders are then introduced under argon flow into a silica tube inserted in a horizontal furnace (Carbolite BGHA12/450B). The tube is evacuated (0.1 mbar) and refilled with anhydrous ammonia (99.99%), nitrogen (99.995%) or argon (99.995%) to atmospheric pressure. Subsequently, the samples are subjected to a cycle of ramping of 5 °C.min^{-1} in the temperature range 200–1000 °C in flowing ammonia, nitrogen or argon (dwelling time of 2 h at the temperature selected among 200, 450, 700 and 1000 °C). A constant flow (120 mL min^{-1}) is passed through the tube during the pyrolysis cycle. After cooling under argon atmosphere, ammonia-, nitrogen- and argon-pyrolyzed samples are stored in the glove-box for characterization. Samples are labeled P-TX with T being the temperature at which the polymer has been exposed and X corresponding to the nature of the atmosphere: X = A for argon, X = N for nitrogen and X = NH for ammonia. For the high temperature (T > 1000 °C) investigation, samples pyrolyzed at 1000 °C (**P-1000NH, P-1000N** and **P-1000A**) are subsequently introduced in a graphite furnace (Gero Model HTK 8 for ammonia-pyrolyzed samples and a Nabertherm VHT-GR for nitrogen- and argon-pyrolyzed samples) for annealing treatments. The furnaces are pumped under vacuum (1.10^{-1} mbar), refilled with nitrogen for ammonia- and nitrogen-pyrolyzed samples or with argon for argon-pyrolyzed samples and maintained under a constant flow of gas (200 mL min^{-1}) during the whole heat treatment. The program consists of a 5 °C.min^{-1} heating ramp up to the maximum temperature fixed in the range 1400–1800 °C, dwelling at the selected temperature for 2 h and cooling down to RT at 5 °C.min^{-1}. Samples are labeled P-TX with T the temperature at which the polymer has been exposed (1400, 1500, 1600 and 1800 °C) and X corresponds to the nature of the atmosphere: X = A for argon, X = N for nitrogen and X = NH for ammonia (200 °C ≤ T ≤ 1000 °C) and then nitrogen T > 1000 °C.

3.4. Material Characterization

As the polymers are reactive towards moisture and oxygen, the following sample preparations were performed within a glove box. The chemical structure of polymers was determined by transmission FTIR spectroscopy using a Nicolet Magna 550 Fourier transform-infrared spectrometer. ^{1}H NMR data of PMVSZ solutions in $CDCl_3$ were obtained using a Bruker AM 300 spectrometer operating at 300 MHz. Tetramethylsilane (TMS) was used as a reference for the NMR data. Solid-state ^{13}C CP MAS, ^{15}N CP MAS and ^{29}Si MAS NMR spectra were recorded on a Bruker AVANCE 300 spectrometer (7.0 T, ν_0(^{1}H) = 300.29 MHz, ν_0(^{13}C) = 75.51 MHz, ν_0(^{15}N) = 30.44 MHz, ν_0(^{29}Si) = 59.66 MHz) using 4 mm and 7 mm Bruker probes and a spinning frequencies of 10 and 5 kHz, respectively. ^{13}C and ^{15}N CP MAS experiments were recorded with ramped-amplitude cross-polarization in the ^{1}H channel to transfer magnetization from ^{1}H to ^{13}C and ^{15}N. (Recycle delay = 3 s, CP contact time = 1 ms, optimized ^{1}H spinal-64 decoupling). Single pulse ^{29}Si MAS NMR spectra were recorded with a recycle delay of 60 s. Chemical shift values were referenced to tetramethylsilane for ^{13}C and ^{29}Si and CH_3NO_2 for ^{15}N. Chemical analyses of the polymers were performed using a combination of several methods at Mikroanalytisches Labor Pascher (Remagen, Germany). Thermogravimetric analyses

(TGA) of samples were performed in flowing ammonia at 5 °C.min^{-1} to 1000 °C using silica crucibles (Setaram TGA 92 16.18, SETARAM Instrumentation (Caluire, France)). In addition, they were done in flowing nitrogen or argon at 5 °C.min^{-1} to 1000 °C using alumina crucibles at ambient atmospheric pressure (STA 449 F3, Netzsch GmbH, Selb, Germany) coupled with mass spectrometer (Omnistar, Pfeiffer Vacuum GmbH, Asslar, Germany) in outlet. The phase composition of ceramic samples was determined by XRD analysis (Bruker AXS D8 Discover, CuK$_\alpha$ radiation, Billerica, MA, USA). The scans were performed in the range of 2θ ∈ ⟨20°; 85°⟩ with a step of 0.015° and an exposure time of 0.7 s. The diffractograms were analyzed using the Diffrac + EVA software with the JCPDS-ICDD database. Crystallite sizes of TiC$_x$N$_y$ crystals were calculated from the FWHM of the (200) diffraction lines using the Scherrer's equation while their chemical composition was measured by considering the interlayer spacing d_{200} and the lattice parameter a considering both the Scherrer equation and the Vegard's law. The chemical analysis of selected powders was determined by scanning electron microscopy equipped with energy dispersive spectroscope (EDS-SEM, JEOL IT 300 LV, Tokyo, Japan) and confirmed for some of them using hot gas extraction techniques with a Horiba EMGA-830 for oxygen, nitrogen and hydrogen contents and using a combustion analysis method for carbon content on a Horiba Emia-321V. Structural and morphological characterizations were conducted by Transmission Electron Microscopy with a JEM-2100F model from JEOL (Japan) for nitrogen- and argon-treated samples and at the Indian Institute of Technology Madras, Chennai using TECNAI 20G^2, FEI instruments (Hillsboro, OR, USA) with an accelerating voltage of 200 kV & JEOL 3010 (Japan) at an accelerating voltage of 300 kV for ammonia-treated samples.

4. Conclusions

Within the present study, a titanium-modified polysilazane has been synthesized by adding tetrakis(dimethylamido)titanium Ti[N(CH$_3$)$_2$]$_4$ (TDMAT as titanium source) to a commercially available poly(vinylmethyl-*co*-methyl)silazane labeled PVMSZ according to an atomic Si:Ti ratio of 2.5. The effect of the nature of the atmosphere including ammonia, nitrogen and argon behind its thermo-chemical conversion and the crystallization behavior of derived amorphous ceramics has been critically investigated. Synthesis reactions involved N-H and Si-H bonds in polysilazanes and N(CH$_3$)-based groups in TDMAT as well as hydrosilylation reactions based on complementary characterization tools including FTIR, solid-state NMR and elemental analysis. Thus, the incorporation of Ti increased the crosslinking degree of the polysilazane and introduced side groups that affect the weight change during the further polymer-to-ceramic conversion. Titanium plays a major role in modifying the mechanisms of the polysilazane-to-ceramic transformation, especially in the low temperature regime of the thermal decomposition by prohibiting the distillation of small polymer fragments occurring with PVMSZ and then releasing amines most probably from side groups and hydrocarbons based on radical reactions. Pyrolysis mechanisms have been identified based on TG experiments, FTIR, solid-state NMR and MS spectroscopies. After pyrolysis to 1000 °C, materials are X-ray amorphous (although nucleation of TiC$_x$N$_y$ (x + y = 1) nanocrystals could be considered) and display a Si:Ti ratio in good agreement with the ratio fixed the polymers. The final nano-/microstructure evolved according to the annealing temperature fixed in the range 1400–1800 °C. Ammonia-treated amorphous ceramics evolved toward nanocrystalline-TiN/amorphous-Si$_3$N$_4$ whereas nitrogen- and argon-treated ceramics led to nanocrystalline-TiC$_x$N$_{1-x}$/amorphous-SiC(N) nanocomposites after annealing at 1400 °C. At 1800 °C, highly crystallized nanocomposites are obtained in the Si-Ti-N (ammonia-treated ceramics) and Si-Ti-C (argon-treated ceramics) systems. The nitrogen-treated samples represent TiC$_x$N$_y$ (x + y = 1)/SiC nanocomposites.

Supplementary Materials: The following are available online, Figure S1: FTIR spectra of PVMSZ, PVMSZ_T and PVMSZTi2.5 samples, Figure S2: liquid-state ^{13}C-NMR spectrum of TDMAT, Figure S3: experimental ^{13}C CP MAS NMR recorded for the PVMSZTi2.5 sample from −20 to 160 ppm, Figure S4: TG curves recorded during decomposition of PVMSZ in flowing ammonia, nitrogen and argon, Figure S5: experimental ^{29}Si CP MAS NMR for the **P-1400NH, P-1400N, P-1400A** and **P-1800A** samples at 7 T, Figure S6: identification of the α and β-Si$_3$N$_4$

phases in the XRD pattern of **P-1500N**, Figure S7: HRTEM micrographs of **P-1500NH** (a), **P-1500N** (b) and **P-1500A** (c) samples with FFT images obtained from HRTEM images of **P-1500N** (b) and **P-1500A** (c) samples as insets, Figure S8: EDS mapping of the **P-1600A** sample, Figure S9: EDS mapping of the **P-1800A** sample.

Author Contributions: Conceptualization, M.B., A.L., V.P., A.V.A.B. and S.B.; methodology, M.B., A.L., V.P., A.V.A.B. and S.B.; validation, R.K., C.G. and S.B.; investigation, M.B., A.L., V.P., A.V.A.B., P.C., C.G. and E.W.A.; formal analysis, E.W.A. and C.G.; writing—original draft preparation, S.B.; writing—review and editing, R.K., R.A.F.M., C.G. and S.B.; visualization, supervision and funding acquisition, R.K., R.A.F.M. and S.B. All authors have read and agreed to the published version of the manuscript.

Funding: This research was funded by the funding agency Agence Nationale de la Recherche (ANR) through the Carapass project (Project number ANR-16-CE08-0026), the funding received from Indo-French Centre for the Promotion of Advanced Research (CEFIPRA, Project No. 5108-1), the financial contribution from COFECUB through the Ph-C 956/19 project.

Acknowledgments: S.B. acknowledges Agence Nationale de la Recherche (ANR) for supporting this work through the Carapass project (PhD thesis of M.B.). S.B. and R.K. would like to thank CEFIPRA to carry out this work and to support the PhD theses of A.L. and E.W.A. S.B. gratefully acknowledge the financial contribution from CNRS and DGA for the PhD thesis of V.P., S.B. and R.A.F.M. acknowledge the CNPq for providing financial support for the PhD thesis of A.V.A.B. as well as COFECUB.

Conflicts of Interest: The authors declare no conflict of interest.

References

1. Hnatko, M.; Hičák, M.; Labudová, M.; Galusková, D.; Sedláček, J.; Lenčéš, Z.; Šajgalík, P. Bioactive silicon nitride by surface thermal treatment. *J. Eur. Ceram. Soc.* **2020**, *40*, 1848–1858. [CrossRef]
2. Sandra, F.; Ballestero, A.; NGuyen, V.L.; Tsampas, M.N.; Vernoux, P.; Balan, C.; Iwamoto, Y.; Demirci, U.B.; Miele, P.; Bernard, S. Silicon Carbide-based Membranes with High Filtration Efficiency, Durability and Catalytic Activity for CO/HC Oxidation and Soot Combustion. *J. Membr. Sci.* **2016**, *501*, 79–92. [CrossRef]
3. Lale, A.; Wasan, A.; Kumar, R.; Miele, P.; Demirci, U.B.; Bernard, S. Organosilicon Polymer-Derived Mesoporous 3D Silicon Carbide, CarboNitride and Nitride Structures as Platinum Supports for Hydrogen Generation by Hydrolysis of Sodium Borohydride. *Int. J. Hydro. Energy* **2016**, *41*, 15477–15488. [CrossRef]
4. Dong, L.; Wang, Y.; Tong, X.; Lei, T. Silicon carbide encapsulated graphite nanocomposites supported Pt nanoparticles as high-performance catalyst for methanol and ethanol oxidation reaction. *Diam. Relat. Mater.* **2020**, *104*, 107739–107744. [CrossRef]
5. Feng, Y.; Yang, Y.; Wen, Q.; Riedel, R.; Yu, Z. Dielectric Properties and Electromagnetic Wave Absorbing Performance of Single-Source-Precursor Synthesized $Mo_{4.8}Si_3C_{0.6}/SiC/C_{free}$ Nanocomposites with an In Situ Formed Nowotny Phase. *ACS Appl. Mater. Int.* **2020**, *12*, 16912–16921. [CrossRef]
6. Lale, A.; Mallmann, M.D.; Tada, S.; Bruma, A.; Özkar, S.; Kumar, R.; Haneda, M.; Machado, R.A.F.; Iwamoto, Y.; Demirci, U.B.; et al. Highly Active, Robust and Reusable Micro-/Mesoporous TiN/Si_3N_4 Nanocomposite-based Catalysts: Understanding the Key Role of TiN Nanoclusters and Amorphous Si_3N_4 Matrix in the Performance of the Catalyst System. *Appl. Catal. B Environ.* **2020**, *272*, 118975–118984. [CrossRef]
7. Mera, G.; Gallei, M.; Bernard, S.; Ionescu, E. Ceramic Nanocomposites from Tailor-Made Preceramic Polymers. *Nanocomposites* **2015**, *5*, 468–540. [CrossRef]
8. Eckardt, M.; Zaheer, M.; Kempe, R. Nitrogen-doped mesoporous SiC materials with catalytically active cobalt nanoparticles for the efficient and selective hydrogenation of nitroarenes. *Sci. Rep.* **2018**, *8*, 2567–2572. [CrossRef]
9. Colombo, P.; Soraru, G.D.; Riedel, R.; Kleebe, H.-J. *Polymer Derived Ceramics: From Nano-Structure to Applications*; DEStech Publications: Lancaster, PA, USA, 2009.
10. Bill, J.; Aldinger, F. Precursor-derived covalent ceramics. *Adv. Mater.* **1995**, *7*, 775–787. [CrossRef]
11. Greil, P. Polymer derived engineering ceramics. *Adv. Eng. Mater.* **2000**, *2*, 339–348. [CrossRef]
12. Colombo, P.; Mera, G.; Riedel, R.; Soraru, G.D. Polymer-derived ceramics: 40 years of research and innovation in advanced ceramics. *J. Am. Ceram. Soc.* **2010**, *93*, 1805–1837. [CrossRef]
13. Ionescu, E.; Kleebe, H.-J.; Riedel, R. Silicon-containing polymer-derived ceramic nanocomposites (PDC-NCs): Preparative approaches and properties. *Chem. Soc. Rev.* **2012**, *41*, 5032–5052. [CrossRef] [PubMed]
14. Zaheer, M.; Schmalz, T.; Motz, G.; Kempe, R. Polymer derived non-oxide ceramics modified with late transition metals. *Chem. Soc. Rev.* **2012**, *41*, 5102–5116. [CrossRef] [PubMed]

15. Bernard, S.; Miele, P. Polymer-derived boron nitride: A review on the chemistry, shaping and ceramic conversion of borazine derivatives. *Materials* **2014**, *7*, 7436–7459. [CrossRef] [PubMed]
16. Stabler, C.; Ionescu, E.; Graczyk-Zajac, M.; Gonzalo-Juan, I.; Riedel, R. Silicon oxycarbide glasses and glass-ceramics: "All-Rounder" materials for advanced structural and functional applications. *J. Am. Ceram. Soc.* **2018**, *101*, 4817–4856. [CrossRef]
17. Viard, A.; Fonblanc, D.; Lopez-Ferber, D.; Schmidt, M.; Lale, A.; Durif, C.; Balestrat, M.; Rossignol, F.; Weinmann, M.; Riedel, R.; et al. Polymer derived Si-B-C-N ceramics: 30 years of research. *Adv. Eng. Mater.* **2018**, *20*, 1800360–1800371. [CrossRef]
18. Ionescu, E.; Bernard, S.; Lucas, R.; Kroll, P.; Ushakov, S.; Navrotsky, A.; Riedel, R. Ultrahigh temperature ceramics (UHTCs) and related materials—Syntheses from polymeric precursors and energetics. *Adv. Eng. Mater.* **2019**, *21*, 1900269–1900292. [CrossRef]
19. Wen, Q.; Yu, Z.; Riedel, R. The fate and role of *in situ* formed carbon in polymer-derived ceramics. *Prog. Mater. Sci.* **2020**, *109*, 100623–100685. [CrossRef]
20. Lale, A.; Schmidt, M.; Mallmann, M.D.; Bezerra, A.V.A.; Diz Acosta, E.; Machado, R.A.F.; Demirci, U.B.; Bernard, S. Polymer-derived ceramics with engineered mesoporosity: From design to application in catalysis. *Surf. Coat. Technol.* **2018**, *350*, 569–586. [CrossRef]
21. Viard, A.; Miele, P.; Bernard, S. Polymer-derived ceramics route toward SiCN and SiBCN fibers: From chemistry of polycarbosilazanes to the design and characterization of ceramic fibers. *J. Ceram. Soc. Jpn.* **2016**, *124*, 967–980. [CrossRef]
22. Flores, O.; Bordia, R.K.; Nestler, D.; Krenkel, W.; Motz, G. Ceramic fibers based on SiC and SiCN systems: Current research, development, and commercial status. *Adv. Eng. Mater.* **2014**, *16*, 621–636. [CrossRef]
23. Barroso, G.; Li, Q.; Bordia, R.K.; Motz, G. Polymeric and ceramic silicon-based coatings—A review. *J. Mater. Chem. A* **2019**, *7*, 1936–1963. [CrossRef]
24. Vakifahmetoglu, C.; Zeydanli, D.; Colombo, P. Porous polymer derived ceramics. *Mater. Sci. Eng. R* **2016**, *106*, 1–30. [CrossRef]
25. Hotza, D.; Nishihora, R.; Machado, R.; Geffroy, P.-M.; Chartier, T.; Bernard, S. Tape casting of preceramic polymers towards advanced ceramics: A review. *Int. J. Ceramic Eng. Sci.* **2019**, *1*, 21–41. [CrossRef]
26. Bechelany, M.C.; Proust, V.; Gervais, C.; Ghisleni, R.; Bernard, S.; Miele, P. In-situ controlled growth of titanium nitride in amorphous silicon nitride: A general route toward bulk non-oxide nitride nanocomposites with very high hardness. *Adv. Mater.* **2014**, *26*, 6548–6553. [CrossRef]
27. Bechelany, M.C.; Proust, V.; Lale, A.; Miele, P.; Malo, S.; Gervais, C.; Bernard, S. Nanocomposites through chemistry of single-source precursors: Understanding the role of chemistry behind the design of monolith-type nanostructured titanium nitride/silicon nitride. *Chem. Eur. J.* **2017**, *23*, 832–845. [CrossRef]
28. Lale, A.; Proust, V.; Bechelany, M.C.; Viard, A.; Malo, S.; Bernard, S. A comprehensive study on the influence of the polyorganosilazane chemistry and material shape on the high temperature behavior of titanium nitride/silicon nitride nanocomposites. *J. Eur. Ceram Soc.* **2017**, *37*, 5167–5175. [CrossRef]
29. Yu, Z.; Min, H.; Zhan, J.; Yang, L. Preparation and dielectric properties of polymer-derived SiCTi ceramics. *Ceram. Int.* **2013**, *39*, 3999–4007. [CrossRef]
30. Yu, Z.; Yang, L.; Min, H.; Zhang, P.; Zhou, C.; Riedel, R. Single-source-precursor synthesis of high temperature stable SiC/C/Fe nanocomposites from a processable hyperbranched polyferrocenylcarbosilane with high ceramic yield. *J. Mater. Chem. C* **2014**, *2*, 1057–1067. [CrossRef]
31. Fonblanc, D.; Lopez-Ferber, D.; Wynn, M.; Lale, A.; Soleilhavoup, A.; Leriche, A.; Iwamoto, Y.; Rossignol, F.; Gervais, C.; Bernard, S. Crosslinking chemistry of poly(vinylmethyl-co-methyl)silazanes toward low-temperature formable preceramic polymers as precursors of functional aluminium-modified Si-C-N ceramics. *Dalton Trans.* **2018**, *47*, 14580–14593. [CrossRef]
32. Viard, A.; Fonblanc, D.; Schmidt, M.; Lale, A.; Salameh, C.; Soleilhavoup, A.; Wynn, M.; Champagne, P.; Cerneaux, S.; Babonneau, F.; et al. Molecular chemistry and engineering of boron-modified polyorganosilazanes as new processable and functional SiBCN precursors. *Chem. Eur. J.* **2017**, *23*, 9076–9090. [CrossRef] [PubMed]
33. Gibson, V.C.; Spitzmesser, S.K. Advances in non-metallocene olefin polymerization catalysis. *Chem. Rev.* **2003**, *103*, 283–316. [CrossRef] [PubMed]
34. Jazzar, R.; Hitce, J.; Renaudat, A.; Sofack-Kreutzer, J.; Baudoin, O. Functionalization of organic molecules by transition-metal-catalyzed C (sp3); H activation. *Chem. A Eur. J.* **2010**, *16*, 2654–2672. [CrossRef] [PubMed]

35. Daugulis, O.; Do, H.-Q.; Shabashov, D. Palladium- and copper-catalyzed arylation of carbon-hydrogen bonds. *Acc. Chem. Res.* **2009**, *42*, 1074–1086. [CrossRef] [PubMed]
36. Giri, R.; Maugel, N.; Li, J.-J.; Wang, D.-H.; Breazzano, S.P.; Saunders, L.B.; Yu, J.-Q. Palladium-catalyzed methylation and arylation of sp2 and sp3 C-H bonds in simple carboxylic acids. *J. Am. Chem. Soc.* **2007**, *129*, 3510–3511. [CrossRef]
37. Schmidt, M.; Durif, C.; Diz Acosta, E.; Salameh, C.; Plaisantin, H.; Miele, P.; Backov, R.; Machado, R.; Gervais, C.; Alauzun, J.G.; et al. Molecular-level processing of Si-(B)-C materials with tailored nano/microstructures. *Chem. Eur. J.* **2017**, *23*, 17103–17117. [CrossRef]
38. Majoulet, O.; Alauzun, J.G.; Gottardo, L.; Gervais, C.; Schuster, M.E.; Bernard, S.; Miele, P. Ordered Mesoporous Silicoboron Carbonitride Ceramics from Boron-Modified Polysilazanes: Polymer Synthesis, Processing and Properties. *Micro. Meso. Mater.* **2011**, *140*, 40–50. [CrossRef]
39. Berger, F.; Müller, A.; Aldinger, F.; Müller, K. Solid-state NMR investigations on Si-B-C-N ceramics derived from boron-modified poly (allylmethylsilazane). *Z. Anorg. Allg. Chem.* **2005**, *631*, 355–363. [CrossRef]
40. Gottardo, L.; Bernard, S.; Gervais, C.; Inzenhofer, K.; Motz, G.; Weinmann, M.; Balan, C.; Miele, P. Chemistry, Structure and Processability of Boron-Modified Polysilazanes as Tailored Precursors of Ceramic Fibers. *J. Mater. Chem.* **2012**, *22*, 7739–7750. [CrossRef]
41. Seitz, J.; Bill, J.; Egger, N.; Aldinger, F. Structural investigations of Si/C/N-ceramics from polysilazane precursors by nuclear magnetic resonance. *J. Eur. Ceram. Soc.* **1996**, *16*, 885–891. [CrossRef]
42. Choong Kwet Yive, N.S.; Corriu, R.J.P.; Leclercq, D.; Mutin, P.H.; Vioux, A. Silicon carbonitride from polymeric precursors: Thermal cross-linking and pyrolysis of oligosilazane model compounds. *Chem. Mater.* **1992**, *4*, 141–146. [CrossRef]
43. Choong Kwet Yive, N.S.; Corriu, R.J.P.; Leclercq, D.; Mutin, P.H.; Vioux, A. Thermogravimetric analysis/mass spectrometry investigation of the thermal conversion of organosilicon precursors into ceramics under argon and ammonia. 2. Poly (silazanes). *Chem. Mater.* **1992**, *4*, 1263–1271. [CrossRef]
44. Allan, J.D.; Delia, A.E.; Coe, H.; Bower, K.N.; Alfarra, M.R.; Jimenez, J.L.; Middlebrook, A.M.; Drewnick, F.; Onasch, T.B.; Canagaratna, M.R.; et al. A generalized method for the extraction of chemically resolved mass spectra from aerodyne aerosol mass spectrometer data. *J. Aerosol Sci.* **2004**, *35*, 909–922. [CrossRef]
45. Lavedrine, A.; Bahloul, D.; Goursat, P.; Choong Kwet Yive, N.S.; Corriu, R.; Leclerq, D.; Mutin, H.; Vioux, A. Pyrolysis of polyvinylsilazane precursors to silicon carbonitride. *J. Eur. Ceram. Soc.* **1991**, *8*, 221–227. [CrossRef]
46. Bahloul, D.; Pereira, M.; Goursat, P.; Choong Kwet Yive, N.S.; Corriu, R.J.P. Preparation of silicon carbonitrides from an organosilicon polymer: I, thermal decomposition of the cross-linked polysilazane. *J. Am. Ceram. Soc.* **1993**, *76*, 1156–1162. [CrossRef]
47. Bahloul, D.; Pereira, M.; Goursat, P. Preparation of silicon carbonitrides from an organosilicon polymer: II, thermal behavior at high temperature under argon. *J. Am. Ceram. Soc.* **1993**, *76*, 1163–1168. [CrossRef]
48. Guilemany, J.; Alcobe, X.; Sanchiz, I. X-Ray Diffraction Analysis of Titanium Carbonitride 30/70 and 70/30 Solid Solutions. *Powder Diffr.* **1992**, *7*, 34–35. [CrossRef]
49. Hering, N.; Schreiber, K.; Riedel, R.; Lichtenberger, O.; Woltersdorf, J. Synthesis of polymeric precursors for the formation of nanocrystalline Ti-C-N/amorphous Si-C-N composites. *Appl. Organomet. Chem.* **2001**, *15*, 879–886. [CrossRef]
50. Cordoba, J.M.; Sayagues, M.J.; Alcalà, M.D.; Gotor, F.J. Synthesis of Titanium Carbonitride Phases by Reactive Milling of the Elemental Mixed Powders. *J. Am. Ceram. Soc.* **2005**, *88*, 1760–1764. [CrossRef]

Sample Availability: Samples of the compounds P1000 to P1800 under argon and nitrogen are available from the authors.

Publisher's Note: MDPI stays neutral with regard to jurisdictional claims in published maps and institutional affiliations.

© 2020 by the authors. Licensee MDPI, Basel, Switzerland. This article is an open access article distributed under the terms and conditions of the Creative Commons Attribution (CC BY) license (http://creativecommons.org/licenses/by/4.0/).

Article

PbS Quantum Dots Decorating TiO₂ Nanocrystals: Synthesis, Topology, and Optical Properties of the Colloidal Hybrid Architecture

Carlo Nazareno Dibenedetto [1,2], Teresa Sibillano [3], Rosaria Brescia [4], Mirko Prato [4], Leonardo Triggiani [2], Cinzia Giannini [3], Annamaria Panniello [2], Michela Corricelli [1], Roberto Comparelli [2], Chiara Ingrosso [2], Nicoletta Depalo [2], Angela Agostiano [1,2], Maria Lucia Curri [1,2], Marinella Striccoli [2,*] and Elisabetta Fanizza [1,2,*]

[1] Dipartimento di Chimica, Università degli Studi di Bari, Via Orabona 4, 70126 Bari, Italy; carlo.dibenedetto@uniba.it (C.N.D); michela.corricelli@gmail.com (M.C.); angela.agostiano@uniba.it (A.A.); marialucia.curri@uniba.it (M.L.C.)
[2] CNR-Istituto per i Processi chimico Fisici (CNR-IPCF), SS Bari, Via Orabona 4, 70126 Bari, Italy; l.triggiani@ba.ipcf.cnr.it (L.T.); a.panniello@ba.ipcf.cnr.it (A.P.); r.comparelli@ba.ipcf.cnr.it (R.C.); c.ingrosso@ba.ipcf.cnr.it (C.I.) n.depalo@ba.ipcf.cnr.it (N.D.)
[3] CNR-Istituto di Cristallografia (CNR-IC, Via Amendola, 122/O, 70126 Bari, Italy; teresa.sibillano@ic.cnr.it (T.S.); cinzia.giannini@ic.cnr.it (C.G.)
[4] IIT- Istituto Italiano di Tecnologia, via Morego 30, 16163 Genova, Italy; Rosaria.Brescia@iit.it (R.B.); Mirko.Prato@iit.it (M.P.)
* Correspondence: m.striccoli@ba.ipcf.cnr.it (M.S.); elisabetta.fanizza@uniba.it (E.F.); Tel.: +39-080-544-2027 (M.S. & E.F.)

Academic Editor: Giuseppe Cirillo
Received: 11 June 2020; Accepted: 24 June 2020; Published: 26 June 2020

Abstract: Fabrication of heterostructures by merging two or more materials in a single object. The domains at the nanoscale represent a viable strategy to purposely address materials' properties for applications in several fields such as catalysis, biomedicine, and energy conversion. In this case, solution-phase seeded growth and the hot-injection method are ingeniously combined to fabricate TiO₂/PbS heterostructures. The interest in such hybrid nanostructures arises from their absorption properties that make them advantageous candidates as solar cell materials for more efficient solar light harvesting and improved light conversion. Due to the strong lattice mismatch between TiO₂ and PbS, the yield of the hybrid structure and the control over its properties are challenging. In this study, a systematic investigation of the heterostructure synthesis as a function of the experimental conditions (such as seeds' surface chemistry, reaction temperature, and precursor concentration), its topology, structural properties, and optical properties are carried out. The morphological and chemical characterizations confirm the formation of small dots of PbS by decorating the oleylamine surface capped TiO₂ nanocrystals under temperature control. Remarkably, structural characterization points out that the formation of heterostructures is accompanied by modification of the crystallinity of the TiO₂ domain, which is mainly ascribed to lattice distortion. This result is also confirmed by photoluminescence spectroscopy, which shows intense emission in the visible range. This originated from self-trapped excitons, defects, and trap emissive states.

Keywords: colloidal heterostructures; seed mediated growth; heterogeneous nucleation; PbS/TiO₂ heterostructure; TiO₂ nanocrystal defects

1. Introduction

The colloidal approach, which is extensively used to prepare nanoparticles (NPs) and nanocrystals (NCs) with a variety of sizes and shapes [1–4], has, in recent decades, made a step forward by fabricating hybrid NPs [5–11] based on the combination of two or more materials in one solid nano-object for application in several fields, including biomedicine, environment, catalysis, and sensing [12–16]. In principle, any desired inorganic material can be purposely assembled to form the hybrid structures, characterized by properties that possibly derive from a simple combination, enhancement, or mitigation of the properties of the individual materials or can bring on completely new physical and chemical properties. Metal-metal [5,6], semiconductor-semiconductor [17–23], metal-semiconductor [24–27], and magnetic-semiconductor [28,29] hybrid structures have been prepared by different solution-phase approaches such as seed mediated epitaxial and non-epitaxial deposition [14,17,30–33], cation exchange [34–36], and coupling promoted by the bifunctional linker [37,38]. With the exception of the last strategy, where pre-synthesized NPs are assembled ex-situ at distances inherently induced by the linker, the other approaches rely on the growth of the new phase on pre-existing NPs with the two domains in intimate contact. Face miscibility and lattice mismatch between the two phases control the feasibility of the hybrid structure and its final topology. Cation exchange, which consists of the kinetically driven replacement of cations of pre-existing NCs with new cations while retaining anion sub-lattice, has been exploited for the fabrication of multi-component materials under a mild temperature condition [34,35]. The new phase, nucleated at the NC surface facet, may topotaxially grow toward the interior of the NCs or form original morphologies due to facet-dependent exchange reaction, nanoscale asymmetry, and anisotropy. Although this strategy shows great potentiality for the easy fabrication of novel heterostructures, it has been mainly limited to the fabrication of hybrid structures belonging to the II–VI, I–III–VI, and IV–VI classes of semiconductors and characterized by the same anion for both domains. Conversely, a huge variety of hybrid structures with each component having a different size, shape, spatial orientation, composition, and crystalline structure, have been synthesized by the multi-step seed-mediated growth [10,39–41]. The reaction scheme consists of the first step of synthesis of the seeds, which is followed by the heterogeneous nucleation and growth of the new phase upon precursor's injection. Even though the presence of seeds in solution provide a thermodynamic and kinetic gain to the nucleation of a new phase by References [8,9,42], and, even though heterogeneous nucleation results in more favoured than homogenous nucleation, the achievement of a high yield of hybrid structures remains challenging, which is often affected by the separate homogenous nucleation of the new phase. According to the colloidal synthesis theory, the Gibbs energy barrier to heterogeneous nucleation is lower than that of homogenous nucleation due to a series of factors including wettability between seeds and nuclei, and further decreases at an increase of a new phase monomer supersaturation and reaction temperature. Similarly, the presence of the seeds affects the nucleation rate. In spite of the fact that heterogeneous and homogeneous nucleation paths follow the same kinetic law, heterogeneous nucleation is generally faster than homogeneous nucleation under an actual supersaturation condition. However, while at low supersaturation, heterogeneous nucleation dominates at high supersaturation. Homogenous nucleation has been reported to be the preferred path [43]. Though temperature and supersaturation define the synthetic condition that make the formation of hybrid structures prevail over homogeneous nucleation, the final hybrid material topology is mainly shaped by the interfacial strain and facet dependent on chemical reactivity [9,23]. A positive strain energy term (γ_{strain}) contributes to the interfacial tension (γ_i) and, together with the solid/solution surface tension of seeds ($\gamma_{seed/solution}$) and nuclei ($\gamma_{nuclei/solution}$), defines the overall surface tension ($\Delta\gamma = \gamma_{nuclei/solution} + \gamma_i - \gamma_{seed/solution}$). The $\Delta\gamma$ value allows prediction of the growth mode of the heterostructure as well as the topology of the hybrid NP. Centrosymmetric core-shell hybrid structures form when the lattice constants of the two components do not differ significantly (Frank-van der Merwe mode, $\Delta\gamma < 0$), while non-centrosymmetric structures are achieved at $\Delta\gamma < 0$, only if interfacial energy between the two materials is large enough and when certain regions on the seed surface are accessible and reactive. At $\Delta\gamma > 0$ and at high lattice mismatch (>5%) between the two phases, the deposition of the new nuclei on the seed occurs by formation of islands-like structures (Volber-Weber mode). In this case, the energy strain is released by atoms rearranging or by interfacial misfit dislocation. However, these theoretical predictions do not take into account the presence in solution phase strategies of organic surfactants, which modulate the solid/solution interfacial tension and enable the formation of heterostructures, even combining dissimilar materials [20].

This work focuses on the synthesis of hybrid nanostructures based on the nucleation of PbS NCs on TiO$_2$ NPs. Colloidal TiO$_2$/PbS hybrid NCs have been prepared following diverse strategies, such as seed-mediated hot-injection [20], ultrasound ionic deposition [44], successive ionic layer adsorption, and a reaction method (SILAR) [45], or by PbS and TiO$_2$ NCs assembling with very short bifunctional linker molecules [38]. Such a type of hybrid structure has gained increasing attention in recent decades for its potential application in solar cell devices [46–48]. Coupling of the TiO$_2$ wide band gap semiconductor (3.2 eV) with PbS NPs with a size dependent narrow band gap that extends to the NIR region may overcome the intrinsic limitation of TiO$_2$, which only absorbs the ultraviolet portion of the solar spectrum [49,50] thanks to PbS-mediated sensitization process [51,52]. Enhanced solar light harvesting and efficient electron transfer have been reported to improve the performances of solar cells [20,44]. However, charge carrier dynamic at the interface strongly depends on the appropriate choice of the two material band offsets. Quantum confined PbS NCs, characterized by a large exciton Bohr radius [44], offer the possibility to well control the band edge alignment with TiO$_2$ by simply tuning their size. Solar light photogenerated electron/hole pairs can be forced either to recombine within the PbS materials, according to type-I heterojunction behaviors or to separate by efficient interparticle electron transfer in a type-II heterojunction [8]. However, as pointed out by Trejo et al. [53], the deposition of PbS over TiO$_2$ suffers from the large lattice mismatch between the two materials. Island-like structures of PbS, grown by chemical vapour deposition on crystalline TiO$_2$ film, show a PbS domain characterized by bond distortion from the bulky crystalline phase, which results in changes in the electronic structure of the PbS and increase of the energy gap. Similarly, as described by Acharya et al. [20], the final topology of the TiO$_2$ nanorods/PbS hybrid structure, grown by the solution phase strategy, and the PbS domain size are dictated by the large interfacial strain at the two-material interface. The work pointed out that, thanks to the use of colloidal NCs as seeds in the synthesis, the interfacial strain can be tuned through the TiO$_2$ ligands, which results in near-epitaxial small island-like or single large PbS structures deposited on TiO$_2$ nanorods [20]. The growth mode and the PbS domain size, lastly, defines the electron transfer properties at the hybrid structure interface [20]. This results in a type-II hybrid heterojunction and efficient electron transfer from PbS to TiO$_2$ domains [20], when PbS domains with sizes below 7 nm decorate the TiO$_2$ nanorods.

The present study aims at synthesizing and extensively characterizing TiO$_2$/PbS hybrid NCs prepared by a seeded growth approach using purposely functionalized TiO$_2$ NCs as seeds and allows the PbS domains growth, in situ, by the hot injection method. The work wants to examine the critical role of interfacial strain in the fabrication of such hybrid TiO$_2$/PbS NCs in directing the heterostructure topology, domain structure, and final optical properties. The effectiveness of PbS deposition on TiO$_2$ seeds, passivated with a native layer of oleic acid (OA), or surface modified with oleylamine (Olam), and the final hybrid structure topology, have been explored by varying the precursor concentration, and, thus, monomer supersaturation as well as injection temperature. The systematic morphological investigation allows defining the synthetic conditions, which brings a high yield of hybrid nanostructures under limited homogeneous nucleation, successfully attained at specific precursor concentration and under controlled temperature conditions, when using Olam-capped TiO$_2$ NCs as seeds. It has been shown that the presence of Olam as a capping agent, loosely bound at the TiO$_2$ NCs surface, allows an easier PbS monomer deposition, which provides the hybrid structure formation. Lastly, the surface modification and interfacial strain between TiO$_2$ and PbS have been found to strongly affect the structural and optical properties of the hybrid NPs and, in particular, the optical properties of the TiO$_2$ domain.

2. Results and Discussion

In this scenario, the synthesis of TiO$_2$/PbS hybrid NCs by seeded-mediated growth and hot injection approaches has been developed. The main aim of this work is to examine how parameters such as injection temperature, precursor concentration, supersaturation, and ligands composition at the TiO$_2$ seed surface can modulate the energetic and kinetic factors that promote colloidal heterogeneous nucleation over homogeneous nucleation, and determine the yield and topology of the heterostructures. The general procedure relies on the hot-injection [54] at different temperatures ranging from 120 °C to 80 °C, into the TiO$_2$ seeds suspension, of the Pb-oleate precursor, prepared by decomposing PbO in oleic acid and ODE. This is followed by the addition of HMDS solution in ODE (Pb:HMDS molar ratio 5:1). Upon injection, Pb-oleate and HMDS readily decompose, which results in the sudden release of monomers that burst the nucleation of PbS [55–58]. Homogenous nucleation represents the upper-limit to heterogeneous nucleation under a high supersaturation condition (Figure 1A), while, at low supersaturation (Figure 1B), the interplay between the lattice mismatch and the solid/solution surface tension for each phase defines the topology of the hybrid structure. In addition, increasing the injection temperature makes the exchange dynamic faster between surfactants bound at the seeds' surface and the free ligands in solution, which renders the seed surface more accessible to the monomer deposition, and, concomitantly, allows faster monomer diffusion [29], which profitably improves heterogeneous nucleation. However, in general, above a certain temperature threshold, which depends on the reaction mixture composition and precursor nature, the higher the temperature, the faster becomes the homogeneous nucleation. In this case, two precursor concentrations have been used with [Pb^{2+}] ranging from 0.01 M to 0.005 M and [HMDS] from 0.002 M and 0.001 M and decreasing injection temperature, namely 120 °C, 100 °C, and 80 °C, have been investigated by fixing the growth temperature at 80 °C.

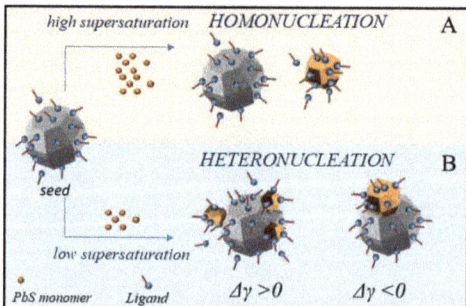

Figure 1. Seeded-growth under high (**A**) and low (**B**) supersaturation condition. Homogenous nucleation represents the upper-limit to heterogeneous nucleation. At low supersaturation, the overall interfacial energy ($\Delta\gamma = \gamma_{nuclei/solution} + \gamma_i - \gamma_{seed/solution}$, with γ_i = interfacial tension) defines the topology of the hybrid structure that, under high interfacial strain, may result in segregated phases ($\Delta\gamma < 0$) or island-like structures ($\Delta\gamma > 0$, Volmer-Weber growth mode). The grey dodecahedral shape NP represents the TiO$_2$ NC seeds while the orange structures represent the PbS NCs.

The reaction has been stopped after 10 min since prolonged heating has been observed to be always accompanied by the reaction mixture discoloration, which is likely ascribed to partial dissolution of the PbS domains.

2.1. Fabrication of TiO$_2$/PbS Hybrid NCs

Monodispersed platelet-like colloidal TiO$_2$ NCs seeds with an average size of nearly 7 nm ($\sigma\% = 13\%$, Figure 2A) have been synthesized by an alcholisis reaction between OA and 1-octadecanol that results in water release in the reaction flask [59] and titanium ethoxide hydrolysis, which is followed by the titanol condensation reaction. The FTIR spectrum of the TiO$_2$ NC sample (Figure 3a and 3(a1) allows disclosing the ligand composition and the type of ligand coordination at the NC surface. The 3300–2800 cm^{-1} range (Figure 3a) shows the vibrational modes of the *cis*-9-octadecenoil hydrocarbon chain of OA (see Figure S1A) used as a coordinating agent during the synthesis with characteristic absorption peaks at 2917 cm^{-1} and 2849 cm^{-1}, ascribed to the intense symmetric and antisymmetric C-H stretching of methylenic groups at 2960 cm^{-1}, attributed to the antisymmetric stretching of the terminal -CH$_3$ group, and, at 3005 cm^{-1}, corresponding to the weak but definite band, which is characteristic of the =C-H symmetric stretching.

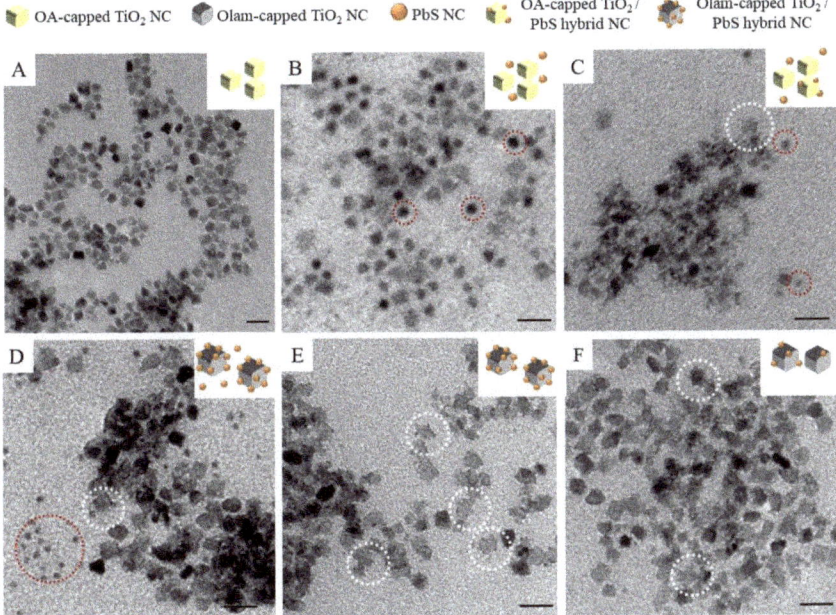

Figure 2. TEM micrographs (scale bar 20 nm) of oleic acid (OA)-capped TiO$_2$ NCs before (**A**) and after (**B-C**) injection of (**B**) [Pb^{2+}] = 0.01M, [HMDS] = 0.002 M at 120 °C and (**C**) [Pb^{2+}] = 0.005M, [HMDS] = 0.001 M at 100 °C. (**D–F**) oleylamine (Olam)-capped TiO$_2$ NCs after injection of [Pb^{2+}] = 0.005 M, [HMDS] = 0.001 M at 120 °C (**D**), 100 °C (**E**), and 80 °C (**F**). Dashed red and white circles in the picture used to highlight homogenously nucleated PbS NCs and TiO$_2$/PbS hybrid NCs, respectively.

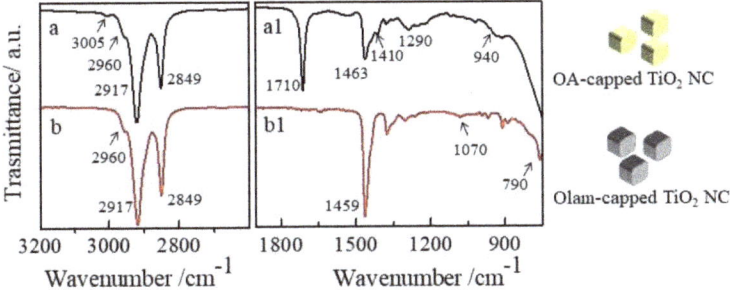

Figure 3. FTIR spectra in the ATR mode of oleic acid (OA)-capped (**a–a1**) and oleylamine (Olam)-capped TiO$_2$ NCs (**b–b1**).

The peaks at 1463, 1410, 1290, and 940 cm^{-1} (Figure 3(a1)) can be ascribed to –CH$_2$– bending of the hydrocarbons chain, –C–O–H in-plane bending, –C–OH stretching vibrations, and –O–H out of the plane mode of the carboxylic acid moiety, respectively. The presence of an intense –C=O stretching mode at 1710 cm^{-1} confirms that OA binds the TiO$_2$ NCs surface through the oxygen of the carbonyl group in a monodentate R–C(=O)–O form rather than forming bidentate RCO$_2$-M structures [60–62]. The FTIR spectrum also shows the characteristic broad and intense stretching of metal–oxygen TiO$_2$ bonds below 800 cm^{-1}. The as-prepared TiO$_2$ NCs have been addressed in the paper as OA-capped TiO$_2$ NCs.

The first set of reactions for the synthesis of hybrid TiO$_2$/PbS NCs has been performed by using seeds as the OA-capped TiO$_2$ NCs suspensions. TEM micrograph of the sample, prepared upon injection at 120 °C of Pb-oleate at [Pb^{2+}] = 0.01M and HDMS at 0.002M is reported in Figure 2B. Small dots of nearly 5 nm in diameter (σ% = 14%), highlighted by the dashed red circles in the picture, are visible together with features clearly ascribed to TiO$_2$ NCs (Figure 2A). Furthermore, isotropic growth and formation of a spherical PbS domain are typically achieved since OA is the sole ligand in the reaction mixture released upon Pb-oleate precursor decomposition and/or desorbed from the TiO$_2$ surface at a high temperature, and, given that OA does not preferentially bind any specific PbS face, and dynamically desorbs and adsorbs at the PbS surface [54].

Homogeneous nucleation of PbS has not been found to be suppressed even by decreasing the injection temperature at 100 °C (Figure S2, [HDMS] = 0.002M, [Pb^{2+}] = 0.01M), which would have been expected to favor heterogeneous nucleation over homogeneous nucleation nor by decreasing the precursor concentration (Figure 2C, [HMDS] = 0.001M, [Pb^{2+}] = 0.005M) while keeping the injection temperature constant. However, in the latter case, even to a lower extent, morphologies characterized by tiny dots decorating the TiO$_2$ NCs seeds have been detected (white dashed circle in Figure 2C), which confirms that depletion of the monomer under low supersaturation condition may improve the yield of heterostructures. The monomer concentration used for the synthesis of the sample reported in Figure 2C has been applied for further experiments.

To optimize reaction conditions toward high yield formation of TiO$_2$/PbS heterostructures without homogeneous nucleation, it should be considered that hybrid structures are difficult to achieve. In fact, since a high energy strain [20,53] has been reported between PbS and TiO$_2$, this, in principle, would not favor the deposition of PbS in intimate contact with the TiO$_2$ domains. However, TiO$_2$/PbS heterostructures have been reported to be successfully achieved when TiO$_2$ {001} facets merge with the cubic PbS (rock-salt) {100} face, even though the two faces show a substantial 6.9% lattice mismatch [20,53]. In this case, OA ligands, which are known to specifically tightly bind to the TiO$_2$ {001} faces [63,64], contribute in limiting the formation of the hybrid structure, which hinders the deposition of PbS on the seed surface. A different scenario is, then, expected by replacing OA with a new capping layer that less strongly passivates the TiO$_2$ NCs {001} faces. After extensive purification steps of the TiO$_2$, NCs to partially remove the physisorbed OA and the excess of OA in solution [65], Olam has been added in large excess for the exchange reaction. Olam is an L-type ligand and it loosely binds by means of nitrogen electron pair the TiO$_2$ NCs [66], preferentially coordinating the {101} faces, which leaves the {001} faces that are highly reactive and available for PbS deposition. According to the ligand-exchange rules [66], since OA binds the TiO$_2$ NC surface through the monodentate oxygen of the carboxylic moieties, acting as an L-type ligand, it can be effectively exchanged by Olam without affecting the electroneutrality of the TiO$_2$ NCs [66]. The FTIR spectrum of the Olam treated TiO$_2$ NCs is reported in Figure 3b and 3(b1). The spectral range between 3100 cm^{-1} and 2800 cm^{-1}, characteristic of the symmetric and antisymmetric C-H stretching modes of saturated and unsaturated hydrocarbons, does not show any appreciable change if compared to the untreated OA-capped TiO$_2$ NCs, since both OA and Olam bring the same alkyl (*cis*-9 octadecenoyl) chain. However, the inspection of 1900–750 cm^{-1} wavenumber range (Figure 3(b1)) shows the complete disappearance of the peaks at 1710 cm^{-1} (v_s –C=O) and at 940 cm^{-1} (O-H out of plane modes) characteristics of the carboxylic group, which confirms the removal of the OA from the TiO$_2$ NC surface. Even though addition of Olam can promote the formation of oleyl ammonium-oleate salt due to acid-base equilibrium, the two bands' characteristics of oleate, centered at 1528 and 1442 cm^{-1}, respectively, are not detected.

The FTIR spectrum of the 1700–800 cm^{-1} region, reported in Figure 3(b1), mainly dominated by the intense peak at 1459 cm^{-1} ascribed to the –CH$_2$ scissoring, shows a peak at 1070 cm^{-1} ascribed to the C–N stretching mode, and a peak at 790 cm^{-1} ascribed to the –NH$_2$ wagging, superimposed to the strong broad stretching of metal–oxygen bonds below 800 cm^{-1}, which is characteristic of the TiO$_2$ NC samples. These active stretching modes and the lack of the peak at 1616 cm^{-1}, which corresponds to the –NH$_2$ scissoring, is visible in the spectrum of the free Olam (Figure S1(B1)). This suggests that Olam coordinates the surface of the TiO$_2$ NCs through the amino groups.

The Olam-capped TiO$_2$ NCs have been used as seeds for preparing a new set of samples. Figure 2D–F reports the morphological characterization of the samples obtained at [HMDS] = 0.001 M, [Pb^{2+}] = 0.005 M, and at an injection temperature of 120 °C (Figure 2D), 100 °C (Figure 2E), and 80 °C (Figure 2F). A high yield (white dashed circle Figure 2D) of TiO$_2$ NCs, decorated with tiny dots (size nearly 1.8 nm and $\sigma_\%$ = 14%), ascribed to TiO$_2$/PbS NCs, are displayed in Figure 2D, which corresponds to the sample prepared at an injection temperature of 120 °C, even though isolated dots of nearly 2.9 nm ($\sigma_\%$ = 9%) are also visible (red dashed circle Figure 2D), which suggests the concomitant occurrence of both homogeneous and heterogeneous nucleation. The decrease of the precursors' injection temperature down to 100 °C (Figure 2E) mainly results in sole nanostructures, based on TiO$_2$ NCs decorated with small spots (nearly 1.7 nm and $\sigma_\%$ = 13%) and a negligible number of isolated dots. Further reduction of the injection temperature at 80 °C causes only a slight darkening of the solution color upon precursors injection. The TEM micrograph (Figure 2F) reveals a low yield of TiO$_2$-decorated structures. Therefore, while in the presence of OA-capped TiO$_2$ seeds, the homogeneous nucleation of PbS could not be eluded, even at the lowest tested supersaturation condition. The exchange of OA with Olam has promoted the formation of structures attributed to heterogeneous NCs and the formation of isolated PbS NCs has been strongly limited by decreasing the reaction temperature. Further chemical, structural, and spectroscopic characterization has been carried out for the sample prepared starting from Olam-capped TiO$_2$ NCs and injection of PbS precursors ([HMDS] = 0.001 M, [Pb^{2+}] = 0.005 M) at 100 °C that shows, by TEM characterization, the limited homogenous nucleation and the higher yield of tiny dots decorated TiO$_2$ NCs.

2.2. Structure and Chemical Composition of TiO$_2$/PbS Hybrid NCs

Olam-capped TiO$_2$ NCs (Figure 4A) and the TiO$_2$/PbS hybrid NC sample (Figure 4B) have been investigated by high-angle annular dark-field-scanning TEM (HAADF-STEM). Based on incoherently scattered electrons, the HAADF-STEM («Z-contrast») imaging mode shows a contrast related to the average atomic number ($\sim Z^{1.7}$). This translates into high sensitivity to variations in the atomic number of atoms in the sample and, thus, enables an easy identification of areas characterized by a different mean atomic number.

Small bright dots, ascribed to a high mean atomic number material, decorate less bright NCs (Figure 4B), which corresponds to TiO$_2$ seeds (Figure 4A). The observed island-like structure of the material decorating the TiO$_2$ NCs retraces the Volmer-Weber growth mode, as expected, from the high lattice mismatch between PbS and TiO$_2$. STEM-EDS analysis (Table 1) and XPS characterization (Figure 4C) have been carried to unveil the chemical composition of the hybrid structure.

Table 1. STEM-EDS analysis (atomic %) of TiO$_2$ NCs and TiO$_2$/PbS hybrid structure.

	Ti%	O%	Pb%	S%
TiO$_2$ NCs	33	66	-	-
TiO$_2$/PbS NCs	20	73	8	low

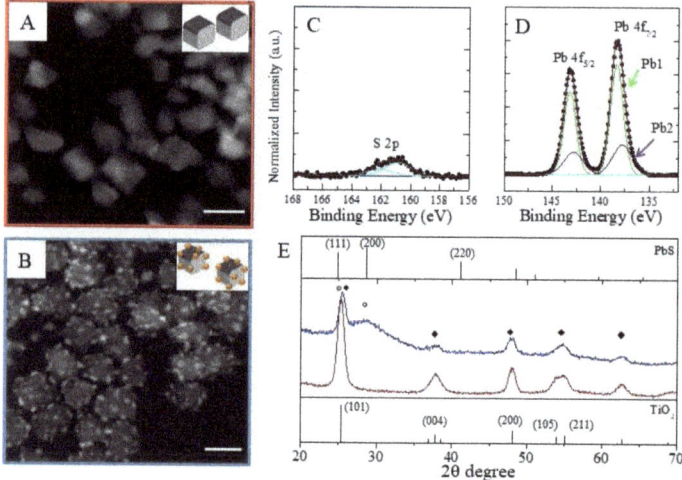

Figure 4. HAADF-STEM images (scale bar 10 nm) of (**A**) TiO$_2$ seed NCs and (**B**) TiO$_2$/PbS hybrid nanostructures. (**C,D**) XPS spectrum of TiO$_2$/PbS hybrid NCs in the 130–170 eV range: raw data (scattered plot) and fitting curve of the S 2p (C, light blue line) and of the Pb 4f (**D**), based on two contributions Pb1 (D, green line) and Pb2 (D, violet line). (**E**) XRD spectra of TiO$_2$ NCs (red line) and TiO$_2$/PbS hybrid (blue line) together with the Bragg hkl reflections positions for TiO$_2$ anatase (bottom markers, crystal system: tetragonal, PDF2-ICDD code: 842186) and PbS (upper panel, crystal system: cubic, PDF2-ICDD code: 00-005-0592) crystal structures. The XRD spectra of the two samples are reported, and shifted for the sake of clarity. Filled square and empty circle symbols are ascribed to TiO$_2$ anatase and PbS peaks, respectively.

The STEM-EDS data confirm the presence Pb and S in the TiO$_2$/PbS hybrid NCs, even though the latter content is lower than that expected to be consistent with the presence of PbS domains. However, it must be mentioned that quantification of sulfur in PbS by EDS is affected by a large uncertainty due to the overlap of the S Kα (2.31 keV) and Pb Mα (2.34 keV) X-ray peaks, given the relatively low energy resolution (~100 eV in the mentioned energy range) of EDS. Therefore, the quantification of the S atomic content relies on fitting the two peaks with a consequent uncertainty. Concomitantly, an increase in O content with respect to Ti (O%: Ti% 3.6:1) is revealed in the hybrid NC sample compared to TiO$_2$ NC seeds (O%: Ti% 2:1), which is in disagreement with the stoichiometry of TiO$_2$. This means that other O sources contribute to the O signal. PbS photo-oxidation [67,68], reasonably taking place in samples stored under air and atmospheric humidity, and residual Pb-oleate, present in excess during the synthesis, can explain the high O content revealed in the hybrid structure. Residual Pb-oleate can also reasonably account for the excess of Pb over S detected by the STEM-EDS analysis. The overall stoichiometry of the TiO$_2$/PbS hybrid system has been assessed by XPS characterization (Figure 4C) [68,69]. The XPS spectrum shows the typical peaks of both S (Figure 4C) and Pb (Figure 4D) in the binding energy range of 130–170 eV. In particular, the Pb 4f signal is composed of two peaks ascribed to Pb4$f_{7/2}$, at 138.4 ± 0.2 eV and Pb 4$f_{5/2}$, at 143.2 ± 0.2 eV. The most prominent feature of the Pb4$f_{7/2}$ peak position agrees with the presence of Pb ions in the Pb^{2+} oxidation state. This peak has been fitted with two major contributions (Figure 4D), which include a prominent Pb1 component centered at 138.6 eV, correlated to binding energy values of oxidized or hydroxylated species (PbSO$_3$, Pb(OH)$_2$) or residual Pb-oleate, expected to have their contributions within the window of 138.4–138.6 eV, according to the NIST database [68] and Pb2, the less intense component, located at 137.8 eV, which is a binding energy value corresponding to Pb in PbS (137.8 ± 0.4eV in the NIST database). The S2p signal (Figure 4C) has been fitted with a single component located at 160.9 ± 0.3 eV that can be attributed to S bound to Pb (PbS rock-salt type formation, 160.6 ± 0.6 eV in the NIST database) [70]. The quantitative analysis, made by comparing the signal of the Pb component at 137.8 eV and the S, with each one attributed to the PbS, shows a Pb: S ratio of 1:1, which reflects the stoichiometry of bulk-like PbS. Therefore, the results of the overall HAADF-STEM, STEM-EDS, and XPS characterization (Figure 4C) suggest the formation of small PbS dots with residual Pb-oleate and other oxidized Pb species [68,69].

X-ray diffraction patterns of TiO_2 (Figure 4E red line) and TiO_2/PbS hybrid NCs (Figure 4E: blue line) together with the Bragg hkl reflections for TiO_2 anatase and PbS crystal structures are reported in Figure 4E. TiO_2 NCs show a pattern characterized by peaks that can be indexed with the crystallographic structure of TiO_2 anatase (crystal system: tetragonal, PDF2-ICDD code: 842186), while the hybrid TiO_2/PbS NCs shows a pattern profile characterized by the presence of both PbS rock-salt (crystal system: cubic, PDF2-ICDD code: 85128) and TiO_2 anatase (crystal system: tetragonal, PDF2-ICDD code: 842186) crystalline phases. In particular, the diffraction peak at nearly $2\theta = 25.4°$, has been indexed as the overlap between the (101) of TiO_2 and (111) of the PbS. The broad peak at $2\theta = 28.6°$ matches the reflection peak (220) of the PbS rock-salt phase, since their broadening accounted for the very small size of the PbS domain, as confirmed by the TEM characterization and XPS chemical analysis. Diffractions peaks at $2\theta \sim 37.9°$, $48.0°$, and $54.6°$ corresponding to the (004), (200), and (211) hkl Bragg reflections of the TiO_2 anatase structure are found both in the hybrid structures and in the TiO_2 NCs.

The apparent difference in relative integral intensity of specific diffraction peaks in TiO_2 NCs compared to TiO_2/PbS hybrid NCs suggests a possible structural change that have been analyzed in the following. The integral intensity ratio for the TiO_2 reflections (004)/(200) has a value of 1.20 in TiO_2 NCs, which decreases to 0.37 in TiO_2/PbS hybrid NCs. Similarly, the integral intensity ratio (101)/(200) is nearly 3.58 in TiO_2 NCs and becomes 1.86 for the NCs' hybrid structures. This latter ratio value should be slightly underestimated, which is the peak (200) at $2\theta = 25.4°$ where the overlap between two contributions arise from the TiO_2 (101) and the PbS (111) reflections. The intensity ratios show that the (004) and (101) TiO_2 reflections are less represented with respect to the (200) one in the TiO_2/PbS hybrid NCs, which indicates a decrease of the crystallinity along the [004] and [101] directions for the TiO_2/PbS NCs hybrid structure with respect to the [200] one.

Bonding distortions in TiO_2/PbS hybrid structures have been demonstrated in literature [53] with the periodic bonds at the TiO_2 surface compromising the crystal structure of PbS during its deposition. In this case, XRD data clearly suggest a structural modification of the TiO_2 seed in the hybrid structure. In order to accommodate the misfit between the two materials' lattice parameters and reduce the interfacial strain, partial disruption of the TiO_2 anatase crystallinity, mainly along the [101] and [004] directions, may occur. However, XRD analysis does not provide any quantitative estimation on the PbS crystalline domain due to the small size of this domain in the hybrid structures.

2.3. Spectroscopic Characterization

Characterizations of colloidal semiconductor NCs by absorption and emission spectra is essential for monitoring the material optical properties in view of their applications in sensing, energy conversion, and optoelectronic devices. In addition, in the case of TiO_2 NCs, photoluminescence spectroscopy represents a powerful technique for tracking the evolution of defective states [71]. Since the valence and conduction bands of TiO_2 are associated with the O-2p and Ti-3d states, respectively, rearrangements of Ti and O atoms can bring modifications in the band edge structures [72].

UV-Vis-NIR absorption spectra (Figure 5A) as well as stationary (Figure 5B, λ_{ex} at 375 nm) and time-resolved emission (Figure S3) in the visible range of Olam-capped TiO_2 seeds (red line, Figure 5A,B) and TiO_2/PbS hybrid nanostructures (blue line, Figure 5A,B) have been recorded. TiO_2 colloidal dispersion (Figure 5A, red line) shows the absorption line profile characteristic of an indirect band gap semiconductor highlighting a steep absorption onset in the UV region, which is consistent with the energy gap of anatase TiO_2 (3.2 eV), while the absorption spectrum of TiO_2/PbS hybrid NCs (Figure 5A, blue line) spans over the UV-Vis-NIR spectral range. This confirms the presence of PbS domains. In particular, the spectrum shows a broad absorption at 545 nm (2.26 eV) consistent with the first exciton transition of small PbS NCs with sizes below 2 nm [73,74]. This is represented by the PbS domain decorating the TiO$_2$ NCs and a week band at 850 nm (1.46 eV) ascribed to exciton transition of homogeneously nucleated PbS NCs, with a size of nearly 3 nm, present in the sample at a very low extent. The band broadening can be attributed to the broad size distribution of PbS domains, which is also revealed by the TEM characterization (Figure 2E).

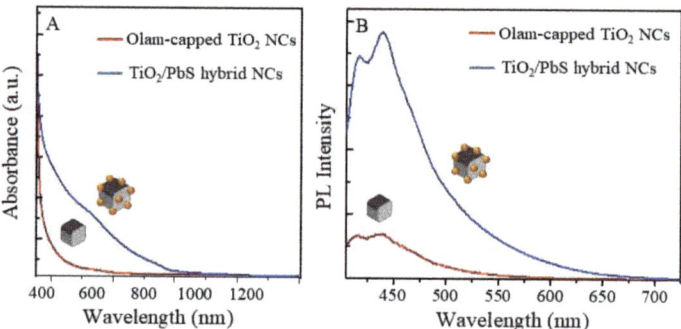

Figure 5. (**A**) UV-Vis-NIR absorbance spectra and (**B**) steady state emission spectra (λ_{ex} = 375 nm) in the visible spectral range of the Olam-capped TiO$_2$ seeds (red line) and TiO$_2$/PbS hybrid structures (blue line). Each sample has been suitably diluted in order to show the same absorbance value at 375 nm.

The PL spectra in the visible region reported in Figure 5B (λ_{ex} = 375 nm) show a structured emission with two main bands peaked at nearly 415 nm and 435 nm for both samples and an emission tail extending up to 600 nm and 700 nm for the TiO$_2$ and TiO$_2$/PbS hybrid NCs, respectively. However, an increased intensity, without evident modifications of the line profile, is detected for the emission spectrum of the TiO$_2$/PbS hybrid structures compared to the TiO$_2$ NC seeds.

Gaussian deconvolution (Figure S4A,B) of the PL spectrum of each sample (TiO$_2$ NC, Figure S4A and TiO$_2$/PbS hybrid NC Figure S4B) performed using four components results in emission bands centered at 411 nm (3 eV), 434 nm (2.86 eV), 445 nm (2,78 eV), and 462 nm (2.68 eV) for the TiO$_2$ NC sample and at 413 nm (3 eV), 439 nm (2.82 eV), 446 nm (2.78 eV), and 488 nm (2.5 eV) for the TiO$_2$/PbS sample, respectively. The component centered at 462 nm, for the TiO$_2$ NC sample, and 488 nm, for the TiO$_2$/PbS sample, are the one that is mainly contributed to the emission tail in the green-red visible spectral region. Early studies [75] attribute blue emission (400–450 nm) in TiO$_2$ NCs to self-trapped exciton (STE), which corresponds to self-localized photogenerated charges, that, in anatase TiO$_2$, can be more easily formed thanks to the long Ti-Ti interionic distances, the low TiO$_6$ octahedral coordination, and the limited symmetry of the structure [75].

Emissive bands mainly located in the green and red spectral region are attributed to structural defects like Ti-interstitials and/or oxygen vacancies, which introduce sub bandgap defect states [76] that contribute to recombination paths of photogenerated electrons with holes trapped on undercoordinated Ti^{3+} or involve shallow states. Even though these emission bands become particularly relevant in high surface area materials such as NCs [75], which are expected to feature a high density of trap/defect states, the PL spectra of the synthesized TiO$_2$ and TiO$_2$/PbS hybrid NCs show only an emission tail in the green and red spectral region. This result can be explained by considering that oxygen vacancies suffer from the presence of O$_2$ when acting as a scavenger of conduction band electrons and induce a strong PL quenching [71,75].

On the basis of these considerations, the fitting contributions of the TiO$_2$ NC PL band centered in the range of 400–450 nm can be mainly attributed to STE while the one centered at 462 nm is ascribed to indirect recombination via oxygen defects [77]. The week emission of the TiO$_2$ NC sample with the tail almost completely quenched at 600 nm suggests a low density of oxygen defects compared to the TiO$_2$/PbS hybrid nanostructure. In the TiO$_2$/PbS hybrid sample, a broad and more intense emissive component centered at 488 nm (2.6 eV) and extending up to 700 nm is displayed. This band is attributed to the charge transfer from Ti^{3+} to the nearby oxygen anion in a TiO_6^{8-} complex structure, and also most likely originated from the recombination of electrons at oxygen-related defect states with the holes in the valence band [71]. Since oxygen vacancies improve the formation of STE, an increased intensity of the emission in the blue spectral region is detected for the TiO_2/PbS hybrid sample. The increase in the surface oxygen vacancy states is also corroborated by the decrease of the average lifetimes of the decay profile. A faster PL decay (Figure S3) at 460 nm has been also measured with averaged lifetimes decreasing from 23 ns for TiO_2 seeds to only 5 ns in the hybrid structures, which indicates the introduction of non-radiative pathways in the recombination of the excited charge carriers. Since oxygen vacancies are more stable, they do not remain as surface defect states, but tend to migrate at subsurface layers, which affects the TiO_2 structures and potentially generates a partially amorphous layer [71], as corroborated by the XRD characterization.

Only a very weak emission of the PbS domain has been observed. The lack of a significant emission signal in the far visible range expected for the PbS domain of 2 nm could be possibly ascribed to a PL quenching induced by electron transfers from PbS to TiO_2 or to a strong red shift of the PL due to energy transfer phenomena among PbS NCs in close proximity with each other. A week and broad band, located in the NIR region at 1095 nm (1.14 eV), is displayed in Figure S4C. However, more in-depth characterization needs to be carried out to confirm the charge transfer process at the interface.

3. Materials and Methods

3.1. Materials

Titanium(IV) ethoxide (TEO, technical grade), 1-decanol (DL, ~97%), 1-octadecene (ODE, technical grade 90%)), oleic acid (OA, technical grade 90%), oleylamine (Olam, technical grade 70%), lead oxide (II) (PbO, ≥99.0%), and hexamethyldisilathiane (HMDS, synthesis grade) were used for NCs synthesis. Acetone (≥99.5%) and ethanol (≥99.8%) were used as non-solvents to recover the colloidal NCs from the reaction mixture. Hexane and tetrachloroethylene (TCE, A.C.S. spectrophotometric grade ≥99%) were used as solvents to disperse the synthesized NCs. All reagents and solvents were purchased by Sigma-Aldrich (Milan, Italy) and used without further purification.

3.2. Synthesis of Organic-Capped TiO_2 NCs

1 mmol of OA and 13 mmol of DL were dissolved in 15 mL of ODE and degassed for 1 h at 120 °C in a three-necked flask. Then, the temperature was set to 290 °C and 1 mmol of TEO was rapidly injected. The reaction was stopped after 1 h by cooling the reaction flask to room temperature. TiO_2 NCs were collected by adding acetone and three cycles of centrifugation/re-dispersion in hexane with the addition of acetone. NCs were dispersed in hexane (OA-capped TiO_2 NCs). Olam-capped TiO_2 NCs were prepared by a ligand exchange reaction on OA-capped TiO_2 NCs with Olam as follows: 1.5 mL of extensively washed native OA-capped TiO_2 NCs were dispersed in 10 mL of ODE in the presence of 0.6 mmol of Olam and sonicated for more than 1 h. The colloidal solution was purified by non-solvent addition, centrifugation, and re-dispersion in hexane.

3.3. Synthesis of TiO_2/PbS Hybrid NPs

The TiO_2/PbS NC hybrid structures were synthesized by a seed-mediated growth reaction. In a three-neck flask, 1.5 mL of TiO_2 NC seeds, 0.2 mmol, (either OA-capped or Olam-capped TiO_2) were dispersed in 10 mL of ODE and degassed at 80 °C for 30 min. In another flask, 0.2 mmol of PbO were dispersed in 10 mL of ODE in the presence of 0.6 mmol OA under inert atmosphere and heated up to 120 °C for 30 min to decompose the PbO. The Pb-oleate precursor was then injected in the flask with the TiO_2 NCs' colloidal dispersion and was stirred under nitrogen for 10 min ([Pb^{2+}] = 0.01 M ÷ 0.005 M), which was followed by the injection of HMDS sulfur precursor (HMDS: Pb molar ratio 1:5) at the different injection temperatures: 120 °C, 100 °C, and 80 °C (T_{inj}) and the sudden decrease of the temperature down to 80 °C (growth temperature T_{growth}) to stop the nucleation. The reaction mixture was left to stir for 10 min and then stopped by cooling down the solution. The NPs were collected by the addition of ethanol as a non-solvent and washed with three cycles of centrifugation/re-dispersion in hexane. The samples were then re-dispersed in 2 mL of TCE for further characterization. Three batches have been prepared for each synthetic approach in order to confirm the reproducibility of the morphological and spectroscopic results.

3.4. Sample Characterization

UV-Vis-NIR absorption spectra were recorded by using a Cary 5000 spectrophotometer (Varian, Agilent Technologies Italia S.p.A, Milano, Italy). Steady-state and time-resolved UV–Vis-NIR photoluminescence experiments (PL) were performed by using a Fluorolog 3 spectrofluorimeter (Horiba Jobin-Yvon, Roma, Italy) equipped with both a continuous wave Xe lamp (450 W) and a ~80 ps pulsed laser source (NanoLED 375 L), which emitted at 375 nm with a repetition rate of 1 MHz, and interfaced with a TBX-PS photon counter for steady-state and time-resolved (TRPL) measurements in the visible range and with a Peltier-cooled InGaAs detector for the NIR range. Fast Fourier transform infrared (FTIR) spectroscopy measurements were carried out in attenuated total reflection (ATR) mode with Spectrum One FTIR spectrometer (Perkin–Elmer, Milan, Italy) equipped with a triglycine sulfate (TGS) detector. The spectral resolution was 4 cm-1. The internal reflection element (IRE) was a three bounce 4 mm diameter diamond microprism. Cast films were prepared directly onto the IRE by depositing the sample solutions (3–5 lL) onto the upper face of the diamond crystal and allowing the solvent to evaporate. For morphological characterization, a JEM-1011 transmission electron microscope (TEM) of JEOL (Tokyo, Japan) was employed, operating at 100 kV acceleration voltage. Samples were prepared by dipping the carbon-coated copper grid in the colloidal dispersion of the NCs and NPs, prepared at a suitable dilution, and let the solvent evaporate. The particle average size and size distribution was obtained by counting at least 150 particles for each sample by means of a freeware Zeiss AxioVision analysis program (Jena, Germany). In particular, the average size was measured and the percentage relative standard deviation (σ%) was calculated in order to define the NC size distributions. High-angle annular dark field-scanning TEM (HAADF-STEM) images and energy-dispersive X-ray spectroscopy (EDS) analyses were acquired using an image C_s-corrected JEM-2200FS TEM (JEOL) with a Schottky emitter, operated at 200 kV, equipped with a Bruker (Berlin, Germany) Quantax 400 STEM system and a XFlash 5060 silicon-drift detector (60 mm^2 active area). The EDS spectra were quantified by the Cliff-Lorimer method applied to the O Kα peak (at 0.52 keV), the S Kα peak (at 2.31 keV), and the Pb Lα peak (at 10.55 keV). For these analyses, the samples were prepared by drop-casting the colloidal suspensions onto a double amorphous carbon film (ultrathin on holey)-coated Cu grid. X-ray photoelectron spectroscopy (XPS) characterizations were performed on an Axis UltraDLD spectrometer (Kratos) using a monochromatic Al Kα source (15 kV, 20 mA). The binding energy was calibrated by setting the main C1s peak (corresponding to C–C bonds) to 284.8 eV. A D8 Discover X-ray powder diffractometer (Bruker AXS Advanced X-ray Solutions GmbH, Karlsruhe, Germany) was used in Bragg-Brentano θ/2θ acquisition geometry using a copper Kα x-ray tube (0.154 nm) and a scintillation detector. The XRD patterns were recorded at a fixed incidence angle of 5° while moving the detector in the range 10–120° with a step size of 0.05°. A qualitative analysis of the crystalline phase content was performed using the QUALX 2.0 program [78]. Samples for XRD characterizations were prepared by drop casting of concentrated NPs' dispersions on silicon substrates.

4. Conclusions

A seeded growth combined with a hot-injection approach has been used to prepare TiO_2/PbS hybrid structures under a controlled experimental condition (reaction mixture composition, seed surface chemistry, and injection/nucleation temperature), suitably defined to limit homogenous nucleation and favor heterogeneous nucleation. The morphological and chemical characterization confirms the formation of small dots ascribed to PbS NCs decorating the oleylamine capped TiO_2 NCs, upon injection at 100 °C of the Pb and S precursors, at the concentrations of 0.005 M and 0.001 M, respectively. The hybrid structure topology has been demonstrated to be strongly affected by the high interfacial strain between the TiO_2 and PbS, which agreed with the Volmer-Weber growth mode.

The XRD characterization suggests a structural modification of the TiO_2 seeds in the hybrid structure with partial disruption of the TiO_2 anatase crystallinity, likely induced to accommodate the misfit between the two materials' lattice parameters and reduce the interfacial strain. The presence of dominant defect states in the TiO_2 domain of the hybrid structure is confirmed by steady state and time resolved photoluminescence. The enhanced intensity of the emission band for the TiO_2/PbS hybrid NCs compared to TiO_2 NCs has been ascribed to a higher density of defect oxygen vacancies with emissive states in the green region and improved formation of STE whose emission falls in the blue region of the visible spectrum.

Supplementary Materials: The following are available online. Figure S1: FTIR spectra of neat oleic acid and oleylamine. Figure S2: TEM micrograph of oleic acid-capped TiO_2 NCs after injection of $[Pb^{2+}]$ = 0.01 M, [HMDS] = 0.002 M at 100 °C. Figure S3: Time-resolved photoluminescence decay at 460 nm (λ_{ex} = 375 nm) of TiO_2 NCs and PbS/TiO_2 hybrid nanostructures. Figure S4: Gaussian deconvolution of photoluminescence spectra (λ_{ex} = 375 nm) in the visible range and in the NIR range of TiO_2 NCs and PbS/TiO_2 hybrid nanostructures.

Author Contributions: Conceptualization, E.F. and M.S. Methodology, E.F., C.N.D., T.S., R.B., M.P., A.P., L.T., M.C., and E.F. Validation, C.N.D., T.S., R.B., M.P., M.S., and E.F. Formal analysis, C.N.D., T.S., R.B., M.P., and E.F. Investigation, C.N.D., T.S., R.B., M.P., and L.T. Resources, E.F., M.S., M.L.C., R.C., and A.A. Data curation, C.N.D., T.S., C.G., R.B., and M.P. Writing—original draft preparation, C.N.D., T.S., R.B., M.P., and E.F. Writing—review and editing, C.G., A.P., C.I., N.D., R.C., M.L.C., and M.S. Visualization, C.N.D., T.S., R.B., M.P., A.A., M.L.C., M.S., and E.F. Supervision, E.F. and M.S. Project administration, E.F. Funding acquisition, E.F., M.S., M.L.C., and R.C. All authors have read and agreed to the published version of the manuscript.

Funding: The PON Project ARS01_00637 "TARANTO", the PON Project ARS01_00519 "BEST-4U", and the MIUR-FIRB Futuro in Ricerca Project (grant number RBFR122HFZ) partially funded this research.

Acknowledgments: The authors are grateful to the National Interuniversity Consortium of Materials Science and Technology (INSTM).

Conflicts of Interest: The authors declare no conflict of interest.

References

1. van Embden, J.; Chesman, A.S.R.; Jasieniak, J.J. The heat-up synthesis of colloidal nanocrystals. *Chem. Mater.* **2015**, *27*, 2246–2285. [CrossRef]
2. Huo, D.; Kim, M.J.; Lyu, Z.; Shi, Y.; Wiley, B.J.; Xia, Y. One-dimensional metal nanostructures: From colloidal syntheses to applications. *Chem. Rev.* **2019**, *119*, 8972–9073. [CrossRef] [PubMed]
3. Nasilowski, M.; Mahler, B.; Lhuillier, E.; Ithurria, S.; Dubertret, B. Two-dimensional colloidal nanocrystals. *Chem. Rev.* **2016**, *116*, 10934–10982. [CrossRef] [PubMed]
4. Carey, G.H.; Abdelhady, A.L.; Ning, Z.; Thon, S.M.; Bakr, O.M.; Sargent, E.H. Colloidal quantum dot solar cells. *Chem. Rev.* **2015**, *115*, 12732–12763. [CrossRef] [PubMed]
5. Ha, M.; Kim, J.-H.; You, M.; Li, Q.; Fan, C.; Nam, J.-M. Multicomponent plasmonic nanoparticles: From heterostructured nanoparticles to colloidal composite nanostructures. *Chem. Rev.* **2019**, *119*, 12208–12278. [CrossRef] [PubMed]
6. Cortie, M.B.; McDonagh, A.M. Synthesis and optical properties of hybrid and alloy plasmonic nanoparticles. *Chem. Rev.* **2011**, *111*, 3713–3735. [CrossRef] [PubMed]
7. Kershaw, S.V.; Susha, A.S.; Rogach, A.L. Narrow bandgap colloidal metal chalcogenide quantum dots: Synthetic methods, heterostructures, assemblies, electronic and infrared optical properties. *Chem. Soc. Rev.* **2013**, *42*, 3033–3087. [CrossRef]
8. Cozzoli, P.D.; Pellegrino, T.; Manna, L. Synthesis, properties and perspectives of hybrid nanocrystal structures. *Chem. Soc. Rev.* **2006**, *35*, 1195–1208. [CrossRef]
9. Casavola, M.; Buonsanti, R.; Caputo, G.; Cozzoli, P.D. Colloidal strategies for preparing oxide-based hybrid nanocrystals. *Eur. J. Inorg. Chem.* **2008**, *2008*, 837–854. [CrossRef]

10. Nag, A.; Kundu, J.; Hazarika, A. Seeded-growth, nanocrystal-fusion, ion-exchange and inorganic-ligand mediated formation of semiconductor-based colloidal heterostructured nanocrystals. *Cryst. Eng. Comm.* **2014**, *16*, 9391–9407. [CrossRef]
11. Ma, D. Chapter 1—Hybrid nanoparticles: An introduction. In *Noble Metal-Metal Oxide Hybrid Nanoparticles*; Mohapatra, S., Nguyen, T.A., Nguyen-Tri, P., Eds.; Woodhead Publishing: Cambridge, UK, 2019; pp. 3–6.
12. Nguyen, K.T.; Zhao, Y. Engineered hybrid nanoparticles for on-demand diagnostics and therapeutics. *Acc. Chem. Res.* **2015**, *48*, 3016–3025. [CrossRef] [PubMed]
13. Yue, S.; Li, L.; McGuire, S.C.; Hurley, N.; Wong, S.S. Metal chalcogenide quantum dot-sensitized 1D-based semiconducting heterostructures for optical-related applications. *Energy Environ. Sci.* **2019**, *12*, 1454–1494. [CrossRef]
14. Selinsky, R.S.; Ding, Q.; Faber, M.S.; Wright, J.C.; Jin, S. Quantum dot nanoscale heterostructures for solar energy conversion. *Chem. Soc. Rev.* **2013**, *42*, 2963–2985. [CrossRef] [PubMed]
15. Zhou, Y.; Zhang, M.; Guo, Z.; Miao, L.; Han, S.-T.; Wang, Z.; Zhang, X.; Zhang, H.; Peng, Z. Recent advances in black phosphorus-based photonics, electronics, sensors and energy devices. *Mater. Horizons* **2017**, *4*, 997–1019. [CrossRef]
16. Wang, Y.; Wu, T.; Zhou, Y.; Meng, C.; Zhu, W.; Liu, L. TiO_2-based nanoheterostructures for promoting gas sensitivity performance: Designs, developments, and prospects. *Sensors* **2017**, *17*, 1971. [CrossRef] [PubMed]
17. Zhao, H.; Vomiero, A.; Rosei, F. Ultrasensitive, biocompatible, self-calibrating, multiparametric temperature sensors. *Small* **2015**, *11*, 5741–5746. [CrossRef] [PubMed]
18. Xu, W.; Niu, J.; Wang, H.; Shen, H.; Li, L.S. Size, shape-dependent growth of semiconductor heterostructures mediated by Ag2Se nanocrystals as seeds. *ACS Appl. Mater. Interfaces* **2013**, *5*, 7537–7543. [CrossRef]
19. Milleville, C.C.; Chen, E.Y.; Lennon, K.R.; Cleveland, J.M.; Kumar, A.; Zhang, J.; Bork, J.A.; Tessier, A.; LeBeau, J.M.; Chase, D.B.; et al. Engineering efficient photon upconversion in semiconductor heterostructures. *ACS Nano* **2019**, *13*, 489–497. [CrossRef]
20. Acharya, K.P.; Hewa-Kasakarage, N.N.; Alabi, T.R.; Nemitz, I.; Khon, E.; Ullrich, B.; Anzenbacher, P.; Zamkov, M. Synthesis of PbS/TiO_2 colloidal heterostructures for photovoltaic applications. *J. Phys. Chem. C* **2010**, *114*, 12496–12504. [CrossRef]
21. Acharya, K.P.; Alabi, T.R.; Schmall, N.; Hewa-Kasakarage, N.N.; Kirsanova, M.; Nemchinov, A.; Khon, E.; Zamkov, M. Linker-free modification of TiO_2 nanorods with PbSe nanocrystals. *J. Phys. Chem. C* **2009**, *113*, 19531–19535. [CrossRef]
22. Hassan, Y.; Chuang, C.-H.; Kobayashi, Y.; Coombs, N.; Gorantla, S.; Botton, G.A.; Winnik, M.A.; Burda, C.; Scholes, G.D. Synthesis and optical properties of linker-free TiO_2/CdSe nanorods. *J. Phys. Chem. C* **2014**, *118*, 3347–3358. [CrossRef]
23. Kwon, K.-W.; Lee, B.H.; Shim, M. Structural evolution in metal oxide/semiconductor colloidal nanocrystal heterostructures. *Chem. Mater.* **2006**, *18*, 6357–6363. [CrossRef]
24. Fenton, J.L.; Hodges, J.M.; Schaak, R.E. Synthetic deconvolution of interfaces and materials components in hybrid nanoparticles. *Chem. Mater.* **2017**, *29*, 6168–6177. [CrossRef]
25. Shaviv, E.; Schubert, O.; Alves-Santos, M.; Goldoni, G.; Di Felice, R.; Vallée, F.; Del Fatti, N.; Banin, U.; Sönnichsen, C. Absorption properties of metal–semiconductor hybrid nanoparticles. *ACS Nano* **2011**, *5*, 4712–4719. [CrossRef] [PubMed]
26. Cozzoli, P.D.; Comparelli, R.; Fanizza, E.; Curri, M.L.; Agostiano, A.; Laub, D. Photocatalytic synthesis of silver nanoparticles stabilized by TiO_2 nanorods: A semiconductor/metal nanocomposite in homogeneous nonpolar solution. *J. Am. Chem. Soc.* **2004**, *126*, 3868–3879. [CrossRef] [PubMed]
27. Cozzoli, P.D.; Fanizza, E.; Curri, M.L.; Laub, D.; Agostiano, A. Low-dimensional chainlike assemblies of TiO_2 nanorod-stabilized Au nanoparticles. *Chem. Commun.* **2005**, 942–944. [CrossRef]
28. He, S.; Zhang, H.; Delikanli, S.; Qin, Y.; Swihart, M.T.; Zeng, H. Bifunctional magneto-optical FePt–CdS hybrid nanoparticles. *J. Phys. Chem. C* **2009**, *113*, 87–90. [CrossRef]
29. Buonsanti, R.; Grillo, V.; Carlino, E.; Giannini, C.; Curri, M.L.; Innocenti, C.; Sangregorio, C.; Achterhold, K.; Parak, F.G.; Agostiano, A.; et al. Seeded growth of asymmetric binary nanocrystals made of a semiconductor TiO_2 rodlike section and a magnetic γ-Fe2O3 spherical domain. *J. Am. Chem. Soc.* **2006**, *128*, 16953–16970. [CrossRef]
30. Santana Vega, M.; Guerrero Martínez, A.; Cucinotta, F. Facile strategy for the synthesis of gold@ silica hybrid nanoparticles with controlled porosity and janus morphology. *Nanomaterials* **2019**, *9*, 348. [CrossRef]

31. Sashchiuk, A.; Yanover, D.; Rubin-Brusilovski, A.; Maikov, G.I.; Čapek, R.K.; Vaxenburg, R.; Tilchin, J.; Zaiats, G.; Lifshitz, E. Tuning of electronic properties in IV–VI colloidal nanostructures by alloy composition and architecture. *Nanoscale* **2013**, *5*, 7724–7745. [CrossRef]
32. Kundu, P.; Anumol, E.A.; Nethravathi, C.; Ravishankar, N. Existing and emerging strategies for the synthesis of nanoscale heterostructures. *PCCP* **2011**, *13*, 19256–19269. [CrossRef] [PubMed]
33. Fan, F.-R.; Liu, D.-Y.; Wu, Y.-F.; Duan, S.; Xie, Z.-X.; Jiang, Z.-Y.; Tian, Z.-Q. Epitaxial growth of heterogeneous metal nanocrystals: from gold nano-octahedra to palladium and silver nanocubes. *J. Am. Chem. Soc.* **2008**, *130*, 6949–6951. [CrossRef] [PubMed]
34. Rivest, J.B.; Jain, P.K. Cation exchange on the nanoscale: An emerging technique for new material synthesis, device fabrication, and chemical sensing. *Chem. Soc. Rev.* **2013**, *42*, 89–96. [CrossRef] [PubMed]
35. De Trizio, L.; Manna, L. Forging colloidal nanostructures via cation exchange reactions. *Chem. Rev.* **2016**, *116*, 10852–10887. [CrossRef] [PubMed]
36. Kriegel, I.; Wisnet, A.; Srimath Kandada, A.R.; Scotognella, F.; Tassone, F.; Scheu, C.; Zhang, H.; Govorov, A.O.; Rodríguez-Fernández, J.; Feldmann, J. Cation exchange synthesis and optoelectronic properties of type II CdTe–Cu_{2-x}Te nano-heterostructures. *J. Mater. Chem. C* **2014**, *2*, 3189–3198. [CrossRef]
37. Dibenedetto, C.N.; Fanizza, E.; Brescia, R.; Kolodny, Y.; Remennik, S.; Panniello, A.; Depalo, N.; Yochelis, S.; Comparelli, R.; Agostiano, A.; et al. Coupling effects in QD dimers at sub-nanometer interparticle distance. *Nano Res.* **2020**, *13*, 1071–1080. [CrossRef]
38. Hyun, B.-R.; Zhong, Y.-W.; Bartnik, A.C.; Sun, L.; Abruña, H.D.; Wise, F.W.; Goodreau, J.D.; Matthews, J.R.; Leslie, T.M.; Borrelli, N.F. Electron injection from colloidal PbS quantum dots into titanium dioxide nanoparticles. *ACS Nano* **2008**, *2*, 2206–2212. [CrossRef]
39. Zhang, H.; Cao, L.; Liu, W.; Su, G.; Gao, R.; Zhao, Y. The key role of nanoparticle seeds during site-selective growth of silver to fabricate core-shell or asymmetric dumbbell heterostructures. *Dalton Trans.* **2014**, *43*, 4822–4829. [CrossRef]
40. Enright, M.J.; Sarsito, H.; Cossairt, B.M. Kinetically controlled assembly of cadmium chalcogenide nanorods and nanorod heterostructures. *Mater. Chem. Front.* **2018**, *2*, 1296–1305. [CrossRef]
41. Corrias, A.; Conca, E.; Cibin, G.; Mountjoy, G.; Gianolio, D.; De Donato, F.; Manna, L.; Casula, M.F. Insights into the structure of dot@rod and dot@octapod cdse@cds heterostructures. *J. Phys. Chem. C* **2015**, *119*, 16338–16348. [CrossRef]
42. Zeng, J.; Wang, X.; Hou, J.G. Colloidal hybrid nanocrystals: Synthesis, properties, and perspectives. In *Nanocrystal*; Masuda, Y., Ed.; IntechOpen: London, UK, 2011.
43. Liu, X.Y. A new kinetic model for three-dimensional heterogeneous nucleation. *J. Chem. Phys.* **1999**, *111*, 1628–1635. [CrossRef]
44. Cai, F.G.; Yang, F.; Xi, J.F.; Jia, Y.F.; Cheng, C.H.; Zhao, Y. Ultrasound effect: Preparation of PbS/TiO_2 heterostructure nanotube arrays through successive ionic layer adsorption and the reaction method. *Mater. Lett.* **2013**, *107*, 39–41. [CrossRef]
45. Hajjaji, A.; Jemai, S.; Rebhi, A.; Trabelsi, K.; Gaidi, M.; Alhazaa, A.N.; Al-Gawati, M.A.; El Khakani, M.A.; Bessais, B. Enhancement of photocatalytic and photoelectrochemical properties of TiO_2 nanotubes sensitized by SILAR-Deposited PbS nanoparticles. *J. Mater.* **2020**, *6*, 62–69. [CrossRef]
46. Díaz-Rodríguez, T.G.; Pacio, M.; Agustín-Serrano, R.; Juárez-Santiesteban, H.; Muñiz, J. Understanding structure of small TiO_2 nanoparticles and adsorption mechanisms of PbS quantum dots for solid-state applications: A combined theoretical and experimental study. *Theor. Chem. Acc.* **2019**, *138*, 92. [CrossRef]
47. Ghadiri, E.; Liu, B.; Moser, J.-E.; Grätzel, M.; Etgar, L. Investigation of interfacial charge separation at PbS QDs/(001) TiO_2 nanosheets heterojunction solar cell. *Part. Part. Syst. Character* **2015**, *32*, 483–488. [CrossRef]
48. Pattantyus-Abraham, A.G.; Kramer, I.J.; Barkhouse, A.R.; Wang, X.; Konstantatos, G.; Debnath, R.; Levina, L.; Raabe, I.; Nazeeruddin, M.K.; Grätzel, M.; et al. Depleted-heterojunction colloidal quantum dot solar cells. *ACS Nano* **2010**, *4*, 3374–3380. [CrossRef]
49. Al Jitan, S.; Palmisano, G.; Garlisi, C. Synthesis and surface modification of TiO_2-based photocatalysts for the conversion of CO_2. *Catalysts* **2020**, *10*, 227. [CrossRef]
50. Dette, C.; Pérez-Osorio, M.A.; Kley, C.S.; Punke, P.; Patrick, C.E.; Jacobson, P.; Giustino, F.; Jung, S.J.; Kern, K. TiO_2 anatase with a bandgap in the visible region. *Nano Lett.* **2014**, *14*, 6533–6538. [CrossRef] [PubMed]

51. Lee, H.; Leventis, H.C.; Moon, S.-J.; Chen, P.; Ito, S.; Haque, S.A.; Torres, T.; Nüesch, F.; Geiger, T.; Zakeeruddin, S.M.; et al. PbS and CdS quantum dot-sensitized solid-state solar cells: "Old concepts, new results". *Adv. Funct. Mater.* **2009**, *19*, 2735–2742. [CrossRef]
52. Fu, H.; Tsang, S.-W. Infrared colloidal lead chalcogenide nanocrystals: Synthesis, properties, and photovoltaic applications. *Nanoscale* **2012**, *4*, 2187–2201. [CrossRef]
53. Trejo, O.; Roelofs, K.E.; Xu, S.; Logar, M.; Sarangi, R.; Nordlund, D.; Dadlani, A.L.; Kravec, R.; Dasgupta, N.P.; Bent, S.F.; et al. Quantifying geometric strain at the PbS QD-TiO$_2$ anode interface and its effect on electronic structures. *Nano Lett.* **2015**, *15*, 7829–7836. [CrossRef] [PubMed]
54. Shrestha, A.; Spooner, N.A.; Qiao, S.Z.; Dai, S. Mechanistic insight into the nucleation and growth of oleic acid capped lead sulphide quantum dots. *PCCP* **2016**, *18*, 14055–14062. [CrossRef]
55. Abel, K.A.; Shan, J.; Boyer, J.-C.; Harris, F.; van Veggel, F.C.J.M. Highly photoluminescent PbS nanocrystals: The beneficial effect of trioctylphosphine. *Chem. Mater.* **2008**, *20*, 3794–3796. [CrossRef]
56. Hines, M.A.; Scholes, G.D. Colloidal PbS nanocrystals with size-tunable near-infrared emission: Observation of post-synthesis self-narrowing of the particle size distribution. *Adv. Mater. (Weinh. Ger.)* **2003**, *15*, 1844–1849. [CrossRef]
57. Corricelli, M.; Enrichi, F.; Altamura, D.; De Caro, L.; Giannini, C.; Falqui, A.; Agostiano, A.; Curri, M.L.; Striccoli, M. Near infrared emission from monomodal and bimodal PbS nanocrystal superlattices. *J. Phys. Chem. C* **2012**, *116*, 6143–6152. [CrossRef]
58. Depalo, N.; Corricelli, M.; De Paola, I.; Valente, G.; Iacobazzi, R.M.; Altamura, E.; Debellis, D.; Comegna, D.; Fanizza, E.; Denora, N.; et al. NIR emitting nanoprobes based on cyclic RGD motif conjugated PbS quantum dots for integrin-targeted optical bioimaging. *ACS Appl. Mater. Interfaces* **2017**, *9*, 43113–43126. [CrossRef] [PubMed]
59. De Trizio, L.; Buonsanti, R.; Schimpf, A.M.; Llordes, A.; Gamelin, D.R.; Simonutti, R.; Milliron, D.J. Nb-doped colloidal TiO$_2$ nanocrystals with tunable infrared absorption. *Chem. Mater.* **2013**, *25*, 3383–3390. [CrossRef]
60. Kuo, M.-S.; Chang, S.-J.; Hsieh, P.-H.; Huang, Y.-C.; Li, C.-C. Efficient dispersants for TiO$_2$ nanopowder in organic suspensions. *J. Am. Ceram. Soc.* **2016**, *99*, 445–451. [CrossRef]
61. Li, H.; Liu, B.; Yin, S.; Sato, T.; Wang, Y. Visible light-driven photocatalytic activity of oleic acid-coated TiO$_2$ nanoparticles synthesized from absolute ethanol solution. *Nanoscale Res. Lett.* **2015**, *10*, 415. [CrossRef]
62. Fanizza, E.; Cozzoli, P.D.; Curri, M.L.; Striccoli, M.; Sardella, E.; Agostiano, A. UV-light-driven immobilization of surface-functionalized oxide nanocrystals onto silicon. *Adv. Funct. Mater.* **2007**, *17*, 201–211. [CrossRef]
63. Cargnello, M.; Gordon, T.R.; Murray, C.B. Solution-phase synthesis of titanium dioxide nanoparticles and nanocrystals. *Chem. Rev.* **2014**, *114*, 9319–9345. [CrossRef]
64. Zhang, Z.; Zhong, X.; Liu, S.; Li, D.; Han, M. Aminolysis route to monodisperse titania nanorods with tunable aspect ratio. *Angew. Chem. Int. Ed.* **2005**, *44*, 3466–3470. [CrossRef] [PubMed]
65. Weir, M.P.; Toolan, D.T.W.; Kilbride, R.C.; Penfold, N.J.W.; Washington, A.L.; King, S.M.; Xiao, J.; Zhang, Z.; Gray, V.; Dowland, S.; et al. Ligand shell structure in lead sulfide–Oleic acid colloidal quantum dots revealed by small-angle scattering. *J. Phys. Chem. Lett.* **2019**, *10*, 4713–4719. [CrossRef]
66. Anderson, N.C.; Hendricks, M.P.; Choi, J.J.; Owen, J.S. Ligand exchange and the stoichiometry of metal chalcogenide nanocrystals: Spectroscopic observation of facile metal-carboxylate displacement and binding. *J. Am. Chem. Soc.* **2013**, *135*, 18536–18548. [CrossRef] [PubMed]
67. Wang, C.; Thompson, R.L.; Ohodnicki, P.; Baltrus, J.; Matranga, C. Size-dependent photocatalytic reduction of CO$_2$ with PbS quantum dot sensitized TiO$_2$ heterostructured photocatalysts. *J. Mater. Chem.* **2011**, *21*, 13452–13457. [CrossRef]
68. Malgras, V.; Nattestad, A.; Yamauchi, Y.; Dou, S.X.; Kim, J.H. The effect of surface passivation on the structure of sulphur-rich PbS colloidal quantum dots for photovoltaic application. *Nanoscale* **2015**, *7*, 5706–5711. [CrossRef]
69. Cant, D.J.H.; Syres, K.L.; Lunt, P.J.B.; Radtke, H.; Treacy, J.; Thomas, P.J.; Lewis, E.A.; Haigh, S.J.; O'Brien, P.; Schulte, K.; et al. Surface properties of nanocrystalline PbS films deposited at the water-oil interface: A study of atmospheric aging. *Langmuir* **2015**, *31*, 1445–1453. [CrossRef] [PubMed]
70. Naumkin, A.V.; Kraut-Vass, A.; Powell, C.J.; Gaarenstroom, S.W. NIST X-ray Photoelectron Spectroscopy Database. Available online: http://srdata.nist.gov/xps/Default.aspx (accessed on 16 April 2019).

71. Saini, C.P.; Barman, A.; Banerjee, D.; Grynko, O.; Prucnal, S.; Gupta, M.; Phase, D.M.; Sinha, A.K.; Kanjilal, D.; Skorupa, W.; et al. Impact of self-trapped excitons on blue photoluminescence in TiO$_2$ nanorods on chemically etched Si pyramids. *J. Phys. Chem. C* **2017**, *121*, 11448–11454. [CrossRef]
72. Saini, C.P.; Barman, A.; Satpati, B.; Bhattacharyya, S.R.; Kanjilal, D.; Kanjilal, A. Defect-engineered optical bandgap in self-assembled TiO$_2$ nanorods on Si pyramids. *Appl. Phys. Lett.* **2016**, *108*, 011907. [CrossRef]
73. Hou, B.; Cho, Y.; Kim, B.-S.; Ahn, D.; Lee, S.; Park, J.B.; Lee, Y.-W.; Hong, J.; Im, H.; Morris, S.M.; et al. Red green blue emissive lead sulfide quantum dots: Heterogeneous synthesis and applications. *J. Mater. Chem. C* **2017**, *5*, 3692–3698. [CrossRef] [PubMed]
74. Miller, E.M.; Kroupa, D.M.; Zhang, J.; Schulz, P.; Marshall, A.R.; Kahn, A.; Lany, S.; Luther, J.M.; Beard, M.C.; Perkins, C.L.; et al. Revisiting the valence and conduction band size dependence of PbS quantum dot thin films. *ACS Nano* **2016**, *10*, 3302–3311. [CrossRef] [PubMed]
75. Pallotti, D.K.; Passoni, L.; Maddalena, P.; Di Fonzo, F.; Lettieri, S. Photoluminescence mechanisms in anatase and rutile TiO$_2$. *J. Phys. Chem. C* **2017**, *121*, 9011–9021. [CrossRef]
76. Zhao, H.; Pan, F.; Li, Y. A review on the effects of TiO$_2$ surface point defects on CO$_2$ photoreduction with H2O. *J. Materiomics* **2017**, *3*, 17–32. [CrossRef]
77. Ruidíaz-Martínez, M.; Álvarez, M.A.; López-Ramón, M.V.; Cruz-Quesada, G.; Rivera-Utrilla, J.; Sánchez-Polo, M. Hydrothermal synthesis of rGO-TiO$_2$ composites as high-performance UV photocatalysts for ethylparaben degradation. *Catalysts* **2020**, *10*, 520. [CrossRef]
78. Altomare, A.; Corriero, N.; Cuocci, C.; Falcicchio, A.; Moliterni, A.; Rizzi, R. QUALX2.0: A qualitative phase analysis software using the freely available database POW_COD. *J. Appl. Crystallogr.* **2015**, *48*, 598–603. [CrossRef]

Sample Availability: Samples of the compounds are available from the authors.

© 2020 by the authors. Licensee MDPI, Basel, Switzerland. This article is an open access article distributed under the terms and conditions of the Creative Commons Attribution (CC BY) license (http://creativecommons.org/licenses/by/4.0/).

Suitable Polymeric Coatings to Avoid Localized Surface Plasmon Resonance Hybridization in Printed Patterns of Photothermally Responsive Gold Nanoinks

Piersandro Pallavicini [1,*], Lorenzo De Vita [1], Francesca Merlin [1], Chiara Milanese [1], Mykola Borzenkov [2], Angelo Taglietti [1] and Giuseppe Chirico [2]

1. Dipartimento di Chimica, Università di Pavia, viale Taramelli, 12–27100 Pavia, Italy; lorenzodevita01@universitadipavia.it (L.D.V.); francesca.merlin01@universitadipavia.it (F.M.); chiara.milanese@unipv.it (C.M.); angelo.taglietti@unipv.it (A.T.)
2. Dipartimento di Fisica "G.Occhialini", Università Milano Bicocca, p.zza della Scienza, 3–20126 Milano, Italy; mykola.borzenkov@unimib.it (M.B.); giuseppe.chirico@mib.infn.it (G.C.)
* Correspondence: piersandro.pallavicini@unipv.it; Tel.: +39-0382-987-336

Academic Editor: Giuseppe Cirillo
Received: 4 May 2020; Accepted: 26 May 2020; Published: 27 May 2020

Abstract: When using gold nanoparticle (AuNP) inks for writing photothermal readable secure information, it is of utmost importance to obtain a sharp and stable shape of the localized surface plasmon resonance (LSPR) absorption bands in the prints. The T increase at a given irradiation wavelength (ΔT_λ) is the retrieved information when printed patterns are interrogated with a laser source. As ΔT_λ is proportional to the absorbance at the wavelength λ, any enlargement or change of the absorbance peak shape in a printed pattern would lead to wrong or unreliable reading. With the aim of preparing AuNP inks suitable for inkjet printing of patterns with stable and reliable photothermal reading, we prepared liquid solutions of spherical AuNP coated with a series of different polymers and with or without additional dispersant. The optical stability of the inks and of the printed patterns were checked by monitoring the shape changes of the sharp LSPR absorption band of AuNP in the visible (λ_{max} 519 nm) along weeks of ageing. AuNP coated with neutral polyethylenglycol thiols (HS-PEG) of mw 2000–20000 showed a strong tendency to rapidly agglomerate in the dry prints. The close contact between agglomerated AuNP resulted in the loss of the pristine shape of the LSPR band, that flattened and enlarged with the further appearance of a second maximum in the Near IR, due to plasmon hybridization. The tendency to agglomerate was found directly proportional to the PEG mw. Addition of the ethylcellulose (EC) dispersant to inks resulted in an even stronger and faster tendency to LSPR peak shape deformation in the prints due to EC hydrophobicity, that induced AuNP segregation and promoted agglomeration. The introduction of a charge on the AuNP coating revelead to be the correct way to avoid agglomeration and obtain printed patterns with a sharp LSPR absorption band, stable with ageing. While the use of a simple PEG thiol with a terminal negative charge, HS-PEGCOO(−) (mw 3000), was not sufficient, overcoating with the positively charged polyallylamine hydrochloride (PAH) and further overcoating with the negatively charged polystyrene sulfonate (PSS) yielded AuNP@HS-PEGCOO(−)/PAH(+) and AuNP@HS-PEGCOO(−)/PAH(+)/PSS(−), both giving stable prints. With these inks we have shown that it is possible to write photothermally readable secure information. In particular, the generation of reliable three-wavelength photothemal barcodes has been demonstrated.

Keywords: gold nanoparticles; photothermal effect; nanoink; inkjet printing; secure writing; anti-counterfeit

1. Introduction

Nanoparticles (NPs) capable of relaxing thermally when irradiated at the wavelength of their absorption bands add interesting photothermal properties to the many peculiarities of matter at the nanoscale. NP of this kind are made of highly absorbing substances such as Prussian Blue [1], copper sulfide [2] or, more frequently, noble metals [3]. In the latter case, the absorption bands responsible of the photothermal effect are due to the well-known phenomenon of localized surface plasmon resonance (LSPR). Excitation at the wavelengths of such bands gives typically a weak fluorescent emission [4] unless pulsed lasers sources are used, generating in the latter case the more intense two photon luminescence (TPL) [5]. Continuous irradiation of LSPR bands with laser or wide spectrum continuous sources leads thus to a largely prevalent thermal relaxation [5]. The maximum of absorption of the LSPR bands (λ_{max}) of Ag and AuNP is a function of their dimensions and shape, that are both tunable during synthesis [6]. In particular, by decreasing the symmetry of Ag and Au nanoparticles from that of a sphere to those of elongated, branched or more complex shapes, such as nanorods, nanostars, nanoplates and nanoshells, the absorption maximum can be easily shifted from the visible to the Near-IR (NIR) range [3,6] including the so-called 'bio-transparent window' (750–900 nm). Following this, many papers have been published proposing antitumoral [7] and antibacterial [8] photothermal through-tissues therapies by using such non-spherical NP. Besides giving a photothermal response in solution and in-vitro or in-vivo wet conditions, dry surfaces coated with patterns of photothermal NP display an even more intense effect [9,10]. With this in mind, we recently proposed a radically different application of the photothermal effect of NIR-absorbing gold nanostars (GNS), i.e. Their use in a new approach in secure writing and anti-counterfeiting applications [11].

Briefly, this approach includes: i) preparing photothermal NP inks, i.e., liquid solutions of NP suitable for inkjet printers; ii) printing patterns on a given bulk substrate (eg paper, glass); iii) interrogating the printed pattern by irradiating it with a laser source at a given wavelength (λ_{exc}); iv) thermally reading the response with a thermocamera, obtaining a thermogram (T vs t, starting from T_0 at t = 0) from which the maximum reached temperature (T_{max}) is determined and the thermal answer is calculated as $\Delta T = T_{max} - T_0$, as sketched in Figure 1A,B.

A simple information is retrieved from this operation, as the ΔT answer may be YES or NO if it is higher or lower than a chosen threshold. In Figure 1, the chosen ΔT threshold is 10 °C, and the exemplificative thermograms of Figure 1A,B give a YES and NO answer, respectively. While we also proposed more sophisticated reading systems to create anticounterfeit secure printed patterns [11] like multiwavelength-interrogation photothermal barcodes, even the simple YES/NO thermal reading on a single irradiated point allows to obtain encoding/reading of secure information, as it requires complex technologies both to be written and to be read, and the correct keys are needed to interpret the thermal response. The keypoint of the present paper is pictorially illustrated in Figure 1C,D. Given a sharp absorption band, the obtained ΔT is strongly dependent on λ_{exc} and once a threshold ΔT has been chosen (10 °C in the exemplifying sketch), a YES answer may become a NO answer even by relatively small changes in λ_{exc} (compare the response of $\lambda_{exc}2$ with that of $\lambda_{exc}1$ and $\lambda_{exc}3$). In this case the knowledge of the correct λ_{exc} (and of the laser irradiance to be used) is one of the keys necessary to a user to verify if a printed point gives the exact expected answer. On the other hand, large absorption bands as in Figure 1D give the same thermal answer (YES, in the Scheme) for a large range of λ_{exc}, eg for the same wavelengths of $\lambda_{exc}1$, $\lambda_{exc}2$ and $\lambda_{exc}3$ of Figure 1C. With such large featureless absorptions the uniqueness of the correct wavelength of the laser source is lost, knocking down the security level of this key.

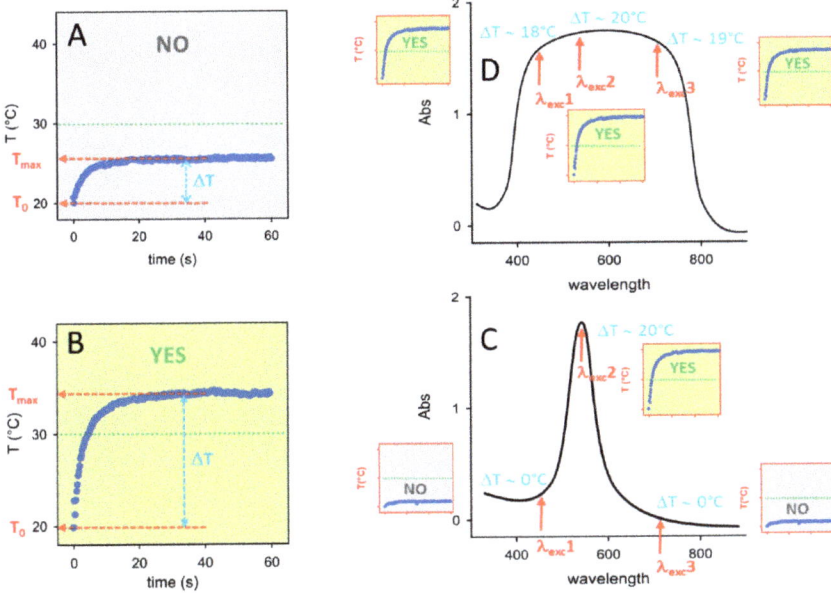

Figure 1. (**A**) and (**B**): examples of typical thermograms, i.e., T vs time profiles obtained irradiating a photothermally responsive printed pattern. ΔT values are calculated as the T differences $T_{max} - T_0$ evidenced by the azure double arrows; T_0 is 20 °C in both examples; a threshold ΔT of 10 °C has been chosen: in panel A ΔT < 10 °C (thermal answer: NO), in panel B ΔT > 10 °C (thermal answer YES). (**C**): visual sketch of how the photothermal answer changes by changing λ_{exc} with a print having a sharp absorption peak: small thermograms in the red-framed squares are what obtained (left to right) with $\lambda_{exc}1$ (ΔT < 10 °C, answer: NO), $\lambda_{exc}2$ (ΔT > 10 °C, answer: YES), $\lambda_{exc}3$ (ΔT < 10 °C, answer: NO). (**D**): same, but with a large featureless absorpion band: at the same three λ_{exc} of panel C all the thermograms (red-framed squares) give ΔT > 10 °C, i.e., a YES answer.

LSPR bands of AuNP are intense and sharp, and, in principle, such NP are ideal for preparing photothermally responsive inks suitable for this approach. However, as we have observed using GNS-containing inks [11], when AuNPs have been printed and the printed pattern is dry, the distance between nanoparticles may get so short that plasmon hybridization takes place [12,13]. This leads to large, featureless absorption bands, completely different from those observed in solution. Moreover, AuNPs are typically prepared in water as colloidal suspensions, while inks for inkjet printers must be liquid mixtures with higher viscosity and lower surface tension than water. Due to this, solvent mixtures have to be used (eg water with alcohols such as isopropanol and ethylene glycol) [11,14] and coatings must be grafted to AuNP to avoid agglomeration and precipitation due to solvent change. In summary, the correct AuNP coating must both grant stability in ink and avoid post-printing hybridization issues, i.e., it must be able to maintain AuNP at a sufficient inter-particle distance once a pattern is printed and dried out. At this regard, ionic polymers have been proposed as optimal coatings for gold nanostars [11] and nanorods [15]. In the present paper we use simple spherical gold nanoparticles (d = 17 nm), whose colloidal solutions have a sharp LSPR absorption in the visible (λ_{max} = 519 nm in water) and we examine a series of different coatings, including neutral or charged thiolated polyethylene glycol polymers (HS-PEG) of increasing length, both in the presence or absence of a typical thickening agent for traditional inks (ethyl cellulose, EC) [16] that may act as a dispersant. After this, we also examine the effect of alternate layers of oppositely charged ionic polymers. With all the coatings we have studied the NP stability in the ink, the sharpness vs flattening (due to hybridization) of the LSPR absorption in the printed patterns and the evolution with time of the LSPR absorption of both inks and prints. Ageing is of course an extremely relevant parameter for any real-life secure writing application. Finally, a proof of concept is presented of how to retrieve a correct or wrong photothermal answer at different λ_{exc} using AuNP inks prepared with optimal or unsuitable coatings.

2. Results and Discussion

2.1. AuNPs

The starting material for all inks preparations was citrate-coated spherical AuNPs prepared by the Turkevich method [17], i.e., by reduction of $HAuCl_4$ with excess sodium citrate in bidistilled water (for details see Materials and Methods). Several 500 mL samples of such AuNP solutions were prepared in the course of the study. In all cases the expected sharp LSPR absorption band typical of small Au nanospheres was observed, with λ_{max} = 519 nm, see Figure 2A, imparting the typical intense purple-red color to the colloidal solutions (Figure 2B).

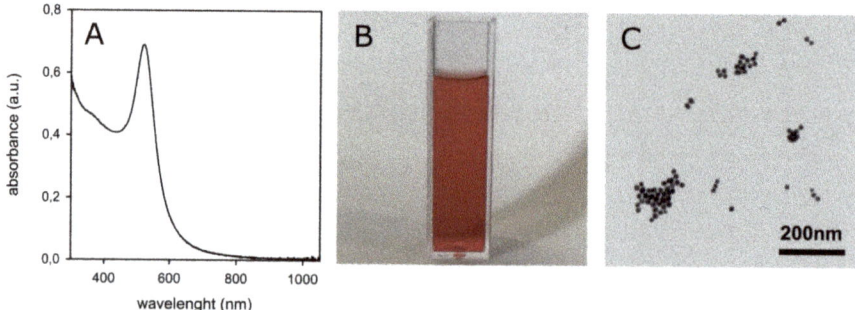

Figure 2. (**A**): absorption spectrum of a freshly prepared solution of citrate-coated AuNP. (**B**): photograph of the same solution. (**C**): TEM image obtained from the same solution.

These aqueous colloidal solutions of citrate-coated AuNP are stable with time (no spectral changes in a 60 days range). Transmission electron microscope (TEM) imaging confirms the expected spheroidal shape of the prepared AuNP, Figure 2C, with average d = 17(±1) nm for all preparations. ζ-potential was −34(±2) mV (average of six preparations), due to the citrate coating. The Au(III) to Au(0) conversion yield can be safely considered ~100%, thanks to the noble nature of gold and to the large excess of the reductant (citrate anion). It has to be noted that following the Turkevich protocol, the total Au concentration in these colloidal solutions is 2.5×10^{-4} M (0.049 mg Au/mL). However, this value is too low for such solutions to be used as an ink component. As an example, we used colloidal solutions with 0.5–0.3 mg Au/mL (corresponding to 16.7–10 nM nanoparticles) for preparing photothermal inks with GNS [11]. Moreover, addition of alcohols is required to tune the viscosity to values suitable for inkjet printers. In this work we adopted a solvent mixture that we have already found to be optimal [9,11], i.e., 70% v/v aqueous AuNP solution, 20% v/v ethylene glycol and 10% v/v 2-propanol, with a viscosity and surface tension 1.92 cP and 40 mN/m, respectively, that is suitable for inkjet printers. Mixing the aqueous colloidal solution to the alcoholic components may induce agglomeration of citrate-coated AuNP. In addition, to prepare inks we needed 10-fold concentrated AuNP solutions (10 × solutions hereinafter), that can be obtained by ultracentrifugation, supernatant removal and pellet redissolution in 1/10 of the initial volume (see Materials and Methods for details). To make AuNP stable during all these procedures we coated them with PEG thiols, obtaining AuNP@HS-PEG.

2.2. AuNPs Coated with Neutral HS-PEG of Different Lengths

We used a series of thiolated PEG of general formula $HS-(CH_2CH_2O)_n-CH_3$, with molecular weights mw = 2000, 5000, 10000 and 20000 (n ~ 44, 112, 226 and 453, respectively). For sake of simplicity, we refer here to such polymers as HS-PEG$_{mw}$. In addition, also the α,ω bifunctional polymer HS-(CH$_2$CH$_2$O)$_n$-CH$_2$COOH was used, with average mw 3000 (n ~ 66), referred to as HS-PEGCOOH in this paper. The –COOH group has typically a pKa of 4–5. Accordingly, in neutral water (pH 7) HS-PEGCOOH is deprotonated, bears a terminal negative charge, and can be referred to as HS-PEGCOO(−). Due to this, the properties of AuNP coated with HS-PEGCOOH are described in the Results and Discussion Section 2.4, that is dedicated to charged coatings, despite of the fact that HS-PEGCOOH has properties similar to those of all the neutral HS-PEG coatings.

The coating step is carried out by adding the chosen HS-PEG in 2×10^{-5} M concentration to a volume of freshly prepared AuNP solution. HS-PEG concentration was chosen with this rationale: a spherical AuNP of 17 nm diameter has a mass of 4.97×10^{-17} g; the Au(0) concentration in the citrate-coated AuNP solutions is 0.049 mg Au/mL; this leads to a 1.64×10^{-9} molar concentration of AuNP; an AuNP of 17 nm has ~ 6×10^3 surface atoms [18], this meaning a concentration of potentially available surface Au atoms in the AuNP solution of ~ 9.8×10^{-6} M. Following also the obvious consideration that, due to steric crowding, not all the Au surface atoms can be coordinated by a thiolate group [19], we considered 2×10^{-5} M as a sufficiently large excess for HS-PEG to saturate the AuNP surface in the coating process. With all the used HS-PEG we observed a ~ 5 nm red-shift of the LSPR band maximum (λ_{max} = 524 nm) on grafting, due to the small local refractive index change when displacing citrate with $^−$S-PEG on the NP surface. As representative of all pegylations, the spectrum of AuNP@HS-PEG$_{5000}$ (10 ×) is compared in Figure 3A (red line) with that of the starting AuNP solution (black line; identical spectra were obtained for all other HS-PEG). It has to be pointed out that the two spectra are recorded in 1 mm and 1 cm cells, respectively, and thus, in principle, their absorbances should be identical. However, pegylation and preparation of 10 × solutions requires repeated ultracentrifugation/redissolution cycles, that slightly decrease the AuNP quantity at each cycle, explaining the difference in Figure 3A. The spectrum of the ink obtained from the 10 × solution is also displayed in Figure 3A (blue line). Ink samples were obtained by adding 100 µL 2-propanol and 200 µL ethylene glycol to 700 µL of a 10 × pegylated AuNP aqueous solution.

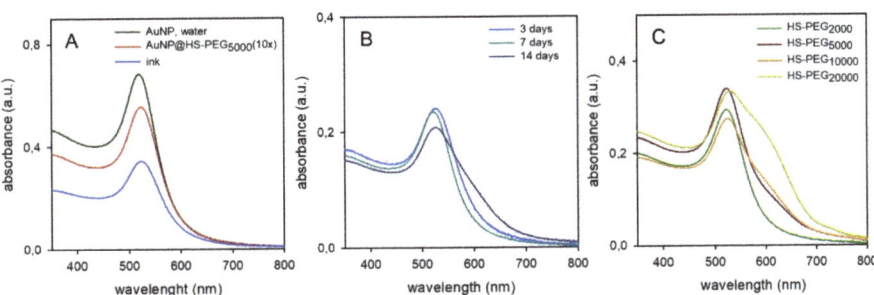

Figure 3. (**A**): absorption spectra of aqueous colloidal solutions of citrate coater AuNP (black) and 10 × AuNP@HS-PEG$_{5000}$ (red) and of the ink obtained from the latter (blue). (**B**): evolution with time (3–14 days) of the absorption spectrum of an ink made with AuNP@HS-PEG$_{2000}$ (time-color correspondances are shown in the palnel). (**C**): absorption spectra after 7 days for inks prepared with AuNP coated with the four different neutral HS-PEG (1 mm cell).

Accordingly, in the just prepared ink the absorbance decrease is due to dilution with alcohols. However, no change in the spectrum shape is observed, indicating stability in the new solvent mixture at least on a short time stint (the spectrum was recorded 1 h after mixing). Also λ_{max} does not shift significantly, as expected from the small refractive index (n_D) differences between water and the additives, (water n_D 1.33, 2-propanol n_D 1.37, ethylene glycol n_D 1.43), from the preponderance of water in the ink and from the small sensitivity to refractive index changes of the LSPR band of gold nanospheres (44 nm/RIU; RIU = refractive index units) [20]. On the other hand, inks show instability on ageing (weeks range), with the ink color changing from red to violet-blue (see SM1). This can be monitored spectroscopically. Figure 3B shows the representative case of AuNP coated with HS-PEG$_{2000}$. While after 3 and 7 days the spectrum was still superimposable on the initial one, after 14 days a shoulder appeared at λ > 650 nm, as a typical indication of AuNP agglomeration [21]. Interestingly, inks prepared with AuNP coated with HS-PEG of increasing molecular weight showed that AuNP with longer PEG coatings undergo more significant AuNP agglomeration, when compared at the same ageing time (Figure 3C, 7 days). This counter-intuitive observation can be rationalized by observing the number of grafted HS-PEG per AuNP, Table 1 (data obtained from thermogravimetric analysis, SM2).

Table 1. Dimensional and chemical-physical data of PEG-coated AuNP.

	Citrate[a]	HS-PEG$_{2000}$	HS-PEG$_{5000}$	HS-PEG$_{10000}$	HS-PEG$_{20000}$
r$_{hyd}$ (nm)[b]	16(1)	23(1)	54	74	93
ζ (mv)	−32(2)	−13(1)	−5(1)	−8(1)	−1(1)
%w/w[c]		10.69(5)	21.95(6)	21.72(3)	29.41(7)
n/NP[d]		1805	1758	890	651

[a] Parent AuNP, with no PEG coating [b] hydrodynamic radius determined in water. [c] % of PEG weight in dry samples; water is adsorbed on the solid samples used for TGA measurements, the % Au is calculated by subtracting the % of PEG and % of water to 100 (see SI for TGA profiles and details); [d] number of polymer chains per NP.

Such a number decreases on increasing the PEG length, most probably due both to hindering of the –SH function of the incoming polymers and to hindering of the available surface on the AuNP by the already grafted chains, two effects that become more significant with increasing the polymer dimensions. The slow agglomeration observed when AuNP coated with high mw PEG are dissolved in the alcohols/water ink mixture corresponds to the situation in which NP with low (i.e., partial) polymer coverage are dissolved in a poor solvent. In this case, we expect to find a minimum with negative free energy for approaching NP that interdigitate their polymer chains at NP-NP distances < 2L (L being the thickness of the coating polymer layer) [22].

Table 1 reports also the hydrodynamic radius and the ζ-potential values of the AuNP with different coatings. Citrate-coated AuNP have a highly negative ζ value (−34 mV) due to the citrate layer. Citrate is displaced when HS-PEGs are added. The latter adhere on the Au surface in the thiolate ⁻S-PEG form and this explains the observed negative ζ values, that however decrease with the increasing coating thickness (the ζ-potential is measured at the slipping plane), erasing any electrostatic contribution to AuNP stability.

Using freshly prepared inks (ageing < 1 day), we printed patterns of ~ 1.0 cm^2 on glass surfaces using the dropcasting protocol described in the Material and Methods Section 3.3.7. Such protocol allowed us to avoid the use of expensive research instruments for inkjet printing (not owned by our laboratories), while also allowing to prepare printed patterns mimicking the actually inkjet-printed ones. We used surface densities similar to those obtained with the optimal parameters standardized on a Dimatix Materials Printer DMP-2800 research inkjet printer (FUJIFILM Dimatix, Inc., Santa Clara (CA), USA) in previous collaborations [9,11]. Such parameters were 10 pL drops with 1681 drops mm^{-2} density and 1–11 printed layers with a 0.42 mg Au/mL ink concentration, corresponding to the 0.71–7.81 µg Au/cm^2 range. In the present paper the inks concentration varied among 0.21 and 0.11 mg Au/mL, depending on the preparation. Using a 40 µL volume of inks, these spread over a ~1 cm^2 surface (see Materials and Methods Section 3.3.7). This allowed us to print surfaces with a Au density between 8.4 µg/cm^2 and 4.4 µg/cm^2. We found a similar behaviour for all prints, almost independently on the mw of the HS-PEG coating (see SM3). Figure 4 shows the representative case of AuNP@HS-PEG$_{5000}$. Printed surfaces were first examined after standard drying (14 h at 40 °C). A λ$_{max}$ red-shift of 20 nm was observed in the absorption spectrum with respect to the liquid ink, blue line in Figure 4A. This is attributable to local refractive index changes, as after drying AuNP are no more dispersed in a water/alcohols mixture but surrounded by the PEG chains, that have an higher refractive index than water (eg n_D = 1.45 for PEG_{200} [23]. AuNP surface wetting by residual ethylene glycol (n_D = 1.43) from the ink formulation should be also taken into account, due to its higher boiling point (197.6 °C) with respect to water and 2-propanol (82.5 °C). A corresponding slight color change is perceivable also to the eye, when a just-dropcasted surface (a 40 µL drop of liquid ink on glass, Figure 4B) is compared with a 14 h-dried out surface, Figure 4C.

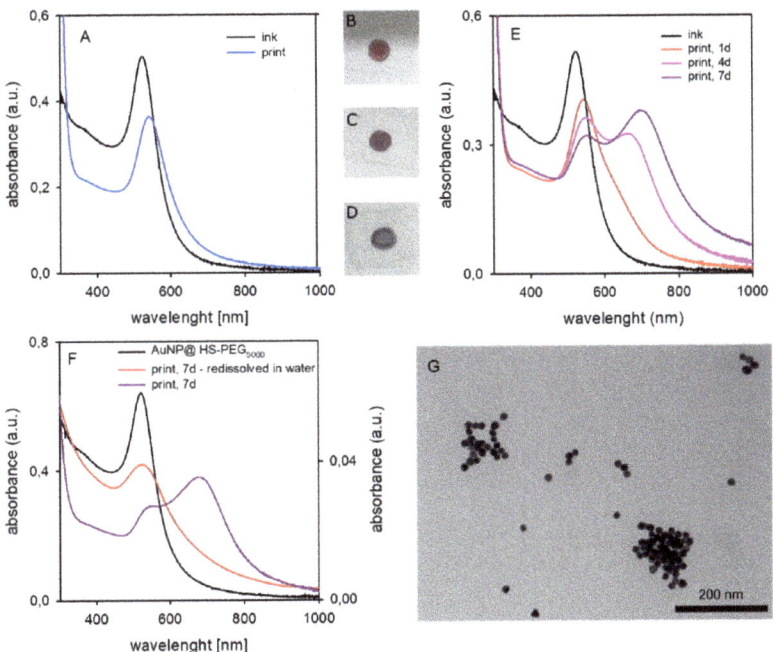

Figure 4. (**A**): absorption spectra of liquid ink containing AuNP@HS-PEG$_{5000}$ (black) and printed on glass (dropcasted ink dried for 14 h, blue). (**B**): a drop of liquid ink just casted on glass. (**C**): same, after 14 h drying (corresponds to the blue spectrum in panel A). (**D**): same, after 7 days ageing (corresponds to the violet spectrum in panel E) (**E**): absorption spectra of the same ink (black) and evolution of the printed pattern after 1–7 days. (**F**): comparison of the absorption spectrum of the AuNP@HS-PEG$_{5000}$ in water (black), of a 7 days aged print (violet), and after redissolution in water of the aged print (red). (**G**): TEM image of the AuNP redissolved from a 7 days aged print.

Following the evolution of the printed pattern with time by absorption spectroscopy (Figure 4E) revealed a progressive enlargement of the LSPR band with the formation and increase of a second maximum at longer wavelengths (λ_{max} = 700 nm after 7 days, dark violet spectrum). This is due to the agglomeration of AuNP, that leads to sufficiently short interparticle distances that LSPR hybridization takes place [13]. In agreement with spectral data, the color of 7 days aged printed patterns turns to blue-violet, Figure 4D. Printed patterns aged 7 days were redissolved in water by prolonged ultrasound treatment. Identically to the parent AuNP@HS-PEG$_{5000}$ aqueous solutions (spectrum added for comparison in Figure 4F, black line), the obtained deep red solution displays an absorption spectrum with λ_{max} at 524 nm, Figure 4F, red line. However, the LSPR absorption is still significant at λ > 600 nm, suggesting a not complete separation into individual AuNP and the persistence of small agglomerates. This is consistent with TEM images obtained from solutions of redissolved AuNP@HS-PEG$_{5000}$ printed patterns, Figure 4G, showing separate and agglomerated AuNP, still maintaining the original dimensions and shape, together with AuNP that have apparently started an authentic aggregation (i.e., fusion) process (see also SM4 for a larger image).

While the general behaviour of printed patterns is similar among all inks made with AuNP coated with neutral HS-PEGs, aggregation and consequent spectral changes parallel what observed for liquid inks, i.e., it is more significant for AuNP coated with the highest molecular weight polymers (see SM3). This clearly states that it is useless to increase the PEG dimensions in the coating to avoid that, in printed patterns, the dry AuNP could come sufficiently close one to the other to give plasmon hybridization. There is an apparent contraddiction with the hydrodynamic radius trend observed in Table 1 for aqueous colloidal solutions, as r_{hyd} increases with PEG mw. However, PEG chains have a good affinity for water, where they tend to outspread, while in the dry printed patterns we can hypothesize that their chains collapse on the AuNP surface, allowing close approach between AuNP.

2.3. Pegylated AuNP Codissolved with EthylCellulose (EC)

In the attempt of obtaining inks that fit the requirements of an inkjet printer while also assuring AuNP separation in the printed patterns, i.e., avoiding LSPR hybridization, we added EC, a non-volatile polymer, to the AuNP liquid inks. The aim was to keep AuNP dispersed in the dry prints, i.e., statistically separated and mechanically immobilized within the dispersant matrix. EC was chosen as it is used as standard dispersant and binder in the formulation of nanoparticle-based inks for inkjet printers [16]. EC scarce solubility in water and aqueous mixtures forced us to use ethanol as the solvent. AuNP@HS-PEG$_{5000}$ were dissolved in ethanol, in which they were stable at least for 24 h, showing a sharp LSPR absorption slightly red shifted with respect to water (λ_{max} 528 nm) due to the refraction index change (see SM5). Addition of 0.1% w/v EC was carried out on 1.5 mL ethanolic AuNP solution, using commercial EC with η (viscosity) of 10, 22, 46 and 100 cp (for all the ECs (ethyl cellulose) samples the nominal viscosity η reported by the seller (Sigma Aldrich, Milano, Italy) in the specification sheets refers to 5% EC solutions in 80:20 toluen/ethanol). The nominal degree of methoxy substitution on the D-glucose units of cellulose is 48% for all the used products, so the differences in η are all attributable to differences in mw, that may be empirically calculated as 243000, 378000, 571000 and 882000 for EC with 10, 22, 46 and 100 cP viscosity, respectively (mw = $k(\eta)^n$, where k and n are empirical constants that depends on the method used to determine mw; mw mentioned in this paper are calculated using typical values reported for EC, $k = 66.96 \times 10^3$, n = 0.56 [24]). In all cases, no significant LSPR shift was observed on EC addition. Ink stability was checked by absorption spectroscopy, observing that after 4 days only the solution with 10 cP EC still showed an acceptable spectrum (comparable to that of the PEG-coated AuNP dissolved in pure ethanol), while solutions with ECs with higher η presented a shoulder at ~ 700 nm, already indicating aggregation (SM6). Nevertheless, glass surfaces were printed using freshly prepared inks with all the four EC additives. A large, shifted absorption band was always observed after the usual 14 h drying process, Figure 5A. We hypothesise that this is due to aggregates of AuNP with different overall shape and dimensions. Further studies were carried out using the EC with lower viscosity (10cp), that was empirically choosen due to the closer similarity of the absorption spectrum of its printed patterns (Figure 5A (i), λ_{max} = 580 nm) with the original AuNP LSPR band. Ethanolic inks containing EC 10cP in different w/v percentages (0.05–0.3%) were prepared and glass surfaces printed. Absorption spectra after an ageing time of 7 days are shown in Figure 5B for 0.1, 0.2 and 0.3% solutions.

A few observations can be made here. First, the use of pure ethanol with added EC caused the spreading of the dropped ink on a larger area (actually all the area delimitated by a PDMS fence, see Materials and Methods) and resulted in a strong coffee stain effect, as it can be seen by the photographs of the printed slides (Figure 5B, insets). Prints changed to a blue color, as expected from the large range of absorption that includes visible and NIR, and a more significant λ_{max} red shift was observed in the spectra for higher EC concentrations. These results discouraged the use of hydrophobic polymeric additives in inks, at least as long as AuNP are coated with hydrophilic polymers. Second, AuNP dramatically agglomerate in the liquid ink and in the dried out printed patterns, with the complete loss of the original sharp LSPR band shape. The longer was the EC polymer chain and the higher was the additive concentration, the more intense was this phenomenon. Although we have discarded and not further investigated these systems, we can hypothesize that the presence of the non-volatile hydrophobic polymer in the printed patterns promotes segregation of the hydrophilic AuNP during the drying process, with the formation of clusters of nanoparticles of different shapes and dimensions. This hypothesis is consistent with the increasing degree of LSPR red shift with increasing EC concentration. SEM images on printed patterns allow to visualize the EC matrix but also the AuNP, see Figure 5C (0.2% EC printed sample). AuNP can be spotted both as isolated particles and as aligned or cropped groups (see in particular the enlarged section framed in red). The different type and degree of aggregation is consistent with the large, featureless absorption band, that is generated by the superimposition of different LSPR hybridizations.

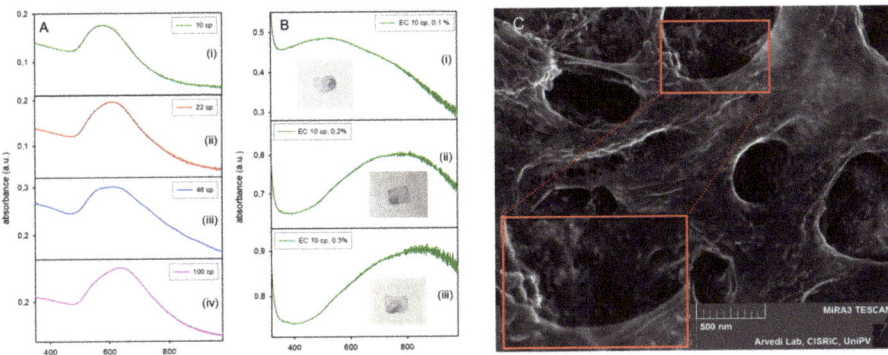

Figure 5. (**A**): absorption spectra of fresh prints (i.e., after 14 h at 40 °C) obtained with AuNP@HS-PEG$_{5000}$ in ethanol with 0.1% EC of different viscosity (viscosity values are indicated in the panel). (**B**): absorption spectra of 7 days aged prints obtained with AuNP@HS-PEG$_{5000}$ in ethanol with 10 cP EC at 0.1–0.3 % concentration. (**C**): SEM image taken on the print (ii) in panel (**B**) (EC 10 cP, 0.2%). The lower red-framed area is a 2 × enlargement of the upper red-framed area.

2.4. Coating with Ionic Polymers

The next step of this study was to coat AuNP with ionic, charged polymers. A first HS-PEGCOOH (mw 3000) layer was grafted on the surface as already described for the neutral HS-PEG, in order to have both sterical stabilisation and a significantly negative ζ-potential for the AuNP. We measured ζ = −22 mV in neutral water. The negative charge is due to the carboxylic acid moieties, that are fully deprotonated at pH 7, as the typical pKa of a –COOH groups is in the 4–5 range (eg the pKa of acetic acid is 4.76). Accordingly, we describe AuNP coated with this polymer as AuNP@HS-PEGCOO(−), see sketch in Figure 6. The ζ value is consistent with those reported in Table 1 for neutral HS-PEG coatings with mw 2000 and 5000 (−13 and −6 mV, respectively) as in the latter cases the weakly negative values are due to the residual influence at the slipping plane of the remote negative thiolate groups grafted on the AuNP surface. From TGA analysis (SM2E-F) we calculated the number of polymers per AuNP as 1532, slightly lower than what found for the longer but neutral HS-PEG$_{5000}$. This is coherent with what we observed with HS-PEGCOOH and GNS [25] and is attributable to the electrostatic repulsive effect between grafted and incoming polymers. Overcoating with PAH, a positively charged ionic polymer, was obtained by electrostatic adhesion of the polymer chains to the negatively charged AuNP@HS-PEGCOO(−). PAH was added in 2 × 10^{-5} M concentration to AuNP@HS-PEGCOO(−) solutions, as this quantity demonstrated sufficient to obtain the maximum PAH coating in the AuNP@HS-PEGCOO(−)/PAH(+) complexes (see Figure 6 for a sketch). These have a ζ-potential of +36(2) mV after 2 h equilibration at room temperature and two cycles of ultracentrifugation-redissolution in bidistilled water to remove non adhering PAH (final pH in the 5–6 range). The use of larger concentrations of PAH did not lead to an higher ζ-potential, suggesting complete coating with the chosen concentration. Finally, these complexes were also further overcoated with the negatively charged polymer PSS using an identical synthetic procedure to obtain the AuNP@HS-PEGCOO(−)/PAH(+)/PSS(−) complex (sketch in Figure 6). This displayed a negative ζ-potential of −27(2) mV (final pH ~ 6). It has to be pointed out that in the slightly acidic pH range of these solutions and up to strongly basic pH values (eg 9 or higher), the ζ-potential values do not change, as all the polymers maintain their ionic state and their typical charge. First, as we have already discussed, a −COOH function is prevalently deprotonated at pH > 5. Then, PAH is a polyamino polymer containing only −NH$_2$ functions, and protonated primary amines have typically pKa values > 10 (eg pKa = 10.60 for n-butylammmonium [26]. Finally, PSS has −SO$_3^-$ groups that are deprotonated up to pH 14 (eg, benzenesulfonic acid is a strong acid [27]). The absorption spectra in water show a sharp LSPR maximum for all complexes, with negligible λ_{max} variations (< 5 nm) with any of the coatings. TGA and DLS measurements (Table 2) gave consistent results, showing an increasing quantity of coating material both as an absolute value and in the ratio with respect to the Au content, and an increase of the hydrodynamic radius on stepping from AuNP@HS-PEGCOO(−) to AuNP@HS-PEGCOO(−)/PAH(+) and to AuNP@HS-PEGCOO(−)/PAH(+)/PSS(−). Ink preparation was carried out using again the standard formulation of 70%(v/v) aqueous component (containing the coated AuNP), 20% v/v ethylene glycol and 10% v/v 2-propanol. Prior to mixing with alcohols, 10 × aqueous solutions of coated AuNP were analyzed by ICP-OES to determine Au concentration (data listed in Table 2). Inks were prepared either with the pure colloidal solution (AuNP@HS-PEGCOO(−)/PAH(+) case) or with the colloidal solutions diluted with a calculated volume of bidistilled water (AuNP@HS-PEGCOO(−) and AuNP@HS-PEGCOO(−)/PAH(+)/PSS(−) cases) so to have all inks with the same Au concentration of 0.11 mg/mL.

Table 2. Dimensional and chemical-physical data of AuNP coated with charged polymers.

	AuNP@HS-PEGCOO(−)	AuNP@HS-PEGCOO(−)/PAH(+)	AuNP@HS-PEGCOO(−)/PAH(+)/PSS(−)
r_{hyd}(nm)[a]	58(2)	178(8)	197(6)
ζ (mv)	−22(2)	+36(2)	−27(3)
%w/w[b]	12.27(4) Au: 86.61(4)	18.17(3) Au: 76.73(3)	20.27(4) Au: 74.46(4)
Au conc[c]	1.07×10^{-3} M (0.211 mg/mL)	7.88×10^{-4} M (0.155 mg/mL)	9.38×10^{-4} M (0.185 mg/mL)
λ_{max}(nm)[d]	524	534	526

[a] Hydrodynamic radius determined in water. [b] % of organic material and % of gold in the dry products, determined by TGA; water is adsorbed on the solid samples, the % Au is calculated by subtracting the % of organic matter and % of water to 100; [c] Au concentration (by ICP-OES) in the 10-fold preconcentrated aqueous solutions, before ink formation; [d] maximum of absorption measured in the ink solvent mixture.

The lower Au concentration in the 10× aqueous solution of AuNP@HS-PEGCOO(−)/PAH(+) is due to the repeated ultracentrifugation cycles required to remove the excess PAH and to the tendency of PAH-coated NP to adhere to the plastic walls of the test tubes, resulting in a less efficient redissolution [25]. Table 2 also reports λ_{max} values of inks absorption spectra recorded 1 h after preparation. Small λ_{max} variations were observed with respect to water, but a sharp LSPR band of identical shape as in water (Figure 6A) was obtained in all cases. However, for the ink containing AuNP@HS-PEGCOO(−), similarly to what described for all the PEG-coated AuNP, ink ageing (7 days) resulted in the formation of a shoulder at longer wavelength, Figure 6B. On the contrary, the spectra of inks with AuNP overcoated with PAH and PSS did not change in 7d (Figure 6B), indicating ink stability. Printing was carried out with the usual procedure using freshly prepared inks, and the stability of the printed patterns was monitored again by absorption spectroscopy. Figure 6C–E compares the spectra recorded 1 day and 14 days after printing for the three inks. While prints with the AuNP@HS-PEGCOO(−) ink showed the growth of a band at longer wavelength (λ_{max} 720 nm), due to AuNP agglomeration, both inks with AuNP overcoated with ionic polymers displayed an excellent peak shape constancy. The insets of Figure 6C–E display the visual aspect of prints after 14 days. Prints with the AuNP@HS-PEGCOO(−) ink turned into a blue-violet color, while the original red-purple color was maintained for prints with the inks containing AuNP further coated with PAH and PSS. To further consolidate this observation, Figure 6F–G display SEM images taken on printed glasses (14 days ageing). A sharp difference was observed between an AuNP@HS-PEGCOO(−) print, Figure 6F, and an AuNP@HS-PEGCOO(−)/PAH(+)/PSS(−) print, Figure 6G. In the former, AuNP are thoroughly aggregated, while in the latter they are all sharply separated. A SEM image for a 14 days aged print from a AuNP@HS-PEGCOO(−)/PAH(+) ink, included in the Supplementary Materials (SM7), showed analogous features as those obtained with the AuNP@HS-PEGCOO(−)/PAH(+)/PSS(−) ink.

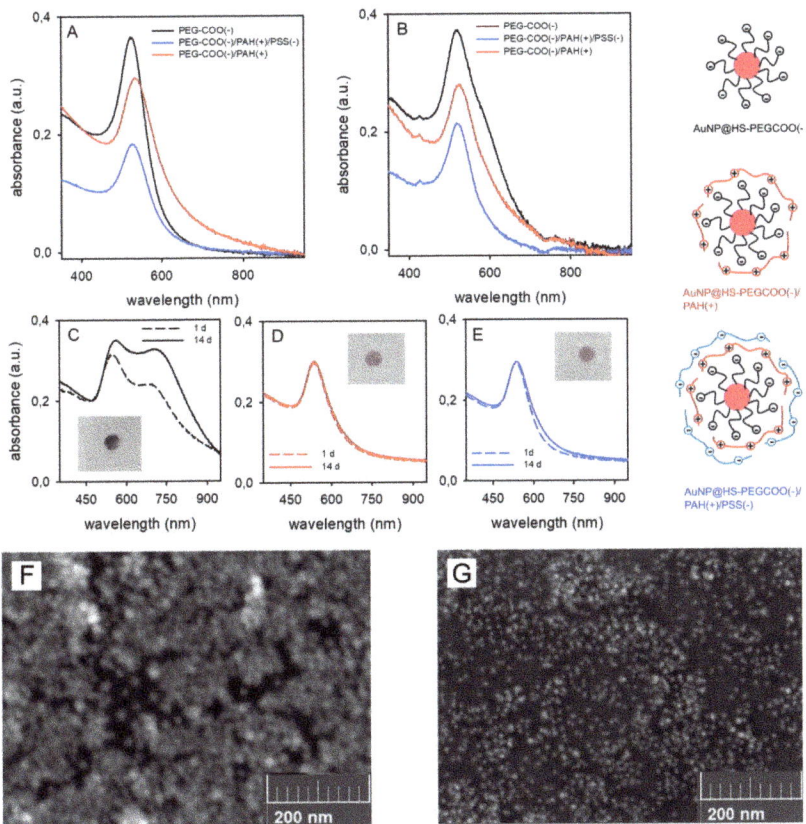

Figure 6. (**A**): absorption spectra of as-prepared inks; (**B**): inks after 7 days ageing; (**C–E**) absorption spectra of dry printed patterns after 1 day (dashed lines) and 14 days ageing (solid lines) for AuNP@HS-PEGCOO(−) (**C**), AuNP@HS-PEGCOO(−)/PAH(+) (**D**), AuNP@HS-PEGCOO(−)/PAH(+)/PSS(−) (**E**); insets are photos taken on 14 days aged prints; F-G: SEM imaged of 14 days aged prints with AuNP@HS-PEGCOO(−) (**F**) and AuNP@HS-PEGCOO(−)/PAH(+)/PSS(−) (**G**).

Noticeably, also in the case of the AuNP@HS-PEGCOO(−) prints, redissolution in water reverted agglomeration, yielding well separated AuNP (TEM image in SM8) with an absorbance identical to that recorded before ink formation (SM8).

2.5. Photothermal Reading of Secure Information

Once we obtained inks that give prints with the desired spectral stability, we carried out a proof-of-concept study on writing secure photothermal information. We used both AuNP@HS-PEGCOO(−)/PAH(+) and AuNP@HS-PEGCOO(−)/PAH(+)/PSS(−) inks, and compared the results with a print of the AuNP@HS-PEGCOO(−) ink. The photothermal response was recorded with interrogation of printed patterns with 1 day and 14 days ageing. The photothermal studies were carried out by irradiating printed patterns with laser sources at three different wavelengths (λ_{irr}), 488, 514 and 720 nm. This is the set of sources that falls inside the absorption range of the prints and that is currently available in our laboratories. We followed a protocol that we have already successfully adopted [10,11]. Briefly, using a E40 thermocamera (FLIR System, Inc., Santa Barbara, CA, USA)) we read a 320 × 240 pixels thermal image, inside which we define a ROI (region of interest) that includes the laser-irradiated area. We run data analysis determining the maximum temperature inside the ROI (± 0.1 °C accuracy). At a λ_{irr} at which the prints display a significant LSPR absorption, a typical steep ascending T vs time profile is observed on irradiation, turning into a plateau in less than 10 s (thermograms sketched in Figure 1 give qualitatively similar examples; SM9 reports actual thermograms for this study). From such data we obtain $\Delta T_\lambda = T_{max} - T_0$, with T_{max} = temperature of the plateau at a given λ of irradiation and T_0 = temperature before irradiation. The ΔT_λ data for prints with the three inks are displayed in Figure 7, superimposed to the absorption spectra of the prints.

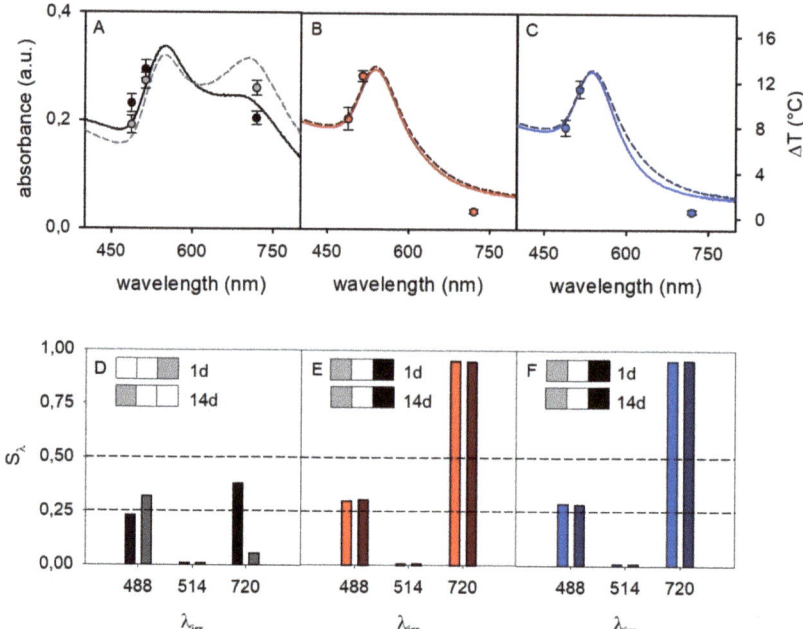

Figure 7. (**A–C**): absorption spectra (left vertical axis; full lines 1 day after print, dashed lines 14 days after print) and ΔT_λ values (circles, right vertical axis) for prints on glass slides with inks containing AuNP@HS-PEGCOO(−), AuNP@HS-PEGCOO(−)/PAH(+) and AuNP@HS-PEGCOO(−)/PAH(+)/PSS(−), respectively; panel A displays ΔT_λ values both for 1 day after print (black circles) and for 14 days after print (grey circles), while in panel (**B**) and (**C**) the ΔT_λ values at 14 days are not displayed as they are superimposable to those at 1 day. (**D–F**): S_λ values and relative three-wavelength barcordes for the same sequence of prints, after 1 day (left bars at a given wavelength) and after 14 days (right bars).

The expected ΔT_λ vs λ_{irr} trend was observed, approximately following the absorbance spectra profile. Noticeably, comparison of spectra at 1 day and 14 days after printing (solid and dashed lines, respectively) evidenced a perfect stability of the prints with inks of AuNP overcoated with PAH and PAH/PSS, while a very significant further spectral change took place for AuNP@HS-PEGCOO(−) ink prints. Among the three inks prints, the most significant differences were observed when irradiating at 720 nm. The absorbance at 720 nm was high for prints with AuNP@HS-PEGCOO(−) inks after 1 day, due to aggregation and plasmon hybridization, and evolved to an even higher absorbance after 14 days. As a consequence $\Delta T_{720} = 8.4\ °C$ (1d) and $11.4\ °C$ (14 days), Figure 7A. On the other hand, in the prints obtained with inks of AuNP@HS-PEGCOO(−)/PAH(+) and AuNP@HS-PEGCOO(−)/PAH(+)/PSS(−) the coatings prevented aggregation and plasmon hybridation, keeping the original shape of the band unchanged after both 1 day and 14 days. Accordingly Abs_{720} is negligible and ΔT_{720} too ($< 1\ °C$). As an example of application of the YES/NO reading scheme highlighted in the Introduction [11], if $\Delta T = 5\ °C$ was chosen as the threshold for a YES answer with $\lambda_{irr} = 720$ nm, the prints with AuNP@HS-PEGCOO(−)/PAH(+) and AuNP@HS-PEGCOO(−)/PAH(+)/PSS(−) inks would give a NO answer, that is the correct one for an unaltered print. On the other hand, prints with the AuNP@HS-PEGCOO(−) ink, would give a YES answer, that is wrong (altered print). Beside the application of this simple concept, a robust way to obtain a more complex information from photothermal data has been introduced in our previous papers [11], that eliminates possible uncertainties due to laser power oscillations or ink concentration. A multiwavelength test with multiple levels can be set by defining a normalized temperature signature:

$$S_\lambda = (\Delta T_{\lambda,max} - \Delta T_\lambda)/ \Delta T_{\lambda,max}$$

where $\Delta T_{\lambda,max}$ is the largest T increase among those obtained with the available set of laser sources (ΔT_{514} in our case) and ΔT_λ is the T increase obtained when irradiating at any wavelength λ. A three level answer generating a three-color barcode can be established by choosing such levels as:

$S_\lambda \leq 0.25$ barcode color: white

$0.25 < S_\lambda \leq 0.5$ barcode color: grey

$S_\lambda > 0.5$ barcode color: black

In this study we have only three available laser sources, so we can generate a barcode with three bars. However, these are enough to establish a clear difference between a correctly printed surface, like those with AuNP@HS-PEGCOO(−)/PAH(+) and AuNP@HS-PEGCOO(−)/PAH(+)/PSS(−) inks, and a print with unsufficiently protected AuNP i.e., that with AuNP@HS-PEGCOO(−) ink.

Figure 7D–F show the S_λ values for the three prints and the corresponding three-wavelengths barcodes. Identical barcodes are obtained for the PAH- and PAH/PSS-overcoated AuNP inks (Figure 7E,F), that do not change after 14 days. On the contrary, the print with the AuNP@HS-PEGCOO(−) ink gives a different barcode at 1 day that, in addition, changes with time (14 days). It has to be noted that this result has been obtained by choosing a given surface concentration of gold, the three irradiation wavelengths and the three S_λ levels. Even considering a very simple printed pattern as in this work, all these parameters are the keys in which information can be hidden, i.e. Their values are all needed both to write and to read the correct multiwavelength barcode. Of course, a stable ink that produces a print with a sharply peaked and time stable absorption is necessary. We have here shown that inks made of AuNP@HS-PEGCOO(−)/PAH(+) and AuNP@HS-PEGCOO(−)/PAH(+)/PSS(−) are suitable for this purpose. Finally, it has to be stressed that periodical check along a 2 months period of the inks and of the prints obtained with the PAH and PAH/PSS coated AuNP gave unmodified spectra.

3. Materials and Methods

3.1. Materials

Tetrachloroauric acid trihydrate (99.99%), sodium citrate dihydrate ($\geq$ 99%), Triton X-100 (laboratory grade), sodium borohydride ($\geq$ 98%), silver nitrate (> 99%), poly(ethylene glycol) methyl ether thiol (mw 2000), poly(allylamine) hydrochloride (PAH, mw 50000), sodium poly(4-styrene sulfonate) (PSS, mw 70000), nitric acid ($\geq$ 65%), nitric acid (1M), hydrochloric acid ($\geq$ 37%), sulfuric acid (95.0–97.0%), hydrogen peroxide (30% w/w), ethanol ($\geq$ 99.8%), 2-propanol (99.5%), and ethyl cellulose (10 cP; 22 cP; 46 cP; 100 cP) were bought from Sigma Aldrich (Milano, Italy); α-mercapto,ω-carboxy poly(ethylene glycol) (mw 3000), poly(ethylene glycol) methyl ether thiol (mw 5000), poly(ethylene glycol) methyl ether thiol (mw 10000), and poly(ethylene glycol) methyl ether thiol (mw 20000) were bought RAPP Polymere (Tübingen, Germany); Ethylene glycol (99.5%) and ammonia solution in water (30% w/w) were bought from Carlo Erba Reagenti S.p.A. (Milano, Italy) Glass coverslides (22 × 26 mm, 0.14 mm thickness) were bought from Delchimica Scientific Gòassware (Napoli, Italy).

3.2. Instrumentation

We used a Sonorex sonicating bath (Bandelin electronic, Berlin, Germany). Ultracentrifugation was carried out with a Z366 ultracentrifuge (Hermle Italia, Rodano, Italy). UV-Vis-NIR absorption spectra were recorded on Cary 50 and Cary 6000 spectrophotometers (Varian, Agilent, Santa Clara (CA), USA); spectra on solutions were taken in 1 cm or 1 mm glass or quartz cuvettes, spectra on glass slides were taken using a dedicated Varian sample holder). Measurement of pH was carried out with a pH-meter (pH 50 model, XS Instruments, Carpi, Italy) with an Orion 91022 BNWP combined glass electrode (Thermo Scientific, Monza, Italy) that was calibrated before measurements with solutions buffered at pH = 4 and pH = 7. Dynamic Light Scattering (nanoparticles dimensions) was performed with a Zetasizer Nano ZS90 instrument (Malvern Panalyticals, Malvern, UK); the same instrument was used for ζ-potential measurements, by using a dedicated dipcell electrode (1 mL samples of colloidal solutions for both techniques). ICP-OES analysis were carried out on an Optima 3000 DW system (Perkin Elmer Italia, Milano, Italy). SEM images were acquired with a Mira XMU series field emission scanning microscope (FEG-SEM) (Tescan, Brno, Czech Republic) at the Arvedi Centre (University of Pavia). Thermogravimetric analyses were carried out on Q5000 instrument (TA Instruments, New Castle (DE), USA). TEM images were taken on JEM-1200 EX II 140 instrument (Jeol Italia SPA, Basiglio, Italy) at the University of Pavia Centro Grandi Strumenti (CGS), on parlodion-coated copper grids, after drying a dropping of 10 microliters of solution. Thermograms were recorded with a FLIR E40 thermocamera driven by the Flir Tools + dedicated software.

3.3. Methods

3.3.1. Glassware and Glass Coverslides Cleaning

Before any synthesis involving nanoparticles, glassware underwent a purification procedure to remove any possible traces of metal ions [28], i.e., it was rinsed with bidistilled water and then cleaned by filling it with aqua regia (3:1 v/v HCl 37% and HNO 3 65%) for 20 min, after which time the oxidant mixture was removed and the glassware filled with bidistilled water and sonicated for 3 min. The last water washing procedure was then repeated 2 further times. Finally glassware was kept at 120 °C in an oven for 2 h. Coverslides were washed with the same procedure, using staining jars filled with aqua regia, in which six or eight coverslides were kept in a vertical position.

3.3.2. AuNP Synthesis

Spherical AuNP were prepared with the standard Turkevich method [17]. Briefly, 87 µL of a 1.44 M tetrachloroauric acid solution in water were added to 500 ml of boiling bidistilled water. Heating was then switched off and under 25 mL of a $1.7 \cdot 10^{-2}$ M sodium citrate dihydrate solution in water were added under magnetic stirring. The synthesis was considered complete after further 2h of stirring. Total Au concentration in the final solution is 2.4×10^{-4} M.

3.3.3. AuNP Coating with HS-PEGs

Pegilation (surface coating with HS-PEGs) was obtained by treating at room temperature a given volume of the citrate-coated AgNP stock colloidal solution with a quantity of the chosen HS-PEG (added as a solid) so to reach a 2.0×10^{-5} M concentration. The obtained solution was then stirred at RT for 3 h, after which time the excess HS-PEG was removed by ultracentrifugation. 10 mL plastic ultracentrifuge test tubes were used, at 13000 rpm for 25 min. After this time the product was found at the bottom of the rest tube a pellet of purple powder. The supernatant was carefully removed with a

Pasteur pipette and the pellet redissolved in 10 mL bidistilled water. The process was repeated two more time. To obtain 10 × concentrated AgNP solution the pellet was redissolved in 1 mL bidistilled water Pegilation (surface coating with HS-PEGs) was obtained by treating at room temperature a given volume of the citrate-coated AgNP stock colloidal solution with a quantity of the chosen HS-PEG (added as a solid) so to reach a 2.0×10^{-5} M concentration. The obtained solution was then stirred at RT for 3 h, after which time the excess HS-PEG was removed by ultracentrifugation. 10 mL plastic ultracentrifuce test tubes were used, at 13000 rpm for 25 min. After this time the product was found at the bottom of the rest tube a pellet of purple powder. The supernatant was carefully removed with a Pasteur pipette and the pellet redissolved in 10 mL bidistilled water. The process was repeated two more time. To obtain 10 × concentrated AgNP solution, at the end of the last cycle the pellet was redissolved in 1mL bidistilled water.

3.3.4. Overcoating with Charged Polymers

The first step required the pegylation of AuNP with HS-PEGCOOH (mw 3000), that was carried out as described in the previous section. After the last ultracentrifugation cycle the AuNP@HS-PEGCOO(−) pellet was redissolved in 10 mL bidistilled water. More 10 mL portions were prepared from the same AuNP stock colloidal solutions, and these were gathered to prepare a 100 mL stock solution of the pegylated particles. After this, a 2×10^{-3} M solution of PAH in bidistilled water was prepared from the commercial product and the quantity needed to obtain a 2×10^{-5} M solution was added with a micropipette to 50 mL of the AuNP@HS-PEGCOO(−) solution. The reaction mixture was alowed to equilibrate for 2 h at RT. After that time, two ultracentrifugation cycles with supernatant discard and redissolution in the same volume of bidistilled water were carried out to remove the PAH excess, obtaining a solution of AuNP@HS-PEGCOO(−)/PAH(+). The same procedure was carried out on the latter solution adding PSS at a 2×10^{-5} M conc, thus preparing AuNP@HS-PEGCOO(−)/PAH(+)/PSS(−). For all ink preparations, a 10 mL volume of a solution of AuNP with the chosen charged coating was ultracentrifuged again, and the pellet redissolved in 1 mL bidistilled water.

3.3.5. Ink Preparation

Inks were prepared by in 1.0 mL volumes by adding 200 µL ethylene glycol and 100 µL 2-propanol to 700 µL of 10 × concentrated AuNP solution, followed by gentle mixing on a reciprocal shaker at RT for 2 min. Larger volumes were prepared maintaining the same volume proportions.

3.3.6. Inks with Ethyl Cellulose.

Due to the negligible solubility of EC in water, a different formulation was used. First, 5 mL of AuNP@HS-PEG$_{5000}$ colloidal solution were added to 5 mL ethanol in ultracentrifuge tubes, followed by 13000 rpm 30 min ultracentrifugation, after which the supernatant was discarded and the pellet redissolved in 10 mL ethanol. A second ultracentrifugation followed, after which the pellet was redissolved in 0.5 mL ethanol to obtain a 10 × AuNP@HS-PEG$_{5000}$ solution. This process was carried out simultaneously on 10 ultracentrifuge tubes and the 10 × portions were gathered to obtain 5 mL of concentrated ethanol solution. 1 mL ink samples were prepared by adding EC as a solid with weight/volume percent of 0.05–0.3% (0.5 to 3 mg/mL).

3.3.7. Printing on Glass Slides

Forty µL of the chosen ink were dropcasted on a glass coverslide that was treated as described in Section 3.3.1. The coverslides still bearing a liquid drop of ink were transferred in an oven thermostatted at 40 °C carefully maintaining a horizontal set up. After 14 h the samples were considered dry and ready for characterization. A circular print of diameter 11–12 mm diameter was obtained (corresponding to a ~ 1cm^2 area; in the rare event of prints with different shape or dimensions, these were discarded). When using the less viscous ethanol inks (EC containing inks), we drop the 40 µL drop inside a polydimethylsiloxane (PDMS) square well (1 cm side) drawn on the glass coverslide. This was done to limit the liquid spreading and maintain the print on the desired surface area (~ 1 cm^2), when using such inks. The square well was drawn by hand with a thin brush and a freshly prepared PDMS solution, obtained by mixing in 10:1 ratio a SYLGARDTM 184 Silicone Elastomer base and its dedicated SYLGARD curing agent (both purchased from Sigma Aldrich, Milano, Italy). PDMS polymerisation was completed with 3 min at 140 °C.

3.3.8. Determination of the Gold Content in Inks

The actual gold content in inks was determined to be sure that it was in the 0.16–0.21 mg/mL range. When lower concentrations were found, the ink was discarded and the colloidal aqueous solution used for its preparation was reconcentrated by ultracentrifugation. For quantitative analysis 200 μL of ink was treated with 500 μL aqua regia, observing immediated decoloration (Au oxidation). After 1 h, the sample was diluted to 10 mL with bidistilled water and then analysed for the gold content with an ICP-OES instrument.

3.3.9. Sample Preparation for Scanning Electon Microscope (SEM) Imaging.

Printed glass coverslided were cut to obtain a 1 cm^2 section, then coated with a 5 nm graphite layer by sputtering before imaging.

3.3.10. Thermogravimetric Analysis (TGA)

One hundred mL of 1 × colloidal solution of AuNP with the chosen coating were subdivided into 10 ultracentrifuge tubes of 10 mL. After the UC, the pellets were redissolved in 1 mL bidistilled water, transferred to 2.5 mL Eppendorf vials and ultracentrifuged for 25 min at 13000 rpm. After discarding most of the supernatant, the 10 samples were carefully gathered using a micropipette in a single Eppendorf vial and dry blown with a nitrogen flow, obtaining 2–3 mg solid samples. Thermogravimetric analysis was performed in a Q5000 system by TA Instruments by heating the obtained powders in a Pt crucible from room temperature up to 600 °C at 5 °C/min under nitrogen flux. The data were elaborated by Universal Analysis v.4.5A by TA Instruments and the mass loss and temperature at mass loss values were evaluated considering the DTG signals

3.3.11. Measurement of the Photothermal Effect

We used a FLIR E40 thermocamera with Flir Tools+ dedicated software for data acquisition and analysis. Thermal images were 320 × 240 pixels. For each thermogram a ROI was defined, that included the laser-irradiated area. Data analysis allowed to determine the maximum temperature inside the ROI (± 0.1 °C accuracy) for each thermal image. In a typical thermogram a thermal image was aquired every 0.25 s for 60–120s. All the laser sources have a beam waist of 1 cm, power was 90 mW for the 488 and 520 nm sources and 116 mW for the 720 nm source (ΔT was consequently normalized).

4. Conclusions

In this research work we have examined different polymeric coatings for spherical AuNPs, with the aim of maintaining their spectral stability both in liquid inks and in dry prints, as this is a mandatory feature for their use in secure writing of photothermally readable information. The traditional linear HS-PEG coatings (mw 2000–20000) were found unsuitable, as they demonstrated uncapable to avoid agglomeration of the coated AuNP both in the liquid ink and in the dry prints. Moreover, we also found a counter-intuitive trend of agglomeration promotion with the increase of the polymer length (i.e., mw), that is due to a less efficient grafting of HS-PEG on AuNP surface with increasing mw. Also the tested dispersant polymer, EC, revealed unefficient in keeping AuNP separated in the prints. On the contrary, the presence of a poorly hydrophilic matrix as EC promotes segregation of the hydrophilic AuNP (coated with HS-PEG$_{5000}$) and their agglomeration. Effective coatings for maintaining a sufficent interparticle separation and avoid LSPR hybridization were instead found when AuNP bearing a first

HS-PEGCOO(−) coating (necessary to introduce an high negative ζ-potential) were overcoated with the positively charged ionic polymer PAH or with PAH and a further overcoating with the negatively charged ionic polymer PSS. In both cases the high ζ-potential and the thickened coating layers contributed to keep AuNP at a sufficiently large distance to avoid the hybridization of their LSPR bands. Accordingly, prints with AuNP@HS-PEGCOO(−) have a poorly efficient protection against AuNP agglomeration and give a photothermal information reading radically different with respect to what expected from their absorbance as inks; morevoer, their photothermal information reading changes with time. Affordable and time-stable prints are instead obtained both with AuNP@HS-PEGCOO(−)/PAH(+) and AuNP@HS-PEGCOO(−)/PAH(+)/PSS(−), from which durable multiwavelength photothermal barcodes can be read after the definition of a number of security keys.

Supplementary Materials: The following are available online at http://www.mdpi.com/1420-3049/25/11/2499/s1, SM1. Inks ageing for AuNP@HS-PEG$_{X000}$ (X = 5, 10, 20), spectra and photographies; SM2. TGA (thermogravimentric analysis) profiles; SM3. Absorption spectra of prints with AuNP@HS-PEGmw; SM4. Larger TEM image of AuNP@HS-PEG$_{5000}$ redissolved in water after printing; SM5. Absorption spectrum of AuNP@HS-PEG$_{5000}$ in ethanol; SM6. Absorption spectra of AuNP@HS-PEG$_{5000}$ in EtOH with added EthylCellulose; SM7. SEM image of a 14 days aged print with AuNP@HS-PEGCOO(−)/PAH(+) ink; SM8. UV-Vis absorbance spectra and TEM images of redissolved AuNP@HS-PEGCOO(−) prints; SM9-Thermograms

Author Contributions: Conceptualization, P.P. and G.C.; methodology, M.B.; investigation, L.D.V., F.M. and C.M.; data curation, A.T.; writing—original draft preparation, P.P.; writing—review and editing, G.C.; funding acquisition, P.P. All authors have read and agreed to the published version of the manuscript.

Funding: This research was funded by University of Pavia–Blue Sky Research 2017, grant number BSR1774514.

Conflicts of Interest: The authors declare no conflict of interest

References

1. Dacarro, G.; Taglietti, A.; Pallavicini, P. Prussian Blue Nanoparticles as a Versatile Photothermal Tool. *Molecules* **2018**, *23*, 1414. [CrossRef] [PubMed]
2. Goel, S.; Chen, F.; Cai, W. Synthesis and Biomedical Applications of Copper Sulfide Nanoparticles: From Sensors to Theranostics. *Small* **2014**, *10*, 631–645. [CrossRef] [PubMed]
3. Jain, P.K.; Huang, X.H.; El-Sayed, I.H.; El-Sayed, M.A. Noble Metals on the Nanoscale: Optical and Photothermal Properties and Some Applications in Imaging, Sensing, Biology, and Medicine. *Acc. Chem. Res.* **2008**, *41*, 1578–1586. [CrossRef] [PubMed]
4. He, H.; Xie, C.; Ren, J. Nonbleaching Fluorescence of Gold Nanoparticles and Its Applications in Cancer Cell Imaging. *Anal. Chem.* **2008**, *80*, 5951–5957. [CrossRef] [PubMed]
5. Sironi, L.; Freddi, S.; Caccia, M.; Pozzi, P.; Rossetti, L.; Pallavicini, P.; Donà, A.; Cabrini, E.; Gualtieri, M.; Rivolta, I.; et al. Gold Branched Nanoparticles for Cellular Treatments. *J. Phys. Chem. C* **2012**, *116*, 18407–18418. [CrossRef]
6. Link, S.; El-Sayed, M.A. Shape and size dependence of radiative, non-radiative and photothermal properties of gold nanocrystals. *Int. Rev. Phys. Chem.* **2000**, *19*, 409–453. [CrossRef]
7. Jain, P.K.; El-Sayed, I.H.; El-Sayed, M.A. Au nanoparticles target cancer. *Nano Today* **2007**, *2*, 18–29. [CrossRef]
8. Pallavicini, P.; Bassi, B.; Chirico, G. Modular approach for bimodal antibacterial surfaces combining photo-switchable activity and sustained biocidal release. *Sci. Rep.* **2017**, *7*, 5259. [CrossRef]
9. Borzenkov, M.; Määttänen, A.; Ihalainen, P.; Collini, M.; Cabrini, E.; Dacarro, G.; Pallavicini, P.; Chirico, G. Fabrication of Inkjet-Printed Gold Nanostar Patterns with Photothermal Properties on Paper Substrate. *ACS Appl. Mater. Inter.* **2016**, *8*, 9909–9916. [CrossRef]
10. Pallavicini, P.; Basile, S.; Chirico, G.; Dacarro, G.; D'Alfonso, L.; Donà, A.; Patrini, M.; Falqui, A.; Sironi, L.; Taglietti, A. Monolayers of gold nanostars with two near-IR LSPRs capable of additive photothermal response. *Chem. Commun.* **2015**, *51*, 12928–12930. [CrossRef]
11. Chirico, G.; Dacarro, G.; O'Regan, C.; Peltonen, J.; Sarfraz, J.; Taglietti, A.; Borzenkov, M.; Pallavicini, P. Photothermally Responsive Inks for Inkjet-Printing Secure Information. *Part. Part. Syst. Charact.* **2018**, *35*, 1800095. [CrossRef]
12. Halas, N.J.; Lal, S.; Chang, W.-S.; Link, S.; Nordlander, P. Plasmons in Strongly Coupled Metallic Nanostructures. *Chem. Rev.* **2011**, *111*, 3913. [CrossRef] [PubMed]

13. Liu, S.-D.; Cheng, M.-T. Linear plasmon ruler with tunable measurement range and sensitivity. *J. Appl. Phys.* **2010**, *108*, 034313. [CrossRef]
14. Sarfraz, J.; Määttänen, A.; Ihalainen, P.; Keppeler, M.; Linden, M.; Peltonen, J. Printed Copper Acetate Based H2S Sensor on Paper Substrate. *Sens. Actuators B* **2012**, *173*, 868–873. [CrossRef]
15. Kang, H.; Lee, J.W.; Nam, Y. Inkjet-Printed Multiwavelength Thermoplasmonic Images for Anticounterfeiting Applications. *ACS Appl. Mater. Interfaces* **2018**, *10*, 6764–6771. [CrossRef]
16. Hill, T.Y.; Reitz, T.L.; Rottmayer, M.A.; Huang, H. Controlling Inkjet Fluid Kinematics to Achieve SOFC Cathode Micropatterns. *ECS J. Solid State Sci.Technol.* **2015**, *4*, 3015–3019. [CrossRef]
17. Turkevich, J. Colloidal Gold. Part I. Historical and preparative aspects, morphology and structure. *Gold Bull.* **1985**, *18*, 86–91. [CrossRef]
18. Poole, C.P., Jr.; Owens, F.J. *Introduction to Nanotechnology*; Wiley Interscience: New York, NY, USA, 2003; pp. 13–14.
19. Borzenkov, M.; Chirico, G.; D'Alfonso, L.; Sironi, L.; Collini, M.; Cabrini, E.; Dacarro, G.; Milanese, C.; Pallavicini, P.; Taglietti, A.; et al. Thermal and Chemical Stability of Thiol Bonding on Gold Nanostars. *Langmuir* **2015**, *31*, 8081–8091. [CrossRef]
20. Chen, H.; Kou, X.; Yang, Z.; Ni, W.; Wang, J. Shape- and Size-Dependent Refractive Index Sensitivity of Gold Nanoparticles. *Langmuir* **2008**, *24*, 5233–5237. [CrossRef]
21. Pallavicini, P.; Preti, L.; De Vita, L.; Dacarro, G.; Diaz Fernandez, Y.A.; Merli, D.; Rossi, S.; Taglietti, A.; Vigani, B. Fast dissolution of silver nanoparticles at physiological pH. *J. Colloid Interf. Sci.* **2020**, *563*, 177–188. [CrossRef]
22. Cao, G. *Nanostructures & Nanomaterials–Synthesis, Properties & Applications*; Imperial College Press: London, UK, 2004; pp. 43–47.
23. Moosavi, M.; Motahari, A.; Omrani, A.; Rostami, A.A. Investigation on some thermophysical properties of poly(ethylene glycol) binary mixtures at different temperatures. *J. Chem. Thermodynamics* **2013**, *58*, 340–350. [CrossRef]
24. Rekhi, G.S.; Jambheka, S.S. Ethyl Cellulose-A Polymer Review. *Drug Dev. Ind. Pharm.* **1995**, *21*, 61–77. [CrossRef]
25. Pallavicini, P.; Cabrini, E.; Cavallaro, G.; Chirico, G.; Collini, M.; D'Alfonso, L.; Dacarro, G.; Donà, A.; Marchesi, N.; Milanese, C.; et al. Gold nanostars coated with neutral and charged polyethylene glycols: A comparative study of in-vitro biocompatibility and of their interaction with SH-SY5Y neuroblastoma cells. *J. Inorg. Biochem.* **2015**, *151*, 123–131. [CrossRef] [PubMed]
26. Weast, R.C.; Astle, M.J.; Beyer, W.H. *CRC Handbook of Chemistry and Physics*, 66th ed.; CRC Press: Boca Raton, FL, USA, 1985; p. 159.
27. Guthrie, J.P. Hydrolysis of esters of oxy acids: pKa values for strong acids. *Can. J. Chem.* **1978**, *56*, 2342–2354. [CrossRef]
28. Turkevich, J.; Stevenson, P.C.; Hillier, J. A study of the nucleation and growth processes in the synthesis of colloidal gold. *Discuss. Faraday Soc.* **1951**, *11*, 55–75. [CrossRef]

© 2020 by the authors. Licensee MDPI, Basel, Switzerland. This article is an open access article distributed under the terms and conditions of the Creative Commons Attribution (CC BY) license (http://creativecommons.org/licenses/by/4.0/).

Article

Nanocrystalline Antiferromagnetic High-κ Dielectric Sr₂NiMO₆ (M = Te, W) with Double Perovskite Structure Type

Jelena Bijelić [1], Dalibor Tatar [1], Sugato Hajra [2], Manisha Sahu [2], Sang Jae Kim [2], Zvonko Jagličić [3,4] and Igor Djerdj [1,*]

1. Department of Chemistry, Josip Juraj Strossmayer Univesity of Osijek, Cara Hadrijana 8/A, HR-31000 Osijek, Croatia; jelena.bijelic@kemija.unios.hr (J.B.); tatar.dalibor42@gmail.com (D.T.)
2. Nanomaterials and System Lab, Major of Mechatronics Engineering, Faculty of Applied Energy Systems, Jeju National University, Jeju 63243, Korea; sugatofl@outlook.com (S.H.); manishafl@outlook.com (M.S.); kimsangj@jejunu.ac.kr (S.J.K.)
3. Institute of Mathematics, Physics and Mechanics, University of Ljubljana, Jadranska 19, SI-1000 Ljubljana, Slovenia; zvonko.jaglicic@imfm.si
4. Faculty of Civil and Geodetic Engineering, University of Ljubljana, Jamova 2, SI-1000 Ljubljana, Slovenia
* Correspondence: igor.djerdj@kemija.unios.hr; Tel.: +385-31-399-975

Academic Editor: Giuseppe Cirillo

Received: 7 August 2020; Accepted: 1 September 2020; Published: 2 September 2020

Abstract: Double perovskites have been extensively studied in materials chemistry due to their excellent properties and novel features attributed to the coexistence of ferro/ferri/antiferro-magnetic ground state and semiconductor band gap within the same material. Double perovskites with Sr₂NiMO₆ (M = Te, W) structure type have been synthesized using simple, non-toxic and costless aqueous citrate sol-gel route. The reaction yielded phase-pure nanocrystalline powders of two compounds: Sr₂NiWO₆ (SNWO) and Sr₂NiTeO₆ (SNTO). According to the Rietveld refinement of powder X-ray diffraction data at room temperature, Sr₂NiWO₆ is tetragonal ($I4/m$) and Sr₂NiTeO₆ is monoclinic ($C12/m1$), with average crystallite sizes of 49 and 77 nm, respectively. Structural studies have been additionally performed by Raman spectroscopy revealing optical phonons typical for vibrations of Te^{6+}/W^{6+}O₆ octahedra. Both SNTO and SNWO possess high values of dielectric constants (341 and 308, respectively) with low dielectric loss (0.06 for SNWO) at a frequency of 1 kHz. These values decrease exponentially with the increase of frequency to 1000 kHz, with the dielectric constant being around 260 for both compounds and dielectric loss being 0.01 for SNWO and 0.04 for SNTO. The Nyquist plot for both samples confirms the non-Debye type of relaxation behavior and the dominance of shorter-range movement of charge carriers. Magnetic studies of both compounds revealed antiferromagnetic behavior, with Néel temperature (T_N) being 57 K for SNWO and 35 K for SNTO.

Keywords: antiferromagnet; double perovskite; high-κ dielectric; nickel; tellurium; tungsten

---

1. Introduction

Complex metal oxides have a great fundamental and practical interest. They attract a great attention of the researchers due to the strong correlation between chemical composition, features of the crystal structure, magnetic, electrical and functional properties [1–3]. Nanostructured magnetic compounds have been investigated thoroughly as they can possess different physical and chemical properties compared to their bulk counterparts [4,5]. For example, many research groups [6–12] have shown that the transition temperature in antiferromagnetic materials (T_N—Néel temperature) increases with the reduction of particle size. Sometimes, in the case of antiferromagnetic materials, additional transition might be observed [13] or spin canting effect might take place [14,15]. This behavior is usually not reported in bulk forms. Alterations can additionally be observed in dielectric properties and usually they are explained by changes in size, shape and particle or grain boundaries [16–18]. Concerning dielectric properties, it is desired to produce a material with high value of dielectric constant (permittivity) and low value of dielectric loss [19–22]. High-κ dielectric materials are materials with high value of dielectric permittivity. They are used in semiconductor manufacturing processes where they replace widely-used silicon dioxide as gate dielectrics. This allows the increase of the gate performance in term of capacitance [23]. However, most of these materials are manufactured as composites with polymers, such as PVA or PVDF, to increase the value of dielectric constant and decrease the value of dielectric loss [24]. High-κ dielectric polymer composites have been reported to be successful in various applications [25–30].

The other thing that can be tailored by particle size control is a value of optical band gap. This has shown a similar trend as transition temperature: It increases with a decrease in particle size [31]. Double perovskite materials are usually reported as semiconductors, with band gap values from 3 to 6 eV [32].

All of the above mentioned properties arise from specific crystal structure and are strongly correlated to it. Perovskite structure is summarized by general formula ABX_3, where A is large cation such as alkali or earth alkali, B is small cation mostly transition metal and X stands for oxide, sulfide or halide anion [33]. Multiple substitutions could be performed at either A, B or X sites. In $A_2B'B''O_6$ double perovskites, half of the B site is occupied by one transition metal cation, such as Ni^{2+} and the other half is occupied by some other transition metal cation, such as W^{6+} or a semimetal like Te^{6+} [34]. Here, Ni^{2+} possess a finite magnetic moment arising from unpaired d electrons, while Te^{6+} and W^{6+} are non-magnetic since their d orbitals are empty. This combination might produce some novel effects in single material, which is why these materials have been of a great interest during the last couple of decades.

Sr_2NiWO_6 (SNWO) and Sr_2NiTeO_6 (SNTO) have been studied since the 1960s and 1970s of the twentieth century and mostly were prepared in the bulk polycrystalline form or in the form of single crystals. SNWO was previously prepared only once in 2016 in nanocrystalline form using the sol-gel method by Xu et al. [35] and its photocatalytic activity was studied. It was first prepared by Fresia et al. [36] in 1959, then it was studied in the 1960s by Brixner [37], and Nomura [38,39], in the 1970s by Köhl [40] and later by Todate [41], Iwanaga [42], Gateshki [43], Tian [44] and more recently by Liu [45], Blum [46] and Rezaei [47]. SNTO was among first studied by Köhl [48], Lentz [49], Rossman [50] and later studied by Todate [41], Iwanaga [42], Ortega San-Martin [51] and Orayech [52].

Mostly the structure and magnetic properties have been investigated for these compounds with lack of knowledge in the dielectric part. Therefore, in this contribution our goal is to clarify dielectric properties and to investigate the effects that occur due to size reduction to nanoscale. For the first time we give a detailed explanation of dielectric properties for both compounds and show effects of size reduction on the magnetic properties.

2. Results and Discussion

2.1. Powder X-ray Diffraction

Powder X-ray diffraction has been performed in order to investigate phase purity, crystal structure as well as microstructural parameters of as synthesized materials. The Rietveld refinement was conducted along with microstructural analysis [53], and the typical Rietveld output plots are given in Figures 1 and 2 whilst the results of the refinement are summarized in Tables 1–4. Both figures reveal phase purity of as synthesized double perovskites since all observed Bragg reflections were correctly described by the calculated curve based on the assumed structural models. Only in Figure 2 are several spikes in the difference curve observed; however, they are most probably attributed to some small preferred orientation.

Table 1. Crystallographic data and Rietveld refinement parameters obtained from the XRD patterns of the synthesized compounds.

Chemical Formula	Sr_2NiTeO_6	Sr_2NiWO_6
Space group	$C12/m1$ (12)	$I4/m$ (87)
Molecular weight	457.53	513.77
Z	2	
Crystal system	Monoclinic	Tetragonal
Lattice parameters (Å)	a = 9.663(1) b = 5.6132(2) c = 5.5833(2) β = 125.32(1)°	a = 5.5644(2) c = 7.9025(4)
Cell volume (Å^3)	247.09(4)	244.68(2)
Calculated density (g/cm^3)	6.15	6.97
Data collection range	10–90°	
No. of parameters refined	24	18
No. of bond lengths restrained	11	0
No. of bond angles restrained	0	0
Average apparent crystallite size (nm)	77	49
Average apparent microstrains (×10^{-4})	17.462	11.828
Phase composition (wt %)	100	
R_B (%)	5.84	6.1
Conventional R_p, R_{wp}, R_e (%)	27.0, 16.5, 11.8	19.9, 14.4, 9.31
χ^2	1.961	2.398

Sr_2NiWO_6 crystallized in tetragonal centrosymmetric space group $I4/m$ with lattice parameters equal to a = 5.5644(2), c = 7.9025(4) Å, while Sr_2NiTeO_6 crystallized in monoclinic centrosymmetric space group $C12/m1$ with lattice parameters: a = 9.663(1), b = 5.6132(2), c = 5.5833(2) Å, β = 125.32(1)°. Although investigated compounds differ significantly in terms of crystal symmetry, their unit cell volume values are quite close.

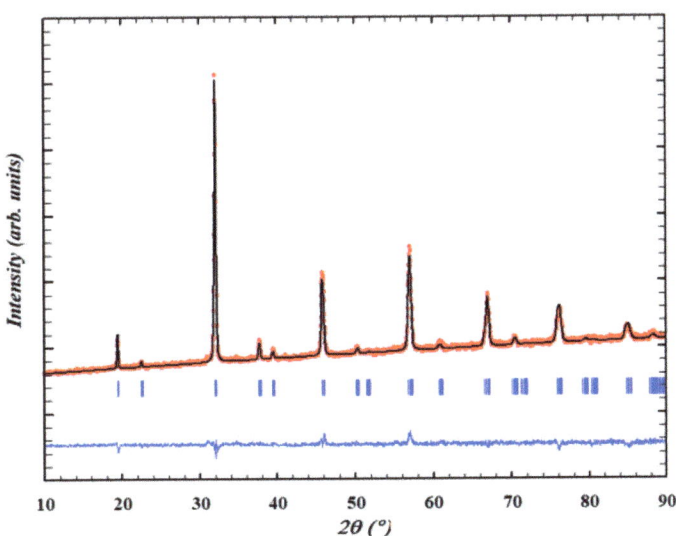

Figure 1. Calculated (black) vs. experimental (red) powder X-ray diffraction pattern for Sr_2NiTeO_6.

Table 2. Structural parameters for Sr_2NiTeO_6 extracted at room temperature (292 K). The site occupancies are expressed in terms of the ratio m:M—site multiplicity:multiplicity of a general position (8).

Atom	Wyckoff Position	x [a]	y [a]	z [a]	B (Å^2) [b]	Occupancy
Sr	4i	0.751(1)	0	0.235(2)	0.9(1)	1/2
Ni	2d	0	1/2	1/2	0.1(1)	1/4
Te	2a	0	0	0	1.8(1)	1/4
O1	8j	0.0264(9)	0.2443(9)	0.266(1)	1.7(2)	1
O2	4i	0.2436(7)	0	0.301(2)	1.7(2)	1/2

[a] Atomic coordinates in 3D space; [b] Debye-Waller factor.

Comparison of atomic positions extracted by Iwanaga et al. [42] for bulk SNTO with this work, presented in Table 2, reveals quite different values, although they correspond to the same space group. As opposed to nanocrystalline SNTO reported in Table 2, in bulk as reported in [42] there are two different positions for magnetically-active Ni^{2+} cation (2a and 2d), which could result in different magnetic properties.

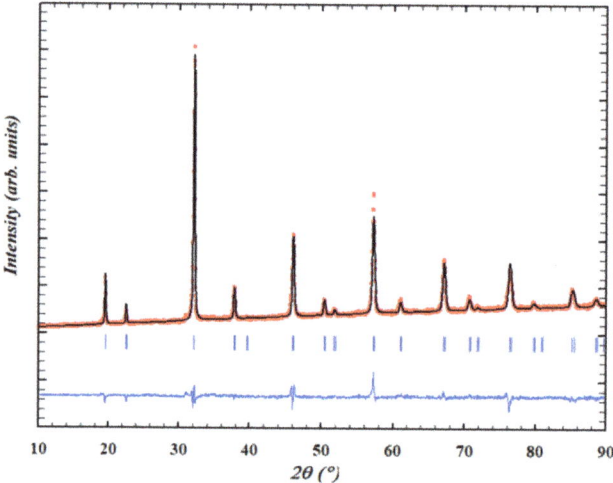

Figure 2. Calculated (black) vs. experimental (red) powder X-ray diffraction pattern for Sr_2NiWO_6.

Table 3. Structural parameters for Sr_2NiWO_6 extracted at room temperature (292 K). The site occupancies are expressed in terms of the ratio m:M—site multiplicity:multiplicity of a general position (16).

Atom	Wyckoff Position	x [a]	y [a]	z [a]	B (Å^2) [b]	Occupancy
Sr	4d	0	1/2	1/4	1.30(7)	1/4
Ni	2a	0	0	0	0.9(1)	1/8
W	2b	0	0	1/2	1.19(5)	1/8
O1	8h	0.269(3)	0.200(3)	0	2.35(4)	1/2
O2	4e	0	0	0.255(2)	2.35(4)	1/4

[a] Atomic coordinates in 3D space; [b] Debye-Waller factor.

Figure 3 shows the crystal structures of as synthesized compounds visualized by VESTA software [54] and Table 4 shows selected interatomic distances calculated by the Rietveld method. SNTO shows structural arrangement consisting of layers of TeO_6 octahedra centered in corners of bc planes, mutually linked with NiO_6 octahedron, centered in the middle of bc rectangle. The 3D structure was further composed by alternate linking of tilted TeO_6-NiO_6 octahedra along the [100] direction. The voids formed with such stacking were filled by SrO_{12} cuboctahedron. SNWO is rock-salt ordered with the arrangements of a mutually slightly-rotated corner-sharing NiO$_6$-WO$_6$ octahedra along the c-axis due to the tetragonal distortion.

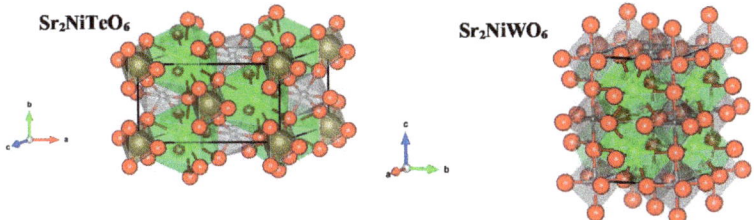

Figure 3. Crystal structures of the synthesized compounds visualized by VESTA [49].

In monoclinic Sr$_2$NiTeO$_6$ there are two types of octahedra that differ according to their lengths: shorter (smaller volume) TeO$_6$ octahedra and slightly larger NiO$_6$ octahedra. Moreover, both octahedra are highly distorted since the apical line is not perpendicular to the equatorial plane. Tetragonal Sr$_2$NiWO$_6$ comprises NiO$_6$ octahedra and WO$_6$ octahedra that differ to each other according to their bond lengths: Equatorial bond lengths in NiO$_6$ are visibly shorter compared to WO$_6$, while the apical lengths of NiO$_6$ are slightly longer than in WO$_6$, as seen from the Table 4. It is noteworthy that SrO$_{12}$ cuboctahedron is more distorted (more different Sn-O bond lengths) for SNTO in comparison with SNWO which is the result of lower symmetry of SNTO. The line broadening analysis performed within the Rietveld refinement reveals that SNTO shows higher crystallinity (average crystallite size equals to 77 nm) compared to SNWO (49 nm). Although the crystallinity is higher for SNTO, its microstrain level is also higher compared to SNWO, which is an unexpected and interesting finding.

Table 4. Selected interatomic distances for synthesized compounds.

Compound	Bond Type	Bond Length (Å)
Sr$_2$NiTeO$_6$	Sr-O1 x2	2.707(7)
	Sr-O1 x2	2.689(1)
	Sr-O1 x2	2.909(1)
	Sr-O1 x2	2.900(8)
	Sr-O2 x1	2.558(2)
	Sr-O2 x1	3.026(1)
	Sr-O2 x2	2.837(3)
	Ni-O1 x4	2.052(7)
	Ni-O2 x2	2.047(6)
	Te-O1 x4	1.929(6)
	Te-O2 x2	1.947(5)
Sr$_2$NiWO$_6$	Sr-O1 x4	2.606(1)
	Sr-O1 x4	2.988(2)
	Sr-O2 x4	2.7825(1)
	Ni-O1 x4	1.873(2)
	Ni-O2 x2	2.0151(1)
	W-O1 x4	2.099(2)
	W-O2 x2	1.9361(1)

2.2. Unpolarized Raman Spectroscopy

Raman spectroscopy was conducted in order to inspect bond length stretching and possible phonon confinement effects that frequently occur in nanocrystalline materials. Figure 4 shows Raman spectra of double perovskites: tetragonal Sr_2NiWO_6 (SNWO in Figure 4a), and monoclinic Sr_2NiTeO_6 (SNTO in Figure 4b). According to Kroumova et al. [55] and based on the space group, Sr_2NiTeO_6 has $7A_g + 5B_g$, 12 first-order Raman active modes (optical modes) in total. Sr_2NiWO_6 has $3A_g + 3B_g + 3^1E_g + 3^2E_g$, which is also 12 first-order Raman active modes in total. Only four bands out of 12 for both SNWO and SNTO appear in Figure 4. Peaks appearing in the range from 100–150 cm^{-1} are assigned to lattice translational modes [51]. They are observed at 134 cm^{-1} for SNWO and at 142 cm^{-1} for SNTO. Peaks in the range 350–450 cm^{-1} are assigned to ν_5 mode, which appears due to oxygen bending in octahedra [56,57]. It is observed at 440 cm^{-1} for SNWO and at 416 cm^{-1} for SNTO. Since the Te^{6+}-O bond is the shortest bond for the studied SNTO, it is also the strongest bond for this compound and higher frequencies in the spectrum should be primarily assigned to ν_1 and ν_2 vibrations of the $Te^{6+}O_6$ octahedron. Hence, for SNTO, the frequencies corresponding to vibrations of $Ni^{2+}O_6$ octahedron are not expected and indeed were not observed. Peaks assigned to ν_2 mode are due to asymmetric stretching that appear in the range 470–650 cm^{-1} according to Silva et al. [56]. This mode is represented by two peaks for both compounds, at 497 and 564 cm^{-1} for SNWO and at 510 and 600 cm^{-1} for SNTO. The highest wavenumber appearing within spectra (850 cm^{-1} for SNWO and 762 cm^{-1} for SNTO) is assigned to ν_1, symmetric stretching of the WO_6 and TeO_6 octahedra, respectively. Optical phonons for both compounds are summarized in Table S1 in the Supporting file. Similar to Silva et al. [56], we also observe a larger difference in the ν_1 vibration between Te^{6+}- (d^{10}) and W^{6+}-based (d^0) compounds. These authors correlated the increase of wavenumber in W-containing compounds compared to Te-containing compounds with the increase of the force constant (bonding energy) of the W-O bond in the WO_6 octahedron. The Te^{6+} cation has a fully-occupied d^{10} orbital configuration, which avoids the formation of π-type Te-O bonds. On the other hand, the W^{6+} cation has a d^0 orbital configuration, allowing the overlap of the t_{2g} orbitals The increase of force constant/bonding energy occurs due to the overlapping of t_{2g} orbitals of the octahedrally-coordinated W^{6+} cation [56]. Similar arguments are valid for the absence of vibrations of $Ni^{2+}O_6$ octahedron for SNWO. Although, the shortest bond in this compound is the equatorial Ni-O bond (1.873(2) Å), the formation of π-type W-O bonds occurred, which resulted in the increase of W-O bonding energy, thus Raman modes of $W^{6+}O_6$ dominate. Ayala et al. [58] studied tetragonal and monoclinic double perovskites and came to conclusion that these spectra resemble to those of cubic Fm-$3m$ perovskites. The similarity comes from the fact that tetragonal and monoclinic structures arise from the small distortions of the cubic cell [56]. When comparing our SNWO spectrum in Figure 7a with the bulk SNWO reported by Manoun et al. [59], there is a progressive shift to lower wavenumbers (lower energies); line shapes are broadened and asymmetric on the lower energy side. Additionally, symmetry breaking becomes more emphasized. These effects occur due to the reduction of size and quantum confinement effect [60–62]. Actually, the impact of the size reduction on the Raman spectra is quite visible, since SNWO contains smaller crystals compared to SNTO and thus its Raman bands are also broader, reflecting a higher disorder in SNWO.

2.3. Scanning Electron Microscopy and Energy Dispersive X-ray Spectroscopy

To investigate the morphology and chemical composition of as-prepared compounds, scanning electron microscopy (SEM) and energy dispersive X-Ray spectroscopy (EDX) measurements were performed. SEM images are shown in Figure 5, EDX spectra for both compounds are shown in Figure S1 and Figure S2 in SI.

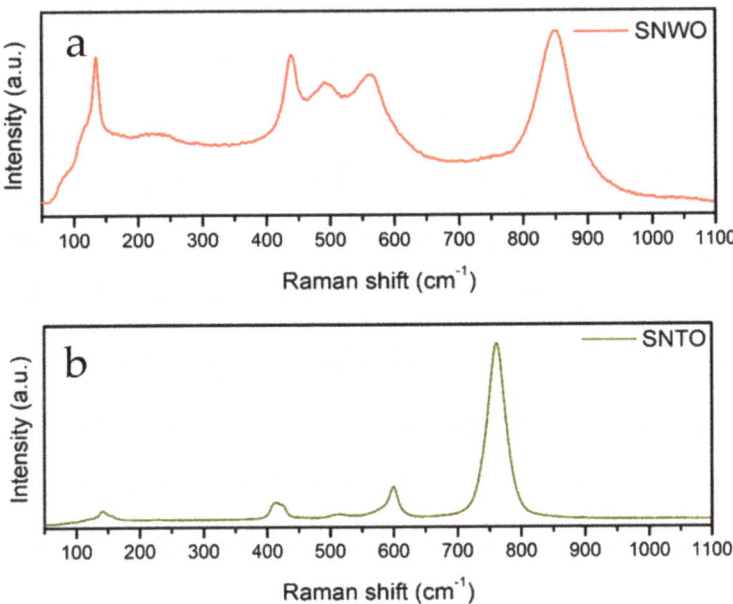

Figure 4. Raman spectra of the Sr_2NiWO_6 (SNWO, (**a**)) and the Sr_2NiTeO_6 (SNTO, (**b**)).

SEM images show irregularly round-shaped particle agglomerates for both SNTO and SNWO at the same magnification (1 µm). It is clearly visible that SNTO particles are larger than SNWO particles, which is in accordance with microstructural analysis conducted using the Rietveld refinement method. The surface morphology of the particles represents the random distribution of the various size of the grains and visible grain boundaries.

Figure 5. SEM images of SNTO (**a**) and SNWO (**b**).

The energy dispersive X-ray spectrum and elemental mapping suggest the presence of all elements is similar to the base composition confirming that both synthesized compounds SNTO and SNWO are free from impurity. Figure S1 shows that Sr_2NiWO_6 consists of 61.2 at % O, 19.7 at % Sr, 10.8 at % Ni and 8.3 at % W, which nearly corresponds to the proposed chemical formula. It is similar with Sr_2NiTeO_6, which consists of 63.4 at % O, 18.1 at % Sr, 9.9 at % Ni and 8.6 at % Te (Figure S2).

2.4. Dielectric Properties

Figure 6a displays dielectric constant and Figure 6b dielectric loss versus frequency for SNTO and SNWO ceramic samples at room temperature. It is observed that the dielectric constant (ε_r)

and dielectric loss (tan δ) are high at lower frequency regime and decreases at higher frequency regime, indicating that the familiar dielectric dispersion phenomenon as observed in the case of normal ferroelectrics. Frequency-temperature dependence of ε_r and tan δ is associated with various polarization effects, such as ionic, dipolar, electronic and space charge that appears at multiple levels of material reaction due to short and long-range movement of mobile charges. At low-frequency regions, the space charge and dipolar polarization are at the peak and interfacial polarizations efficiently add to the upper value of ε_r. In the high-frequency region, the electronic polarizations become most significant compared to other polarizations, leading to invariant dielectric constant. The decrease in the value of dielectric constant with the rise in frequency may be due to the dipoles unable to follow the rapid oscillating field [63]. The dielectric loss has a similar type of trend as ε_r in the low-frequency regime, where the loss is high due to the influence of compositional disorder, leading to a relaxation phenomenon [64]. Mutual comparison between SNTO and SNWO reveals that SNWO has a lower value of ε_r (308 compared to 341 at 1 kHz), while its dielectric loss is significantly lower (0.06 compared to 0.23). Moreover, frequency-dependent quantities show milder decrease for SNWO, indicating its better perspective for electronic applications.

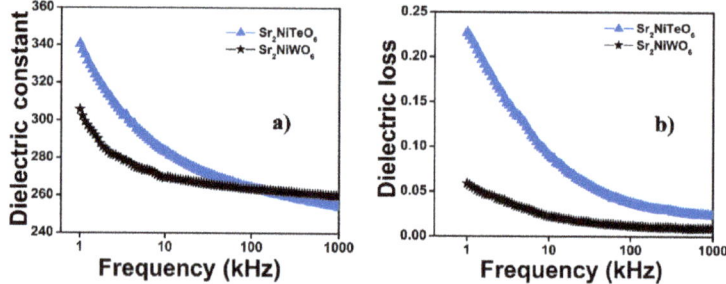

Figure 6. (a) Dielectric constant and (b) dielectric loss of Sr_2NiWO_6 and Sr_2NiTeO_6 ceramics at room temperature.

2.5. Impedance Spectroscopy

The Nyquist plot in general has several semicircular arcs, whereby the first circle (higher frequency) represents the grain/bulk effect, the second circle (intermediate frequency) represents the grain boundary effect and the third circle (lower frequency) represents the electrode effect. Normally, depressed semicircular arcs are obtained with the center below the abscissa implying departure from Debye type behavior. There are several factors, such as atomic defect distribution, stress–strain phenomena, grain boundary, and grain orientation, which can be correlated to the above-described non-ideal behavior. The depressed semicircles also provide evidence of polarization phenomena with the allocation of relaxation times. In Debye-type behavior, the center of the circle lies exactly on the real Z-axis [65]. Figure 7 shows the depressed semicircle and the non-Debye type of relaxation phenomena are occurring for both synthesized materials. The commercial Z Smipwin software is used to fit the experimental impedance data to an equivalent circuit model bearing a constant phase element (CPE). The parameters derived from fitting are grain resistance, grain capacitance, grain boundary resistance, and grain boundary capacitance and corresponding values are listed in Table 5. Figure 8 shows the frequency-dependent M″ and Z″ plot of (a) SNTO and (b) SNWO ceramics. It represents the effect of longer- and shorter-range movement of charge carriers on the relaxation process. The shorter-range charge carrier dominancy is interpreted by the M″ and Z″ peaks mismatch, whereas if the M″ and Z″ peaks exactly match then it is related to the long-range motion of charge carriers [66]. In Figure 8a,b, it is depicted that the peaks of the M″ and Z″ of SNWO and SNTO coincides in the same position, suggesting the dominance of shorter-range movement of charge carriers.

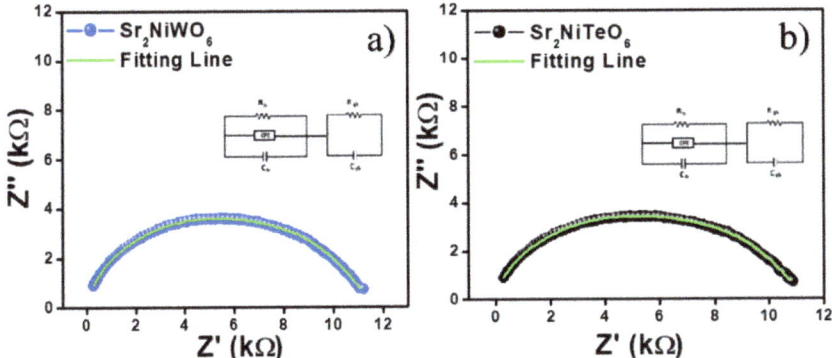

Figure 7. Nyquist plot of (**a**) Sr$_2$NiWO$_6$ and (**b**) Sr$_2$NiTeO$_6$ ceramics at room temperature. In the inset, the equivalent electrical circuits are displayed.

Table 5. The values of grain resistance and grain capacitance from the Nyquist plot fitting.

Composition	R_g (Ω) [a]	C_g (F) [b]	R_{gb} (Ω) [c]	C_{gb} (F) [d]
Sr$_2$NiWO$_6$	1.054×10^4	1.270×10^{-10}	1102	1.083×10^{-8}
Sr$_2$NiTeO$_6$	1.016×10^4	1.268×10^{-10}	1179	1.057×10^{-8}

[a] Grain resistance; [b] Grain capacitance; [c] Grain boundary resistance; [d] Grain boundary capacitance.

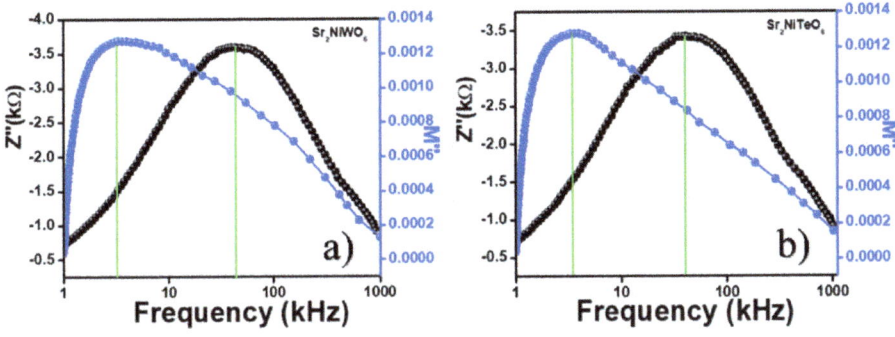

Figure 8. Frequency dependent imaginary part of impedance and modulus: (**a**) Sr$_2$NiWO$_6$ and (**b**) Sr$_2$NiTeO$_6$ ceramics.

2.6. Magnetic Properties

Temperature-dependent susceptibility of SNTO measured in a magnetic field of 1000 Oe is shown in Figure 9. The susceptibility increases as temperature decreases from room temperature down to 35 K. At Néel temperature (T_N) = 35.0 K, the susceptibility exhibits a maximum, in agreement with already reported antiferromagnetic ground state of SNTO [42]. No difference between zero-field-cooled (ZFC) and field-cooled (FC) susceptibility was observed. The measured susceptibility in high temperature region (T > 100 K) was analyzed using a Curie-Weiss law $\chi = C/(T-\vartheta)$. The result, presented as a full line in χ^{-1} vs. T plot (inset in Figure 9), gave the Curie constant C = 1.4 emu K/mol and Curie–Weiss temperature ϑ = −210 K. The effective magnetic moment $\mu = \sqrt{8\,C} = 3.3\ \mu B$ in agreement with the expected value for Ni(II) ions with non-zero orbital contribution L [67]. The magnetization curve M(H) at 2 K (Figure S3) is linear up to the maximal magnetic field of 50 kOe with very small magnetization (0.04 μB at 50 kOe), as expected for the antiferromagnetic ground state.

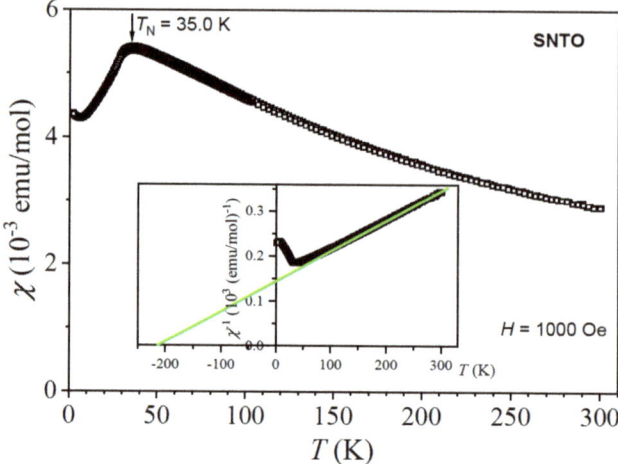

Figure 9. Temperature-dependent susceptibility and inverse susceptibility (inset) of SNTO.

In Figure 10a the temperature-dependent magnetic susceptibility is shown for SNWO. A local maximum of susceptibility (Néel temperature) appears here at a higher Néel temperature (T_N) = 56.6 K. The magnetization M(H) curve at 2 K is linear as in SNTO (Figure S3), with a very small value of the magnetization in a maximal field of 50 kOe. Both results speak in favor of antiferromagnetic ground state in SNWO, as already reported in the literature [42,46]. A slightly higher Néel temperature (56.6 K) than the literature data for bulk material (53 K in [42], 54 K in [46]) might be ascribed to a nanosized material in our case. Various researchers have already described that Néel temperature increases with the decrease in crystallite size [6–12].

Above T_N, the susceptibility follows the Curie–Weiss law (inset in Figure 10). The same analyses as performed in the case of SNTO (Curie-Weiss fit), which gave the Curie–Weiss temperature $\vartheta = -125$ K and effective magnetic moment of 3.2 µB. These values are similar to the literature data [42,46].

However, below T_N, the susceptibility starts to increase once again. The ZFC/FC splitting can be observed below 6 K, which was, to our knowledge, not reported yet in literature. In order to investigate this low temperature signal in more detail, we performed additional AC susceptibility measurements (susceptibility in alternating magnetic field) shown in Figure 10b. The AC susceptibility has been measured at three different frequencies of the applied magnetic field—1, 10, and 100 Hz. The amplitude of AC magnetic field was 6 Oe. The maximum of AC susceptibility is at approximately $T_{max} = 6$ K for the frequency of the applied field of 1 Hz, and shifts to higher temperatures with frequency. This feature is typical for a frustrated magnetic system, where both ferromagnetic and antiferromagnetic interaction are present in the system simultaneously, and (or) there is some degree of disorder in a distribution of magnetic moments. [68] The relative shift of T_{max} with frequency (inset in Figure 10b), $G = \frac{\Delta T_{max}}{T_{max(1\ Hz)} \Delta \log_\nu} = 0.07$, falls somewhere between the values of typical spin glasses and superparamagnetic systems [69]. A very similar signal can be found also in NiO nanoparticles [68]. We tentatively ascribe this low temperature signal to the magnetism of uncompensated nickel ions on the surface of nanoparticles, where the perfect antiferromagnetically-ordered environment around the magnetic ions, as established in bulk compounds, is corrupted.

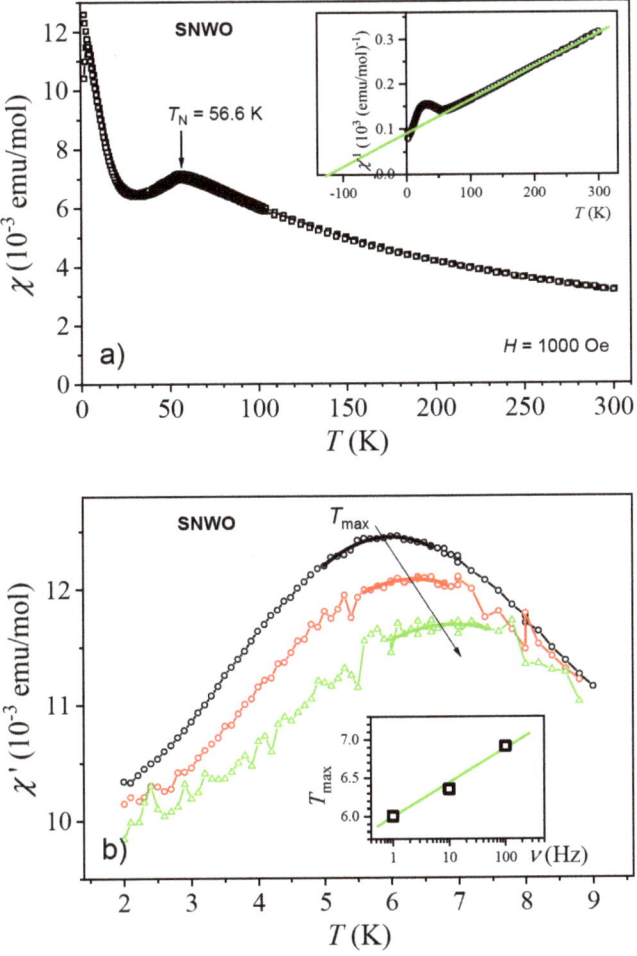

Figure 10. Temperature-dependent susceptibility and inverse susceptibility (inset) of SNWO (**a**). AC susceptibility (susceptibility in alternating magnetic field) around the low temperature peak (**b**) and shift of the maximum of AC susceptibility with frequency (inset in (**b**)).

3. Materials and Methods

Materials. The following commercially available chemicals were used without further purification: Ammonium tungsten oxide hydrate and ammonium tellurate from Alfa Aesar, Germany; strontium(II) nitrate, and nickel(II) nitrate hexahydrate from Sigma Aldrich, Germany; citric acid monohydrate 99.9% from T.T.T., Croatia; and concentrated ammonia solution 25% from Gram-Mol, Croatia.

Synthesis. Double perovskites were synthesized using aqueous sol-gel citrate method, previously reported for triple perovskites [57,70]. Stoichiometric amounts were dissolved in 10% citric acid solution (10 g of citric acid in 100 mL of MiliQ water) for synthesis of SNTO: A total of 2 mmol of strontium(II) nitrate, 1 mmol of nickel(II) nitrate hexahydrate and 1 mmol of ammonium tellurate. For synthesis of SNWO, instead of ammonium tellurate, 1/12 mmol of ammonium tungsten oxide (molecular formula: $H_{40}N_{10}O_{41}W_{12} \bullet xH_2O$) was used. The value of pH was adjusted to 5 using concentrated ammonia solution (pH-meter 211, HANNA, Zagreb, Croatia). Reaction mixture was then evaporated at 95 °C, with constant stirring on a magnetic stirrer (IKA RCT Basic, Staufen, Germany), until black resin was formed. Black resin was further dried in a drying oven (Instrumentaria ST-05, Sesvete, Croatia) at 120 °C for 24 h and grinded in a mortar with a pestle. Obtained black powder was first calcined at 600 °C for 8 h and then at 950 °C (SNTO) or 1000 °C (SNWO) for 12 h (Furnace SN 342689, Nabertherm GmbH, Lilienthal, Germany). Heating rate for both calcination steps was 2 °C per minute. Synthesis procedure is represented in Scheme 1 below.

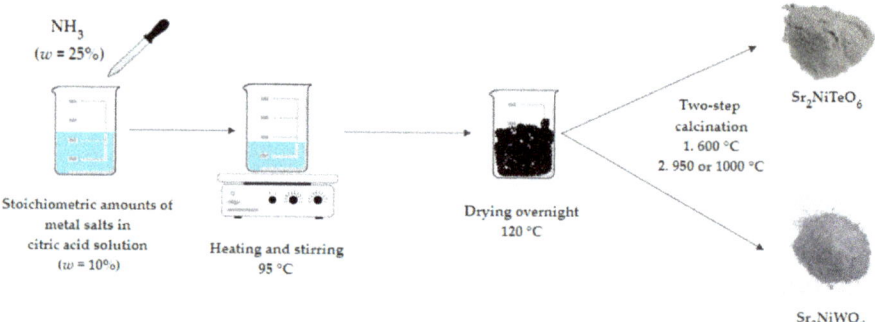

Scheme 1. Schematic representation of sol-gel synthesis of double perovskites Sr_2NiTeO_6 and Sr_2NiWO_6.

Preparation of pellets for dielectric measurements. The calcined powder was converted into fine particles and 5 wt % of binder polyvinyl alcohol (PVA) was mixed. Further, disk-shape pellets comprising a diameter of 10 mm and a thickness of 1.2 mm were sieved employing a hydraulic press (Model number: 220-93614-01, Shimadzu, Kyoto, Japan) with cold pressure of 4.0×10^6 Nm^{-2}. Subsequently, the green pellets were sintered at 1250 °C for 4 h at a heating rate of 8 °C/min in a tube furnace (Model number: SH-FU-50TS, SH Scientific, Daejeon, Korea). One of the pellets of each composition was taken and rubbed with zero emery paper to make the opposite sides smooth and parallel. The high-quality silver paint was painted on both opposite sides of each pellet that served as an electrode and further electrical characterization was performed.

Methods. Powder X-Ray diffraction patterns were collected on a Panalytical XPert Pro Diffractometer with θ–2θ geometry (Malvern Panalytical, Malvern, UK), using monochromatized CuKα radiation (40 kV, 40 mA) at 292(2) K, in the range of 2θ from 10°–90° with the step size of 0.02.

Unpolarized Raman spectroscopy was performed at room temperature in backscattering geometry using a Renishaw inVia Raman microscope system (Reinshaw GmbH, Pliezhausen, Germany) with a HeNe laser (633 nm) for excitation, in the range 50–1100 cm^{-1} with 2 mW laser power. Samples were placed on the microscope slides in a powder form.

The surface morphology was investigated using a scanning electron microscope (M/S TESCAN Mira 3, Brno, Czech Republic).

The resistive and capacitive parameters were fetched from a computer-coupled phase-sensitive meter (Hioki IM3570, Nagano, Japan). The electrical properties of the prepared pellets were examined over a frequency sweep (1 kHz–1 MHz) and at room temperature.

Magnetic properties were investigated using a MPMS-XL-5 magnetometer (Quantum Design, San Diego, CA, USA). All presented data were corrected for diamagnetic contribution estimated from Pascal's constants [71].

4. Conclusions

This study was performed to clarify the impact of size reduction on magnetic properties and to investigate dielectric properties of synthesized Ni-based double perovskites. Most of the reported literature deal with bulk materials and their magnetic properties along with detailed crystal structure analysis. Since miniaturizing and enhancing the performance has become a trend in the production of electronic devices, our motivation was to decrease crystallite size of already known compounds and to shed some light on their magnetic and electrical properties.

Double perovskites Sr_2NiTeO_6 and Sr_2NiWO_6 both possess magnetically active Ni^{2+} cation and magnetically inactive d^{10} Te^{6+} (diameter of 0.56 Å for coordination number 6 [72]) and d^0 W^{6+} (diameter of 0.60 Å for coordination number 6 [72]) cations, respectively. Coexistence of these two species in a single-phase material is one of the conditions for coupling of magnetic and dielectric properties within the same material, which leads to multifunctionality.

So far, we have successfully synthesized, using modified sol-gel route phase pure Ni-based perovskites, Sr_2NiTeO_6 and Sr_2NiWO_6. Since the ionic radii values of octahedrally-coordinated Te^{6+} and W^{6+} are similar [72], it would be expected that their crystal structures are also similar. The value of their Goldschmidt tolerance factors are the same up to fourth decimal digit (t = 0.9889) [33]. However, even the small difference in ionic radii can produce different structures. Thus, Sr_2NiWO_6 is crystalized in the tetragonal crystal system ($I4/m$) with an average crystallite size of 49 nm, while Sr_2NiTeO_6 is monoclinic ($C12/m1$) with 77 nm the average crystallite size. On the other hand, monoclinic and tetragonal structures are both formed because of small distortions of the cubic cell, so their similarity is still preserved.

Both compounds are antiferromagnetically ordered, as it has already been reported before for their bulk and single crystal forms. Sr_2NiTeO_6 behaves the same as nanocrystalline and bulk material, having the same value of Néel temperature (T_N = 35 K). However, Sr_2NiWO_6 possesses a slightly larger value of T_N (56.6) than single crystal (53 K) and bulk (54 K) forms, which often occurs due to size reduction.

For the first time we reported the results of dielectric measurements for these compounds. Both compounds have high values of dielectric constants (341 for Sr_2NiTeO_6 and 308 for Sr_2NiWO_6) at room temperature and low frequency (1 kHz), which classifies them as high-κ dielectrics. Moreover, the W-containing compound is promising for application in electronic devices, since its dielectric loss of 0.06 at 1 kHz is significantly lower compared to its Te counterpart. Impedance spectroscopy results point out that the non-Debye type of relaxation phenomena are occurring for both compounds, suggesting also the dominance of the shorter-range movement of charge carriers.

These materials could be promising candidates to use in electronic devices because they possess both dielectric and antiferromagnetic behavior. Additionally, the size reduction to nanoscale favors the production of smaller devices with enhanced functionality.

Supplementary Materials: The following are available online, Table S1: Optical phonons (in cm^{-1}) of phase pure SNWO and SNTO, Figure S1: EDX spectrum of SNWO, Figure S2: EDX spectrum of SNTO, Figure S3: Magnetization curves of SNTO and SNWO at 2 K.

Author Contributions: Conceptualization, supervision, funding acquisition, review and editing and project administration, I.D.; methodology, formal analysis and original draft preparation, J.B.; synthesis and investigation, D.T.; the study of dielectric properties and supervision, S.H., M.S. and S.J.K.; the study of magnetic properties, Z.J. All authors have read and agreed to the published version of the manuscript.

Funding: This research has been fully funded by the Croatian Science Foundation under the project number IP-2016-06-3115, entitled "Solution chemistry routes towards complex multiferroic materials" partially by the Slovenian Research Agency (Grant No. P2-0348). S.H., M.S., S.J.K. were supported by the Basic Science Research Program through the National Research Foundation of Korea (NRF) grant funded by the Korean government (MSIT) (2018R1A4A1025998, 2019R1A2C3009747).

Acknowledgments: Authors are thankful to Pascal Cop and Sebastian Werner from the AG Smarsly, Institute for Physical Chemistry, Justus Liebig University of Giessen for fruitful discussions, useful advices and great support with powder XRD measurements. J.B., D.T. and I.D. are also thankful to their master student Marina Sekulić for helping with various sample preparations.

Conflicts of Interest: Authors declare no competing financial interests. The funders had no role in the design of the study; in the collection, analyses, or interpretation of data; in the writing of the manuscript, or in the decision to publish the results.

References

1. Ketsko, V.A.; Beresnev, E.N.; Kop'Eva, M.A.; Elesina, L.V.; Baranchikov, A.E.; Stognii, A.I.; Trukhanov, A.V.; Kuznetsov, N.T. Specifics of pyrohydrolytic and solid-phase syntheses of solid solutions in the $(MgGa_2O_4)_x(MgFe_2O_4)_{1-x}$ system. *Russ. J. Inorg. Chem.* **2010**, *55*, 427–429. [CrossRef]
2. Vinnik, D.; Klygach, D.; Zhivulin, V.; Malkin, A.; Vakhitov, M.; Gudkova, S.; Galimov, D.M.; Zherebtsov, D.A.; Trofimov, E.; Knyazev, N.; et al. Electromagnetic properties of $BaFe_{12}O_{19}$:Ti at centimeter wavelengths. *J. Alloys Compd.* **2018**, *755*, 177–183. [CrossRef]
3. Almessiere, M.; Slimani, Y.; Tashkandi, N.; Baykal, A.; Saraç, M.; Trukhanov, A.; Ercan, I.; Belenli, I.; Ozçelik, B. The effect of Nb substitution on magnetic properties of $BaFe_{12}O_{19}$ nanohexaferrites. *Ceram. Int.* **2019**, *45*, 1691–1697. [CrossRef]
4. Lead, J.R.; Wilkinson, K.J. Aquatic Colloids and Nanoparticles: Current Knowledge and Future Trends. *Environ. Chem.* **2006**, *3*, 159–171. [CrossRef]
5. Hough, R.; Noble, R.; Reich, M. Natural gold nanoparticles. *Ore Geol. Rev.* **2011**, *42*, 55–61. [CrossRef]
6. Tang, Z.X.; Hadjipanayis, G.C.; Sorensen, C.M.; Klabunde, K.J. Size-dependent Curie temperature in nanoscale $MnFe_2O_4$ particles. *Phys. Rev. Lett.* **1991**, *67*, 3602–3605. [CrossRef]
7. Kulkarni, G.U.; Kannan, K.R.; Arunarkavalli, T.; Rao, C.N.R.; Joshi, S.K.; Mashelkar, R.A. Particle-size effects on the value of Tc of $MnFe_2O_4$: Evidence for finite-size scaling. *Linus Pauling—Selected Sci. Pap.* **1995**, *49*, 528–531. [CrossRef]
8. van der Zaag, P.J.; Noordermeer, A.; Johnson, M.T.; Bongers, P.F. Comment on "Size-dependent Curie temperature in nanoscale $MnFe_2O_4$ particles". *Phys. Rev. Lett.* **1992**, *68*, 3112. [CrossRef]
9. van der Zaag, P.J.; Brabers, V.A.M.; Johnson, M.T.; Noordermeer, A.; Bongers, P.F. Comment on "Particle-size effects on the value of TC of $MnFe_2O_4$: Evidence for finite-size scaling.". *Phys. Rev. B* **1995**, *51*, 12009–12011. [CrossRef]
10. Chen, J.-P.; Sorensen, C.M.; Klabunde, K.J.; Hadjipanayis, G.C.; Devlin, E.; Kostikas, A. Size-dependent magnetic properties of $MnFe_2O_4$ fine particles synthesized by coprecipitation. *Phys. Rev. B* **1996**, *54*, 9288–9296. [CrossRef]
11. Yang, A.; Chinnasamy, C.N.; Greneche, J.M.; Chen, Y.; Yoon, S.D.; Hsu, K.; Vittoria, C.; Harris, V.G. Large tunability of Néel temperature by growth-rate-induced cation inversion in Mn-ferrite nanoparticles. *Appl. Phys. Rev.* **2009**, *94*, 113109. [CrossRef]
12. Gajbhiye, N.; Balaji, G.; Ghafari, M. Magnetic Properties of Nanostructured $MnFe_2O_4$ Synthesized by Precursor Technique. *Phys. Status Solidi A* **2002**, *189*, 357–361. [CrossRef]
13. Yao, Q.; Tian, C.; Lu, Z.; Wang, J.; Zhou, H.; Rao, G. Antiferromagnetic-ferromagnetic transition in Bi-doped $LaFeO_3$ nanocrystalline ceramics. *Ceram. Int.* **2020**, *46*, 20472–20476. [CrossRef]
14. Li, D.; Han, Z.; Zheng, J.G.; Wang, X.L.; Geng, D.; Li, J.; Zhang, Z.D. Spin canting and spin-flop transition in antiferromagnetic Cr_2O_3 nanocrystals. *J. Appl. Phys.* **2009**, *106*, 053913. [CrossRef]
15. Javed, Q.-U.-A.; Feng-Ping, W.; Rafique, M.Y.; Toufiq, A.M.; Iqbal, M.Z. Canted antiferromagnetic and optical properties of nanostructures of Mn_2O_3 prepared by hydrothermal synthesis. *Chin. Phys. B* **2012**, *21*, 117311. [CrossRef]
16. Dhaouadi, H.; Ghodbane, O.; Hosni, F.; Touati, F. Mn_3O_4 Nanoparticles: Synthesis, Characterization, and Dielectric Properties. *ISRN Spectrosc.* **2012**, *2012*, 1–8. [CrossRef]
17. Gopalan, E.V.; Malini, K.A.; Saravanan, S.; Kumar, D.S.; Yoshida, Y.; Anantharaman, M.R. Evidence for polaron conduction in nanostructured manganese ferrite. *J. Phys. D Appl. Phys.* **2008**, *41*, 185005. [CrossRef]
18. Shenoy, S.; Joy, P.; Anantharaman, M.R. Effect of mechanical milling on the structural, magnetic and dielectric properties of coprecipitated ultrafine zinc ferrite. *J. Magn. Magn. Mater.* **2004**, *269*, 217–226. [CrossRef]
19. Sahu, M.; Hajra, S.; Choudhary, R.N.P. Structural, Bulk Permittivity, and Magnetic Properties of Lead-Free Electronic Material: $Ba_1Bi_1Cu_1Fe_1Ni_1Ti_3O_{12}$. *J. Supercond. Nov. Magn.* **2019**, *32*, 2613–2621. [CrossRef]
20. Prasad, B.V.; Rao, G.N.; Chen, J.-W.; Babu, D. Abnormal high dielectric constant in $SmFeO_3$ semiconductor ceramics. *Mater. Res. Bull.* **2011**, *46*, 1670–1673. [CrossRef]
21. Mahmoud, A.E.-R.; Afify, A.S.; Parashar, S.K.S. Dielectric, tunability, leakage current, and ferroelectric properties of $(K_{0.45}Na_{0.55})_{0.95}Li_{0.05}NbO_3$ lead free piezoelectric. *J. Mater. Sci. Mater. Electron.* **2019**, *30*, 2659–2668. [CrossRef]

22. Babeer, A.M.; El-Sadek, M.S.A.; Mahmoud, A.E.-R.; Afify, A.S.; Tulliani, J.-M.; El-Sadek, M.S.A. Structural, humidity sensing and dielectric properties of Ca-modified Ba(Ti$_{0.9}$Sn$_{0.1}$)O$_3$ lead free ceramics. *J. Mater. Sci. Mater. Electron.* **2016**, *27*, 7622–7632. [CrossRef]
23. Hoefflinger, B. ITRS: The International Technology Roadmap for Semiconductors. In *CHIPS 2020, New Vistas in Nanoelectronics*, 2nd ed.; The Frontiers Collection; Springer: Cham, Switzerland, 2016.
24. Wang, D.; Bao, Y.; Zha, J.-W.; Zhao, J.; Dang, Z.-M.; Hu, G.-H. Improved Dielectric Properties of Nanocomposites Based on Poly(vinylidene fluoride) and Poly(vinyl alcohol)-Functionalized Graphene. *ACS Appl. Mater. Interfaces* **2012**, *4*, 6273–6279. [CrossRef] [PubMed]
25. Niu, Y.; Wang, H. Dielectric Nanomaterials for Power Energy Storage: Surface Modification and Characterization. *ACS Appl. Nano Mater.* **2019**, *2*, 627–642. [CrossRef]
26. Jiang, J.; Shen, Z.; Qian, J.; Dan, Z.; Guo, M.; He, Y.; Lin, Y.; Nan, C.-W.; Chen, L.; Shen, Y. Synergy of Micro-/Mesoscopic Interfaces in Multilayered Polymer Nanocomposites Induces Ultrahigh Energy Density for Capacitive Energy Storage. *Nano Energy* **2019**, *62*, 220–229. [CrossRef]
27. Liu, F.; Li, Q.; Cui, J.; Li, Z.; Yang, G.; Liu, Y.; Dong, L.; Xiong, C.; Wang, H.; Wang, Q. High-Energy-Density Dielectric Polymer Nanocomposites with Trilayered Architecture. *Adv. Funct. Mater.* **2017**, *27*, 1606292. [CrossRef]
28. Huang, X.; Sun, B.; Zhu, Y.; Li, S.; Jiang, P. High-κ Polymer Nanocomposites with 1D Filler for Dielectric and Energy Storage Applications. *Prog. Mater. Sci.* **2019**, *100*, 187–225. [CrossRef]
29. Dang, Z.-M.; Zha, J.-W.; Zheng, M.-S. 1D/2D Carbon Nanomaterial-Polymer Dielectric Composites with High Permittivity for Power Energy Storage Applications. *Small* **2016**, *12*, 1688–1701. [CrossRef]
30. Zeraati, A.S.; Arjmand, M.; Sundararaj, U. Silver Nanowire/MnO$_2$ Nanowire Hybrid Polymer Nanocomposites: Materials with High Dielectric Permittivity and Low Dielectric Loss. *ACS Appl. Mater. Interfaces* **2017**, *9*, 14328–14336. [CrossRef]
31. Deotale, A.J.; Nandedkar, R. Correlation between Particle Size, Strain and Band Gap of Iron Oxide Nanoparticles. *Mater. Today Proc.* **2016**, *3*, 2069–2076. [CrossRef]
32. Sun, Q.; Wang, J.; Yin, W.-J.; Yan, Y. Bandgap Engineering of Stable Lead-Free Oxide Double Perovskites for Photovoltaics. *Adv. Mater.* **2018**, *30*, e1705901. [CrossRef]
33. West, A.R. *Crystal Structures and Crystal Chemistry. Solid State Chemistry and its Applications*, 2nd ed.; Wiley: Chichester, UK, 2014; pp. 54–63.
34. Vasala, S.; Karppinen, M. A2B'B"O6 perovskites: A review. *Prog. Solid State Chem.* **2015**, *43*, 1–36. [CrossRef]
35. Xu, L.; Qin, C.; Wan, Y.; Xie, H.; Huang, Y.; Qin, L.; Seo, H.J. Sol–gel preparation, band structure, and photochemical activities of double perovskite A$_2$NiWO$_6$ (A = Ca, Sr) nanorods. *J. Taiwan Inst. Chem. Eng.* **2017**, *71*, 433–440. [CrossRef]
36. Fresia, E.J.; Katz, L.; Ward, R. Cation Substitution in Perovskite-like Phases1,2. *J. Am. Chem. Soc.* **1959**, *81*, 4783–4785. [CrossRef]
37. Brixner, L.H. Preparation and Structure Determination of Some New Cubic and Tetragonally-Distorted Perovskites. *J. Phys. Chem.* **1960**, *64*, 165–166. [CrossRef]
38. Nomura, S.; Kawakubo, T.T. Phase Transition in Sr(NiW)$_{0.5}$O$_3$-Ba(NiW)$_{0.5}$O$_3$ System. *J. Phys. Soc. Jpn.* **1962**, *17*, 1771–1776. [CrossRef]
39. Nomura, S.; Nakagawa, T. Magnetic Properties and Optical and Paramagnetic Spectra of Divalent Nickel in Sr$_2$(NiW)O$_6$. *J. Phys. Soc. Jpn.* **1966**, *21*, 1679–1684. [CrossRef]
40. Köhl, P. Die Kristallstruktur von Perowskiten A$_2^{II}$Ni^{II}M^{VI}O$_6^{II}$. Das Sr$_2$NiWO$_6$. *Z. Anorg. Allg. Chem.* **1973**, *401*, 121–131. [CrossRef]
41. Todate, Y. Exchange interactions in antiferromagnetic complex perovskites. *J. Phys. Chem. Solids* **1999**, *60*, 1173–1175. [CrossRef]
42. Iwanaga, D.; Inaguma, Y.; Itoh, M. Structure and magnetic properties of Sr$_2$NiAO$_6$ (A = W, Te). *Mater. Res. Bull.* **2000**, *35*, 449–457. [CrossRef]
43. Gateshki, M.; Igartua, J.M.; Hernandez-Bocanegra, E. X-ray powder diffraction results for the phase transitions in Sr$_2$MWO$_6$ (M = Ni, Zn, Co, Cu) double perovskite oxides. *J. Phys. Condens. Matter* **2003**, *15*, 6199–6217. [CrossRef]
44. Tian, S.; Zhao, J.; Qiao, C.; Ji, X.; Jiang, B. Structure and properties of the ordered double perovskites Sr$_2$MWO$_6$ (M=Co, Ni) by sol-gel route. *Mater. Lett.* **2006**, *60*, 2747–2750. [CrossRef]

45. Liu, Y.P.; Fuh, H.R.; Wang, Y.K. Expansion research on half-metallic materials in double perovskites of Sr$_2$BB'O$_6$ (B = Co, Cu, and Ni; B' = Mo, W, Tc, and Re; and BB' = FeTc). *Comp. Mater. Sci.* **2014**, *92*, 63–68. [CrossRef]
46. Blum, C.; Holcombe, A.; Gellesch, M.; Sturza, M.I.; Rodan, S.; Morrow, R.; Maljuk, A.; Woodward, P.; Morris, P.; Wolter, A.; et al. Flux growth and characterization of Sr$_2$NiWO$_6$ single crystals. *J. Cryst. Growth* **2015**, *421*, 39–44. [CrossRef]
47. Rezaei, N.; Hashemifar, T.; Alaei, M.; Shahbazi, F.; Hashemifar, S.J.; Akbarzadeh, H. Ab initio investigation of magnetic ordering in the double perovskite Sr$_2$NiWO$_6$. *Phys. Rev. B* **2019**, *99*, 104411. [CrossRef]
48. Schultze-Rhonhof, E.; Reinen, D. Die Kristallstruktur von Perowskiten A$_2^{II}$Ni^{II}M^{VI}O$_6^I$. Das Sr$_2$[NiTe]O$_6$. *Z. Anorg. Allg. Chem.* **1970**, *378*, 129–143. [CrossRef]
49. Lentz, A. Schwingungsspektroskopische Untersuchungen an Hexaoxotelluraten vom Perowskittyp. *Z. Anorg. Allg. Chem.* **1972**, *392*, 218–226. [CrossRef]
50. Rossman, G.R.; Shannon, R.D.; Waring, R.K. Origin of the yellow color of complex nickel oxides. *J. Solid State Chem.* **1981**, *39*, 277–287. [CrossRef]
51. Ortega-San-Martin, L.; Chapman, J.P.; Cuello, G.; Gonzalez-Calbet, J.M.; Arriortua, M.I.; Rojo, T. Crystal Structure of the Ordered Double Perovskite, Sr$_2$NiTeO$_6$. *Zeitschrift für Anorganische und Allgemeine Chemie* **2005**, *631*, 2127–2130. [CrossRef]
52. Orayech, B.; Ortega-San-Martín, L.; Urcelay-Olabarria, I.; Lezama, L.; Rojo, T.; Arriortua, M.I.; Igartua, J.M. The effect of partial substitution of Ni by Mg on the structural, magnetic and spectroscopic properties of the double perovskite Sr$_2$NiTeO$_6$. *Dalton Trans.* **2016**, *45*, 14378–14393. [CrossRef]
53. Rodriguez-Carvajal, J. Recent advances in magnetic-structure determination by neutron powder diffraction. *Phys. B* **1993**, *192*, 55–69. [CrossRef]
54. Momma, K.; Izumi, F. VESTA 3 for three-dimensional visualization of crystal, volumetric and morphology data. *J. Appl. Crystallogr.* **2011**, *44*, 1272–1276. [CrossRef]
55. Kroumova, E.; Aroyo, M.I.; Mato, J.M.; Kirov, A.; Capillas, C.; Ivantchev, S.; Wondratschek, H. Bilbao Crystallographic Server: Useful Databases and Tools for Phase-Transition Studies. *Phase Transit.* **2003**, *76*, 155–170. [CrossRef]
56. Silva, E.N.; Guedes, I.; Ayala, A.P.; López, C.A.; Augsburger, M.S.; del Viola, M.C.; Pedregosa, J.C. Optical-active phonons in A$_3$Fe$_2$B''O$_9$ (A=Ca, Sr; B''=Te, W) double perovskites. *J. Appl. Phys.* **2010**, *107*, 043512. [CrossRef]
57. Bijelic, J.; Stanković, A.; Medvidović-Kosanović, M.; Markovic, B.; Cop, P.; Sun, Y.; Hajra, S.; Sahu, M.; Vukmirovic, J.; Markovic, D.; et al. Rational Sol–Gel-Based Synthesis Design and Magnetic, Dielectric, and Optical Properties Study of Nanocrystalline Sr$_3$Co$_2$WO$_9$ Triple Perovskite. *J. Phys. Chem. C* **2020**, *124*, 12794–12807. [CrossRef]
58. Ayala, A.P.; Guedes, I.; Silva, E.N.; Augsburger, M.S.; Viola, M.D.C.; Pedregosa, J.C. Raman investigation of A$_2$CoBO$_6$ (A=Sr and Ca, B=Te and W) double perovskites. *J. Appl. Phys.* **2007**, *101*, 123511. [CrossRef]
59. Manoun, B.; Igartua, J.M.; Lazor, P. High temperature Raman spectroscopy studies of the phase transitions in Sr$_2$NiWO$_6$ and Sr$_2$MgWO$_6$ double perovskite oxides. *J. Mol. Struct.* **2010**, *971*, 18–22. [CrossRef]
60. Kumar, C.S.S.R. *Raman Spectroscopy for Nanomaterials Characterization*; Springer: Berlin/Heidelberg, Germany, 2012; pp. 379–387.
61. Arora, A.K.; Rajalakshmi, M.; Ravindran, T.R.; Sivasubramanian, V. Raman spectroscopy of optical phonon confinement in nanostructured materials. *J. Raman Spectrosc.* **2007**, *38*, 604–617. [CrossRef]
62. Campbell, I.; Fauchet, P. The effects of microcrystal size and shape on the one phonon Raman spectra of crystalline semiconductors. *Solid State Commun.* **1986**, *58*, 739–741. [CrossRef]
63. Hajra, S.; Sahu, M.; Purohit, V.; Panigrahi, R.; Choudhary, R.N.P. Investigation of structural, electrical and magnetic characterization of erbium substituted lead free electronic materials. *Mater. Res. Express* **2019**, *6*, 096319. [CrossRef]
64. Dai, Z.; Akishige, Y. Electrical properties of multiferroic BiFeO$_3$ ceramics synthesized by spark plasma sintering. *J. Phys. D Appl. Phys.* **2010**, *43*, 445403. [CrossRef]
65. Hajra, S.; Sahu, M.; Purohit, V.; Choudhary, R. Dielectric, conductivity and ferroelectric properties of lead-free electronic ceramic: $_{0.6}$Bi(Fe$_{0.98}$Ga$_{0.02}$)O$_3$-$_{0.4}$BaTiO$_3$. *Heliyon* **2019**, *5*, e01654. [CrossRef] [PubMed]
66. Setter, N.; Waser, R. Electroceramic materials. *Acta Mater.* **2000**, *48*, 151–178. [CrossRef]

67. Mabbs, F.E.; Machin, D.J. *Magnetism and Transition Metal Complexes*; Dover Publications, Inc. Mineola: New York, USA, 1973.
68. Jagodic, M.; Jagličić, Z.; Jelen, A.; Lee, J.B.; Kim, Y.M.; Kim, H.J.; Dolinšek, J. Surface-spin magnetism of antiferromagnetic NiO in nanoparticle and bulk morphology. *J. Phys. Condens. Matter* **2009**, *21*, 215302. [CrossRef] [PubMed]
69. Mydosh, J.A. *Spin Glasses: An Experimental Introduction*; Taylor & Francis: London, UK, 1993.
70. Bijelić, J.; Stankovic, A.; Matasović, B.; Marković, B.; Bijelić, M.; Skoko, Ž.; Popović, J.; Stefanic, G.; Jagličić, Z.; Zellmer, S.; et al. Structural characterization and magnetic property determination of nanocrystalline $Ba_3Fe_2WO_9$ and $Sr_3Fe_2WO_9$ perovskites prepared by a modified aqueous sol–gel route. *CrystEngComm* **2019**, *21*, 218–227. [CrossRef]
71. Bain, G.A.; Berry, J.F. Diamagnetic Corrections and Pascal's Constants. *J. Chem. Educ.* **2008**, *85*, 532. [CrossRef]
72. Shannon Database of Ionic Radii. Available online: http://abulafia.mt.ic.ac.uk/shannon/ptable.php (accessed on 14 April 2020).

Sample Availability: Samples of the compounds Sr2NiTeO6 and Sr2NiWO6 are available on demand from the authors.

© 2020 by the authors. Licensee MDPI, Basel, Switzerland. This article is an open access article distributed under the terms and conditions of the Creative Commons Attribution (CC BY) license (http://creativecommons.org/licenses/by/4.0/).

Review

Environment-Induced Reversible Modulation of Optical and Electronic Properties of Lead Halide Perovskites and Possible Applications to Sensor Development: A Review

Maria Luisa De Giorgi [1,*], Stefania Milanese [1], Argyro Klini [2] and Marco Anni [1]

1. Dipartimento di Matematica e Fisica "Ennio De Giorgi", Università del Salento, Via per Arnesano, 73100 Lecce, Italy; stefania.milanese@studenti.unisalento.it (S.M.); marco.anni@unisalento.it (M.A.)
2. Institute of Electronic Structure and Laser, Foundation for Research and Technology-Hellas, P.O. Box 1385, Heraklion, 71110 Crete, Greece; klini@iesl.forth.gr
* Correspondence: marialuisa.degiorgi@unisalento.it

Abstract: Lead halide perovskites are currently widely investigated as active materials in photonic and optoelectronic devices. While the lack of long term stability actually limits their application to commercial devices, several experiments demonstrated that beyond the irreversible variation of the material properties due to degradation, several possibilities exist to reversibly modulate the perovskite characteristics by acting on the environmental conditions. These results clear the way to possible applications of lead halide perovskites to resistive and optical sensors. In this review we will describe the current state of the art of the comprehension of the environmental effects on the optical and electronic properties of lead halide perovskites, and of the exploitation of these results for the development of perovskite-based sensors.

Citation: De Giorgi, M.L.; Milanese, S.; Klini, A.; Anni, M. Environment-Induced Reversible Modulation of Optical and Electronic Properties of Lead Halide Perovskites and Possible Applications to Sensor Development: A Review. *Molecules* **2021**, *26*, 705. https://doi.org/10.3390/molecules26030705

Academic Editor: Giuseppe Cirillo
Received: 9 January 2021
Accepted: 23 January 2021
Published: 29 January 2021

Publisher's Note: MDPI stays neutral with regard to jurisdictional claims in published maps and institutional affiliations.

Copyright: © 2021 by the authors. Licensee MDPI, Basel, Switzerland. This article is an open access article distributed under the terms and conditions of the Creative Commons Attribution (CC BY) license (https://creativecommons.org/licenses/by/4.0/).

Keywords: lead halide perovskites; sensors; photoluminescence; amplified spontaneous emission; thin films

1. Introduction

The field of the development of chemical sensors has undergone a significant expansion in the last decade and it is currently one of the most active research area in nanoscience. The possible applications of gas sensors are extremely broad, and include, for example, the detection of air pollutants for environmental monitoring, of explosive vapors for security control and of toxic gases hazardous to human health [1–5].

Among the possible active materials, thin films of semiconductors are particularly promising, due to the possibility to tune electrical and/or optical properties when interacting with the gas analyte. The main properties of ideal sensors are the sensitivity, the selectivity and the reversibility which permit one to detect very low concentrations of gases (target gases), discriminate among them and come back to the pristine stage when the exposition is over. Another crucial issue is a fast response to the stimulus induced by the interaction of the gas with the sensing element and a high recovery speed. The possibility to use the sensor in simple working conditions, moreover, requires the capability to operate at room temperatures without the need for an external trigger, and good stability to external conditions; low-cost and facile fabrication processes are also important for the development of cheap devices with a long operational lifetime.

To date, sensing has been achieved almost entirely by examining changes in the electrical conductance of properly prepared nanostructures of semiconducting materials exposed to different environments [6]. The typical active materials are metal oxides, such as SnO_2, ZnO, TiO_2, Co_3O_4 and FeO, which need elevated temperatures (200–400 °C) to operate [7]. This feature limits the application of these materials for the detection in flammable or explosive atmospheres and the investigation of biomaterials. A further drawback of metal oxide resistive sensors is the typical need for pre-treatments in order to induce the sensing properties and the need for expensive deposition techniques for the realization of the sensing devices [8].

For these reasons, the research aiming to develop novel active materials for sensing, able to combine cheap deposition, high sensitivity and the capability to operate at room temperature is still running.

Among the different classes of innovative semiconducting materials, lead halide perovskites recently started to receive enormous interest thanks to their unique capability to combine simple film deposition from a solution, active properties typical of semiconductors and wide chemical flexibility [9–12]. The general chemical formula of lead halide perovskites is $APbX_3$, where X = Cl, Br or I, or mixed Cl/Br and Br/I systems, and A is an organic group in hybrid organic–inorganic perovskites or an inorganic cation in totally inorganic ones.

The chemical and physical properties of the perovskites strongly depend on the preparation procedures. $APbX_3$ are typically obtained by two methods: synthesis from solution and deposition from vapor phase. Alternative procedures, even if less common, are through solid-state reactions [13] or atomic-layer-deposition (ALD) [14]. In the precipitation-from-solution procedure, the precursors (PbX_2 and AX) are simultaneously dissolved in a solution; the subsequent evaporation of the solvent, by heating or spin-coating, leads to the formation of the perovskite crystals (one step method). In the two-step procedure, instead, a solution containing the lead halide salt is spin-coated onto a substrate and dried in order to obtain a PbX_2 film. Then the film is exposed to AX (e.g., dipped in a AX-based solution) to convert the PbX_2 film into the perovskite. In the vapor-phase deposition the perovskite is obtained by the coevaporation of the lead halide salt and the AX halide. This method generally yields more uniform films if compared with those synthesized from solution. In the solid state reactions, equimolar amounts of PbX_2 and AX are loaded in a tube, shaken to homogenize the mixture and then heated to obtain a homogeneous solid (in some cases through a melted phase). Alternatively, PbX_2 and AX are placed in a mortar and ground with a pestle to get a uniform powder. Finally, multistep synthesis based on the conversion of a seed PbS film, deposited by ALD, into PbX_2 and the subsequent formation of the perovskite through a dip in an AX solution has been explored by Sutherland et al. [14]. This procedure yields thin films with highly controlled thickness and excellent purity. Under appropriate synthesis conditions, colloidal lead halide nanocrystals can be obtained in solutions and nanocrystal (NC) films can be deposited onto substrates by means of drop-casting or spin-coating. [15,16].

The first synthesis of metal halide perovskites dates back to the late 1970s [17,18], but their physical and chemical properties have not been deeply investigated until the pioneering research of Miyasaka's group in 2009 [19], when they used hybrid perovskite as a light absorbing material in a solar cell, reaching less than 4% efficiency. Since then, thanks to their large absorption coefficient, low defect density and high carrier mobility [12,20,21], perovskites have attracted more and more attention, and in a short time all the research efforts have brought to the actual solar cells' power a conversion efficiency of 25.5% [22].

Together with the development of solar cells, a wide variety of active properties of lead halide perovskites have been demonstrated, such as electroluminescence, efficient photoluminescence and optical gain, thereby showing perovskites to be candidate active materials for other optoelectronic devices.

Metal halide perovskites indeed have also been exploited for the realization of LEDs. Starting from the first reports of the electroluminescence in organic–inorganic perovskites [23,24], lots of advances have been made, leading to bright emission devices [25]. The possibility to deposit perovskite films through solution-based techniques and tune the emission wavelength by varying the chemical composition of the perovskite allowed researches to obtain light emitting diodes with increased efficiency and wide color gamuts [10,26–28].

Concurrently, the first report of room temperature amplified spontaneous emission (ASE) has been provided [29], which cleared the path for the implementation of lead halide perovskites as active materials for lasers. Great efforts have been made to improve the ASE properties [16,30–35] and lasers have been realized with different geometries, such as microcavity lasers [11], distributed feedback lasers [36], whispering gallery mode lasers [37] and random lasers [38].

Moreover, due to perovskite attractive electrical and optical properties, a new route for the application of perovskites in the detection of high-energy radiation, such as X-ray, gamma and deep-UV photons, has been opened up [39–48]. In particular, the ability to convert high-energy photons into lower-energy, direct bandgap emissions recently made Cs-based perovskites ideal candidates for implementation in deep-UV photodetectors. As an example, inorganic $CsPbBr_3$ quantum dots (QDs) have been integrated into perovskite photodiodes as fluorophors playing two roles: they act as a down-conversion layer, converting incident UV into emitted 510 nm light, and protect hybrid perovskite from UV degradation [46].

However, the lack of long term chemical stability to date still prevents applications in commercial devices.

In particular, many experiments demonstrated that lead halide perovskite films are sensitive to the environment chemical composition, and depending on the chemical nature of the interacting species, the material can show irreversible interactions, leading to degradation or reversible interactions thereby preserving the active properties when the external gas source is removed. The reversible interactions could be potentially exploited for sensing applications.

Another interesting aspect of lead halide perovskites for sensors development is related to their tendency to form films with pores on the surface, potentially allowing an efficient superficial interaction with gas species and improving gas adsorption and evacuation. These features could be important to allow fast sensor response and recovery times and high sensitivity even at room temperature [49].

Recently, various review articles debating the use of perovskite as sensing materials have been published [50–52], that deal with different classes of perovskites (such as metal oxide and metal halide perovskites), concern the interaction of the sensing material with metal ions, solvents and gases, and focus on the chemical point of view of the interaction between the perovskite and the analyte [53].

In this paper, differently from what already done, we want to focus on a particular class of perovskites, i.e., lead halide ones, and review the actual state of art of the research on the effects of interaction with the environment. We will initially investigate the main processes inducing perovskite degradation, due to the interaction with light, surrounding atmosphere (oxygen, moisture, NH_3 or other gases or volatile organic compounds (VOCs)), and due to heating. Then we will focus on the reversible modulation of electrical and optical properties of perovskites, that can be exploited for the development of perovskite gas sensors giving an almost complete overview of the tested devices.

2. Environmental Stability of Perovskites

The environmental stability is a particularly relevant property for any material aiming to be used in any application, as it allows the realization of devices with a long operative life. In particular an active material for a solar cell should be stable when operating in air and under sun light, and should also be stable against heating induced by the sunlight exposure. For these reasons, given the impressive increase of the power conversion efficiency of perovskite solar cells, many groups investigated the stability of perovskite thin films under exposure to different external perturbations, with a particular focus on the perovskite irreversible degradation channels.

In this section we will describe the main results in literature about perovskite degradation due to the exposure to light, heating, and environmental species, such as oxygen, moisture or other gases (such as NH_3, H_2 and volatile organic compounds).

Even if most of the referenced papers below report stability tests on perovskite for the realization of solar cells, exploring both the variations of structure/morphology and of the electro-optical properties, the results about the intrinsic stability of the material are expected to be of general validity and independent of the specific application.

2.1. Light and Temperature Effects

Most of the experiments about lead halide perovskites have been performed on iodide perovskites, and in particular, on MAPbI$_3$, which thanks to its absorption edge in the near infrared, has been widely used as active material in solar cells.

A preliminary interesting result on the MAPbI$_3$ stability has been obtained on active films stored in dark and vacuum condition [54], surprisingly demonstrating that the material shows irreversible degradation even if not exposed to light and to external chemical species. The self degradation was ascribed to a sequence of four different processes: decomposition of MAPbI$_3$ in $[MA]^+$, $[PbI_2]^-$ and $[I]^-$ ions; dissociation of $[PbI_2]^-$ into Pb^0 and I_2^- at the perovskite surface; regeneration of I_2^- ions at the perovskite interface and formation of AuI$_2$ at the interface with the gold electrode of the solar cell device.

Starting from this evidence of intrinsic instability it is easy to understand that additional factors, such as light exposure, heating or interaction with the environment can induce further degradation processes.

The effects of sunlight exposition are reported by Misra et al. [55] who investigated the material stability by measuring the evolution of the absorption spectra of MAPbI$_3$ thin films, deposited and encapsulated in a glove box (thus exposed to light in an inert environment), and exposed to concentrated sunlight of 100 suns. The authors observed that light exposure for times up to 1 h has no effects on the absorption spectra if the sample was kept at 25 °C (see Figure 1a). On the contrary, a clear variation of the absorption is observed when the film is exposed to concentrated sunlight with a film temperature increase to only 45–55 °C (see Figure 1b). In particular, beyond a general reduction of the total absorption (see inset of Figure 1b), a clear variation of the absorption lineshape is observed, with a reduction of the intrinsic MAPbI$_3$ absorption and an increase of absorption in the range 400–460 nm, ascribed to MAPbI$_3$ decomposition with crystallization of its inorganic component PbI$_2$. This effect is not observed when the sample is heated at the same temperature without light exposure, evidencing that a combination of heating and lighting is required to determine the degradation. In the same experiment the authors also demonstrated that no degradation takes place in MAPbBr$_3$ films, both at 25 °C and at 45–55 °C, clearly evidencing that the light-induced degradation is highly dependent on the hybrid perovskite composition (see Figure 1c). The higher stability of Br-based hybrid lead halide perovskite with respect to the I-based one was ascribed to differences in the ionic ratio of Br$^-$ and I$^-$ ions that influences the perovskite crystal structure. In particular the cubic crystalline phase of MAPbBr$_3$ results denser and thus less prone to environmental molecules attacks than the tetragonal structure of MAPbI$_3$.

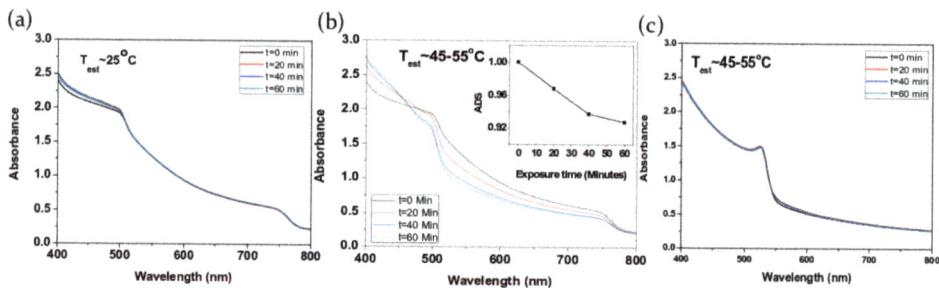

Figure 1. UV–Vis absorption spectra of encapsulated MAPbI$_3$ films exposed for various times at (a) ≃25 °C and (b) ≃45–55 °C: spectral modifications appear at high temperature, whereas (c) the MAPbBr$_3$ film remains stable even at ≃45–55 °C. Adapted with permission from [55]. Copyright © 2015, American Chemical Society.

The higher stability of MAPbBr$_3$ with respect to MAPbI$_3$ also comes from the higher bromine electronegativity and to the stronger Pb-Br bond compared to the Pb-I one, and of the H-Br bond compared to the H-I one [54,55].

The details of photo-degradation and thermal decomposition of MAPbI$_3$ and MAPbBr$_3$ have been also investigated by Juarez-Perez [56]. Beyond evidencing that light and heat can, in proper conditions, determine the perovskite degradation, the experiment allowed to conclude that parts of the reaction determining the release of gas compounds are reversible and cannot be considered real degradation channels. In particular the released CH$_3$NH$_2$ cannot be considered as degradation products of perovskites because they can resynthesize MAPbI$_3$ if the film is in a closed environment (for example, an encapsulated film). On the contrary, the back formation to MAX (X = Br, I) or MAPbX$_3$ from the released CH$_3$X + NH$_3$ molecules is thermodynamically unfavorable and prone to form non-primary ammonium salts, and it is thus an authentic detrimental pathway for perovskite degradation.

Further insight in the light induced MAPbI$_3$ degradation processes was obtained by X-ray photoelectron spectroscopy (XPS) on films irradiated in vacuum by a blue laser [57]. A two step degradation process was proposed, with an initial decomposition of MAPbI$_3$ leading to the formation of PbI$_2$ and volatile compounds, such as NH$_3$ and HI. Then, PbI$_2$ further decomposes into metallic lead (Pb^0) and iodine (I$_2$). A saturation of the degradation was observed after 480 min of laser irradiation, when the ratio of metallic Pb to total Pb was about 33%.

The thermal stability of MAPbI$_3$ perovskite films has been also investigated by Conings et al. [58] in various ambient conditions (pure N$_2$, O$_2$ and ambient air), corroborating the idea that bringing perovskite films to a temperature of 85 °C, which is close to the temperature at which perovskite is formed, can induce surface decomposition, even in an inert atmosphere. The presence of oxygen and water molecules further accelerates the disintegration process since they react with the methylammonium iodide (MAI) cation, leading to the breakage of the perovskite unit cell.

A nice example of the correlation between the active material degradation and the variation of its optical properties has been obtained by Motti et al. [59] who analyzed the effects of the photoexcitation energy density on the photoluminescence (PL) and the transient absorption of polycrystalline thin films, of both hybrid (MAPbBr$_3$ and MAPbI$_3$) and fully inorganic (CsPbBr$_3$) perovskites.

In all the investigated materials they demonstrated that the laser pumping results in the formation, even in vacuum, of sub-bandgap defect states, quenching the PL. This effect is evidenced by PL measurements as a function of the excitation density showing a progressive increase of the PL intensity as the excitation density increases, but also a systematically lower PL intensity when the excitation density is decreased back to the initial value (see Figure 2a). This effect is also present when the samples are pumped in air, but the relative decrease of the PL intensity is, surprisingly, much weaker. This suggests that the presence of oxygen in the atmosphere, even in small amounts, attenuates the effect as it acts as passivating agent for such defect states.

The role of the excitation wavelength on MAPbI$_3$ perovskite thin films has instead been described by Quitsch et al. [60]. They demonstrated that an increase of the photoluminescence intensity (called photobrightening) occurs under green laser illumination ($\lambda > 520$ nm), while the decrease of the emitted PL is obtained by excitation with a blue led ($\lambda < 520$ nm), as shown in Figure 2b. In particular, the threshold wavelength, separating photobrigthening from photodegradation, is determined by the PbI$_2$ energy band gap. The quenching of the photoluminescence is indeed related to a photodegradation process, ascribed to the formation of I$_2$ by photolysis of PbI$_2$ and/or hole transfer between the photoexcited species under blue light exposure. Moreover, similarly to the previous report, oxygen has been demonstrated to have a positive role in inducing the photobrightening, since oxygen-based species act as passivating agents for perovskite halide vacancies.

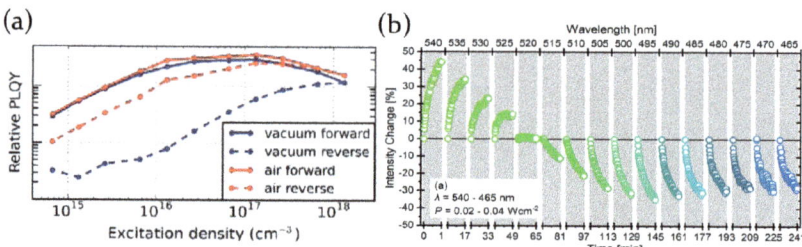

Figure 2. (a) Relative photoluminescence quantum yield (PLQY) curves of MAPbBr$_3$ polycrystalline thin films in air (red) and in a vacuum (blue). Adapted with permission from [59]. Copyright © 2016, American Chemical Society. (b) Percentage PL intensity change of a MAPbI$_3$ film for 60 s of illumination in air, for excitation wavelength between 540 and 465 nm (gray shades). Between each measurement the sample was kept for 15 min in the dark in order to reduce the overall stress (white columns); the emission was detected at the PL peak position (778 nm). Adapted with permission from [60]. Copyright © 2018, American Chemical Society.

The evidence of several degradation channels in hybrid organic–inorganic perovskites stimulated the research of novel chemical compositions allowing a stability increase. In this frame a particularly interesting class of materials are fully inorganic perovskites, in which the organic cation is replaced by an inorganic atom (typically Cs) [61].

A couple of interesting experiments have been performed in order to investigate the stability of CsPbBr$_3$ perovskite nanocrystals films [62,63], evidencing that also these materials show several degradation processes induced by light and depending on the environment chemical composition.

In particular it has been shown that under illumination of 450 nm light-emitting diode they suffer degradation, resulting in a sample color change from green to yellow, both in solutions and in films.

As highlighted in Figure 3c–f, the pristine green nanocrystal (NC) solution is formed by highly monodispersed nanocrystals with an average length of about 5–10 nm; as the exposition time goes on, NCs tend to aggregate forming clusters, up to 50 nm after about 2 h of illumination, and the solution color turns to yellow (Figure 3a,b). This effect has been observed both in solution and in films and is ascribed to the light source which removes bonding ligands from the surface of the nanocrystals, allowing them to merge and form bigger aggregates. Concurrently, the PL emission intensity decreases as a function of the illumination time, coming with a redshift of the PL peak. The surface decomposition of the nanocrystals determines indeed an increase of the surface charge trap states, which quenches the sample light emission, whereas the shift toward longer wavelengths is attributed to the greater crystals dimensions. Moreover, the degradation process has been demonstrated to be dependent on the power illumination, since a 89.4% PL loss has been obtained at 175 mW/cm^2 and it reaches higher values (99.3%) at 350 mW/cm^2 for CsPbBr$_3$ films.

In addition to illumination, other factors have been reported to influence the sample degradation. The color variation is indeed not present in encapsulated samples, evidencing the effects is not simply due to photodegradation. On the other hand, no degradation is also observed if the samples are stored in dark, even in oxygen atmosphere, suggesting that a combination of light exposure and oxygen is necessary to induce the degradation.

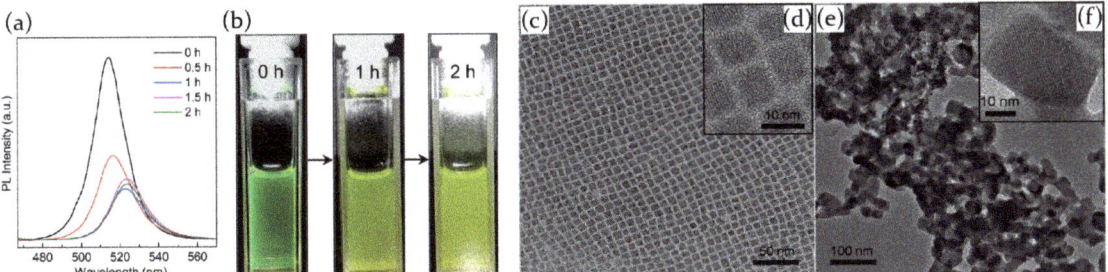

Figure 3. (a) PL emission spectra of the colloidal CsPbBr$_3$ toluene solution and (b) the optical images of the cuvette as a function of the illumination time (175 mW/cm^2). (c) Transmission electron microscope (TEM) and (d) high resolution transmission electron microscope (HRTEM) images of CsPbBr$_3$-NCs. (e) TEM and (f) HRTEM images of the a CsPbBr$_3$ sample after 2 h of illumination. Adapted with permission from [62]. Copyright © 2017, American Chemical Society.

The role on the NCs size for CsPbBr$_3$ films under UV light illumination has also been investigated in [63], evidencing that the morphology of the film remains more stable (quite unchanged) for samples composed of bigger grains. The authors deposited two CsPbBr$_3$ NCs films with an average crystal dimensions of 9.7 nm and 11.4 nm, respectively, emitting at 495 nm (NC495) and 520 nm (NC520). The experimental results evidenced that under UV light exposure NC520 NCs remain rather unchanged, whereas smaller crystals in the NC495 sample tend to merge and form bigger oval-shaped aggregates, related to the formation of Cs$_4$PbBr$_6$ products, as evidenced in XRD spectra.

However, both samples (NC495 and NC520) suffer from the combined detrimental action of light irradiation and temperature, since both show a PL intensity decrease for increasing temperature values from 80 K to 400 K, under UV light illumination. The observed quenching of the photoluminescence is probably related to the formation of trap states which limit radiative recombination processes, and are ascribed to the photooxidation mechanism of lead atoms in perovskite films.

2.2. Air, Oxygen and Humidity Effects

There is strong evidence that in presence of air, oxygen and/or humidity the perovskites can modify their properties, and the irradiation and temperature effects are enhanced. With the aim to understand the ambient induced degradation, numerous investigations have been carried on by studying the effects of oxygen and wet or dry air.

For example, close attention has been payed to the role of humidity in the atmosphere in contact with a perovskite sample, even if opposing results have been reported. Some of them point out a detrimental action of water molecules on the morphological stability of the perovskite structure, others evidence that it does not inevitably lead to deleterious effects.

The ascertained intrinsic thermal instability of MAPbI$_3$ [58] is boosted in ambient air. Even if the samples annealed at 85 °C after 24 h show structure variation regardless the atmosphere (N$_2$, dry O$_2$ or ambient air), the authors found that the major changes are induced when the samples are treated in air. Water and oxygen molecules can interact with the methylammonium cations, leading to the formation of new carbon based species and the release of nitrogen. Moreover, moisture accelerates the formation of metallic lead (Pb^0) clusters from Pb^{2+} ions, in addition to the most commonly known formation of PbI$_2$ as degradation product of MAPbI$_3$ perovskite.

All these degradation pathways find confirmation in XPS measurements, as reported in Figure 4. The left panel of Figure 4 shows the carbon compounds XPS peaks of the as-prepared sample (black curve) and for the sample stored in different atmospheric conditions (air, oxygen and nitrogen, up to 120 h of storage). The pristine sample is characterized by two carbon contributions, one at higher binding energy (286.7 eV) and the other one at lower binding energy (285.3 eV). The first one is ascribed to the C-N bound of the MAI cation (indicated as "perov." in the figure), whereas the second one refers to C-C and C-H bounds of hydrocarbons formed on the perovskite surface at the deposition stage.

Heating the sample supports the release of the methylammonium volatile compounds, leading to a gradual disappearance of the C-N bound peak, in favor of the hydrocarbon ones. This effect results to be more prominent in presence of oxygen and water molecules, and is corroborated by a concurrent decrease of the N1s core level peak intensity, at 402 eV, sign of the release of N_2 molecules from the perovskite surface (central panel of Figure 4). Besides, the panel on the right evidences the appearance of metallic lead with the Pb^0 peak at about 137 eV at increasing exposition times in air.

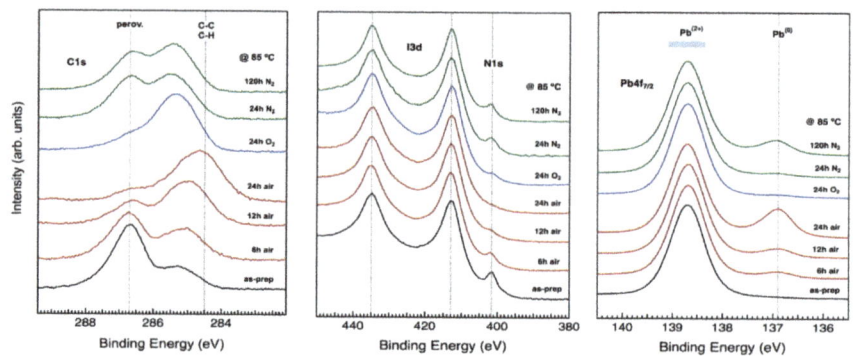

Figure 4. XPS measurements on MAPbI$_3$ perovskite samples treated at 85 °C for up to 24 h in different atmospheres. Adapted with permission from [58]. Copyright © 2015, WILEY-VCH Verlag GmbH and Co. KGaA, Weinheim.

Humidity is also revealed to be a critical factor in the performance of MAPbI$_3$ solar cells due to the deleterious effects of H_2O vapor on the perovskite [64]. Morphology and crystal structure measurements have been performed for perovskite films exposed to different atmospheric conditions in the dark (0–90% RH). It has been observed that before humidity exposure all the perovskite samples are characterized by a rough surface whereas, after being in contact with water molecules, they undergo a recrystallization process, becoming smooth and highly ordered; moreover, this process is more emphasized for higher humidity content. X-ray diffraction (XRD) measurements then prove the formation of an hydrate product, similar to $(CH_3NH_3)_4PbI_6 \cdot 2H_2O$, after the exposure of perovskite to humid air in the dark. Anyway, the hydration process seems to be reversible and the perovskite can partially recover its pristine structure after being stored in vacuum or dry atmosphere. In addition, time-resolved absorption spectra highlighted that there are not radical changes in perovskite typical peaks for samples stored up to 14 days in different humidity conditions meaning that, despite the morphological changes observed, the perovskite excited state properties do not change.

In agreement with the previous report, Galisteo-Lopez et al. [65] rule out the irreversible detrimental action of moisture on MAPbI$_3$ perovskite film chemical composition. Optically pumping the sample, the PL maximum registered at 775 nm shows an initial photoactivation stage followed by a radiation quenching (Figure 5a). Measurements taken in different atmospheres (ambient air, N_2 and O_2; Figure 5b) allow one to identify moisture and oxygen as the responsible molecules for photodarkening and photoactivation processes, respectively. Water molecules, in particular, induce the formation of hydrates species on the surface of perovskite, determining a quenching of the photoluminescence. Switching the atmosphere from ambient air to a pure oxygen one allows, on the other hand, to partially recover the sample emission, evidencing that O_2 acts as a trap states filler.

Even if the presence of moisture in the atmosphere surrounding the sample influences its photoluminescence emission, it has been verified that water molecules do not cause permanent degradation of the film. Under prolonged optical pumping in humid conditions, it has been indeed demonstrated that the film does not suffer any color change in its aspect, proving that the interaction of perovskite with water molecules does not induce the formation of PbI$_2$ molecules on the sample surface.

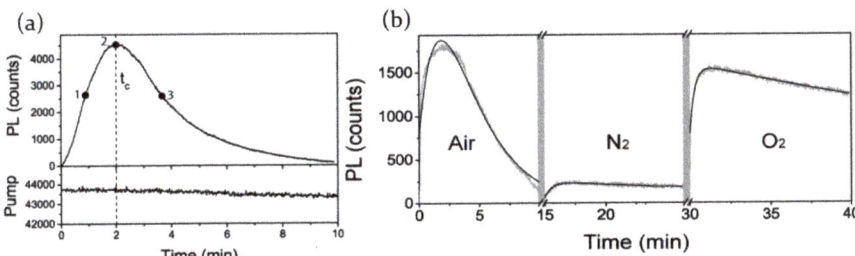

Figure 5. (a) Temporal evolution of PL intensity maximum at 775 nm for MAPbI$_3$ perovskite films, under monochromatic light excitation (500 nm), showing an initial photoactivation stage followed by intensity quenching after t_c. (b) Evolution of the PL peak intensity when switching the atmosphere from air to nitrogen and then to pure oxygen. Adapted with permission from [65]. Copyright © 2015, American Chemical Society. https://pubs.acs.org/doi/10.1021/acs.jpclett.5b00785. Further permission should be directed to the ACS.

On the contrary, other works support the idea of moisture induced permanent degradation, both in MAPbI$_3$ [66,67], FAPbI$_3$ [68] and mixed halide perovskites [69].

Concerning the interaction between the active mean and the environment (in particular, moisture), film morphology seems to play a fundamental role. Being a phenomenon related to uppermost part (surface) of the perovskite, the morphology and the dimensions of the grains composing the deposited film, the presence of grain boundaries and their chemical composition can deeply influence the interaction process, and as a consequence, limit or accelerate the degradation of the film.

The morphology of FAPbI$_3$ perovskite films, for example, has been analyzed through atomic force microscope (AFM) topography images, which evidenced that the sample remains stable when stored in controlled atmosphere below 30% of relative humidity (RH) at room temperature, but suffers moisture action when the atmosphere reaches 50% RH, and completely decomposes in PbI$_2$ when soaked in water.

In particular, grain boundaries result to be the starting point for perovskite decomposition, representing a pathway for the interaction with water molecules; the degradation then proceeds deeper into the grains. Once the degradation process is started, the morphology of the film changes, resulting in a merging of neighbour grains which form bigger clusters and the appearance of a non perovskite phase, as evidenced in Figure 6a–c [68].

As a consequence, a scaling behavior with grain size has been revealed [67]: since degradation process starts from grain boundaries (Figure 6d), the perovskite films composed of bigger grains result more stable than the films with smaller grains. Another peculiarity relates to the fact that the degradation process, starting from grain boundaries, expands along in-plane direction when the sample is annealed after deposition; otherwise, when the sample does not receive any post-deposition treatment, chemical modifications result more uniform. An annealing process, indeed, facilitate the release of MA ions or MAI from the surface of the perovskite, leaving a PbI$_2$-terminated surface. As PbI$_2$ is less soluble in water than MA ions, the top plane surface of the annealed film results less prone to humidity degradation. Moreover, grain boundaries have been seen to be composed of an amorphous region, caused by an excess of MAI during the film formation phase, which facilitates moisture permeation and then induces degradation (Figure 6f).

The chemical composition of the active material has been proved to play a fundamental role in the response to humidity, as evidenced through the experiment of Howard et al. [69] on thin films of Cs$_x$FA$_{1-x}$Pb(I$_y$Br$_{1-y}$)$_3$ mixed halide perovskites. Different concentrations of Cs/FA cations and I/Br halides have been chosen in order to show which is the photoluminescence response of the films at different humidity concentrations (with RH ranging from 5 to 55%).

At low humidity levels, for RH ranging from 5% to 35%, all the samples analyzed, regardless their chemical composition, show an enhancement of the emitted light intensity for increasing humidity content. Water molecules, indeed, remove defect states not already passivated from O_2 molecules, determining a PL intensity increase.

Switching the atmosphere from 35% to 55% of RH, on the other hand, the chemical composition of the active material results to be crucial. Low (17%) Br content samples, indeed, maintain the PL intensity at the same level; high (38%) Br content samples, instead, show a PL intensity decrease combined with a PL peak red-shift, both independently from the Cs percentage.

The presence of high humidity content, moreover, is responsible for the degradation of the perovskite. 55% of RH is indeed sufficient to form a thin layer of monohydrate species on the perovskite surface. As the water adsorption proceeds, dihydrate species are formed, leading to perovskite degradation with PbI_2, $PbBr_2$ and $Cs_xFA_{1-x}I_yBr_{1-y}$ formation and the presence of new surface trap states. Since moisture presence influences the halide ions mobility in the perovskite crystal structure, major Br-content samples results to be more affected by degradation and suffers PL intensity quenching processes to a greater extent.

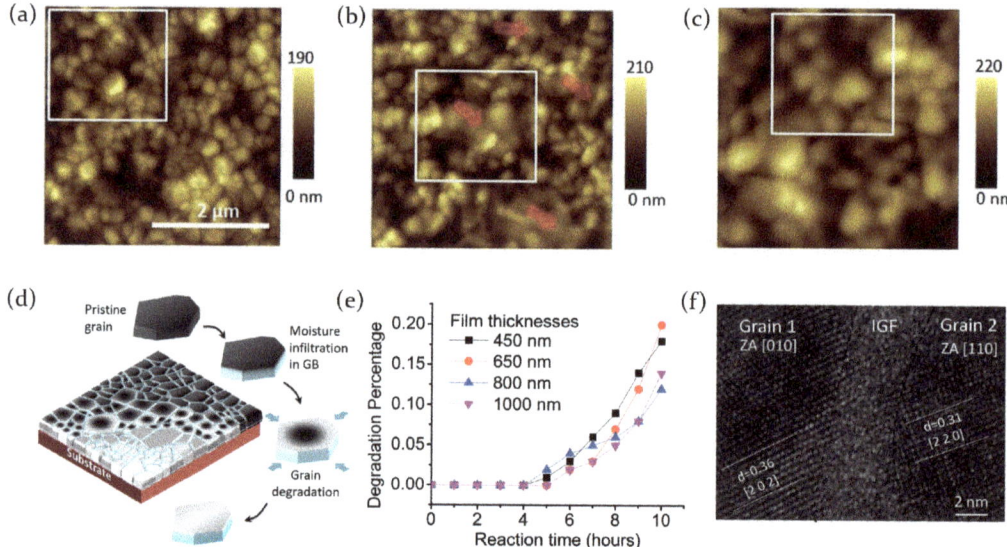

Figure 6. (a–c) AFM images of $FAPbI_3$ film, showing the different stages of the degradation process, from the fresh sample (dark black film) to the degraded one (pale yellow). Adapted with permission from [68]. Copyright 2018, WILEY-VCH Verlag GmbH and Co. KGaA, Weinheim. (d) Schematic representation of the degradation of $MAPbI_3$ perovskite film in moisture; the degradation mechanism starts from grain boundaries and extends towards the grain center along the in-plane direction. (e) Degradation percentage as a function of the moisture exposure time for different film thicknesses. (f) A HRTEM cross-sectional image highlighting the presence of an intergranular film (IGF) between two neighboring grains. Republished with permission of the Royal Society of Chemistry, from [67], © 2017.

Another chemical species deeply investigated is oxygen; metal halide perovskites have been revealed indeed to be highly sensitive to the presence of O_2 molecules, both in a pure oxygen atmosphere and in ambient air.

Fang et al. [70], for example, showed how the atmosphere surrounding a $MAPbBr_3$ perovskite single crystal can modulate its photoluminescence emission. Pumping the sample with a 400-nm excitation laser, with a pulse repetition of 1.4 MHz, the effect of different gases has been studied, monitoring the PL emission every 10 s. Is has been verified that the presence of dry N_2, CO_2 and Ar does not affect perovskite light emission in time; air, dry O_2 and wet N_2 instead induce PL changes, in particular determining photoactivation processes. The major PL enhancement is observed in presence of air, suggesting that oxygen and water molecules are the real responsible for the observed emission variations. Switching the chamber atmosphere to vacuum condition, an almost immediate photoluminescence quenching is observed, and the process results to be reversible when going back to air atmosphere.

The experimental results thus lead to think that the interaction between perovskite and water/oxygen molecules is a physisorption, rather than a chemisorption. Oxygen and water molecules role is to act as electron donors, neutralizing an excess of positive Pb^{2+} charges on perovskite. Since those trap states have been demonstrated to be located mostly on the surface rather than in the perovskite bulk, a control of the surface trap states density through sample exposure to different gases is possible. As a consequence, a reversible PL intensity modulation can be obtained and $MAPbBr_3$ single crystals become suitable for new potential applications in the detection of oxygen and water vapor.

Similarly to what reported above, also Zhang et al. [71] observed a PL intensity modulation from $MAPbBr_3$ perovskite single crystals. Interestingly, the analysis evidenced that the photoluminescence intensity decrease, obtained passing from air to vacuum atmosphere, was accompanied by a corresponding modulation of the electrical properties. In particular, both dark current and photocurrent, measured through the gold electrodes deposited on the $MAPbBr_3$ film surface, have been observed to increase in vacuum conditions. This suggests the formation of shallow trap states in ambient air exposure; since those trap states are close to the perovskite band edges, photogenerated carriers can be trapped in, forming radiative recombination centers and then leading to a PL enhancement. As a consequence, in presence of air charge carriers are trapped and radiatively recombine, and can not be available for photocurrent in the perovskite device; viceversa, in vacuum conditions, the density of trap states is reduced and the current increases.

Air, moisture and light stimuli have then been also studied together, in order to evaluate the possible synergistic effects arising from their combined presence.

As an example, Tian and its coworkers [72] investigated the luminescence properties of a spin coated $MAPbI_3$ perovskite polycrystalline film, studying both the PL intensity variation by CW 514 nm excitation laser and the PL decay dynamics under picosecond pulsed laser, when the sample is exposed to different external stimuli. In particular, they focused on the curing action induced by the combined presence of light and oxygen, needed to deactivate quenching sites in the perovskite crystal structure. As a consequence, initial low PL intensity can increase for more than three order of magnitudes as the light exposition goes on. Along with the PL intensity enhancement, also a PL lifetime increase, from several nanoseconds to several hundreds of nanoseconds, has been observed. Excitons and free carriers (electrons and holes) created from the absorption of photons can be trapped by perovskite defect sites, thereby inducing non radiative recombinations. The combined presence of oxygen and light can, however, reduce the concentration of trapping sites and increase the charge carrier diffusion length. As a consequence, if initially the process is confined to the upper region of the perovskite layer, as the curing action goes on, trapping sites in bulk perovskite can be filled.

Other works confirming the competing action of photobrightening and photodarkening are reported in literature.

As an example, Godding et al. [73] used $MAPbI_3$ films excited with a picosecond pulsed laser emitting at 504 nm, in order to explore the effects of the combined action of light excitation and air exposure on the photoluminescence properties of the sample. In particular, they observed an initial photobrightening phase, followed by a worsening stage of the emission properties in which the PL intensity starts to saturate. Light excitation is thought to determine the modification of the perovskite stoichiometry, through the generation of atomic lead and the loss of iodine, increasing surface state density responsible for charge non-radiative recombination. Molecular oxygen then diffuses into the film forming superoxide species (O_2^-), whose interaction with methylammonium cations leads to peroxide H_2O_2 formation. The latter, in turn, interacts with atomic lead, forming lead oxide (PbO). This oxidation process improves the emission properties since it contributes to passivate trapping sites. On the other hand, the simultaneous loss of MA ions and the formation of PbI_2 overcomes the beneficial effect of passivation and brings to perovskite degradation.

Motti et al. [59] repeated the study of the irradiation effect on perovskite polycrystalline films in atmosphere and showed that PL emission greatly increases when passing from vacuum to air. Comparison between the results obtained in wet and in dry air leads to conclude that oxygen is the molecule responsible of the observed PL enhancement. As stated above, light irradiation creates sub band-gap states which trap carrier charges and induce an instant PL quenching. Filling the chamber with air, however, allows one to restore the photoluminescence, thanks to the combined action of light and oxygen, which passivates defects. PL decay and transient absorption (TA) measurements, moreover, allow one to shed light on the PL decay dynamics, showing the presence of a fast component, on the order of few nanoseconds, and a slow one, which extends on microsecond time scale. The first one is associated with the carrier trapping mechanism, whereas the second one is related to the recombination of trapped carriers. Exposing the sample to oxygen reduces the trap sites density; as a consequence, the fast decay component shows a longer lifetime compared to inert atmosphere, whereas the slow component almost disappears.

Starting from the first evidence of the effect of the excitation wavelength on $MAPbI_3$ perovskite thin films (which showed photobrightening induced by green excitation light and photodegradation by blue light), Quitsch et al. [60] further went into the issue in order to reveal which is the role of the atmosphere. For this purpose, photostability measurements have been performed both in vacuum and in ambient air.

The experimental results evidenced that under green laser illumination a photobrightening effect is always observed and appears to be promoted in air. On the other hand, the presence or absence of humidity in the atmosphere reveals to be crucial for a blue light illumination. In vacuum or dry air condition, indeed, the PL intensity initially increases and then suffers a decrease with ongoing illumination time, sign of the degradation of the material. Differently, in presence of humid atmosphere, the photodegradation process, responsible for the PL quenching, is boosted and completely hides the photobrightening effect. These results suggest the coexistence of a wavelength-independent photobrightening and a wavelength-dependent photodegradation effect in $MAPbI_3$ upon excitation above PbI_2 band gap.

According to the authors, the reason behind the increase of the photoluminescence is attributable to the passivation of iodine vacancies on the perovskite surface. These vacancies indeed act as non-radiative trap states located in energy immediately below the conduction band minimum. Once they are passivated (for example, by superoxide ions formed due to the presence of light and oxygen), sub-bandgap states shift into the valence band and trap states density is reduced, determining an increase of the PL intensity [74].

On the other hand, exciting the sample with a wavelength shorter than 520 nm allows the formation of PbI_2, detrimental for the stability of the perovskite, which therefore brings to a PL decrease.

The presence of a light source, combined with an oxygen rich atmosphere, is thought to be responsible for the formation of superoxide species (O_2^-) through charge transfer from perovskite to molecular oxygen, thereby activating degradation processes of the active mean.

The photo-oxidative degradation mechanism has been hypothesized by Ouyang et al. [75] for $MAPbI_3$ perovskite and includes three steps, as schematically represented in Figure 7a–c.

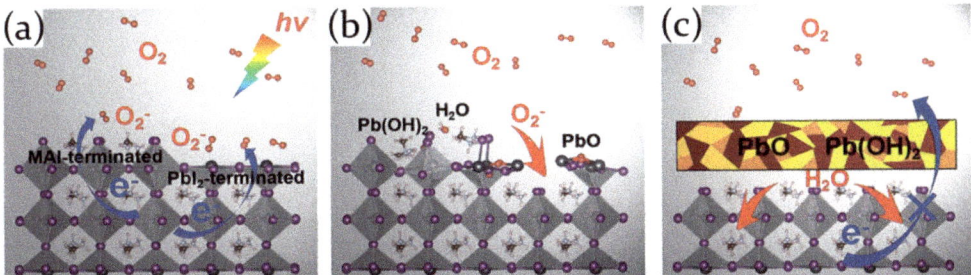

Figure 7. Schematic representation of the photo-oxidative degradation process of the MAPbI$_3$ surface. (**a**) O$_2$ molecules close to the surface of MAPbI$_3$ get the photo-excited electrons from perovskite, forming superoxide anions (O$_2^-$). (**b**) O$_2^-$ reacts with MA$^+$ and Pb, yielding H$_2$O and Pb(OH)$_2$ on the MAI-terminated surface. Oxidation of lead on the PbI$_2$-terminated surface leads to the disintegration of the Pb–I bond, producing PbO and leaving the underlying MAI-terminated surface exposed. (**c**) The oxidation products PbO and Pb(OH)$_2$ form a protection layer, preventing the oxidation of inner MAPbI$_3$. However, the fresh water molecules produced in previous steps cause hydration of the inner perovskite. Used with permission from the Royal Society of Chemistry from [75], © 2019; permission conveyed through Copyright Clearance Center, Inc.

First, superoxide anions (O$_2^-$) are created through a charge transfer from perovskite surface to molecular oxygen. The second step depends on the surface composition; indeed, since MAPbI$_3$ perovskite surface can be PbI$_2$-terminated or MAI-terminated, two different degradation mechanisms are proposed. For a PbI$_2$-terminated surface, due to the higher electronegativity of oxygen compared to iodine, oxygen atoms replace I atoms on the surface, forming Pb-O bonds. The formation of lead oxide results in the release of iodine molecules (I$_2$), with a consequent breakage of the surface and the unveiling of the MAI-terminated layer. The exposed MAI-terminated surface is then further oxidized and PbO, H$_2$O, and the unstable Pb(OH)$_2$ are formed. Third, the oxidation products PbO and Pb(OH)$_2$ form a protection layer to prevent a further oxidation of inner perovskite. Nevertheless, the fresh H$_2$O molecules produced in the previous steps act as active species, causing hydration of the inner perovskite and ultimately destroying the MAPbI$_3$.

The formation of negatively charged oxygen molecules influencing the PL properties of the sample is reported also in [76]. The negative oxygen molecules form a layer on the perovskite surface activating, by electrostatic repulsion, the migration of interstitial halide anions toward the perovskite bulk. This mechanism favors the formation of halide vacancy/interstitial Frenkel pairs reducing the density of non-radiative trap states in the bulk of the material and thus inducing the increase of the PL intensity. On the other hand, light exposition in an oxygen atmosphere concurrently induces Pb^0, Br$_2$ and methylammine formation, leading to MAPbBr$_3$ degradation and PL decrease. When light and oxygen sources are removed, oxygen species (O$_2^-$ and water molecules, in particular) remain on the perovskite surface, as inferred from XPS measurements; Br signal is partially recovered, but C and N are irreversibly lost, confirming the hypothesis of volatile methylammine formation and the disintegration of the perovskite lattice (Figure 8a–c) [76].

A combination of experimental measurements and theoretical density fuctional theory (DFT) calculations allowed Aristidou et al. [77,78] to evaluate the effect of the exposure of perovskite films to oxygen and identify the route of superoxide species formation. The authors focused their attention on a couple of samples, namely MAPbI$_3$ and MAPbI$_3$(Cl), which are identical in chemical composition but differ for the deposition stage. The MAPbI$_3$(Cl) has been synthesized by using chloride as precursor, even if it does not influence the final chemical composition of the film. Chlorine ions in the precursor mixture is known to slow down the perovskite formation process, allowing to obtain grains bigger than those formed in MAPbI$_3$.

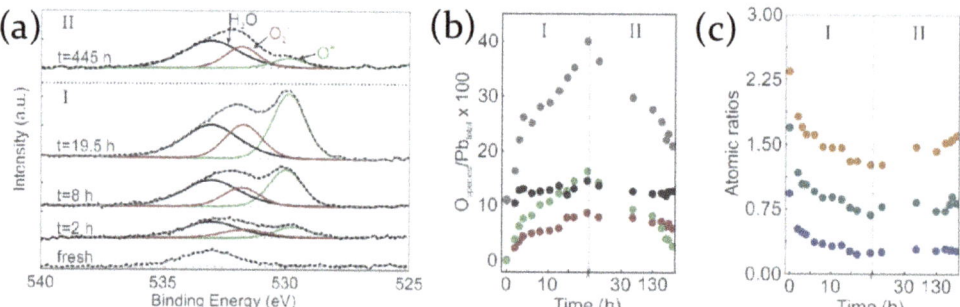

Figure 8. (a) Photoelectron energy spectra for (**I**) a photoexcited MAPbBr$_3$ sample at different times and (**II**) after keeping the sample stored in the dark and in vacuum, with a deconvolution of different oxygen signals. (b) Evolution of the O$_{total}$/Pb$_{total}$ (gray circles), H$_2$O/Pb$_{total}$ (black circles), O$_2^-$/Pb$_{total}$ (red circles) and O^{2-}/Pb$_{total}$ (green circles) atomic ratios for the whole experiment, for phases I and II. (c) Evolution of the Br/Pb (orange circles), C/Pb (dark green circles) and N/Pb (blue circles) atomic ratios for the whole experiment, for phases I and II. Adapted with permission from [76]. Copyright © 2018, American Chemical Society. https://pubs.acs.org/doi/10.1021/acs.jpclett.8b01830. Further permission should be directed to the ACS.

First of all, the oxygen diffusion mechanism in MAPbI$_3$ and MAPbI$_3$(Cl) perovskite films has been analyzed with isothermal gravimetric analysis (IGA) and time of flight-secondary ion mass spectrometry (ToF-SIMS) measurements, evidencing that the diffusion of O$_2$ into the lattice is rapid, reaching the saturation after only 5–10 min, and this represents the reason of the high instability of the material. Then they analyzed through DFT simulations the combined effect of light illumination and oxygen presence, leading to perovskite degradation. Iodine vacancies are formed under light illumination and act as non radiative recombination centers; then, due to an electron transfer mechanism, O$_2^-$ ions are formed. The entire mechanism is shown to be influenced by the particle size of the film: smaller grains show higher defect density and thus, minor stability, whereas films composed of larger crystals result to be less prone to degradation. As a consequence, MAPbI$_3$(Cl) film is more stable than MAPbI$_3$ under atmosphere exposure. In order to hinder O$_2^-$ ion formation and stop the degradation of the material, authors propose a treatment with iodine salts. They, indeed, substitute oxygen molecules in the process of recombination with iodine vacancies, suppressing the formation of the superoxide species and reducing the number of non-radiative trap sites. Moreover, iodine salts treatment does not modify the perovskite structure, leading the process of oxygen diffusion unchanged.

Recently, fully inorganic perovskites have been discovered to be excellent alternatives to their hybrid counterparts since, by substituting the organic cation with a cesium ion into the perovskite, the structure gains major stability and suffers less long term degradation phenomena.

Nevertheless, to date, there are very few reports in literature which examine in depth the effects of the interaction between the inorganic perovskite and the ambient air (such as oxygen and water molecules, alone or combined), and provide details about the morphology of the film, the photoluminescence and electrical properties of the sample before and after the exposure to the ambient gas. As a consequence, lots of questions about the topic remain unsolved.

Photoluminescence measurements of CsPbBr$_3$ perovskite nanocrystals in controlled oxygen atmosphere reveal strong quenching, which is in contrast to the photobrightening effect observed in both MAPbX$_3$ and CsPbX$_3$ perovskite films and single crystals [59,65,70,72].

In order to investigate the optical response to environmental gases of CsPbBr$_3$ perovskite nanocrystals, the effects of the atmosphere in the exciton recombination processes and the nature of trap states, Lorenzon et al. [79] conducted PL measurements combined with spectro-electrochemical analysis.

The application of electrochemical potentials induces the variation of the Fermi level position, which modulates the emission intensity by altering the occupancy of defect states without degrading the nanocrystals. It has been observed that when a negative reductive potential is applied, the PL emission is strongly quenched, whereas upon the application of a positive oxidative potential the emission slightly increases. In particular the reductive potential leads to the raising of the Fermi level, and the emission mechanism is mostly influenced by trapping of photogenerated holes, whereas electron trapping has a negligible role in nonradiative PL quenching; on the contrary, an oxidative potential corresponds to the lowering of the Fermi level leading to the suppression of hole trapping in defect states. The difference between the PL emission, upon negative and positive electrochemical potential, and the relatively high PL quantum efficiency in unperturbed conditions suggest that the number of active intragap traps is very small and, as a consequence, the oxygen molecules directly interact with the photogenerated electrons in the conduction band without the mediation of the structural defects.

This behavior confirms that the interaction mechanism between $CsPbBr_3$ perovskite NCs and molecular oxygen is clearly different from the process leading to the photobrightening observed in perovskites films and single crystals when exposed to oxygen, which is ascribed to the passivation of defect states by adsorption of O_2 molecules on the sample surface.

The ambient effects on the degradation processes of fully inorganic perovskite nanocrystals (NCs) when illuminated with UV radiation have been deeply investigated also in [62,63].

As an example, Huang et al. [62] performed several experiments on $CsPbBr_3$ NCs both in toluene solution and as thin films, testing the role of illumination power density, moisture and oxygen, and concluding that the NCs surface decomposition is due to a synergistic combination of the three elements (Figure 9a–f). It has been demonstrated indeed that in presence of light, humidity sustains the degradation process (Figure 9e), whereas in dark conditions the effect of oxygen and RH is negligible, as evidenced in Figure 9f.

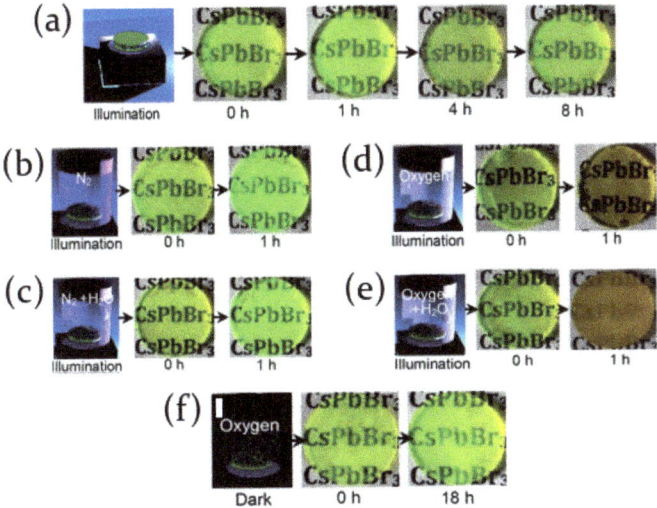

Figure 9. Optical images of the $CsPbBr_3$ film (**a**) sandwiched between two quartz coverslips as a function of the illumination time (175 mW/cm^2, RH 60%). Optical images of the $CsPbBr_3$ film exposed to (**b**) pure N_2 in an illuminated environment, (**c**) N_2 + 0.5 µL H_2O in an illuminated environment, (**d**) oxygen in an illuminated environment, (**e**) oxygen + 0.5 µL H_2O in an illuminated environment and (**f**) oxygen in a dark environment. Adapted with permission from [62]. Copyright © 2017, American Chemical Society.

Indeed, oxygen and moisture can reach perovskite surface and the NCs interfaces thanks to their hydrophilic properties, so that $CsPbBr_3$ hydrates species can be formed. As an effect of illumination, surface bonding ligands are removed through photons absorption. The energetic barrier among crystals becomes weaker, and thus less stable, and NCs tend to aggregate.

Oxygen molecules can erode the NCs, while moisture presence enhances ion migration leading to the formation of PbO and $PbCO_3$ species on the perovskite surface. Surface modification contribute to the generation of trap states, resulting in decreased PL emission [62,63].

2.3. Gas Atmosphere Effects (NH_3, NO_2, VOC, O_3, H_2)

The evidence of possible reversible modulation of the active properties of perovskite thin films stimulated experiments aiming to exploit this effect, in order to detect gas analytes in the sample environment, with particular interest toward pollutants and gases hazardous for human health and environment.

In this section we will resume the main results on the interaction between perovskites and several gas species.

2.3.1. Ammonia

The detection of NH_3 molecules is of high importance since they represent one of the most harmful pollutants; checking the level of ammonia in the atmosphere thus represents the first step to avoid problems on human health. In this regard, metal halide perovskites turned out to be fundamental since their high sensitivity to ammonia presence has been demonstrated both for hybrid and fully inorganic active materials, becoming candidates for the development of a sensing device.

Maity and his coworkers deeply investigated the interaction of hybrid lead I-based perovskite with NH_3 molecules, concluding than the main effect is the decomposition of the active material into PbI_2 [80,81]. Interestingly, scanning electron microscopy (SEM) images of $MAPbI_3$ films deposited on a paper substrate prove that after the interaction with the ammonia the perovskite morphology is more similar to the PbI_2 than to the pristine $MAPbI_3$, as evidenced in Figure 10a–d [80]. Moreover, also the visual aspect of the film evidently suffers from the interaction, since a color change is observed from perovskite natural dark brown/black to a pale yellow, whose shade is influenced from NH_3 gas concentration.

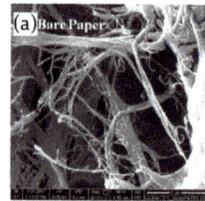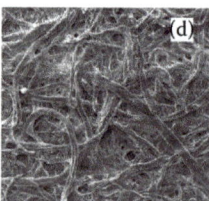

Figure 10. Field emission scanning electron microscope (FESEM) images of (**a**) bare paper and (**b**) $MAPbI_3$-coated paper: perovskite exploits the fibril structure of the paper substrate to grow with a nanorod-like structure. FESEM images of (**c**) PbI_2 and (**d**) $MAPbI_3$ after NH_3 exposure: the morphology of the $MAPbI_3$ film after interacting with ammonia looks like PbI_2 film more than the pristine perovskite film. Adapted with permission from [80], under the terms of the Creative Commons Attribution 4.0 International License. Copyright © 2018, Springer Nature.

In this regard, gas concentration resulted fundamental for the sensing properties of the sample. Considering the color change of the sample, the response time of the active mean, that is the time needed to reach the 90% of the maximum of the response, exponentially decreases with higher gas concentrations. At low NH_3 concentration (10 ppm) the film changes its color with a response time of about 12 s, and reaches 3 s for 30 ppm of gas content. Moreover, it has been demonstrated that for low NH_3 concentration (10 ppm) the transformation of MAPbI$_3$ perovskite in PbI$_2$ molecules is reversible upon removal of the gas source, whereas it becomes irreversible for higher concentration (30 ppm), leading to a detrimental perovskite disintegration, thereby imposing an upper limit to the sensor application [80].

However, the effects of the interaction of hybrid lead halide perovskite with ammonia are not totally clear yet, since several authors support the idea that the exposition to NH$_3$ induces the formation of new unknown non-luminescent phases, not attributable to PbI$_2$ molecules [82–85].

Bao et al. [84], for example, found that MAPbI$_3$ perovskite films interacting with NH$_3$ transform in a new unknown compound, named (MAPbI$_3$ + NH$_3$), which visually appears as a color change from typical perovskite dark brown to a light yellow. Interestingly, they found that, even if the phase transition becomes irreversible after only 50 s, the permanent modification of the film appearance is not accompanied by a modification of the electrical behavior of the sample. The new (MAPbI$_3$ + NH$_3$) compound seems indeed to respond to ammonia presence identically to pristine perovskite, changing its resistance as a function of the on/off NH$_3$ cycles.

Structural modifications of perovskite samples in presence of ammonia have also been explored by Singh and his coworkers [82], who showed that both bare and PMMA-covered MAPbBr$_3$ quantum dots (QDs) lost their photoluminescence emission when interacting with NH$_3$ molecules. After exposing MAPbBr$_3$ QDs to ammonia, NH$_4^+$ cations are thought to replace methylammonium cations (CH$_3$NH$_3^+$) in perovskite, leading to the formation of NH$_4$PbBr$_3$ species. However, on the removal of the NH$_3$ source, the reaction does not turn back alone and the injection of CH$_3$NH$_2$ molecules is needed to restore pristine perovskite structure and its PL emission.

Similarly, Ruan et al. [83] found that metal halide perovskite crystals and films interacting with ammonia create a new composite, starting from the oxidation process of NH$_3$ to NH$_4^+$. In this case, however, reduced methylammonium ion (MA$^+$ to MA) is thought to remain inside the perovskite crystal structure, forming a weakly coordinated complex (NH$_4$PbX$_3 \cdot$ MA); the weak bound with MA can be thermally broken, restoring pristine perovskite crystal structure. Interestingly, experimental XRD measurements evidenced that this is only partially true, since the chemical composition of the perovskite influences its stability to external stimuli. Only MAPbCl$_3$ single crystals have been found to be completely restored when the NH$_3$ source is removed, whereas MAPbBr$_3$ and MAPbI$_3$ showed traces of the NH$_4$PbX$_3 \cdot$ MA complex in their XRD patterns after ammonia treatment.

Contrary to hybrid perovskites, recently it has been demonstrated that fully inorganic CsPbBr$_3$ perovskite QDs do not suffer from structural modification induced by the interaction with ammonia [86]. XRD patterns of perovskite before and after NH$_3$ exposure have indeed been shown to be quite similar (Figure 11c) and measurements on morphology additionally confirm that the size ($\simeq$10 nm) and the cubic phase of the quantum dots remain unaltered (Figure 11a,b). As a consequence, the interaction between inorganic perovskite and ammonia is physical rather than chemical, with NH$_3$ assuming the role of passivating agent for surface defects through the interaction with Pb ions. The reduction of trap states density is confirmed by TA measurements and leads to the enhancement of the sample PL emission in presence in ammonia.

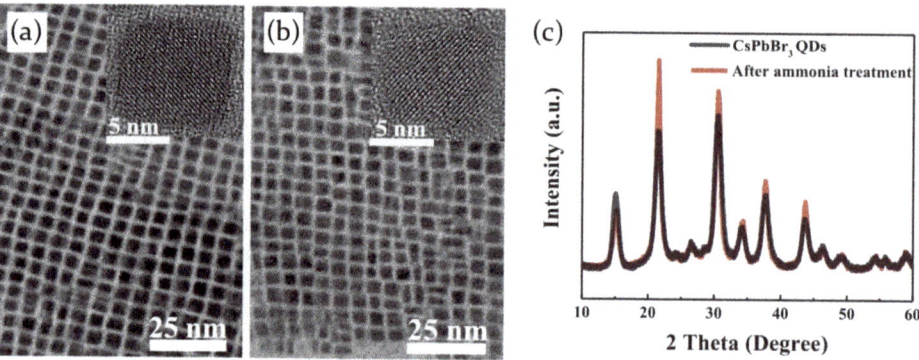

Figure 11. TEM images of CsPbBr$_3$ QDs (**a**) before and (**b**) after ammonia gas exposure. (**c**) XRD pattern of CsPbBr$_3$ QDs before and after NH$_3$ treatment. Adapted with permission from [86]. Copyright © 2020, WILEY-VCH Verlag GmbH and Co. KGaA, Weinheim.

2.3.2. Nitrogen Dioxide

MAPbI$_3$ perovskite is demonstrated to be highly sensitive also to NO$_2$ [87], one of the most common harmful chemicals for human health. In presence of NO$_2$, both at atmospheric pressure and high pressure, the MAPbI$_3$ films increase their conductivity. Then, the conductivity decreases back to the initial value when the films are exposed to an inert gas. On the basis of computational simulations and experimental Fourier transform infrared spectroscopy (FTIR) results, a simple model is proposed to describe the interaction between the gas and the surface atoms, ascribed to electron transfer from perovskite to NO$_2$ molecules (Figure 12a–c). Nitrogen dioxide molecules, in their approach to perovskite surface, are thought to interact with the methyl group -CH$_3$ rather than the -NH$_3$ one.

This effect is confirmed by computational analysis in which Radial Distribution Functions (RDF) of NO$_2$ molecules adsorbed on the (110) face of MAPbI$_3$ have been calculated and reported in Figure 12a,b. In particular, Figure 12a evidences the RDFs ascribed to the bond between NO$_2$ and -CH$_3$ group of MA$^+$ ion through the O atom (black line) and the N atom (red line), whereas in Figure 12b the RDFs between N and O atoms of NO$_2$ and the -NH$_3$ group of MA$^+$ are reported in black and red, respectively. A better overlap of the RDFs first peaks at 2.50 Å and 2.63 Å (Figure 12a) obtained for the interaction between NO$_2$ molecules and -CH$_3$ group reveals a stronger hydrogen bond than that obtained from the interaction between NO$_2$ and -NH$_3$ group, evidenced by peaks at 2.94 Å and 3.53 Å. Once the hydrogen bond is formed, a charge transfer can occur. Starting from the perovskite Pb-I skeleton, which is known to represents the major channel of charge transport [88], some electrons are transferred to MA$^+$ ions and then out of perovskite, leaving the MAPbI$_3$ crystal structure with a lack of negative charge. The hole concentration increase induces a resistance decrease, and subsequently, a current enhancement. As a consequence, the higher the NO$_2$ pressure is, the more efficient the electron transfer becomes, resulting in an increase of current in the device.

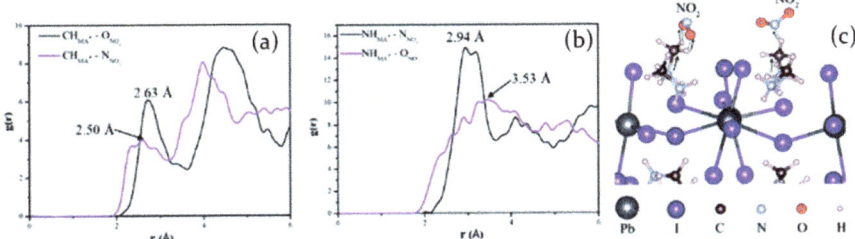

Figure 12. Radial distribution functions (RDF) of adsorbed NO$_2$ molecules on the (110) face of MAPbI$_3$, including both hydrogen bond interactions with (**a**) −CH$_3$ (peaks 2.50 Å and 2.63 Å) and (**b**) −NH$_3$ (peaks at 2.94 Å and 3.53 Å). (**c**) Schematic illustration of the electron transfer process from perovskite to the NO$_2$ molecules via the hydrogen bond channel. Adapted with permission from [87] under the terms of the Creative Commons Attribution 3.0 Unported license. Copyright © 2018, Royal Society of Chemistry.

2.3.3. Volatile Organic Compound

Volatile organic compound (VOC) gases contribute to ambient air pollution [89], in particular in indoor space, leading to health problems such as respiratory diseases [90], eye irritation [91], skin allergy [92], fatigue [93], and central nervous system disorder [94]. For this reason finding VOC sensors with high sensitivity and fast response, able to operate at room temperature, would be very important [7].

Nur'aini et al. [7] have shown that MAPbI$_3$ perovskite thin films are very sensitive to the presence of VOC gases in the atmosphere and then are good active materials for sensors operating at room temperature, with good reversibility and repeatability. In particular, electrical properties of perovskite films are tested by applying a voltage bias to the interdigitated electrode and measuring the current-time response of the film. Samples are tested toward a range of typical organic VOCs, both polar gases, such as ethanol, acetone, isopropanol, acetonitrile and methanol, and non-polar gases, such as toluene, benzene, chloroform, and hexane, always obtaining an increase of the current in presence of gas and a subsequent current decrease when the atmosphere is switched back to inert gas.

A sensing mechanism based on charge trap state passivation is proposed. MAPbI$_3$ is well known to contain iodine vacancies, which could be located in MA-I and Pb–I layers [95]. As MA-I layer is a termination layer of MAPbI$_3$, most of iodine vacancies are located on the surface, interfacial sites, and grain boundaries. During the film deposition, the solvent evaporation leads to the formation of a high density of crystal defects which act as trap states. In pristine film, electrons can fall from the conduction band into the iodine vacancies trap states, and as a consequence, the film is characterized by low conductivity. In presence of an ambient gas, the gas molecules passivate the iodine vacancies and the trapped electrons can be restored into the conduction band, determining an increase of the conductivity. If an inert gas is then injected to recover the perovskite, the absorbed inert gas molecules expel the VOC gas ones and recreate vacancies, re-establishing the crystal defects of the perovskite film and reducing its conductivity. This sensing mechanism is supported also by the photoluminescence measurements, which highlight a PL increase after the exposition to the gas and a return to lower PL intensity on the removal of the gas.

The effects of volatile organic compounds (VOCs) on the photoluminescence properties of lead bromide perovskite have also been investigated by Kim et al. [96], who focused on the role of aliphatic amines. MAPbBr$_3$ thin films have been exposed to three different amine vapors with different aliphatic chains: monoethylamine (EtNH$_2$), diethylamine (Et$_2$NH) and trimethylamine (Et$_3$N). In response to the amine vapor exposure the perovskite films suffered photoluminescence quenching. Anyway MAPbBr$_3$ showed full recovery only after exposure to Et$_3$N, whereas the Et$_2$NH exposed film was characterized by a slower and only partial PL recovery, and EtNH$_2$ vapor determined an irreversible quenching of the photoluminescence. XRD measurements evidenced that fluorescence quenching originates from the structural changes in the perovskite films (Figure 13a) caused by the interaction with the aliphatic amines. In particular, the films exposed to $EtNH_2$ and Et_2NH clearly show full and partial degradation of the crystal structure, while no structural changes are observed in the film exposed to Et_3N, thereby explaining the different behavior in terms of reversibility.

The alteration of the perovskite crystal structure is due to the intercalation of polar molecules into the lattice, which breaks the hydrogen bonds of the perovskite and forms new ones with the halides. Then, in order to investigate the origin of the observed reversible/irreversible changes in emission signals, density functional theory (DFT) and molecular electrostatic potential (MEP) calculations have been performed. The hydrogen bonding distances between perovskite MA^+ ions and aliphatic amines are calculated to be 1.07 Å, 2.25 Å and 3.58 Å for $EtNH_2$, Et_2NH and Et_3N, respectively (Figure 13b); it suggests that, due to the longer hydrogen bonding distance, the interaction strength between Et_3N and MA^+ is lower, allowing reversible photoluminescence quenching. This behavior is confirmed by the MEP calculations that indicate a significant decrease of negative potential of nitrogen atom in Et_3N and confirm its relatively lower hydrogen bonding strength with perovskite. Vice versa the shorter distance, and then the higher interaction strength, between $EtNH_2$ and MA^+ results in irreversibility.

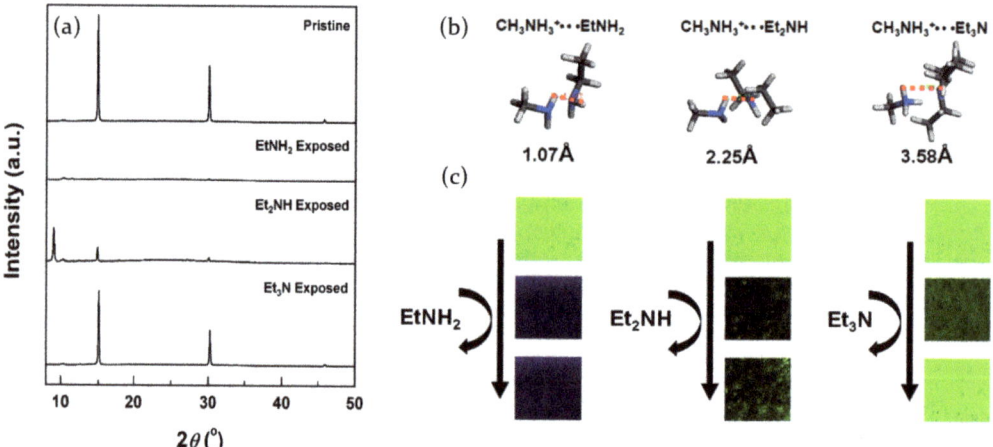

Figure 13. (a) X-ray diffraction (XRD) patterns of $CH_3NH_3PbBr_3$ films before amine exposure (pristine), after exposure to $EtNH_2$, Et_2NH and Et_3N. (b) Interaction distances between MA^+ of perovskite and aliphatic amines obtained through DFT calculation. (c) Photographs of $MAPbBr_3$ perovskite films recorded under 365 nm UV light after exposure to $EtNH_2$, Et_2NH and Et_3N. Reprinted from [96]. Copyright © 2017, with permission from Elsevier.

The possibility to develop metal halide gas sensors has also been investivated for fully inorganic perovskites. Porous layers of $CsPbBr_3$ have been indeed demonstrated to be highly sensitive to the presence of volatile organic compounds, showing a modulation of electrical properties under visible light excitation. In particular, the samples have been analyzed under exposure to acetone and ethanol, and also to oxygen, showing a clear enhancement of the photocurrent in presence of the target gas and a decrease when, subsequently, is exposed to inert atmosphere. Moreover, the choice of using different gas sources allowed to demonstrate that the perovskite is able to respond with an equal current change to both oxidizing and reducing gas, due to the perovskite ambipolar charge transport due to the similarity of electron effective mass and hole effective mass [9,49,97].

Lead bromide perovskites in pristine conditions—in an inert atmosphere—are characterized by a high density of bromide vacancies, which act as trap states for photoexcited charges. As a consequence, the conductivity of the sample results unavoidably low. On the contrary, when the sample is exposed to an external gas source, such as oxygen, ethanol or acetone, the gas molecules act as trap fillers for bromide vacancies. Decreasing the density of trap states means a higher number of photoexcited charges available for electrical transport, which translates into a photocurrent increase. The overall process is reversible, so when the chamber is evacuated, gas molecules desorb from the perovskite surface, restoring the initial trap states density condition.

2.3.4. Ozone

A gas-sensing mechanism based on the surface trap passivation has also been hypothesized to explain the interaction between ozone molecules and the surface of lead mixed halide perovskite $MAPbI_{3-x}Cl_x$ thin films, resulting in a decrease of the electrical resistance of the sample [98]. The gas molecules, adsorbed within the perovskite lattice close to the surface, passivate the traps (unpaired Pb^{2+} ions), and as a result, the sensing films resistance is reduced. In detail, the increase of the current through the perovskite is due to the electron transfer from the O_3 to the Pb^{2+} cations. Consequently, the excess of positive charges is neutralized, the surface recombination rate in the film is modulated and the surface trap passivation leads to the enhancement of the conductivity, due to a lower hole-electron carrier trapping. Such a change in the conductivity depends on the amount of ozone concentration molecules adsorbed onto the surface and consequently on the surface morphology of the film (grain size, porosity, roughness). When the sample chamber is evacuated, the ozone molecules are quickly desorbed, and thus the perovskite film restores its initial electrical conductivity [99] in very short response times.

The hypothesized sensing process is confirmed by the photoluminescence measurements which demonstrate that, after the exposure of the film to ozone, PL intensity increases significantly. Indeed, due to the O_3 passivation of the surface traps in the perovskite films, the rate of the non-radiative recombination [99] decreases. Additionally, the absence of shift of the peak PL wavelength indicates the absence of substantial crystal phase change during the ozone exposure (Figure 14b).

Moreover, it has been verified that a prolonged exposure to ozone at high gas concentration can induce detrimental effect in the perovskite. As evidenced in Figure 14a, the exposure of the $MAPbI_{3-x}Cl_x$ thin films to 2500 ppb ozone concentration leads to a uniform drop of the UV-Vis absorption spectra intensity after about 60 min. However, such high levels of ozone exposure are quite drastic, so in standard conditions (gas concentration lower than 75 ppb, which is the safety limit imposed by International Agencies) [100] the sensing process can be considered fully reversible.

2.3.5. Hydrogen

Lead mixed halide perovskite $MAPbI_{3-x}Cl_x$ thin films are sensitive also to hydrogen (H_2) [8]. Hydrogen is a reducing gas, and since the $MAPbI_{3-x}Cl_x$ is a p-type semiconductor, the adsorption of the H_2 molecules causes the lowering of the current. The interaction between the perovskite and the gas is typically physical rather than chemical: the H_2 molecules, adsorbed through the porous of the perovskite film, bond loosely close to the surface and leave the film after the removal of the gas, without inducing any structure modification. This is confirmed by the XRD patterns of the sample taken before and after the exposure to H_2. They indeed show the same peaks without substantial differences in intensity (Figure 14c).

Under H_2 exposure the film resistance increases because, being the hydrogen a reducing gas, it releases electrons that recombine with the holes (majority charge carriers in the p-type semiconductor) resulting in the lowering of the current through the film. When hydrogen is desorbed the resistance of the material returns to the its initial value.

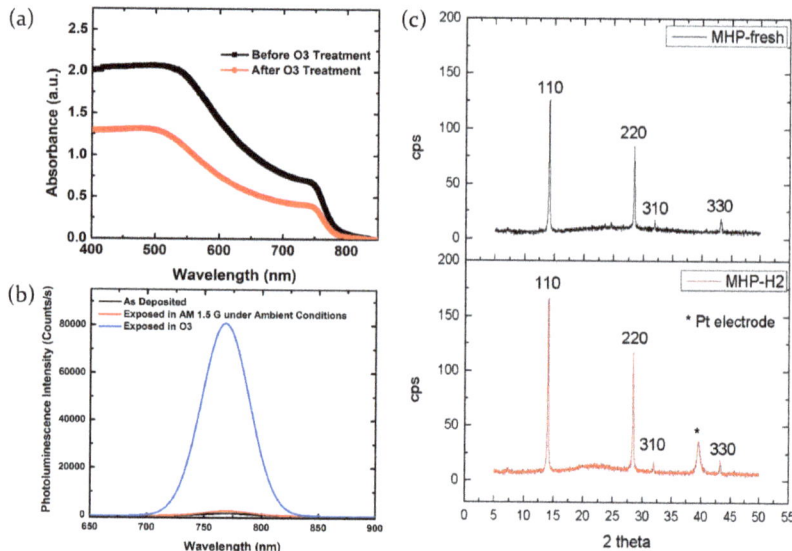

Figure 14. (a) UV–Vis absorption measurements before and after O_3 exposure for $MAPbI_{3-x}Cl_x$ perovskite films. (b) PL spectra of the pristine (black line), Solar A.M. 1.5G exposed (red line) and O_3 exposed (blue line) $MAPbI_{3-x}Cl_x$ perovskite films. Adapted with permission from [98]. Copyright © 2017, American Chemical Society. https://pubs.acs.org/doi/10.1021/acssensors.7b00761. Further permission should be directed to the ACS. (c) XRD patterns of $MAPbI_{3-x}Cl_x$ films before and after exposure to 100 ppm hydrogen gas, which demonstrate that the perovskite does not change crystal structure after interacting with H_2. The only different feature noticeable in H_2-exposed perovskite film relates to the Pt electrode; it is absent in the pristine sample spectra, since the XRD of the non-exposed sample was done on a glass substrate without electrodes. Adapted with permission from [8] under the terms of the Creative Commons Attribution 3.0 license. Copyright © 2020, IOP Science.

3. Sensing Application of Perovskites

The reversible response to the environment makes the perovskite-based films attractive sensing elements in the detection of harmful gases. It has been observed that sometimes the exposition to the surrounding atmosphere does not induce any phase transformation, indicating a non-chemical interaction between the target gas and the perovskite, and it has been demonstrated that in these cases the surface traps play an important role in determining the gas sensing mechanism both in the optical and the electrical response.

The conductivity of the perovskite layers is greatly influenced by the environmental gases and the electrical response to both oxidizing and reducing gases, as already evidenced, is symmetrical. This behavior indicates a sensing mechanism in perovskites different from that of semiconductors typically employed in MOS technology. Indeed, because of the similarity of electron and hole effective masses in halide perovskites, the charge transport is bipolar leading to analogous variation of photocurrent with increasing concentrations of oxidizing or reducing gases. For this reason these materials are able to operate as sensors for both the types of analytes.

To date the most diffused approach reported in literature is the realization of hybrid lead halide perovskites gas sensors based on the modulation of the electrical properties, induced by the atmosphere chemical composition variation [81,84].

On the contrary, the development of lead halide perovskites optical sensors, exploiting the reversible photoluminescence alterations due to the interaction with the analyte molecules, is much less developed, with limited examples in literature.

3.1. Resistive Sensors

The first experimental demonstration of the use of perovskite-based device as a resistive sensor has been provided by Stoeckel et al. [66], who exploited the interaction of MAPbI$_3$ polycrystalline films with oxygen. A more-than-three order of magnitude increase in current has been measured changing the atmosphere from pure N$_2$ to O$_2$. Oxygen molecules diffuse inside the perovskite crystal structure and reversibly fill iodine vacancies, which act as sensing sites. This determines a reduced carrier trapping resulting in a conductivity, and thus current, increase. The trap healing mechanism does not involve the creation of a strong covalent bond between O$_2$ molecules and perovskite, but it is just a physical interaction guided by the gas concentration in the chamber. It then leads to a fast response of the device (400 ms) and a full reversibility of the process (Figure 15c,d).

The analysis, moreover, revealed that the morphology, depending on the deposition technique, strongly influences the response to an external stimulus. Two samples were prepared through solution-based deposition methods, in one (1S) or two steps (2S). In the 2S approach, a PbI$_2$ solution was spin coated on a Si/SiO$_2$ substrate, followed by the deposition of a MAI solution on the previously prepared PbI$_2$ film; on the other hand, the 1S technique directly involved the deposition of an equimolar mixture of the two precursors. The sample 2S is characterized by a more uniform coverage (Figure 15b), with very small crystals (about 500 nm in length and 400 nm in thickness), which maximizes the diffusion of oxygen molecules in the film, and as a consequence, the sensing response. In a 95% O$_2$ atmosphere almost all traps in the film result to be filled. On the contrary, fibril-like structure obtained in the 1S sample (Figure 15a) does not allow oxygen to diffuse properly in the sample, resulting in a less efficient sensing response.

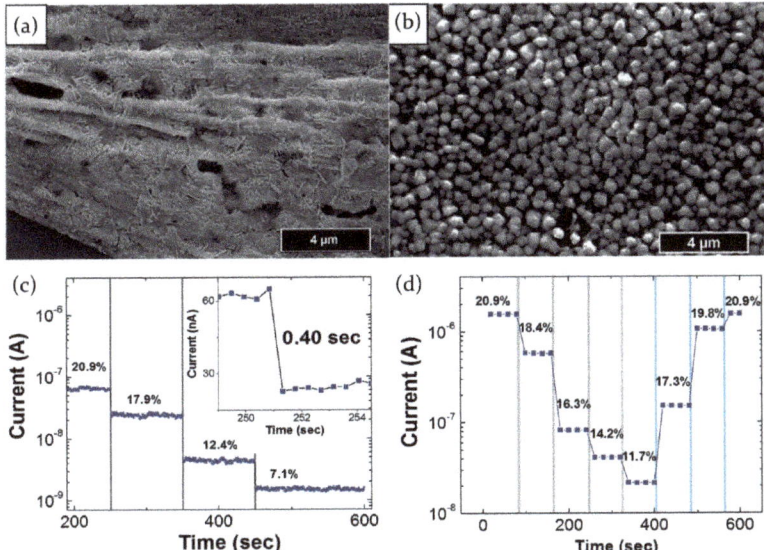

Figure 15. SEM images of (**a**) 1S sample and (**b**) 2S sample of MAPbI$_3$ polycrystalline films, showing the different morphologies coming from the deposition technique used. (**c**) Current change at decreasing oxygen levels. Inset: magnification of the current plot showing the response time of the device. (**d**) Current plot highlighting the reversibility of the sensor for decreasing (from 20.9% to 11.7%) and then increasing (from 11.7% to 20.9%) O$_2$ gas concentration. Adapted with permission from [66]. Copyright © 2017, WILEY-VCH Verlag GmbH and Co. KGaA, Weinheim.

Furthermore, the device has also been tested with other gases in order to investigate the selectivity of the response, that is the ability to respond only to a single gas source, that represents one of the main properties of a correctly working sensor. SO_2 and NH_3 have been shown to induce a current change much lower than that induced by oxygen molecules; the presence of moisture, instead, leads to the formation of electrically insulating species, detrimental for the sensor.

The process of physical adsorption of atmosphere molecules in $MAPbI_3$ perovskite crystal lattice has been also verified by Bao et al. [84] in presence of NH_3 as target gas. The exposition to the chosen gas has been found to induce an immediate current increase, with a reversible behavior in case of gas evacuation. Besides, authors interestingly noticed how this reversible current modulation was accompanied by an irreversible phase transformation, showing a resistance higher than the pristine perovskite, but similar sensing ability. As a results, switching on and off the NH_3 flow, current plot is characterized by an overall gradual decrease in the first few minutes of the measurement. Once the perovskite is completely transformed in the new phase, the current sets up on a constant level, following the NH_3 on/off cycles with a corresponding current increase/decrease, with a response time and a recovery time, respectively, of 3 s and 4.5 s (Figure 16a,b).

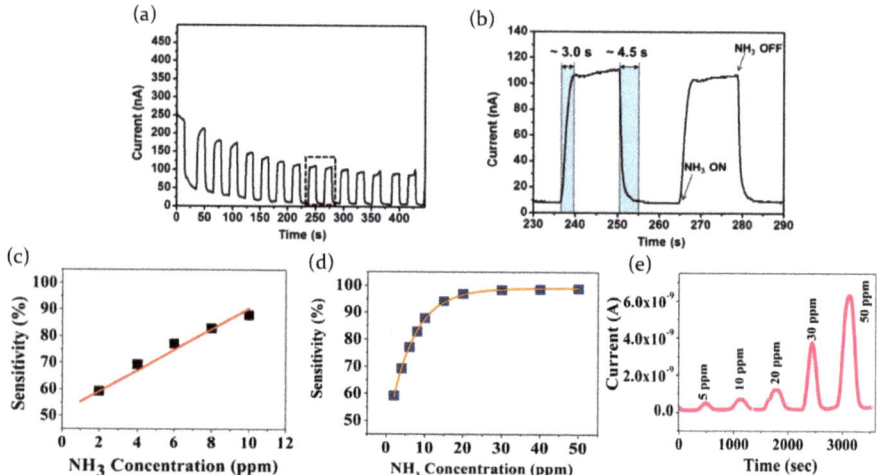

Figure 16. (a) I–t curve for a $MAPbI_3$ perovskite film for NH_3 on/off cycles; the current plot shows an initial overall gradual decrease due to the phase transformation, then settling at a constant level. (b) Expanded view of the on and off switches of the gas, with the corresponding response and recovery times (blue shade). Used with permission from the Royal Society of Chemistry [84]. Copyright © 2015. Permission conveyed through Copyright Clearance Center, Inc. (c) Linear dependence of the sensitivity as a function of NH_3 concentration in a 1–10 ppm range for a $MAPbI_3$ film deposited on a paper substrate; (d) sensitivity as a function of higher values, ammonia concentration 1–50 ppm ; (e) current modulation resulting from NH_3 on/off switches at increasing gas concentration. Adapted with permission from [81] under the terms of the Creative Commons Attribution 4.0 International License. Copyright © 2019, Springer Nature.

Current variation after the interaction of $MAPbI_3$ perovskite with NH_3 gas has also been investigated by Maity et al. [81], and the electrical response has been related to the formation of PbI_2 on the surface of the active layer. The perovskite film, deposited on a paper substrate, consists of nano and microrods, with an average length of 30 μm and a diameter ranging from 0.7 to 1.6 μm. The exposure to NH_3 leads to perovskite transformation into PbI_2 that has a higher conductivity, thereby causing a current increase (Figure 16e). However, the gas concentration plays an important role in the process, since only for low values (<10 ppm) the reaction mechanism is fully reversible. At increasing gas pressures, the transformation of perovskite into PbI_2 goes on, and allows reversible current switches only for short time exposure. The process then becomes totally irreversible for NH_3 concentration higher than 50 ppm.

The sensitivity S of the sample (defined as the ratio $\Delta R/R_0$, where $\Delta R = R_0 - R_g$ is the variation of resistance with R_g the resistance when the sensor is exposed to the gas (NH_3) and R_0 the resistance in inert atmosphere) reflects this behavior. It indeed linearly depends on the gas concentration for low values (<10 ppm), and saturates after reaching 30 ppm, as shown in Figure 16c,d. However, the device has shown to be able to detect low gas concentration down to 1 ppm with a 55% sensitivity and reaches sensitivity of almost 90% at 10 ppm. The response and recovery times have also been investigated, both being of about 120–130 s at a NH_3 concentration of 10 ppm.

This experiment investigated also the selectivity of the response by testing the device with other volatile compounds, such as ethanol, methanol, TCE (trichloro ethelyne) and IPA (2 propanol). All of them have shown to induce a small negative sensitivity response, which means that the resistance of the film increases when exposed to those gases. Only acetone induced a current increase, even if the response, compared to NH_3 one, is negligible.

The quick room-temperature reversible behavior of $MAPbI_3$ films when exposed to nitrogen dioxide indicates their potential application as NO_2 sensors, showing a fast decrease of the resistance in presence of the external stimulus [87]. The NO_2-sensing properties of $MAPbI_3$ thin films have been investigated both at ambient and high pressure (with argon from 1 to 8 MPa). Interestingly, the sensor presents a very low detection limit with a response even under extremely low NO_2 concentrations (about 1 ppm) (Figure 17d) and an average sensitivity as high as 0.62 ppm^{-1} (inset of Figure 17d). In addition, the $MAPbI_3$-based NO_2 sensor exhibits a very quick-responsive character with average response and recovery times of about 5 s and 25 s, respectively (Figure 17e), good selectivity versus other gases, such as SO_2, HCHO, CH_4, CO, NH_3, $(CH_3)_3N$, O_2 and H_2O, and interesting reproducibility, as the sensor response remains stable still after more than 12 cycles.

In order to enhance $MAPbI_3$ gas sensing performances toward NO_2 and acetone vapors, a doping process has been used, introducing thiocyanate ions (SCN-) into the perovskite lattice and obtaining $CH_3NH_3PbI_{3-x}(SCN)_x$ with x in the range 0.016–0.053 [49]. As in other reports, also Zhuang's research points out how the film morphology results fundamental in the sensing process. A film thickness of 120 nm maximizes the electrical response compared to 170 and 210 nm thick layers (Figure 17a). Thinner layers indeed have an higher roughness and provide a larger surface to volume ratio, inducing a stronger sensing response. On the other hand, it is demonstrated how layers thinner than 120 nm do not uniformly cover the substrate, creating non conductive channels.

The detection limits of the 120 nm-film are found to be 20 ppm and 200 ppb for acetone and NO_2, respectively, whereas its sensitivity is $5.6 \cdot 10^{-3}$ ppm^{-1} for acetone and $5.3 \cdot 10^{-1}$ ppm^{-1} for NO_2 (Figure 17b,c). Recovery times also show differences between the two active means, obtaining 4 min for acetone and 1.5 min for NO_2. Differences in detection limit and recovery times arise from a stronger (for acetone) or weaker (for NO_2) interaction with the perovskite surface. However, it is interesting to notice how perovskite positively responds to the presence of both an electron-withdrawing (NO_2) and an electron-donating (acetone) gas, resulting in both cases in a conductivity enhancement. It is related to the ambipolar charge transport nature of hybrid metal halide perovskite, due to the similar effective masses of electrons and holes [97]. This sensor also exhibits excellent reproducibility and greatly improved environmental stability thanks to the chemical bound between SCN- ions and Pb atoms in perovskite lattice, compared to the non doped $MAPbI_3$ perovskite.

The performance of $MAPbI_3$ films as sensors has been tested with ethanol and other polar and non-polar organic species at room temperature [7]. The resistance of the film substantially decreases when it is exposed to VOC vapor and recovers back to high resistance when the VOC gas is removed and the surface is exposed to an inert gas. Ethanol exposure does not damage the $MAPbI_3$ structure and the interaction between the perovskite and the gas is assumed to be mainly due to physisorption and limited on the perovskite grain boundaries. From the dynamic response–time plot of $MAPbI_3$ film (Figure 17f), when it is alternately exposed to ethanol and inert gas, a response time of 66 s and a recovery time of 67 s have been inferred, which are faster than other type gas sensors [101]. In addition, the films are characterized by a sensitivity of $3 \cdot 10^{-4}$ ppm^{-1} and a limit of detection (LOD) of 1300 ppm, lower than 3300 ppm, immediately dangerous for life or health.

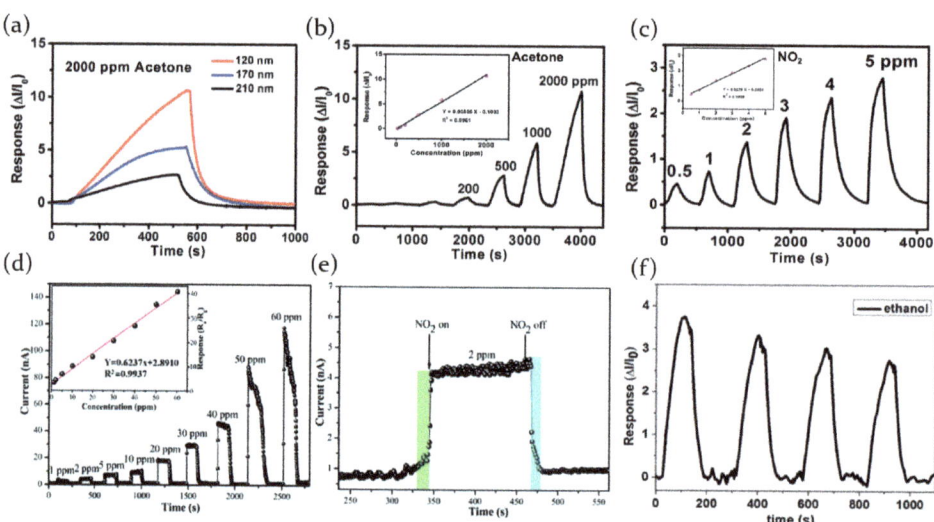

Figure 17. (**a**) Response of the SCN-organometal halide perovskites-based sensor with a 120, 170 or 210 nm-thick sensing layer when exposed to 2000 ppm of acetone. (**b**) Current response of the 120 nm thick perovskite layer to increasing acetone concentrations. Inset: linear dependence of the current response of the sensor to the gas concentration, showing an average sensitivity of $5.6 \cdot 10^{-3}$ ppm^{-1}. (**c**) Current response of the 120 nm thick perovskite layer to NO$_2$ concentrations ranging from 500 ppb to 5 ppm. Inset: linear increase of the device response to the gas concentration, resulting in an average sensitivity of $5.3 \cdot 10^{-1}$ ppm^{-1}. Used with the permission of the Royal Society of Chemistry, from [49]. Copyright © 2017. Permission conveyed through Copyright Clearance Center, Inc. (**d**) Response curves of the MAPbI$_3$ film sensor at different NO$_2$ concentrations (1–60 ppm). Inset: linear dependence of the resistive response to the gas concentration, resulting in an average sensibility of 0.62 ppm^{-1}. (**e**) Expanded view of the on and off switches of the gas, with the corresponding response and recovery processes (green and blue shades). Adapted with permission from [87] under the terms of the Creative Commons Attribution 3.0 Unported license. Copyright © 2018, Royal Society of Chemistry. (**f**) Response–time graph of MAPbI$_3$ film when exposed to on/off switches of ethanol gas. Adapted with permission from [7] under a Creative Commons Attribution-NonCommercial 3.0 Unported license. Copyright © 2020, Royal Society of Chemistry.

Moreover, tests with polar and non-polar organic VOCs indicate that generally the perovskite sensors are more sensitive to the former, resulting in higher responses. This behavior could be ascribed to the fact that the polar gas molecules attach stronger to MAPbI$_3$ defect sites, since MAPbI$_3$ perovskite itself is a polar ionic crystal.

Hybrid lead halide MAPbI$_{3-x}$Cl$_x$ films have been successfully tested as portable, flexible, self-powered, and ultrasensitive ozone sensors operating at room temperature [98].

Ozone (O$_3$) is considered one of the principal pollutants harmful to the human respiratory system, causing inflammation and congestion of the respiratory tract. Therefore it is necessary to control the ozone concentration in the atmosphere and in confined environments through the continuous monitoring of the gas. For this reason, the development of both effective and inexpensive methods and devices are required.

The electrical resistance of the film promptly decreases when exposed to ozone with a response time between 188 and 225 s, depending on the O$_3$ concentration (with response time getting lower values as the concentration of the O$_3$ increases), and recovers to the pristine values within 40–60 s, after the complete removal of gas. In addition, tests have shown that the films are able to detect, with good repeatability, even ultralow ozone concentrations ranging from 2500 to 5 ppb, characterized by a reduction of the response magnitude with ozone concentration decrease (Figure 18a).

Exploiting the reversible interaction of mixed MAPbI$_{3-x}$Cl$_x$ perovskite with H$_2$ molecules, inducing a decrease of current in the film, Gagaoudakis et al. [8] have realized portable, flexible, self-powered, fast and sensitive hydrogen sensing elements, operating at room temperature. Concentrations of H$_2$ gas down to 10 ppm could be detected with a sensitivity depending on the concentration and included in the range from 0.3% at 10 ppm to 5.2% at 100 ppm (Figure 18b,c). Moreover, the sensing element film shows a good reversibility and quick response and recovery times with average values of 45 and 35 s, respectively.

As regarding the MAPbBr$_3$ sensing properties, its interaction with air (O$_2$ and H$_2$O molecules) has been found to induce a decrease of both photocurrent and dark current (Figure 18d,e), compared to vacuum atmosphere, accompanied in this case by a PL enhancement. Air infiltration in perovskite lattice is thought to be responsible for the formation of shallow trap states. Photo-generated carriers can then be trapped and form radiative recombination centers, inducing the decrease of both dark and photo current; on the other hand, charge carrier trapping is seen to reversibly enhance the sample PL emission [71].

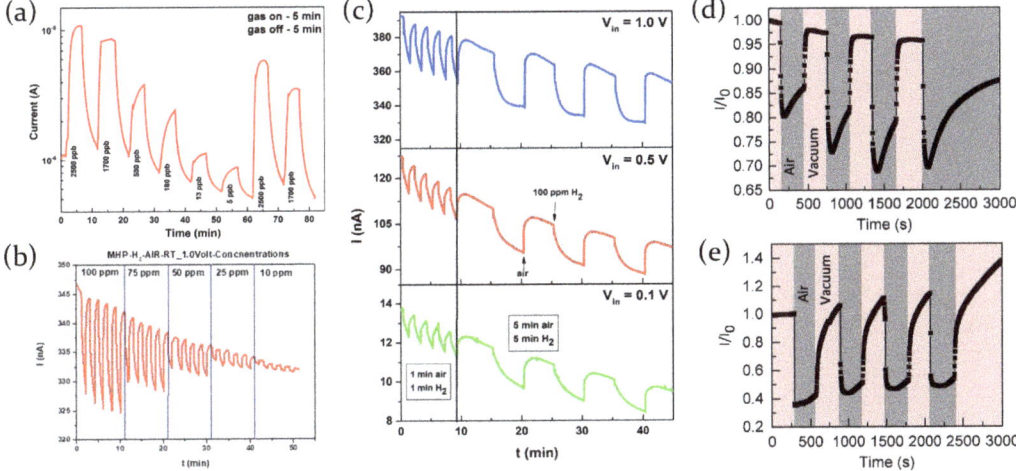

Figure 18. (**a**) Electrical response of MAPbI$_{3-x}$Cl$_x$ films under various ozone concentrations (5–2500 ppb). Adapted with permission from [98]. Copyright © 2017, American Chemical Society. https://pubs.acs.org/doi/10.1021/acssensors.7b00761. Further permission should be directed to the ACS. (**b**) Reversible current modulation for 300 nm thick mixed MAPbI$_{3-x}$Cl$_x$ perovskite films under exposure (for five minutes) in various hydrogen concentrations, revealing a detection limit of 10 ppm of H$_2$. (**c**) Current modulation of mixed MAPbI$_{3-x}$Cl$_x$ perovskite layers under H$_2$ and air flow, for various exposure times (one and five minutes) and under different forward biases (0.1, 0.5 and 1 Volt). Adapted with permission from [8] under the terms of the Creative Commons Attribution 3.0 license. Copyright © 2020, IOP Science. (**d**) Dark current and (**e**) photocurrent modulation for MAPbBr$_3$ perovskite measured under vacuum and standard atmosphere. Reprinted from [71], with permission of AIP Publishing; Copyright © 2017.

Due to their superior stability at ambient conditions [102,103], totally inorganic perovskites have been explored as self-powered oxygen and VOCs sensors [104]. Porous CsPbBr$_3$ layers with an average thickness of about 350 nm, constituted of interconnected grains of 100–200 nm in diameter, have been tested. It has been demonstrated that at room temperature and under visible light illumination, the photocurrent generated by these devices is highly sensitive to both oxidizing and reducing gases with very quick response and recovery time.

The dynamic responsivity of the CsPbBr$_3$ devices when exposed to O$_2$, acetone and ethanol (Figure 19a–c) indicates that the photocurrent rapidly increases and stabilizes upon gas injection, and quickly recovers its initial value after switching back to the pure N$_2$

atmosphere. Response time and recovery time of 17 and 128 s, respectively, have been estimated for oxygen, while 9.8 s and 5.8 s for acetone. Moreover the CsPbBr$_3$ devices could easily detect down to 1 per cent, and possibly lower content, of O$_2$ in N$_2$ and concentrations as low as 1 ppm of acetone and ethanol. Finally, the sensor reveals a good repeatability when exposed to consecutive cycles of gas exposition, and after two weeks storage, confirming that they are compelling alternatives to oxide-based sensors.

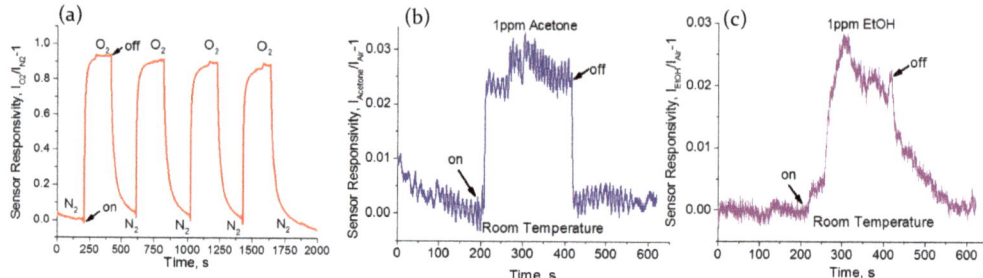

Figure 19. Dynamic CsPbBr$_3$ sensor responsivity to (**a**) O$_2$, (**b**) acetone and (**c**) ethanol exposure at room temperature under visible-light irradiation. Adapted with permission from [104]. Copyright © 2017, WILEY-VCH Verlag GmbH and Co. KGaA, Weinheim.

In the Table 1, the main sensing properties of the described perovskite devices are reported with some information about the active material (chemical composition and structure) and the target gas which induces the electrical response.

Table 1. Sensing properties for perovskite-based resistive sensors described in this review. Data are grouped taking into account the perovskite active mean. The morphology of the film, the target gas used as analyte and the main sensing properties are reported.

Perovskite	Structure	Target Gas	T_{res}	T_{rec}	Sensitivity	LOD	Ref.
MAPbI$_3$	Polycrystalline	O$_2$	0.4 s		3000		[66]
MAPbI$_3$	Thin film	NH$_3$	3 s	4.5 s			[84]
MAPbI$_3$	Nanorods	NH$_3$	≃130 s	≃120 s	≃90%	1 ppm	[81]
MAPbI$_3$	Thin film	NO$_2$	5 s	25 s	0.62 ppm^{-1}	1 ppm	[87]
SCN – MAPbI$_3$	Thin film	NO$_2$	3.7 min	6 min	$5.3 \cdot 10^{-3}$ ppm^{-1}	200 ppb	[49]
		Acetone	4.5 min	4 min	$5.6 \cdot 10^{-3}$ ppm^{-1}	20 ppm	
MAPbI$_3$	Thin film	Ethanol	66 s	67 s	$3 \cdot 10^{-4}$ ppm^{-1}	1300 ppm	[7]
MAPbI$_{3-x}$Cl$_x$	Thin film	O$_3$	188 s	40 s	9.69	5 ppb	[98]
MAPbI$_{3-x}$Cl$_x$	Thin film	H$_2$	39 s	61 s	0.3%	10 ppm	[8]
MAPbBr$_3$	Single crystal	Air					[71]
CsPbBr$_3$	Nanocrystals	O$_2$	17 s	128 s	0.93	10,000 ppm	[104]
		Acetone	9.8 s	5.8 s	0.03	1 ppm	
		Ethanol			0.025	1 ppm	

T_{res} and T_{rec} refer to the response and recovery time, defined as the time needed to reach the 90% of the final value and the time to fall to 10% of the final value after removing the gas source, respectively. The last column shows instead the limit of detection (LOD), i.e., the minimum gas quantity the sensor is able to detect.

Concerning the sensitivity (S), in literature there is not a common definition, so some disorder can be generated and a clarification is needed. Some authors refer to S as the ratio $\frac{\Delta I}{I_0} = \frac{I - I_0}{I_0}$ where I represents the current value obtained in presence of the target analyte and I_0 is the reference value in inert atmosphere [8,66,104]. Similarly, S can be considered as the ratio $\frac{\Delta R}{R_0} = \frac{R - R_0}{R_0}$, where the resistance change is evaluated [81]. So they provide S value with a reference gas concentration. Others distinguish between R_S (the sensor response) and S, specifying that the first one refers to the above mentioned ratio $\Delta I/I_0$ [7,49] or the resistance ratio R/R_0 [87], whereas the S represents the sensor response (R_S) divided by gas concentration, expressed in ppm^{-1}. In these cases, sensitivity values are not dimentionless and often refer to average values [49,87]. A last sensitivity definition comes from [98] where S is defined as the ratio I_{max}/I_{min} where the terms represents the maximum and the minimum values the current reaches in the on/off gas cycle. Finally, some authors do not report any sensitivity value [71,84].

3.2. Optical Sensors

Even if the metal halide perovskite resistive response to the presence of a particular gas is the most studied thus far, recently some reports have highlighted how the presence or absence of a particular analyte can induce the activation or deactivation of some relaxation processes, and consequently, determine a change in the sample photoluminescence emission. Having a deeper insight into the issue can represent the starting point for the application of lead halide perovskite as luminescent-based gas sensor. The interaction between perovskite films and gases of different nature has been proved to induce modification both of the visual aspect of the film, noticeable as a color change, and a PL modulation, which returns to the initial value once the gas source is removed.

3.2.1. Color Change Gas Sensor

The first report of a reversible color change from perovskite films interacting with the surrounding atmosphere dates back to 2013 with Zhao's experimental analysis [85]. The exposition of MAPbI$_3$ perovskite films to ammonia gas was found to lead to a rapid (<1 s) change in its color from brown to transparent, and then turn back to its pristine condition upon removing the NH$_3$ gas. XRD measurements highlighted structural changes in perovskite crystals, likely due to the formation of an intercalation compound or a coordination complex from perovskite and NH$_3$ interaction. Since similar results have been found also for MAPbI$_{3-x}$Cl$_x$ and MAPbBr$_3$ perovskites, the discovery opened a route for a potential use of lead halide perovskite as optical sensor application.

Later, Maity's group developed a paper-based gas sensor in which MAPbI$_3$ perovskite film changes its color from pristine black to yellow when exposed to NH$_3$ vapors (Figure 20a) [80]. Contrary to what previously stated, color change has been ascribed to PbI$_2$ molecules formation, confirmed by XRD measurements.

Unfortunately, both studies highlighted the presence of an exposition limit in terms of time and gas concentration, above which the color change process becomes irreversible due to the perovskite film degradation (Figure 20b).

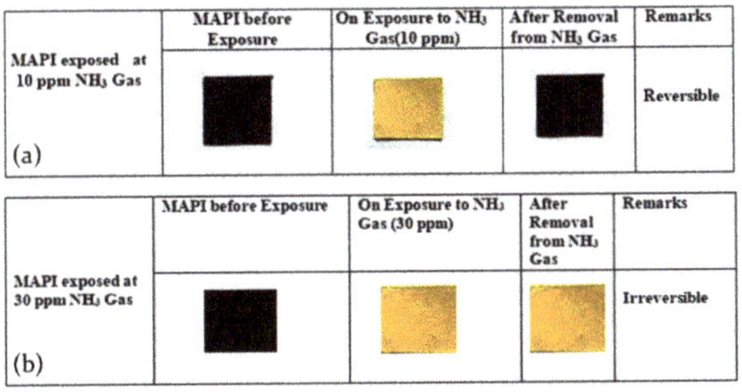

	MAPI before Exposure	On Exposure to NH$_3$ Gas(10 ppm)	After Removal from NH$_3$ Gas	Remarks
MAPI exposed at 10 ppm NH$_3$ Gas (a)	■	▩	■	Reversible

	MAPI before Exposure	On Exposure to NH$_3$ Gas (30 ppm)	After Removal from NH$_3$ Gas	Remarks
MAPI exposed at 30 ppm NH$_3$ Gas (b)	■	▩	▩	Irreversible

Figure 20. Photographs of the MAPbI$_3$ film exposed to (**a**) 10 ppm NH$_3$ gas, which allows a reversible process, and to (**b**) 30 ppm, which induces an irreversible film transformation. Adapted with permission from [80] under the terms of the Creative Commons Attribution 4.0 International License. Copyright © 2018, Springer Nature.

3.2.2. PL Emission-Based Gas Sensors

The atmosphere around a perovskite sample has been proved to be crucial for the modulation of the photoluminescence emission. Oxygen, water molecules or other volatile compounds are indeed verified to be responsible, for example, for trap states passivation processes, leading to an increase of the photoluminescence intensity, or modification of the chemical structure of the perovskite, inducing a worsening of the optical properties of the film.

In case of controlled atmosphere inside a test chamber, the effect of a single source can be evaluated, as it happened for the experimental analysis of Howard et al. [69], who demonstrated how humidity alone can induce perovskite degradation through the formation of hydrated species or facilitate ion migration. Besides, the combined effect of two analytes has also been evaluated.

The role of oxygen and water molecules on the photoluminescence properties of hybrid $MAPbI_3$ perovskite films has been subject matter of the research for Galisteo-Lopez et al. [65], evidencing that the PL activation is related to the passivation of trap states by O_2 molecules, whereas a humid atmosphere is responsible for the photodarkening process. This feature is evident considering the PL maximum at 775 nm as a function of illumination time, for a sample exposed to air under continuous optical 500 nm excitation. As shown in Figure 5a, the PL peak shows an initial increase, reaches its maximum value at t_c and then decreases for higher illumination times. Moreover, both photoactivation and photodarkening processes are demonstrated to be dependent on the pump intensity; for increasing excitation intensities, indeed, the rate of photoactivation process increases and t_c becomes smaller, reaching a stationary behavior after 0.7 W/cm^2 pump intensity. However, the emission spectral shape has been evaluated at different excitation times, for a fixed pump intensity, evidencing that it remains unchanged. This feature, together with the fact that no color change of the film has been observed, allows one to exclude any permanent degradation process of the film.

A possible explanation of the observed experimental results involves the presence of Pb^{2+} ions on the pristine perovskite surface, which act as trap sites for photoexcited charges, reducing the number of charges available for radiative recombination processes. Once oxygen molecules are injected in the test chamber, they passivate such trap states with the formation of lead oxide, leading to the PL intensity increase. On the other hand, moisture in atmosphere induces the formation of hydrated species, which weaken the hydrogen bonds between organic cations and PbI_6 octahedra and support the formation of complexes with water molecules. Since it does not completely destroy the perovskite crystal structure, a recovery of the photoluminescence is possible switching the atmosphere from air to pure oxygen under illumination (Figure 5b).

Air molecules have been demonstrated to act as passivating agents also for bromide based hybrid lead halide perovskites, affecting the photoluminescence emission [70,71]. The result is a PL enhancement when the external atmosphere is switched from vacuum to air. However, differently from the above mentioned research [65], O_2 and water contributions are not separated.

In particular, Fang et al. [70] observed a drastic photoluminescence decrease passing from air to vacuum atmosphere, the emission going down to 10% of the initial maximum value after 120 s. Since the effect of the exposure to the gas can be reset by evacuating the test chamber, it implies the interaction is governed by physical rather than chemical processes. With the aim of checking the selectivity of the response, the $MAPbBr_3$ single crystal samples have been exposed to other gas sources (dry N_2, CO_2 and Ar); however it was found that they do not produce any effects on the PL intensity. Afterwards, the authors investigated the trap healing mechanism in order to have a deeper knowledge about the position of the traps inside the crystal.

In this regard, both 400 nm- and 800 nm-laser excitations have been used, the latter allowing to obtain a two-photon excitation and interaction with the innermost regions of the crystal below the surface. Unlike in the 400-nm excitation experiment (Figure 21d), the

PL intensity modulation under atmosphere change has been found to be much smaller under two-photon excitation (Figure 21b), concluding that the photocarrier recombination is strongly affected by the surface properties (Figure 21a–d). Additional evidence comes from 2D pseudoplots of the time resolved photoluminescence (TRPL) under 800 nm and 400 nm excitation, as shown in Figures 21a,c. The latter indeed evidences that the emission peak wavelength suffers a redshift with ongoing illumination time, whereas in the 800 nm excitation case the emission wavelength remain unaltered, proving that the traps are located near the surface of the film rather than in the bulk.

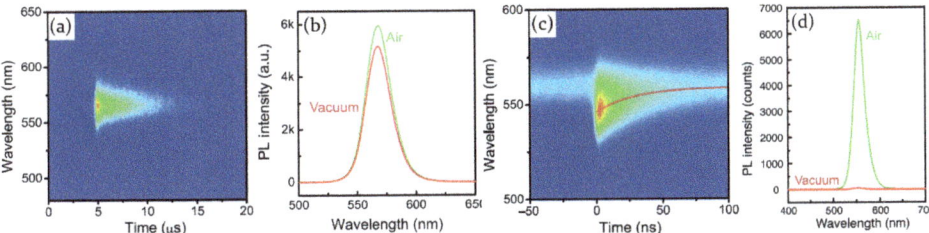

Figure 21. (**a**) 2D pseudocolor plot of TRPL under 800 nm femtosencond laser excitation in air. (**b**) Two-photon (800 nm) excited PL spectra measured in air and in a vacuum. (**c**) 2D pseudocolor plot of TRPL under 400 nm excitation; the red line indicates the emission peak wavelength. (**d**) PL spectra registered in a vacuum and in air, under 400 nm laser excitation. Adapted with permission from [70] under the terms of the Creative Commons Attribution-NonCommercial license. Copyright © 2016, American Association for the Advancement of Science.

Moreover, the PL quenching observed from the exposition of MAPbBr$_3$ single crystals from air to vacuum has been noticed to have a corresponding current modulation, revealing a positive correlation between the resistance of the sample and the air pressure [71]. As a consequence, both dark current and photocurrent have been shown to increase in vacuum. This behavior relates to the fact that air infiltration in perovskite lattice can form shallow trap states close to the perovskite band edges. Those states can trap photo generated carriers, acting as radiative recombination centers. Therefore, in presence of air the PL intensity enhances and a smaller quantity of carriers is available for the conduction, resulting in a current decrease.

The interaction with ammonia, on the contrary, induces structural changes in hybrid lead halide perovskite, regardless of the chemical composition. Ruan et al. [83], for example, realized MAPbX$_3$ (X = I, Br, Cl) single crystals and washed them with a 30 wt.% ammonia solution, revealing an instantaneous color change for all the samples, which indicates that a new phase on the film is formed. Despite XRD measurements indicated that MAPbX$_3$ structure is recovered after an annealing process post-treatment, it evidences how the halogen size influences the stability of the material. Cl presence in MAPbCl$_3$ single crystals, indeed, not only leads to a contracted crystal unity cell, but also determines a more stable perovskite structure compared with the larger unit cell of MAPbI$_3$, which shows a complete phase change after ammonia treatment.

Among the three samples, MAPbBr$_3$ showed an almost complete restoration of the XRD peaks after the ammonia treatment; as a consequence, it has been chosen as active material for the realization of the NH$_3$ sensor. In particular, its photoluminescence decreases of 60% in 2 s when exposed at low ammonia concentration (vapor from 0.3 wt.% NH$_3$ solution), and recovers the initial level in 20 s after removing the NH$_3$ source. However, for higher gas concentrations the sensing response has been found to be slower, reaching 45 s and 1300 s as recovery time for 1 wt.% and 3 wt.% NH$_3$ solution, respectively. The sensor has then been tested with other volatile compounds, such as water, methanol, ethanol and acetone and the response has been measured as the ratio $\Delta I / I_0$, where ΔI represents the variation of PL intensity with and without the external stimulus. Methanol, ethanol and acetone interaction resulted in a PL increase of 25%, 5% and 8%, respectively, whereas water vapor determined a 32% luminescence quenching.

Hybrid lead halide perovskites have also been tested for the detection of volatile organic compounds, exploring the role of the chemical nature of the target gas and its influence on the response of the active mean.

The sensing mechanism of the $MAPbI_3$ perovskite devices proposed by Nur'aini et al. [7] for VOCs is supported by the photoluminescence (PL) measurements (Figure 22a). Adsorption of VOC gas molecules to vacancies in $MAPbI_3$ film leads to trap state passivation. Before ethanol exposure, high density of crystal defects is indicated by low PL intensity, whereas during the exposition the defect density becomes lower and crystallinity of perovskite improves, leading to the PL intensity increases.

From the study of Kim et al. [96], focused on the interaction between $MAPbBr_3$ and aliphatic amine gases ($EtNH_2$, Et_2NH and Et_3N) (Figure 13c), effective naked-eye fluorescence sensors based on $MAPbBr_3$ nanoparticle films have been realized. The exposition to all amines has been observed to lead to PL quenching with changes in intensity of strong green fluorescence; nevertheless the process is fully reversible only for Et_3N exposure with very fast recovery time (<1 s).

$MAPbBr_3$ nanoparticles are used as excellent optical sensors for the detection of picric acid (2,4,6-trinitrophenol, TNP), potent explosive for lethal weapons, with high selectivity and sensitivity both in solution and vapor state [105]. The nanoparticles in toluene, yellow-colored under room light, emit bright green fluorescence when irradiated with UV light at 364 nm (inset of Figure 22b) and the solution shows quenching of the fluorescence (up to 97%) almost instantaneously when TNP is added (Figure 22c,d). Moreover the perovskite nanoparticles are very sensitive and are able to detect TNP with a very low limit of detection, down to femtomolar (10^{-15} M) concentrations. The response is due to the properties of TNP that, being a good electron acceptor, quenches the perovskite fluorescence efficiently. A key role on the overall detection mechanism is played by the hydroxyl group which interacts with the perovskite nanoparticles through the formation of very stable hydrogen bonding.

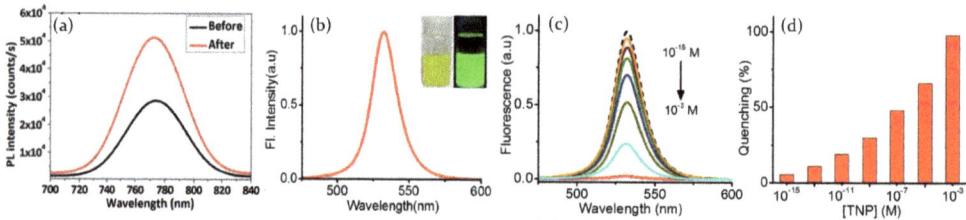

Figure 22. (a) Photoluminescence intensity of $MAPbI_3$ perovskite film before and after ethanol gas exposure. Adapted with permission from [7] under a Creative Commons Attribution-NonCommercial 3.0 Unported license. Copyright © 2020, Royal Society of Chemistry. (b) Emission spectrum of $MAPbBr_3$ nanoparticles. Inset: photographs of the nanoparticle suspension in toluene under room light (left, yellow colored) and 364 nm UV light (right, green colored). (c) Fluorescence spectra of $MAPbBr_3$ nanoparticles at an increasing concentration of TNP, ranging from 10^{-3} M to 10^{-15} M, with the (d) corresponding quenching percentage for the increasing TNP concentration. Used with the permission of the Royal Society of Chemistry, from [105]. Copyright © 2014. Permission conveyed through Copyright Clearance Center, Inc.

Recently, also fully inorganic perovskites have been explored as optical sensors because they exhibit higher long term stability with respect to the hybrid perovskites.

Based on the results inferred from the investigations on the interaction between fully inorganic perovskites and ambient air, $CsPbBr_3$ perovskite nanocrystals have been tested as oxygen sensing materials [79] and PL measurements have revealed a strong emission quenching. Actually, the photoluminescence emission modulation in presence of O_2 rich atmosphere is due to the direct extraction of photogenerated electrons from the conduction band of the NCs causing the PL quenching (Figure 23).

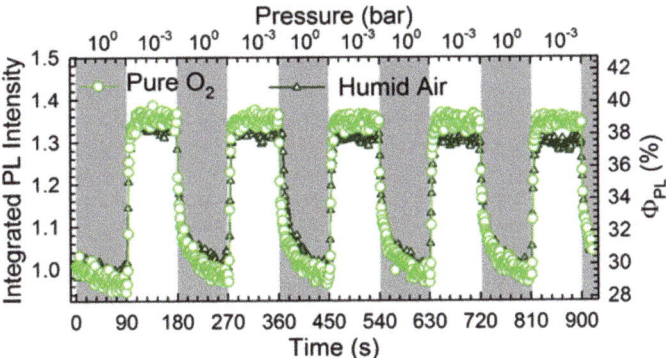

Figure 23. Integrated PL intensity and corresponding PL quantum yield (Φ_{PL}) of CsPbBr$_3$ NCs for O$_2$/vacuum and humid air/vacuum cycles between P = 1 bar (gray shades) and P = 10^{-3} bar (white). Adapted with permission from [79]. Copyright © 2017, American Chemical Society. https://pubs.acs.org/doi/10.1021/acs.nanolett.7b01253. Further permission should be directed to the ACS.

Differently, the photoluminescence modulation in CsPbBr$_3$ single crystals in presence of oxygen [106,107] is an effect of the trap state passivation and through theoretical calculations the nature of these trap states has been investigated. The presence of bromide vacancies on the perovskite surface determines the formation of shallow energy levels at the bottom of the conduction band minimum (CBM) that could trap charge carriers and give rise to non-radiative recombination. When O$_2$ molecules are adsorbed at the location of bromide vacancies, a charge redistribution happens and the shallow trap states at the bottom of the CBM are removed, inducing an increase in the photoluminescence emission.

Recent researches report fully inorganic perovskites as sensing material for the detection of ammonia and explosive vapors, confirming the idea that these materials could represent the ideal candidates for the realization of optical sensors.

Contrary to results obtained with hybrid perovskites, CsPbBr$_3$ based sensor works in a PL turn-on mode rather than in a PL quenching one [86]. The exposition of the sample to ammonia gas reveals an incredible photoluminescence intensity increase of the QDs (about three times than the original one) and the pristine condition is restored when the gas source is removed (Figure 24a). Moreover, TEM and XRD measurements reveal that the size and morphology of the QDs remain the same after treatment, and the crystalline structure is retained after being exposed to ammonia gas. It suggests the interaction between perovskite and NH$_3$ is physical rather than chemical, where ammonia molecules passivate surface defects of perovskite QDs, reducing the non radiative component from 18.78% to 12.73%. The sensor is shown to have a low detection limit of 8.85 ppm and a response and recovery times of about 10 s and 30 s, respectively, for a 50 ppm ammonia concentration (Figure 24b). The selectivity of the active material has also been tested, first evaluating the effect of O$_2$ molecules, then a variety of other gases (such as acetone, water, isopropanol, HCl, ethanol, CO$_2$), all of them inducing just a slightly decrease of the photoluminescence intensity.

The effect of the dimensionality of the sensing material has instead been analyzed by Harwell et al. [108], evidencing how by tuning the perovskite dimensionality from 3D (bulk CsPbBr$_3$) to 2D (layered phenethylammonium (PEA)$_2$Cs$_2$Pb$_3$Br$_{10}$) and finally to 0D (CsPbBr$_3$ nanocrystals) structures can dramatically enhance the response of the films to external stimuli. CsPbBr$_3$ nanocrystals (Figure 24c), thanks to their low dimensionality, have a high surface to volume ratio, which makes them efficient for the detection of 2,4-dinitrotoluene (DNT) vapors through photoluminescence quenching. Contrary to what observed for 3D and 2D perovskite structures, CPB nanocrystals have been indeed seen to respond to the presence of DNT vapors with a rapid drop in PL in about 1 min, and recover about 50% of the emission when returning to pure nitrogen atmosphere (Figure 24d). The reversibility of the PL response to the exposure to DNT vapors is probably due to the presence of ligands on the perovskite NC surface, which prevent DNT molecules from strongly binding to the perovskite. As a consequence, when the gas source is removes, the analyte molecules can easily desorb and the pristine condition is restored.

Contrary to what we have seen for perovskite resistive sensors, the study on photoluminescence modulation based sensors, both with hybrid and fully inorganic perovskites, is not furthered yet. Physical and chemical mechanisms behind the interaction between perovskite and different analytes are not clear so far. Literature lacks of a systematic and extensive study of sensing properties, allowing to identify a good sensing material thanks to parameters such as response and recovery times, sensitivity and limit of detection. Up to now, most authors have provided qualitative researches, showing the potential response of the active mean to a determined analyte in terms of increase or quenching of the photoluminescence. However, it could represent a starting point for a future research involving a complete study of the sensing properties of both hybrid and promising fully inorganic perovskites, with the aim to have a better insight into the question and develop an efficient gas sensor.

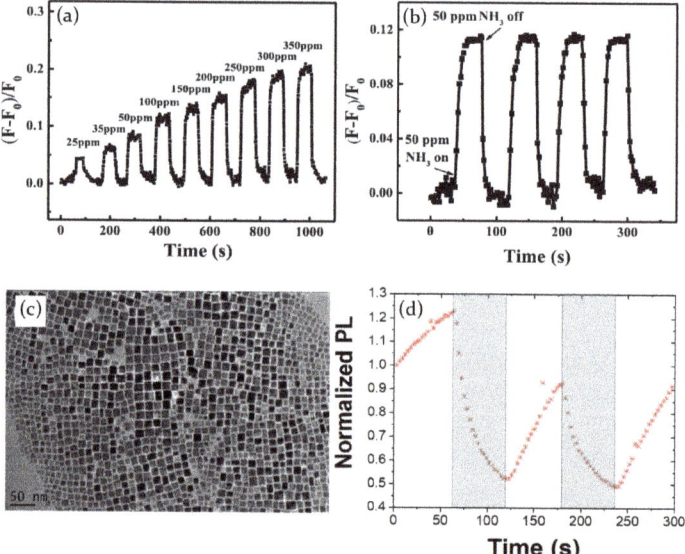

Figure 24. (**a**) Dynamic sensing response of $CsPbBr_3$ quantum dots (QDs) for different NH_3 concentrations (from 25 to 350 ppm). (**b**) Four successive sensing cycles of the $CsPbBr_3$ QDs at 50 ppm. Adapted with permission from [86]. Copyright © 2020, WILEY-VCH Verlag GmbH and Co. KGaA, Weinheim. (**c**) Transmission electron microscopy images of $CsPbBr_3$ nanocrystals. (**d**) Photoluminescence emission as a function of time for a $CsPbBr_3$ nanocrystal film under repeated cycles of exposure to pure nitrogen (white areas) and DNT vapors in N_2 (gray areas). Adapted with permission from [108] under a Creative Commons Attribution (CC BY) license. Copyright © 2020, AIP Publishing.

4. Perspectives

As we have extensively discussed, the interaction of the perovskites with the ambient, when does not lead to reversible effects, is detrimental to their sensing properties. So improving the performance of gas sensing devices requires specific expedients to reduce their degradation over time and when subjected to external stimuli.

To this aim, perovskites with different chemical compositions, dimensionality and surface morphology, or passivation strategies could be employed for the optimization and the stabilization of the sensing devices.

Chemical action on the perovskite composition can improve the sensing ability of the active material—for example, through a doping process [49] or by simply modifying the halogen, which has been demonstrated to influence both the stability and the response speed [83].

Moreover, the deposition process can change the morphology properties of the resulting film (such as the thickness and the substrate coverage) [49,66] affecting, in turn, the device sensitivity. Additionally, the dimensionality of the perovskite can strongly influence the response of the device and it has been shown that switching from 3D to 0D structures allows one to obtain an higher sensitivity to the external stimuli thanks to the 0D greater surface to volume ratio [108].

Furthermore, post-deposition processes, such as annealing treatments, have been seen to improve the stability of the perovskite against moisture, and consequently, enhance the sensitivity of the perovskite toward environmental gases [70].

Operational stability improvement of lead halide perosvkites has been proved also for drop cast $CsPbBr_3$ NC thin films deposited on hexamethyldisilazane (HMDS) hydrophobic functionalized substrates. These results are ascribed to a closer NC packing in the films on treated substrate, allowing the reduction of the film interaction with external moisture [16].

In the research of new sensing systems, with the aim of the device sensitivity enhancement, another route is open.

In PL-based optical sensors, the photoluminescence of the active mean is sensitive to the surrounding gas partial pressure with quenching or enhancement of emission in presence of a target gas. It is proven that in some materials (mostly polymers) exhibiting amplified spontaneous emission (ASE), the sensing sensitivity can be further boosted by harnessing the ASE and the lasing action.

Therefore, novel methods of gas sensing based on ASE or lasing in perovskites should be explored and proposed.

It has been demonstrated that metal-lead halide perovskites exhibit good light amplification emission. Evidence of high optical gains and efficient amplified spontaneous emissions (ASE) is reported for bulk polycrystalline thin films of organic-inorganic [29,31,109,110] and fully inorganic perovskites [33,35] and for perovskite nanocrystal films [16,38,111–114].

However, it has been observed that laser irradiation in vacuum of some perovskites thin films could result in ASE degradation [30]. The correlation between the observed ASE operational stability and the irradiation-induced variations in the local morphology and emission properties, observed at high excitation energy density, suggests that the main process leading to ASE quenching is related to the film melting induced by localized heating. Instead, the photodegradation lessens at lower excitation energy density. Surprisingly, ASE properties, such as intensity and stability, are improved when the film is irradiates in air, indicating that the interaction with oxygen has overall positive effects on the emission properties.

In spite of that, the influence of ambient gas on the perovskite active material that leads to the ASE intensity modulation is to date almost unexplored. Otherwise, the gas-induced modulation of ASE intensity has been already observed and studied in other materials.

ASE sensitivity to an external stimulus was demonstrated since 2005 [115] in semiconducting organic polymer. Trace vapors of the explosives 2,4,6-trinitrotoluene (TNT) and 2,4-dinitrotoluene (DNT) induce ASE attenuation during irradiation of thin films due to the introduction of non-radiative deactivation pathways competing with stimulated emission. Attenuation of ASE intensity is observed when the organic polymer thin films are irradiated in presence of TNT vapors down to 5 ppb, showing a sensitivity to the explosives, which is higher when films are pumped at energy densities near the lasing threshold, up to more than 30 times higher than that observed from spontaneous emission.

Analogously, also oligofluorene based compounds, such as Ter(9,9-diarylfluorene) (TDF) films, are demonstrated to be sensitive to the surrounding oxygen partial pressure with fast response, reversibility and high efficiency [116]. The PL emission intensity is quenched by the presence of O_2, while the emission spectral features, such as peak wavelength, vibronic progression and bandwidth, are unaffected. Beside it has been shown that sensitivity of a chemical sensor can be enhanced in both amplified spontaneous emission and lasing action. In particular, ASE attenuation is found to be 10 times higher than spontaneous emission quenching and enhanced up to 20-fold in lasing action.

The experimental evidence of the enhanced sensitivity of an optical gas sensor when the ASE or the lasing action are monitored, instead of the PL emission, stimulates the exploration of the response of perovskites at excitation densities overcoming the ASE threshold. Indeed, similar sensing improvements could be expected in perovskites optical sensors, but such results are still missing in recent literature.

To obtain high-performance sensing devices it is mandatory to have active materials with high operation stability. For this reason, the stability improvement of light emission, amplification and lasing is an issue that is receiving growing attention. To enhance the ASE operational stability, the deposition of encapsulating layers, such a poly(methyl methacrylate) (PMMA) layer, followed by epoxy resin and glass encapsulation [109], or a commercial fluoropolymer layer (CYTOP) [36], has been proposed for $MAPbBr_3$ bulk polycrystalline thin films. However, these expedients are not suitable for sensing devices, since the top layers prevent the interaction of the active material with the surrounding gases.

An alternative strategy is the treatment of the surface or, in some cases, of the substrate with functionalization chemicals.

As an example, concerning NC films an ASE stability improvement has been reported in films of $MAPbBr_3$ nanocrystals treated with benzyl alcohol [113]. The addition of benzyl alcohol during the synthesis positively affects the structural and photophysical properties of the $MAPbBr_3$ nanocrystals resulting in very stable nanocrystals and nanocrystal thin films, even after 4 months storage under ambient conditions, exhibiting near-unity PLQY, high optical gain (520 cm^{-1}) and ultralow ASE thresholds (~13.9 $\mu J\, cm^{-2}$) under femtosecond excitation.

It has been demonstrated also that a hydrophobic functionalization of the substrates with HMDS improves the ASE properties of drop cast $CsPbBr_3$ nanocrystal thin films with a decrease of threshold down to 45%, an optical gain increase of up to 1.5 times and an ASE operational stability increase of up to 14 times [16].

In future, the employment of functionalized perovskites with good ASE properties, high stability and remarkable sensitivity to ambience, could lead to the realization of high-performance devices, by monitoring the ASE and the laser emission in the presence of target gases.

Author Contributions: M.L.D.G and M.A., conceptualization; M.L.D.G. and S.M. wrote and prepared the original draft; all the authors (M.L.D.G., S.M., A.K. and M.A.) contributed to the preparation, revision and editing of the manuscript; M.A. supervised the work. All authors have read and agreed to the published version of the manuscript.

Funding: This research received no external funding.

Institutional Review Board Statement: Not applicable.

Informed Consent Statement: Not applicable.

Conflicts of Interest: The authors declare no conflict of interest.

References

1. Sun, Y.F.; Liu, S.B.; Meng, F.L.; Liu, J.Y.; Jin, Z.; Kong, L.T.; Liu, J.H. Metal Oxide Nanostructures and Their Gas Sensing Properties: A Review. *Sensors* **2012**, *12*, 2610–2631. [CrossRef] [PubMed]
2. Chiu, S.W.; Tang, K.T. Towards a Chemiresistive Sensor-Integrated Electronic Nose: A Review. *Sensors* **2013**, *13*, 14214–14247. [CrossRef] [PubMed]
3. Gupta Chatterjee, S.; Chatterjee, S.; Ray, A.K.; Chakraborty, A.K. Graphene–metal oxide nanohybrids for toxic gas sensor: A review. *Sens. Actuators B* **2015**, *221*, 1170–1181. [CrossRef]
4. Zhu, L.; Zeng, W. Room-temperature gas sensing of ZnO-based gas sensor: A review. *Sens. Actuators A* **2017**, *267*, 242–261. [CrossRef]
5. Nazemi, H.; Joseph, A.; Park, J.; Emadi, A. Advanced Micro- and Nano-Gas Sensor Technology: A Review. *Sensors* **2019**, *19*. [CrossRef] [PubMed]
6. Arafat, M.M.; Dinan, B.; Akbar, S.A.; Haseeb, A.S.M.A. Gas Sensors Based on One Dimensional Nanostructured Metal-Oxides: A Review. *Sensors* **2012**, *12*, 7207–7258. [CrossRef] [PubMed]
7. Nur'aini, A.; Oh, I. Volatile organic compound gas sensors based on methylammonium lead iodide perovskite operating at room temperature. *RSC Adv.* **2020**, *10*, 12982–12987. [CrossRef]
8. Gagaoudakis, E.; Panagiotopoulos, A.; Maksudov, T.; Moschogiannaki, M.; Katerinopoulou, D.; Kakavelakis, G.; Kiriakidis, G.; Binas, V.; Kymakis, E.; Petridis, K. Self-powered, flexible and room temperature operated solution processed hybrid metal halide p-type sensing element for efficient hydrogen detection. *J. Phys. Mater.* **2020**, *3*, 014010. [CrossRef]
9. Zhao, Y.; Zhu, K. Organic-inorganic hybrid lead halide perovskites for optoelectronic and electronic applications. *Chem. Soc. Rev.* **2016**, *45*, 655–689. [CrossRef]
10. Protesescu, L.; Yakunin, S.; Bodnarchuk, M.I.; Krieg, F.; Caputo, R.; Hendon, C.H.; Yang, R.X.; Walsh, A.; Kovalenko, M.V. Nanocrystals of Cesium Lead Halide Perovskites ($CsPbX_3$, X = Cl, Br, and I): Novel Optoelectronic Materials Showing Bright Emission with Wide Color Gamut. *Nano Lett.* **2015**, *15*, 3692–3696. [CrossRef]
11. Deschler, F.; Price, M.; Pathak, S.; Klintberg, L.E.; Jarausch, D.D.; Higler, R.; Hüttner, S.; Leijtens, T.; Stranks, S.D.; Snaith, H.J.; et al. High Photoluminescence Efficiency and Optically Pumped Lasing in Solution-Processed Mixed Halide Perovskite Semiconductors. *J. Phys. Chem. Lett.* **2014**, *5*, 1421–1426. [CrossRef] [PubMed]
12. Brenner, T.; Egger, D.; Kronik, L.; Hodes, G.; Cahen, D. Hybrid Organic-Inorganic Perovskites: Low-Cost Semiconductors with Intriguing Charge-Transport Properties. *Nat. Rev. Mater.* **2016**, *1*, 15007. [CrossRef]
13. Stoumpos, C.C.; Malliakas, C.D.; Kanatzidis, M.G. Semiconducting tin and lead iodide perovskites with organic cations: phase transitions, high mobilities, and near-infrared photoluminescent properties. *Inorg. Chem.* **2013**, *52*, 9019–9038. [CrossRef] [PubMed]
14. Sutherland, B.R.; Hoogland, S.; Adachi, M.M.; Kanjanaboos, P.; Wong, C.T.; McDowell, J.J.; Xu, J.; Voznyy, O.; Ning, Z.; Houtepen, A.J.; et al. Perovskite thin films via atomic layer deposition. *Adv. Mater.* **2015**, *27*, 53–58. [CrossRef] [PubMed]
15. Motti, S.G.; Krieg, F.; Ramadan, A.J.; Patel, J.B.; Snaith, H.J.; Kovalenko, M.V.; Johnston, M.B.; Herz, L.M. $CsPbBr_3$ Nanocrystal Films: Deviations from Bulk Vibrational and Optoelectronic Properties. *Adv. Funct. Mater.* **2020**, *30*, 1909904. [CrossRef]
16. De Giorgi, M.L.; Krieg, F.; Kovalenko, M.; Anni, M. Amplified Spontaneous Emission Threshold Reduction and Operational Stability Improvement in $CsPbBr_3$ Nanocrystals Films by Hydrophobic Functionalization of the Substrate. *Sci. Rep.* **2019**, *9*. [CrossRef] [PubMed]
17. Weber, D. $CH_3NH_3PbX_3$, a Pb(II)-System with Cubic Perovskite Structure. *Z. Naturforschung B* **1978**, *33*, 1443–1445. [CrossRef]
18. Weber, D. $CH_3NH_3SnBr_xI_{3-x}$ (x=0-3), a Sn(II)-System with Cubic Perovskite Structure. *Z. Naturforschung B* **1978**, *33*, 862–865. [CrossRef]
19. Kojima, A.; Teshima, K.; Shirai, Y.; Miyasaka, T. Organometal Halide Perovskites as Visible-Light Sensitizers for Photovoltaic Cells. *J. Am. Chem. Soc.* **2009**, *131*, 6050–6051. [CrossRef]
20. Sun, S.; Salim, T.; Mathews, N.; Duchamp, M.; Boothroyd, C.; Xing, G.; Sum, T.C.; Lam, Y.M. The origin of high efficiency in low-temperature solution-processable bilayer organometal halide hybrid solar cells. *Energy Environ. Sci.* **2014**, *7*, 399–407. [CrossRef]
21. Stranks, S.D.; Snaith, H. Metal-halide perovskites for photovoltaic and light-emitting devices. *Nat. Nanotechnol.* **2015**, *10*, 391–402. [CrossRef] [PubMed]
22. National Renewable Energy Laboratory. Research Cell Record Efficiency Chart. Available online: https://www.nrel.gov/pv/cell-efficiency.html (accessed on 15 December 2020).
23. Hattori, T.; Taira, T.; Era, M.; Tsutsui, T.; Saito, S. Highly efficient electroluminescence from a heterostructure device combined with emissive layered-perovskite and an electron-transporting organic compound. *Chem. Phys. Lett.* **1996**, *254*, 103–108. [CrossRef]
24. Chondroudis, K.; Mitzi, D.B. Electroluminescence from an Organic-Inorganic Perovskite Incorporating a Quaterthiophene Dye within Lead Halide Perovskite Layers. *Chem. Mater.* **1999**, *11*, 3028–3030. [CrossRef]
25. Kojima, A.; Ikegami, M.; Teshima, K.; Miyasaka, T. Highly Luminescent Lead Bromide Perovskite Nanoparticles Synthesized with Porous Alumina Media. *Chem. Lett.* **2012**, *41*, 397–399. [CrossRef]
26. Tan, Z.; Moghaddam, R.; Lai, M.; Docampo, P.; Higler, R.; Deschler, F.; Price, M.; Sadhanala, A.; Pazos, L.; Credgington, D.; et al. Bright light-emitting diodes based on organometal halide perovskite. *Nat. Nanotechnol.* **2014**, *9*, 687–692. [CrossRef]

27. Li, G.; Tan, Z.K.; Di, D.; Lai, M.L.; Jiang, L.; Lim, J.H.W.; Friend, R.H.; Greenham, N.C. Efficient Light-Emitting Diodes Based on Nanocrystalline Perovskite in a Dielectric Polymer Matrix. *Nano Lett.* **2015**, *15*, 2640–2644. [CrossRef]
28. Zhang, F.; Zhong, H.; Chen, C.; Wu, X.g.; Hu, X.; Huang, H.; Han, J.; Zou, B.; Dong, Y. Brightly Luminescent and Color-Tunable Colloidal $CH_3NH_3PbX_3$ (X = Br, I, Cl) Quantum Dots: Potential Alternatives for Display Technology. *ACS Nano* **2015**, *9*, 4533–4542. [CrossRef]
29. Xing, G.; Mathews, N.; Lim, S.S.; Yantara, N.; Liu, X.; Sabba, D.; Grätzel, M.; Mhaisalkar, S.; Sum, T.C. Low-temperature solution-processed wavelength-tunable perovskites for lasing. *Nat. Mater.* **2014**, *13*, 476–480. [CrossRef]
30. De Giorgi, M.L.; Lippolis, T.; Jamaludin, N.F.; Soci, C.; Bruno, A.; Anni, M. Origin of Amplified Spontaneous Emission Degradation in $MAPbBr_3$ Thin Films under Nanosecond-UV Laser Irradiation. *J. Phys. Chem. C* **2020**, *124*, 10696–10704. [CrossRef]
31. Stranks, S.D.; Wood, S.M.; Wojciechowski, K.; Deschler, F.; Saliba, M.; Khandelwal, H.; Patel, J.B.; Elston, S.J.; Herz, L.M.; Johnston, M.B.; et al. Enhanced Amplified Spontaneous Emission in Perovskites Using a Flexible Cholesteric Liquid Crystal Reflector. *Nano Lett.* **2015**, *15*, 4935–4941. [CrossRef]
32. Ngo, T.T.; Suarez, I.; Antonicelli, G.; Cortizo-Lacalle, D.; Martinez-Pastor, J.P.; Mateo-Alonso, A.; Mora-Sero, I. Enhancement of the Performance of Perovskite Solar Cells, LEDs, and Optical Amplifiers by Anti-Solvent Additive Deposition. *Adv. Mater.* **2017**, *29*, 1604056. [CrossRef] [PubMed]
33. Pourdavoud, N.; Haeger, T.; Mayer, A.; Cegielski, P.J.; Giesecke, A.L.; Heiderhoff, R.; Olthof, S.; Zaefferer, S.; Shutsko, I.; Henkel, A.; et al. Room-Temperature Stimulated Emission and Lasing in Recrystallized Cesium Lead Bromide Perovskite Thin Films. *Adv. Mater.* **2019**, *31*, 1903717. [CrossRef] [PubMed]
34. Brenner, P.; Bar-On, O.; Jakoby, M.; Allegro, I.; Richards, B.; Paetzold, U.; Howard, I.; Scheuer, J.; Lemmer, U. Continuous wave amplified spontaneous emission in phase-stable lead halide perovskites. *Nat. Commun.* **2019**, *10*. [CrossRef] [PubMed]
35. De Giorgi, M.L.; Perulli, A.; Yantara, N.; Boix, P.P.; Anni, M. Amplified Spontaneous Emission Properties of Solution Processed $CsPbBr_3$ Perovskite Thin Films. *J. Phys. Chem. C* **2017**, *121*, 14772–14778. [CrossRef]
36. Harwell, J.; Whitworth, G.; Turnbull, G.; Samuel, I. Green Perovskite Distributed Feedback Lasers. *Sci. Rep.* **2017**, *7*, 11727. [CrossRef]
37. Zhang, Q.; Ha, S.T.; Liu, X.; Sum, T.C.; Xiong, Q. Room-Temperature Near-Infrared High-Q Perovskite Whispering-Gallery Planar Nanolasers. *Nano Lett.* **2014**, *14*, 5995–6001. [CrossRef]
38. Yakunin, S.; Protesescu, L.; Krieg, F.; Bodnarchuk, M.; Nedelcu, G.; Humer, M.; De Luca, G.; Fiebig, M.; Heiss, W.; Kovalenko, M. Low-threshold amplified spontaneous emission and lasing from colloidal nanocrystal of caesium lead halide perovskites. *Nat. Commun.* **2015**, *6*, 8056. [CrossRef]
39. Tian, W.; Zhou, H.; Li, L. Hybrid Organic–Inorganic Perovskite Photodetectors. *Small* **2017**, *13*, 1702107. [CrossRef]
40. Yao, F.; Gui, P.; Zhang, Q.; Lin, Q. Molecular engineering of perovskite photodetectors: Recent advances in materials and devices. *Mol. Syst. Des. Eng.* **2018**, *3*, 702–716. [CrossRef]
41. Yakunin, S.; Dirin, D.N.; Shynkarenko, Y.; Morad, V.; Cherniukh, I.; Nazarenko, O.; Kreil, D.; Nauser, T.; Kovalenko, M.V. Detection of gamma photons using solution-grown single crystals of hybrid lead halide perovskites. *Nat. Photonics* **2016**, *10*, 585–589. [CrossRef]
42. Wei, H.; Fang, Y.; Mulligan, P.; Chuirazzi, W.; Fang, H.H.; Wang, C.; Ecker, B.; Gao, Y.; Loi, M.; Cao, L.; et al. Sensitive X-ray detectors made of methylammonium lead tribromide perovskite single crystals. *Nat. Photonics* **2016**, *10*, 333–339. [CrossRef]
43. Xu, Q.; Zhang, H.; Nie, J.; Shao, W.; Wang, X.; Zhang, B.; Ouyang, X. Effect of methylammonium lead tribromide perovskite based-photoconductor under gamma photons radiation. *Radiat. Phys. Chem.* **2021**, *181*, 109337. [CrossRef]
44. Yang, J.; Kang, W.; Liu, Z.; Pi, M.; Luo, L.B.; Li, C.; Lin, H.; Luo, Z.; Du, J.; Zhou, M.; et al. High-Performance Deep Ultraviolet Photodetector Based on a One-Dimensional Lead-Free Halide Perovskite $CsCu_2I_3$ Film with High Stability. *J. Phys. Chem. Lett.* **2020**, *11*, 6880–6886. [CrossRef] [PubMed]
45. Kang, C.H.; Dursun, I.; Liu, G.; Sinatra, L.; Sun, X.; Kong, M.; Pan, J.; Maity, P.; Ooi, E.N.; Ng, T.K.; et al. High-speed colour-converting photodetector with all-inorganic $CsPbBr_3$ perovskite nanocrystals for ultraviolet light communication. *Light Sci. Appl.* **2019**, *8*, 1–12. [CrossRef] [PubMed]
46. Zou, T.; Liu, X.; Qiu, R.; Wang, Y.; Huang, S.; Liu, C.; Dai, Q.; Zhou, H. Enhanced UV-C Detection of Perovskite Photodetector Arrays via Inorganic $CsPbBr_3$ Quantum Dot Down-Conversion Layer. *Adv. Opt. Mater.* **2019**, *7*, 1801812. [CrossRef]
47. Tong, G.; Li, H.; Zhu, Z.; Zhang, Y.; Yu, L.; Xu, J.; Jiang, Y. Enhancing Hybrid Perovskite Detectability in the Deep Ultraviolet Region with Down-Conversion Dual-Phase ($CsPbBr_3$–Cs_4PbBr_6) Films. *J. Phys. Chem. Lett.* **2018**, *9*, 1592–1599. [CrossRef]
48. Nguyen, T.M.H.; Kim, S.; Bark, C.W. Solution-processed and self-powered photodetector in vertical architecture using mixed-halide perovskite for highly sensitive UVC detection. *J. Mater. Chem. A* **2021**. [CrossRef]
49. Zhuang, Y.; Yuan, W.; Qian, L.; Chen, S.; Shi, G. High-performance gas sensors based on a thiocyanate ion-doped organometal halide perovskite. *Phys. Chem. Chem. Phys.* **2017**, *19*, 12876–12881. [CrossRef]
50. Lou, S.; Xuan, T.; Wang, J. (INVITED) Stability: A desiderated problem for the lead halide perovskites. *Opt. Mater. X* **2019**, *1*, 100023. [CrossRef]
51. Shellaiah, M.; Sun, K.W. Review on Sensing Applications of Perovskite Nanomaterials. *Chemosensors* **2020**, *8*. [CrossRef]

52. Kymakis, E.; Panagiotopoulos, A.; Stylianakis, M.M.; Petridis, K. 5-Organometallic hybrid perovskites for humidity and gas sensing applications. In *2D Nanomaterials for Energy Applications*; Zafeiratos, S., Ed.; Micro and Nano Technologies; Elsevier: Amsterdam, The Netherlands, 2020; pp. 131–147. [CrossRef]
53. Halali, V.V.; Sanjayan, C.G.; Suvina, V.; Sakar, M.; Balakrishna, R.G. Perovskite nanomaterials as optical and electrochemical sensors. *Inorg. Chem. Front.* **2020**, *7*, 2702–2725. [CrossRef]
54. Gunasekaran, R.K.; Chinnadurai, D.; Selvaraj, A.R.; Rajendiran, R.; Senthil, K.; Prabakar, K. Revealing the Self-Degradation Mechanisms in Methylammonium Lead Iodide Perovskites in Dark and Vacuum. *Chem. Phys. Chem.* **2018**, *19*, 1507–1513. [CrossRef] [PubMed]
55. Misra, R.K.; Aharon, S.; Li, B.; Mogilyansky, D.; Visoly-Fisher, I.; Etgar, L.; Katz, E.A. Temperature- and Component-Dependent Degradation of Perovskite Photovoltaic Materials under Concentrated Sunlight. *J. Phys. Chem. Lett.* **2015**, *6*, 326–330. [CrossRef] [PubMed]
56. Juarez-Perez, E.J.; Ono, L.K.; Maeda, M.; Jiang, Y.; Hawash, Z.; Qi, Y. Photodecomposition and thermal decomposition in methylammonium halide lead perovskites and inferred design principles to increase photovoltaic device stability. *J. Mater. Chem. A* **2018**, *6*, 9604–9612. [CrossRef]
57. Li, Y.; Xu, X.; Wang, C.; Ecker, B.; Yang, J.; Huang, J.; Gao, Y. Light-Induced Degradation of $CH_3NH_3PbI_3$ Hybrid Perovskite Thin Film. *J. Phys. Chem. C* **2017**, *121*, 3904–3910. [CrossRef]
58. Conings, B.; Drijkoningen, J.; Gauquelin, N.; Babayigit, A.; D'Haen, J.; D'Olieslaeger, L.; Ethirajan, A.; Verbeeck, J.; Manca, J.; Mosconi, E.; et al. Intrinsic Thermal Instability of Methylammonium Lead Trihalide Perovskite. *Adv. Energy Mater.* **2015**, *5*, 1500477. [CrossRef]
59. Motti, S.G.; Gandini, M.; Barker, A.J.; Ball, J.M.; Srimath Kandada, A.R.; Petrozza, A. Photoinduced Emissive Trap States in Lead Halide Perovskite Semiconductors. *ACS Energy Lett.* **2016**, *1*, 726–730. [CrossRef]
60. Quitsch, W.A.; de Quilettes, D.W.; Pfingsten, O.; Schmitz, A.; Ognjanovic, S.; Jariwala, S.; Koch, S.; Winterer, M.; Ginger, D.S.; Bacher, G. The Role of Excitation Energy in Photobrightening and Photodegradation of Halide Perovskite Thin Films. *J. Phys. Chem. Lett.* **2018**, *9*, 2062–2069. [CrossRef]
61. Kulbak, M.; Gupta, S.; Kedem, N.; Levine, I.; Bendikov, T.; Hodes, G.; Cahen, D. Cesium Enhances Long-Term Stability of Lead Bromide Perovskite-Based Solar Cells. *J. Phys. Chem. Lett.* **2016**, *7*, 167–172. [CrossRef]
62. Huang, S.; Li, Z.; Wang, B.; Zhu, N.; Zhang, C.; Kong, L.; Zhang, Q.; Shan, A.; Li, L. Morphology Evolution and Degradation of $CsPbBr_3$ Nanocrystals under Blue Light-Emitting Diode Illumination. *ACS Appl. Mater. Interfaces* **2017**, *9*, 7249–7258. [CrossRef]
63. Li, Y.; Wang, L.; Yuan, X.; Bo, B.; Li, H.; Zhao, J.; Gao, X. Ultraviolet light induced degradation of luminescence in $CsPbBr_3$ perovskite nanocrystals. *Mater. Res. Bull.* **2018**, *102*, 86–91. [CrossRef]
64. Christians, J.A.; Miranda Herrera, P.A.; Kamat, P.V. Transformation of the Excited State and Photovoltaic Efficiency of $CH_3NH_3PbI_3$ Perovskite upon Controlled Exposure to Humidified Air. *J. Am. Chem. Soc.* **2015**, *137*, 1530–1538. [CrossRef] [PubMed]
65. Galisteo-López, J.F.; Anaya, M.; Calvo, M.E.; Míguez, H. Environmental Effects on the Photophysics of Organic-Inorganic Halide Perovskites. *J. Phys. Chem. Lett.* **2015**, *6*, 2200–2205. [CrossRef] [PubMed]
66. Stoeckel, M.A.; Gobbi, M.; Bonacchi, S.; Liscio, F.; Ferlauto, L.; Orgiu, E.; Samorì, P. Reversible, Fast, and Wide-Range Oxygen Sensor Based on Nanostructured Organometal Halide Perovskite. *Adv. Mater.* **2017**, *29*, 1702469. [CrossRef]
67. Wang, Q.; Chen, B.; Liu, Y.; Deng, Y.; Bai, Y.; Dong, Q.; Huang, J. Scaling behavior of moisture-induced grain degradation in polycrystalline hybrid perovskite thin films. *Energy Environ. Sci.* **2017**, *10*, 516–522. [CrossRef]
68. Yun, J.S.; Kim, J.; Young, T.; Patterson, R.J.; Kim, D.; Seidel, J.; Lim, S.; Green, M.A.; Huang, S.; Ho-Baillie, A. Humidity-Induced Degradation via Grain Boundaries of $HC(NH_2)_2PbI_3$ Planar Perovskite Solar Cells. *Adv. Funct. Mater.* **2018**, *28*, 1705363. [CrossRef]
69. Howard, J.M.; Tennyson, E.M.; Barik, S.; Szostak, R.; Waks, E.; Toney, M.F.; Nogueira, A.F.; Neves, B.R.A.; Leite, M.S. Humidity-Induced Photoluminescence Hysteresis in Variable Cs/Br Ratio Hybrid Perovskites. *J. Phys. Chem. Lett.* **2018**, *9*, 3463–3469. [CrossRef]
70. Fang, H.H.; Adjokatse, S.; Wei, H.; Yang, J.; Blake, G.R.; Huang, J.; Even, J.; Loi, M.A. Ultrahigh sensitivity of methylammonium lead tribromide perovskite single crystals to environmental gases. *Sci. Adv.* **2016**, *2*. [CrossRef]
71. Zhang, H.; Liu, Y.; Lu, H.; Deng, W.; Yang, K.; Deng, Z.; Zhang, X.; Yuan, S.; Wang, J.; Niu, J.; et al. Reversible air-induced optical and electrical modulation of methylammonium lead bromide ($MAPbBr_3$) single crystals. *Appl. Phys. Lett.* **2017**, *111*, 103904. [CrossRef]
72. Tian, Y.; Peter, M.; Unger, E.; Abdellah, M.; Zheng, K.; Pullerits, T.; Yartsev, A.; Sundström, V.; Scheblykin, I.G. Mechanistic insights into perovskite photoluminescence enhancement: Light curing with oxygen can boost yield thousandfold. *Phys. Chem. Chem. Phys.* **2015**, *17*, 24978–24987. [CrossRef]
73. Godding, J.S.; Ramadan, A.J.; Lin, Y.H.; Schutt, K.; Snaith, H.J.; Wenger, B. Oxidative Passivation of Metal Halide Perovskites. *Joule* **2019**, *3*, 2716–2731. [CrossRef]
74. Brenes, R.; Guo, D.; Osherov, A.; Noel, N.; Eames, C.; Hutter, E.; Pathak, S.; Niroui, F.; Friend, R.; Islam, M.S.; et al. Metal Halide Perovskite Polycrystalline Films Exhibiting Properties of Single Crystals. *Joule* **2017**, *1*, 155–167. [CrossRef]
75. Ouyang, Y.; Li, Y.; Zhu, P.; Li, Q.; Gao, Y.; Tong, J.; Shi, L.; Zhou, Q.; Ling, C.; Chen, Q.; et al. Photo-oxidative degradation of methylammonium lead iodide perovskite: Mechanism and protection. *J. Mater. Chem. A* **2019**, *7*, 2275–2282. [CrossRef]

76. Anaya, M.; Galisteo-López, J.F.; Calvo, M.E.; Espinós, J.P.; Míguez, H. Origin of Light-Induced Photophysical Effects in Organic Metal Halide Perovskites in the Presence of Oxygen. *J. Phys. Chem. Lett.* **2018**, *9*, 3891–3896. [CrossRef]
77. Aristidou, N.; Sanchez-Molina, I.; Chotchuangchutchaval, T.; Brown, M.; Martinez, L.; Rath, T.; Haque, S.A. The Role of Oxygen in the Degradation of Methylammonium Lead Trihalide Perovskite Photoactive Layers. *Angew. Chem. Int. Ed.* **2015**, *54*, 8208–8212. [CrossRef]
78. Aristidou, N.; Eames, C.; Sanchez-Molina, I.; Bu, X.; Kosco, J.; Islam, M.S.; Haque, S.A. Fast oxygen diffusion and iodide defects mediate oxygen-induced degradation of perovskite solar cells. *Nat. Commun.* **2017**, *8*, 15218. [CrossRef]
79. Lorenzon, M.; Sortino, L.; Akkerman, Q.; Accornero, S.; Pedrini, J.; Prato, M.; Pinchetti, V.; Meinardi, F.; Manna, L.; Brovelli, S. Role of Nonradiative Defects and Environmental Oxygen on Exciton Recombination Processes in $CsPbBr_3$ Perovskite Nanocrystals. *Nano Lett.* **2017**, *17*, 3844–3853. [CrossRef]
80. Maity, A.; Ghosh, B. Fast response paper based visual color change gas sensor for efficient ammonia detection at room temperature. *Sci. Rep.* **2018**, *8*. [CrossRef]
81. Maity, A.; Raychaudhuri, A.; Ghosh, B. High sensitivity NH_3 gas sensor with electrical readout made on paper with perovskite halide as sensor material. *Sci. Rep.* **2019**, *9*, 7777. [CrossRef]
82. Singh, A.K.; Singh, S.; Singh, V.N.; Gupta, G.; Gupta, B.K. Probing reversible photoluminescence alteration in $CH_3NH_3PbBr_3$ colloidal quantum dots for luminescence-based gas sensing application. *J. Colloid Interface Sci.* **2019**, *554*, 668–673. [CrossRef]
83. Ruan, S.; Lu, J.; Pai, N.; Ebendorff-Heidepriem, H.; Cheng, Y.B.; Ruan, Y.; McNeill, C.R. An optical fibre-based sensor for the detection of gaseous ammonia with methylammonium lead halide perovskite. *J. Mater. Chem. C* **2018**, *6*, 6988–6995. [CrossRef]
84. Bao, C.; Yang, J.; Zhu, W.; Zhou, X.; Gao, H.; Li, F.; Fu, G.; Yu, T.; Zou, Z. A resistance change effect in perovskite $CH_3NH_3PbI_3$ films induced by ammonia. *Chem. Commun.* **2015**, *51*, 15426–15429. [CrossRef]
85. Zhao, Y.; Zhu, K. Optical bleaching of perovskite $(CH_3NH_3)PbI_3$ through room-temperature phase transformation induced by ammonia. *Chem. Commun.* **2014**, *50*, 1605–1607. [CrossRef] [PubMed]
86. Huang, H.; Hao, M.; Song, Y.; Dang, S.; Liu, X.; Dong, Q. Dynamic Passivation in Perovskite Quantum Dots for Specific Ammonia Detection at Room Temperature. *Small* **2020**, *16*, 1904462. [CrossRef]
87. Fu, X.; Jiao, S.; Dong, N.; Lian, G.; Zhao, T.; Lv, S.; Wang, Q.; Cui, D. A $CH_3NH_3PbI_3$ film for a room-temperature NO_2 gas sensor with quick response and high selectivity. *RSC Adv.* **2018**, *8*, 390–395. [CrossRef]
88. Zhang, Z.J.; Xiang, S.C.; Guo, G.C.; Xu, G.; Wang, M.S.; Zou, J.P.; Guo, S.P.; Huang, J.S. Wavelength-Dependent Photochromic Inorganic–Organic Hybrid Based on a 3D Iodoplumbate Open-Framework Material. *Angew. Chem. Int. Ed.* **2008**, *47*, 4149–4152. [CrossRef]
89. Huang, H.; Xu, Y.; Feng, Q.; Leung, D.Y.C. Low temperature catalytic oxidation of volatile organic compounds: A review. *Catal. Sci. Technol.* **2015**, *5*, 2649–2669. [CrossRef]
90. Shuai, J.; Kim, S.; Ryu, H.; Park, J.; Lee, C.; Kim, G.B.; Ultra, V.U., Jr.; Yang, W. Health risk assessment of volatile organic compounds exposure near Daegu dyeing industrial complex in South Korea. *BMC Public Health* **2018**, *18*. [CrossRef]
91. Hempel-Jorgensen, A.; Kjaergaard, S.K.; Molhave, L.; Hudnell, K.H. Sensory Eye Irritation in Humans Exposed to Mixtures of Volatile Organic Compounds. *Arch. Environ. Health* **1999**, *54*, 416–424. [CrossRef]
92. Ahn, K. The role of air pollutants in atopic dermatitis. *J. Allergy Clin. Immunol.* **2014**, *5*, 993–1000. [CrossRef]
93. Otto, D.A.; Hudnell, H.K.; House, D.E.; Mølhave, L.; Counts, W. Exposure of Humans to a Volatile Organic Mixture. I. Behavioral Assessment. *Arch. Environ. Health* **1992**, *47*, 23–30. [CrossRef] [PubMed]
94. Genc, S.; Zadeoglulari, Z.; Fuss, S.; Genc, K. The Adverse Effects of Air Pollution on the Nervous System. *J. Toxicol.* **2012**, *2012*, 782462. [CrossRef] [PubMed]
95. Wei, L.Y.; Ma, W.; Lian, C.; Meng, S. Benign Interfacial Iodine Vacancies in Perovskite Solar Cells. *J. Phys. Chem. C* **2017**, *121*, 5905–5913. [CrossRef]
96. Kim, S.H.; Kirakosyan, A.; Choi, J.; Kim, J.H. Detection of volatile organic compounds (VOCs), aliphatic amines, using highly fluorescent organic-inorganic hybrid perovskite nanoparticles. *Dyes Pigm.* **2017**, *147*, 1–5. [CrossRef]
97. Giorgi, G.; Fujisawa, J.I.; Segawa, H.; Yamashita, K. Cation Role in Structural and Electronic Properties of 3D Organic-Inorganic Halide Perovskites: A DFT Analysis. *J. Phys. Chem. C* **2014**, *118*, 12176–12183. [CrossRef]
98. Kakavelakis, G.; Gagaoudakis, E.; Petridis, K.; Petromichelaki, V.; Binas, V.; Kiriakidis, G.; Kymakis, E. Solution Processed $CH_3NH_3PbI_{3-x}Cl_x$ Perovskite Based Self-Powered Ozone Sensing Element Operated at Room Temperature. *ACS Sens.* **2018**, *3*, 135–142. [CrossRef]
99. Wen, X.; Wu, J.; Gao, D.; Lin, C. Interfacial engineering with amino-functionalized graphene for efficient perovskite solar cells. *J. Mater. Chem. A* **2016**, *4*, 13482–13487. [CrossRef]
100. Department of Health and Human Services. US Food and Drug Administration, Federal Register. 1997. Available online: https://www.govinfo.gov/content/pkg/FR-1997-04-17/pdf/97-9706.pdf (accessed on 15 December 2020).
101. Acharyya, D.; Bhattacharyya, P. Alcohol sensing performance of ZnO hexagonal nanotubes at low temperatures: A qualitative understanding. *Sens. Actuators B* **2016**, *228*, 373–386. [CrossRef]
102. Liang, J.; Wang, C.; Wang, Y.; Xu, Z.; Lu, Z.; Ma, Y.; Zhu, H.; Hu, Y.; Xiao, C.; Yi, X.; et al. All-Inorganic Perovskite Solar Cells. *J. Am. Chem. Soc.* **2016**, *138*, 15829–15832. [CrossRef]
103. Beal, R.E.; Slotcavage, D.J.; Leijtens, T.; Bowring, A.R.; Belisle, R.A.; Nguyen, W.H.; Burkhard, G.F.; Hoke, E.T.; McGehee, M.D. Cesium Lead Halide Perovskites with Improved Stability for Tandem Solar Cells. *J. Phys. Chem. Lett.* **2016**, *7*, 746–751. [CrossRef]

104. Chen, H.; Zhang, M.; Bo, R.; Barugkin, C.; Zheng, J.; Ma, Q.; Huang, S.; Ho-Baillie, A.W.Y.; Catchpole, K.R.; Tricoli, A. Superior Self-Powered Room-Temperature Chemical Sensing with Light-Activated Inorganic Halides Perovskites. *Small* **2018**, *14*, 1702571. [CrossRef] [PubMed]
105. Muthu, C.; Nagamma, S.R.; Nair, V.C. Luminescent hybrid perovskite nanoparticles as a new platform for selective detection of 2,4,6-trinitrophenol. *RSC Adv.* **2014**, *4*, 55908–55911. [CrossRef]
106. Wang, Y.; Ren, Y.; Zhang, S.; Wu, J.; Song, J.; Li, X.; Jiayue, X.; Sow, C.; Zeng, H.; Sun, H. Switching excitonic recombination and carrier trapping in cesium lead halide perovskites by air. *Commun. Phys.* **2018**, *1*. [CrossRef]
107. Lu, D.; Zhang, Y.; Lai, M.; Lee, A.; Xie, C.; Lin, J.; Lei, T.; Lin, Z.; Kley, C.S.; Huang, J.; et al. Giant Light-Emission Enhancement in Lead Halide Perovskites by Surface Oxygen Passivation. *Nano Lett.* **2018**, *18*, 6967–6973. [CrossRef]
108. Harwell, J.R.; Glackin, J.M.E.; Davis, N.J.L.K.; Gillanders, R.N.; Credgington, D.; Turnbull, G.A.; Samuel, I.D.W. Sensing of explosive vapor by hybrid perovskites: Effect of dimensionality. *APL Mater.* **2020**, *8*, 071106. [CrossRef]
109. Li, J.; Si, J.; Gan, L.; Liu, Y.; Ye, Z.; He, H. Simple Approach to Improving the Amplified Spontaneous Emission Properties of Perovskite Films. *ACS Appl. Mater. Interfaces* **2016**, *8*, 32978–32983. [CrossRef]
110. Qin, L.; Lv, L.; Li, C.; Zhu, L.; Cui, Q.; Hu, Y.; Lou, Z.; Teng, F.; Hou, Y. Temperature dependent amplified spontaneous emission of vacuum annealed perovskite films. *RSC Adv.* **2017**, *7*, 15911–15916. [CrossRef]
111. Balena, A.; Perulli, A.; Fernandez, M.; De Giorgi, M.L.; Nedelcu, G.; Kovalenko, M.V.; Anni, M. Temperature Dependence of the Amplified Spontaneous Emission from $CsPbBr_3$ Nanocrystal Thin Films. *J. Phys. Chem. C* **2018**, *122*, 5813–5819. [CrossRef]
112. Papagiorgis, P.; Manoli, A.; Protesescu, L.; Achilleos, C.; Violaris, M.; Nicolaides, K.; Trypiniotis, T.; Bodnarchuk, M.I.; Kovalenko, M.V.; Othonos, A.; et al. Efficient Optical Amplification in the Nanosecond Regime from Formamidinium Lead Iodide Nanocrystals. *ACS Photonics* **2018**, *5*, 907–917. [CrossRef]
113. Veldhuis, S.A.; Tay, Y.K.E.; Bruno, A.; Dintakurti, S.S.H.; Bhaumik, S.; Muduli, S.K.; Li, M.; Mathews, N.; Sum, T.C.; Mhaisalkar, S.G. Benzyl Alcohol-Treated $CH_3NH_3PbBr_3$ Nanocrystals Exhibiting High Luminescence, Stability, and Ultralow Amplified Spontaneous Emission Thresholds. *Nano Lett.* **2017**, *17*, 7424–7432. [CrossRef]
114. Dey, A.; Rathod, P.; Kabra, D. Role of Localized States in Photoluminescence Dynamics of High Optical Gain $CsPbBr_3$ Nanocrystals. *Adv. Opt. Mater.* **2018**, *6*, 1800109. [CrossRef]
115. Rose, A.; Zhu, Z.; Madigan, C.F.; Swager, T.M.; Bulovic, V. Sensitivity gains in chemosensing by lasing action in organic polymers. *Nature* **2005**, *434*, 876–879. [CrossRef] [PubMed]
116. Lin, H.W.; Huang, M.H.; Chen, Y.H.; Lin, W.C.; Cheng, H.C.; Wu, C.C.; Chao, T.C.; Wang, T.C.; Wong, K.T.; Tang, K.C.; et al. Novel oxygen sensor based on terfluorene thin-film and its enhanced sensitivity by stimulated emission. *J. Mater. Chem.* **2012**, *22*, 13446–13450. [CrossRef]

Quasi-3D Hyperbolic Shear Deformation Theory for the Free Vibration Study of Honeycomb Microplates with Graphene Nanoplatelets-Reinforced Epoxy Skins

Hossein Arshid [1], Mohammad Khorasani [2], Zeinab Soleimani-Javid [3], Rossana Dimitri [4] and Francesco Tornabene [4,*]

1. Department of Mechanical Engineering, Qom Branch, Islamic Azad University, Qom 3749113191, Iran; Arshid@ut.ac.ir
2. Department of Basic and Applied Sciences for Engineering, Faculty of Civil and Industrial Engineering, Sapienza University, 00161 Rome, Italy; Mohammad.Khorasani94@hotmail.com
3. Department of Solid Mechanics, Faculty of Mechanical Engineering, University of Kashan, Kashan 8731753153, Iran; As.Javid71@gmail.com
4. Department of Innovation Engineering, Università del Salento, 73100 Lecce, Italy; rossana.dimitri@unisalento.it
* Correspondence: francesco.tornabene@unisalento.it; Tel./Fax: +39-0832297275

Academic Editors: Giuseppe Cirillo and Long Y Chiang
Received: 23 September 2020; Accepted: 30 October 2020; Published: 2 November 2020

Abstract: A novel quasi-3D hyperbolic shear deformation theory (QHSDT) with five unknowns is here employed, together with the Hamilton's principle and the modified couple stress theory (MCST) to analyze the vibrational behavior of rectangular micro-scale sandwich plates resting on a visco-Pasternak foundation. The sandwich structure features a Nomex or Glass phenolic honeycomb core, and two composite face sheets reinforced with graphene nanoplatelets (GPLs). The effective properties of both face sheets are evaluated by means of the Halpin-Tsai and extended rule of mixture (ERM) micromechanical schemes. The governing equations of the problem are derived by applying the Hamilton's principle, whose solutions are determined theoretically according to a classical Navier-type procedure. A parametric study checks for the effect of different material properties, length-scale parameters, foundation parameters and geometrical properties of the honeycomb cells, and the reinforcing GPLs, on the vibration response of the layered structure, which can be of great interest for many modern engineering applications and their optimization design.

Keywords: graphene nanoplatelets; honeycomb structures; modified couple stress theory; quasi-3d hyperbolic shear deformation theory; sandwich structures; vibration analysis

1. Introduction

In the last decades, lightweight mechanical components and layered structures have increased the attention of many researchers and scientists, due to the increased demand in modern engineering, together with a possible reduction in their production cost. Among them, sandwich structures can be regarded as subset of multilayered composite structures consisting of outer facings and a soft core in-between, including foam, honeycomb, corrugated core, various bio-inspired cores, etc. The choice of sandwich materials depends on the structural functionality as well as on the lifetime loading, availability and cost. For example, Graphite-epoxy and carbon-epoxy multilayered facings are typically used in aerospace applications, whereas glass-epoxy or glass-vinyl ester are adopted in civil and marine layered structures. At the same time, the core of sandwich aerospace structures is often made of aluminum or

Nomex honeycomb, whereas, a closed-cell or open-cell foam represents the typical core choice in civil engineering, instead of a balsa core, usually applied in ship sandwich structures. As far as the honeycomb sandwich-type plate is concerned, the adhesive bonding between the honeycomb core and face sheets is the responsible for the load transferring among the sandwich constitutive parts. In such a context, one of the pioneering works on the topic is represented by Ref. [1], where the authors studied the vibrational behavior of sandwich beams with a honeycomb core [1]. In 2004, different vibration tests were performed experimentally by Yanfeng and Jinghui [2] to study the vibration transmissibility and shock-absorbing properties of the honeycomb thin plates, while computing their damping ratios and highest frequencies of vibration. From a theoretical and numerical perspective, a comprehensive review of studies on sandwich structures is mentioned in the following, covering the more recent developments on the topic. Li and Jin [3] applied a third-order shear deformation plate theory (TSDT) and classical plate theory (CPT) to examine the free vibration of rectangular plates with a honeycomb structure, whereas a semi-analytical approach was suggested in [4] for the bending, buckling and free vibration analysis of sandwich panels with square-honeycomb cores. At the same time, the influence of the skin/core debonding phenomena on the overall vibrational behavior of sandwich plates was analyzed by Burlayenko and Sadowski [5], whose results helped to address sandwich plates non-destructive damages. In line with this work, a wavelet analysis has been recently applied by Katunin [6], to detect and identify possible damages in sandwich structures and their effect on the global vibrational structural response. The sensitivity of the vibration response of a honeycomb core structure to random geometrical or mechanical irregularities was also outlined theoretically by Mukhopadhyay and Adhikari [7]. A novel method was proposed by Duc et al. [8] to study the vibrational response of sandwich cylindrical panels with a honeycomb core, based on the first-order shear deformation theory (FSDT), fourth-order Runge-Kutta method and Galerkin method. Among the most recent solutions of increasing the intrinsic damping properties of sandwich structures, Piollet et al. [9] proposed the use of entangled cross-linked fibers as core materials within sandwich beams and performed different steady-state tests for different excitation levels to study their high-damping and nonlinear vibration response. Moreover, Kumar and Renji [10] studied the acceleration response and natural modes of sandwich panels with a honeycomb core subjected a diffused acoustic field, developing a methodology to estimate their strain field in low frequency modes, based on the acceleration response. A novel model based on the differential quadrature method (DQM) was successfully proposed by Sobhy [11] to study the coupled hygrothermal bending response of functionally graded (FG) graphene platelets/aluminum sandwich-curved beams equipped by a honeycomb core. A numerical and experimental investigation based on a classical finite element approach and imaging correlation method was also performed by Li et al. [12] for the study of the dynamic response of shallow sandwich arches with aluminum face sheets and auxetic reentrant hexagonal metallic honeycomb core under a localized impulsive loading, providing useful data and results for the honeycomb cells deformation [12]. In 2017, Chen et al. [13] examined the nonlinear mechanical behavior of a sandwich structure. Their model was made of FG porous layer reinforced by graphene nanoplatelets (GPLs). Moreover, Karimiasl et al. [14] studied the nonlinear vibration behavior of multiscale nanocomposites nanoshells, resting on an elastic foundation, and subjecting to a hygrothermal environment. Furthermore, in 2019, the instability characteristics of a magnetorheological (MR) fluid core patched to two piezoelectric FG-GPLRC face sheets were investigated by Eyvazian et al. [15], while proving the positive effect of magnetic field on the system's mechanical behavior. More recently in 2020, Torabi and Ansari [16] hired the Mindlin's plate model and the phase-field approach to have a throughout comprehension of the vibration response for cracked FG GPL-RC plates with stationary cracks.

The large benefits of sandwich structures and their mechanical performances, have increased the interest of the scientific community to develop even more accurate theories for their study. For example, an improper definition of a mechanical parameter, even at small scales, can cause a meaningful variation in the acquisition of results, with deleterious effects on the overall performance of sensitive systems, as aircraft and space vehicles. This makes extremely important the use of accurate theories, where the proper definition of the mechanical parameters is mandatory to obtain reliable results. In such a context, many works from the literature have applied CPTs, FSDTs, or higher-order-theories (HSDTs) for the study of plate and shell structures even with complicated materials and geometries. For example, Khoa et al. [17] applied a HSDT to examine the vibration response of FG carbon nanotubes (CNTs) reinforced composites cylindrical shells in thermal environment. The same problem was also studied by Ibrahim et al. [18], according to FSDT, and coupled with thermal conditions. Li et al. [19] used CPT to model clamped honeycomb sandwich panels to study the nonlinear forced vibrational response. Many further applications of the HSDT to coupled problems of sandwich panels and shell structures can be found in [20–28]. A valid theoretical alternative to handle the plate structures is represented by the quasi-3D hyperbolic shear deformation theory (QHSDT) which accounts for both transverse shear and normal deformations and satisfies the zero traction boundary conditions on the plate surfaces without using any shear correction factor. In QHSDT the number of unknown functions involved in displacement field is only equal to five, instead of six or more unknowns required by the other shear and normal deformation theories. The computational efficiency of this method was recently verified in Refs. [29–31]. Inspired by these few pioneering works from the literature, in the present paper we propose a QHSDT to study the free vibration response of sandwich structures with a honeycomb core resting on a visco-Pasternak foundation. The governing equations of the problem are derived from the Hamilton's principle and solved in closed form via the Navier's method. The analytical solutions from our formulation are verified with those reported in literature, where a parametric investigations aims at determining the effect of the material variation, GPLs gradient index and dispersion patterns, geometry, internal cells angle, or thickness of layers on the natural frequencies for the selected sandwich structure.

2. Theoretical Formulation

Consider a rectangular sandwich plate with thickness h, length a, width b, as illustrated in Figure 1, together with the reference coordinate system (x, y, z). The sandwich structure is immersed within a visco-Pasternak elastic foundation, and it is made of a honeycomb core with thickness h_c and two composite face sheets with thickness h_t and h_b at the top and bottom side, respectively. This means that the total thickness of the structure is $h = h_c + h_t + h_b$.

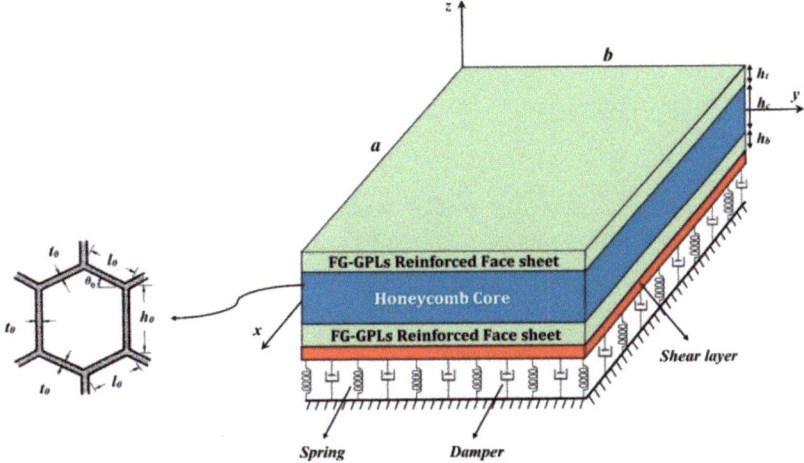

Figure 1. Geometrical model of sandwich structure.

In the current study, a QHSDT is adopted to define the position of an arbitrary point in the micro-model. The major advantage of using such a displacement field is that the problem is not limited to plane-strain conditions (i.e., $\varepsilon_{zz} \neq 0$), as typically occurs in the other 2-D theories such as FSDT, that could cause possible discrepancies between the theoretical and experimental results. Based on a QHSDT, the displacement field is defined as [32]

$$U(x,y,z,t) = u(x,y,t) - z\frac{\partial}{\partial x}w_b(x,y,t) - f(z)\frac{\partial}{\partial x}w_s(x,y,t),$$
$$V(x,y,z,t) = v(x,y,t) - z\frac{\partial}{\partial y}w_b(x,y,t) - f(z)\frac{\partial}{\partial y}w_s(x,y,t), \quad (1)$$
$$W(x,y,z,t) = w_b(x,y,t) - w_s(x,y,t) - w_{st}(x,y,z,t)$$

where u and v stand for the displacement components along the x and y directions, respectively; w_s, w_b and w_{st} are the transverse displacement components due to bending, shear and stretching effects, respectively, with

$$w_{st}(x,y,z,t) = g(z)\varphi(x,y,t) \quad (2)$$

In the last relation, φ is an additional displacement variable that accounts for the effect of normal stress; $g(z)$ and $f(z)$ are expressed by the following functions [29]

$$f(z) = ((h/\pi)\sinh(\pi z/h) - z)/(\cosh(\pi/2) - 1), \quad (3)$$

$$g(z) = 1 - f'(z) \quad (4)$$

where $f'(z)$ denotes the first derivative of function f with respect to z. The strain-displacement relations follow the von-Karman's assumptions [31]

$$\begin{aligned}
\varepsilon_{xx} &= \frac{\partial u(x,y,t)}{\partial x} - z\frac{\partial^2 w_b(x,y,t)}{\partial x^2} - f(z)\frac{\partial^2 w_s(x,y,t)}{\partial x^2}, \\
\varepsilon_{yy} &= \frac{\partial v(x,y,t)}{\partial y} - z\frac{\partial^2 w_b(x,y,t)}{\partial y^2} - f(z)\frac{\partial^2 w_s(x,y,t)}{\partial y^2}, \\
\varepsilon_{zz} &= -\frac{\partial^2(f(z)\varphi(x,y,t))}{\partial z^2}, \\
\gamma_{xy} &= \frac{\partial v(x,y,t)}{\partial x} - 2z\frac{\partial^2 w_b(x,y,t)}{\partial x \partial y} - 2f(z)\frac{\partial^2 w_s(x,y,t)}{\partial x \partial y} + \frac{\partial u(x,y,t)}{\partial y}, \\
\gamma_{xz} &= \frac{\partial w_s(x,y,t)}{\partial x} + (1-f'(z))\frac{\partial \varphi(x,y,t)}{\partial x} - f'(z)\frac{\partial w_s(x,y,t)}{\partial x}, \\
\gamma_{yz} &= \frac{\partial w_s(x,y,t)}{\partial y} + (1-f'(z))\frac{\partial \varphi(x,y,t)}{\partial y} - f'(z)\frac{\partial w_s(x,y,t)}{\partial y},
\end{aligned} \quad (5)$$

whereas the constitutive equations for the honeycomb core and FG-GPLs face sheets, read as follows [33]

$$\left\{\begin{array}{c}\sigma_{xx}\\\sigma_{yy}\\\sigma_{zz}\\\sigma_{xy}\\\sigma_{yz}\\\sigma_{xz}\end{array}\right\}^{c,f} = \left[\begin{array}{cccccc}C_{11}&C_{12}&C_{13}&0&0&0\\C_{12}&C_{22}&C_{23}&0&0&0\\C_{13}&C_{23}&C_{33}&0&0&0\\0&0&0&C_{44}&0&0\\0&0&0&0&C_{55}&0\\0&0&0&0&0&C_{66}\end{array}\right]\left\{\begin{array}{c}\varepsilon_{xx}\\\varepsilon_{yy}\\\varepsilon_{zz}\\\gamma_{xy}\\\gamma_{yz}\\\gamma_{xz}\end{array}\right\} \quad (6)$$

where C_{ij} are the elastic constants for each part of the sandwich structure. More specifically, for the honeycomb core, the elastic constants read as follows [29]

$$\begin{aligned}
C_{11c} &= \frac{E_{11}(-\nu_{23}\nu_{32}+1)}{\delta}, & C_{22c} &= \frac{E_{22}(-\nu_{13}\nu_{31}+1)}{\delta}, \\
C_{33c} &= \frac{E_{33}(-\nu_{12}\nu_{21}+1)}{\delta}, & C_{12c} &= C_{21c} = \frac{E_{11}(\nu_{23}\nu_{31}+\nu_{21})}{\delta}, \\
C_{13c} &= C_{31c} = \frac{E_{11}(\nu_{21}\nu_{32}+\nu_{31})}{\delta}, & C_{23c} &= C_{32c} = \frac{E_{22}(\nu_{12}\nu_{31}+\nu_{32})}{\delta}, \\
C_{44c} &= G_{23}, & C_{55c} &= G_{13}, & C_{66c} &= G_{12}
\end{aligned} \quad (7)$$

where

$$\delta = 1 - 2\nu_{12}\nu_{13}\nu_{32} - \nu_{12}\nu_{21} - \nu_{13}\nu_{31} - \nu_{23}\nu_{32} \quad (8)$$

and

$$E_{11} = E_h \frac{\cos\theta_0(1-\gamma_0^2\cot^2\theta_0)}{\sin^2\theta_0(\phi_0+\sin\theta_0)}\gamma_0^3, \quad (9)$$

$$E_{22} = E_h \frac{(1-\gamma_0^2(\phi_0\sec^2\theta_0+\tan^2\theta_0))(\phi_0+\sin\theta_0)}{\cos^3\theta_0}\gamma_0^3, \quad (10)$$

$$E_{33} = E_h \frac{2+\phi_0}{2\cos\theta_0(\phi_0+\sin\theta_0)}\gamma_0, \tag{11}$$

$$G_{12} = E_h \frac{(\phi_0+\sin\theta_0)}{\phi_0^2(1+2\phi_0)\cos\theta_0}\gamma_0^3, \tag{12}$$

$$G_{13} = G_h \frac{\cos\theta_0}{(\phi_0+\sin\theta_0)}\gamma_0, \tag{13}$$

$$G_{23} = G_h \left(\frac{(\phi_0+\sin\theta_0)}{(1+2\phi_0)\cos\theta_0} + \frac{(\phi_0+2\sin^2\theta_0)}{2(\phi_0+\sin\theta_0)} \right) \frac{\gamma_0}{2\cos\theta_0}, \tag{14}$$

$$\rho^c = \rho_h \frac{2+\phi_0}{2\cos\theta_0(\phi_0+\sin\theta_0)}\gamma_0, \tag{15}$$

$$\nu_{12} = \frac{\cos^2\theta_0(1-\gamma_0^2\csc^2\theta_0)}{\sin\theta_0(\phi_0+\sin\theta_0)}, \tag{16}$$

$$\nu_{21} = \frac{(1+\phi_0)(1-\gamma_0^2\sec^2\theta_0)\cos^2\theta_0}{\sin\theta_0(\phi_0+\sin\theta_0)}, \tag{17}$$

$$\nu_{31} = \nu_{32} = \nu_h \tag{18}$$

In the relations above, the Young's modulus, shear modulus, density, and Poisson's ratio are defined in a homogenized form by means of the mechanical properties E_h, G_h, ρ_h and ν_h of the honeycomb material [34]. Besides, ϕ_0 is the internal cells angle of the honeycomb structure; $\varphi_0 = h_0/l_0$ and $\gamma_0 = t_0/l_0$ stand for the internal aspect ratio and dimensionless cells thickness, respectively, in which h_0, l_0 and t_0 are the geometrical parameters defining the hexagonal cells as represented in Figure 1. For FG-GPLs reinforced face sheets, the elastic constants C_{ijf} are given by

$$\begin{aligned} C_{11f} &= C_{22f} = C_{33f} = \frac{(1-\nu(z))E(z)}{(1+\nu(z))(1-2\nu(z))}, \\ C_{12f} &= C_{13f} = C_{23f} = \frac{\nu(z)E(z)}{(1+\nu(z))(1-2\nu(z))}, \\ C_{44f} &= C_{55f} = C_{66f} = \frac{E(z)}{2(1+\nu(z))} \end{aligned} \tag{19}$$

The mechanical properties for both face sheets vary throughout the thickness, and they are clearly function of the effective material properties, defined, in turn, by means of the Halpin-Tsai micromechanical model, as follows [35]

$$E(z) = \frac{3}{8}\frac{1+\zeta_L\eta_L V_{GPL}}{1-\eta_L V_{GPL}}E_M + \frac{5}{8}\frac{1+\zeta_W\eta_W V_{GPL}}{1-\eta_W V_{GPL}}E_M \tag{20}$$

In the last relation, E_M denotes the Young's modulus of the matrix; V_{GPL} refers to the volume fraction of GPLs; ζ_L, ζ_W, η_L and η_W are the geometrical properties of GPLs, i.e.,

$$\begin{aligned} \zeta_L &= 2\frac{L_{GPL}}{h_{GPL}}, & \eta_W &= (E_{GPL}/E_M - 1)/(E_{GPL}/E_M + \zeta_W), \\ \zeta_W &= 2\frac{w_{GPL}}{h_{GPL}}, & \eta_L &= (E_{GPL}/E_M - 1)/(E_{GPL}/E_M + \zeta_L), \end{aligned} \tag{21}$$

L_{GPL}, h_{GPL}, w_{GPL} and E_{GPL} being the length, thickness, width and Young's modulus of GPLs, respectively. It is noteworthy that the summation of GPLs and matrix volume fractions equals one, where the GPLs volume fraction is determined as

$$V_{GPL} = \frac{g_{GPL}(z)}{g_{GPL}(z) + (\frac{\rho_{GPL}}{\rho_M})(1-g_{GPL}(z))} \tag{22}$$

where ρ_{GPL} and ρ_M refer to the density of the reinforcement phase and matrix, respectively. Moreover, g_{GPL} is the weight fraction of GPLs that obey the following relations for three different dispersion patterns though the face sheets thicknesses [36]

For a parabolic pattern

$$g_{GPL}(z) = \frac{4}{h_f^2}\lambda_P W_{GPL} z^2 \tag{23}$$

For a linear pattern

$$g_{GPL}(z) = \lambda_L W_{GPL}\left(\frac{1}{2} \pm \frac{z}{h_f}\right) \tag{24}$$

in which the positive and negative signs are related to the top and bottom face sheets, respectively. For a uniform pattern

$$g_{GPL}(z) = \lambda_U W_{GPL} \tag{25}$$

In Equations (23)–(25), λ_P, λ_L and λ_U are the gradient index of GPLs for their parabolic, linear, and uniform dispersion patterns, referred to the total GPLs content, as reported in Table 1.

Table 1. Graphene nanoplatelets (GPLs) gradient index for different values of their total content [36].

Total GPLs Content (Percentage)	λ_U	λ_L	λ_P
0	0	0	0
1/3	1/3	2/3	1
1	1	2	3

The further properties for the face sheets are the Poisson's ratio and density, which are determined via the ERM as [37]

$$\rho(z) = \rho_{GPL} V_{GPL} + \rho_M V_M, \tag{26}$$

$$\nu(z) = \nu_{GPL} V_{GPL} + \nu_M V_M \tag{27}$$

3. Governing Equations of the Problem

The Hamilton's principle is here applied to gain the governing equations of the problem [38]

$$\int_{t_1}^{t_2} \delta(\Lambda - K - \Pi)dt = 0 \tag{28}$$

where Π, Λ and K denote the applied external work, the strain energy, and the kinetic energy for the sandwich structure, respectively. The strain energy of the system consists of two parts: the classical strain energy and the energy component from the MCST. The following relation is used to define the total strain energy for the selected sandwich structure [39]

$$\Lambda = \frac{1}{2}\left(\iiint\limits_{x\ y\ core} \left(\sigma_{ij}^c \varepsilon_{ij} + m_{ij}^c \chi_{ij}\right) dz\, dy\, dx + \iiint\limits_{x\ y\ faces} \left(\sigma_{ij}^f \varepsilon_{ij} + m_{ij}^f \chi_{ij}\right) dz\, dy\, dx \right); i,j = x,y,z \tag{29}$$

where m_{ij} and χ_{ij} stand for the higher-order stresses and symmetric rotation gradient tensor, respectively, defined in the following

$$m_{ij} = 2l_m^2 \mu \chi_{ij}; \qquad (i,j = x,y,z) \tag{30}$$

where l_m is the MCST material length scale parameter, and μ is the Lame's parameter. Moreover, the components of the symmetric rotation gradient tensor can be determined using the following compact relation

$$\chi_{ij} = \frac{1}{2}\left(\Theta_{i,j} + \Theta_{j,i}\right) \tag{31}$$

This means that

$$\chi_{xx} = \frac{\partial}{\partial x}\Theta_x, \quad \chi_{yy} = \frac{\partial}{\partial y}\Theta_y, \quad \chi_{zz} = \frac{\partial}{\partial z}\Theta_z,$$
$$\chi_{xy} = \frac{1}{2}\left(\frac{\partial}{\partial y}\Theta_x + \frac{\partial}{\partial x}\Theta_y\right), \quad \chi_{yz} = \frac{1}{2}\left(\frac{\partial}{\partial z}\Theta_y + \frac{\partial}{\partial y}\Theta_z\right), \quad \chi_{xz} = \frac{1}{2}\left(\frac{\partial}{\partial z}\Theta_x + \frac{\partial}{\partial x}\Theta_z\right) \tag{32}$$

in which, the infinitesimal rotation vector Θ is defined as

$$\Theta_i = \frac{1}{2}(curl(u))_{,i} \tag{33}$$

which means

$$\Theta_x = \frac{1}{2}\left(\frac{\partial}{\partial y}W(x,y,z,t) - \frac{\partial}{\partial z}V(x,y,z,t)\right),$$
$$\Theta_y = \frac{1}{2}\left(\frac{\partial}{\partial z}U(x,y,z,t) - \frac{\partial}{\partial x}W(x,y,z,t)\right), \tag{34}$$
$$\Theta_z = \frac{1}{2}\left(\frac{\partial}{\partial x}V(x,y,z,t) - \frac{\partial}{\partial y}U(x,y,z,t)\right)$$

In addition, the kinetic energy for the whole microstructure can be defined as [40].

$$K = \frac{1}{2}\int_x\int_y\int_{-h/2}^{+h/2}\rho_{c,f}(z)\left[\left(\frac{\partial U}{\partial t}\right)^2 + \left(\frac{\partial V}{\partial t}\right)^2 + \left(\frac{\partial W}{\partial t}\right)^2\right]dzdydx \tag{35}$$

where U, V and W refer to the displacement components introduced in Equation (1).

For a structure resting on a visco-Pasternak elastic foundation, the external work due to the substrate can be defined as follows [41]

$$\Pi = \int_x\int_y \frac{1}{2}\left(\begin{array}{l} K_W(w_b+w_s)^2 - K_G(w_b+w_s)\frac{\partial^2(w_b+w_s)}{\partial x^2} - \\ K_G(w_b+w_s)\frac{\partial^2(w_b+w_s)}{\partial y^2} + \\ C_d(w_b+w_s)\frac{\partial(w_b+w_s)}{\partial t} \end{array}\right)dxdy \tag{36}$$

where K_W is the Winkler parameter, K_G is the shear layer parameter, and C_d denotes the damping parameter, respectively. By substitution of Equations (29), (35), (36) into the Hamilton's principle (28), after a mathematical manipulation we get the following governing equations of motion in terms of displacement field

$$\delta u:$$

$$-C_{110}\frac{\partial^2 u(x,y,t)}{\partial x^2} + C_{111}\frac{\partial^3 w_b(x,y,t)}{\partial x^3} + F_{110}\frac{\partial^3 w_s(x,y,t)}{\partial x^3} +$$
$$+C_{121}\frac{\partial^3 w_b(x,y,t)}{\partial x\partial y^2} + F_{120}\frac{\partial^3 w_s(x,y,t)}{\partial x\partial y^2} - E_{130}\frac{\partial \varphi(x,y,t)}{\partial x} +$$
$$-C_{440}\frac{\partial^2 v(x,y,t)}{\partial x\partial y} - \frac{1}{4}K\frac{\partial^4 v(x,y,t)}{\partial x\partial y^3} - C_{120}\frac{\partial^2 v(x,y,t)}{\partial x\partial y} +$$
$$+2C_{441}\frac{\partial^3 w_b(x,y,t)}{\partial x\partial y^2} + 2F_{440}\frac{\partial^3 w_s(x,y,t)}{\partial x\partial y^2} - C_{440}\frac{\partial^2 u(x,y,t)}{\partial y^2} + \tag{37}$$
$$+\frac{1}{4}K\frac{\partial^4 u(x,y,t)}{\partial y^4} - \frac{1}{4}K\frac{\partial^4 v(x,y,t)}{\partial x^3\partial y} + \frac{1}{4}K\frac{\partial^4 u(x,y,t)}{\partial x^2\partial y^2} +$$
$$-I_0\frac{\partial^2 u(x,y,t)}{\partial t^2} + I_1\frac{\partial^3 w_b(x,y,t)}{\partial t^2\partial x} + I_3\frac{\partial^3 w_s(x,y,t)}{\partial t^2\partial x} = 0$$

$\delta v:$

$$\begin{aligned}
&-C_{120}\frac{\partial^2 u(x,y,t)}{\partial x \partial y} + C_{121}\frac{\partial^3 w_b(x,y,t)}{\partial x^2 \partial y} + F_{120}\frac{\partial^3 w_s(x,y,t)}{\partial x^2 \partial y} - C_{220}\frac{\partial^2 v(x,y,t)}{\partial y^2} + \\
&+ C_{221}\frac{\partial^3 w_b(x,y,t)}{\partial y^3} + F_{220}\frac{\partial^3 w_s(x,y,t)}{\partial y^3} - E_{230}\frac{\partial \varphi(x,y,t)}{\partial y} - C_{440}\frac{\partial^2 v(x,y,t)}{\partial x^2} + \\
&+ 2C_{441}\frac{\partial^3 w_b(x,y,t)}{\partial x^2 \partial y} + 2F_{440}\frac{\partial^3 w_s(x,y,t)}{\partial x^2 \partial y} - C_{440}\frac{\partial^2 u(x,y,t)}{\partial y \partial x} + \tfrac{1}{4}K\frac{\partial^4 v(x,y,t)}{\partial x^4} + \\
&- \tfrac{1}{4}K\frac{\partial^4 u(x,y,t)}{\partial x^3 \partial y} + \tfrac{1}{4}K\frac{\partial^4 v(x,y,t)}{\partial x^2 \partial y^2} - \tfrac{1}{4}K\frac{\partial^4 u(x,y,t)}{\partial x \partial y^3} + \\
&- I_0\frac{\partial^2 v(x,y,t)}{\partial t^2} + I_1\frac{\partial^3 w_b(x,y,t)}{\partial t^2 \partial y} + I_3\frac{\partial^3 w_s(x,y,t)}{\partial t^2 \partial y} = 0
\end{aligned} \qquad (38)$$

$\delta w_b:$

$$\begin{aligned}
&-C_{121}\frac{\partial^3 u(x,y,t)}{\partial x \partial y^2} + C_{122}\frac{\partial^4 w_b(x,y,t)}{\partial x^2 \partial y^2} + F_{121}\frac{\partial^4 w_s(x,y,t)}{\partial x^2 \partial y^2} - C_{221}\frac{\partial^3 v(x,y,t)}{\partial y^3} + \\
&+ C_{222}\frac{\partial^4 w_b(x,y,t)}{\partial y^4} + F_{221}\frac{\partial^4 w_s(x,y,t)}{\partial y^4} - E_{231}\frac{\partial^2 \varphi(x,y,t)}{\partial y^2} - 2C_{441}\frac{\partial^3 v(x,y,t)}{\partial x^2 \partial y} + \\
&+ 4C_{442}\frac{\partial^4 w_b(x,y,t)}{\partial x^2 \partial y^2} + 4F_{441}\frac{\partial^4 w_s(x,y,t)}{\partial x^2 \partial y^2} - 2C_{441}\frac{\partial^3 u(x,y,t)}{\partial y^2 \partial x} + \\
&- C_{111}\frac{\partial^3 u(x,y,t)}{\partial x^3} + C_{112}\frac{\partial^4 w_b(x,y,t)}{\partial x^4} + F_{111}\frac{\partial^4 w_s(x,y,t)}{\partial x^4} - C_{121}\frac{\partial^3 v(x,y,t)}{\partial x^2 \partial y} + \\
&+ C_{122}\frac{\partial^4 w_b(x,y,t)}{\partial x^2 \partial y^2} + 2K\frac{\partial^4 w_b(x,y,t)}{\partial x^2 \partial y^2} + K\frac{\partial^4 w_s(x,y,t)}{\partial x^2 \partial y^2} + \\
&+ K_0\frac{\partial^4 \varphi(x,y,t)}{\partial x^2 \partial y^2} - K_1\frac{\partial^4 w_s(x,y,t)}{\partial y^4} + K\frac{\partial^4 w_b(x,y,t)}{\partial y^4} + \tfrac{1}{2}K\frac{\partial^4 w_s(x,y,t)}{\partial y^4} + \\
&+ \tfrac{1}{2}K_0\frac{\partial^4 \varphi(x,y,t)}{\partial y^4} + \tfrac{1}{2}K_1\frac{\partial^4 w_s(x,y,t)}{\partial y^4} + K\frac{\partial^4 w_b(x,y,t)}{\partial y^4} + \tfrac{1}{2}K_1\frac{\partial^4 w_s(x,y,t)}{\partial x^4} + \\
&+ \tfrac{1}{2}K\frac{\partial^4 w_s(x,y,t)}{\partial x^4} + \tfrac{1}{2}K_0\frac{\partial^4 \varphi(x,y,t)}{\partial x^4} + F_{121}\frac{\partial^4 w_s(x,y,t)}{\partial x^2 \partial y^2} - E_{131}\frac{\partial^4 \varphi(x,y,t)}{\partial x^2} + \\
&- C_d\frac{\partial w_b(x,y,t)}{\partial t} - C_d\frac{\partial w_s(x,y,t)}{\partial t} + K_G\frac{\partial^2 w_b(x,y,t)}{\partial x^2} + K_G\frac{\partial^2 w_s(x,y,t)}{\partial x^2} + \\
&+ K_G g\frac{\partial^2 \varphi(x,y,t)}{\partial x^2} + K_G\frac{\partial^2 w_b(x,y,t)}{\partial y^2} + K_G\frac{\partial^2 w_s(x,y,t)}{\partial y^2} + K_G g\frac{\partial^2 \varphi(x,y,t)}{\partial y^2} + \\
&- C_d g\frac{\partial \varphi(x,y,t)}{\partial t} - K_W g\varphi(x,y,t) - K_W w_s(x,y,t) - K_W w_b(x,y,t) - I_1\frac{\partial^3 u(x,y,t)}{\partial t^2 \partial x} + \\
&+ I_2\frac{\partial^4 w_b(x,y,t)}{\partial t^2 \partial x^2} + I_5\frac{\partial^4 w_s(x,y,t)}{\partial t^2 \partial x^2} - I_1\frac{\partial^3 v(x,y,t)}{\partial t^2 \partial y} + I_2\frac{\partial^4 w_b(x,y,t)}{\partial t^2 \partial y^2} + \\
&+ I_5\frac{\partial^4 w_s(x,y,t)}{\partial t^2 \partial y^2} - I_0\frac{\partial^2 w_b(x,y,t)}{\partial t^2} - I_0\frac{\partial^2 w_s(x,y,t)}{\partial t^2} - I_4\frac{\partial^2 \varphi(x,y,t)}{\partial t^2} = 0
\end{aligned} \qquad (39)$$

$\delta w_s:$

$$\begin{aligned}
&\tfrac{1}{2}K_1\tfrac{\partial^4 w_b(x,y,t)}{\partial x^4} + \tfrac{1}{4}K_3\tfrac{\partial^4 w_s(x,y,t)}{\partial x^4} + \tfrac{1}{4}K_2\tfrac{\partial^4 \varphi(x,y,t)}{\partial x^4} - F_{111}\tfrac{\partial^4 w_b(x,y,t)}{\partial x^4} + \\
&+\tfrac{1}{2}K_1\tfrac{\partial^4 w_b(x,y,t)}{\partial y^4} + \tfrac{1}{4}K_3\tfrac{\partial^4 w_s(x,y,t)}{\partial x^4} + \tfrac{1}{4}K_2\tfrac{\partial^4 \varphi(x,y,t)}{\partial x^4} - \tfrac{1}{4}K_5\tfrac{\partial^2 w_s(x,y,t)}{\partial x^2} + \\
&-\tfrac{1}{4}K_5\tfrac{\partial^2 w_s(x,y,t)}{\partial y^2} - \tfrac{1}{4}K_4\tfrac{\partial^2 \varphi(x,y,t)}{\partial x^2} - \tfrac{1}{4}K_4\tfrac{\partial^2 \varphi(x,y,t)}{\partial y^2} - F_{110}\tfrac{\partial^3 u(x,y,t)}{\partial x^3} + \\
&-F_{120}\tfrac{\partial^3 v(x,y,t)}{\partial x^2 \partial y} + F_{121}\tfrac{\partial^4 w_b(x,y,t)}{\partial y^2 \partial x^2} + F_{122}\tfrac{\partial^4 w_s(x,y,t)}{\partial y^2 \partial x^2} - E_{132}\tfrac{\partial^2 \varphi(x,y,t)}{\partial x^2} + \\
&-F_{120}\tfrac{\partial^3 u(x,y,t)}{\partial y^2 \partial x} + F_{121}\tfrac{\partial^4 w_b(x,y,t)}{\partial x^2 \partial y^2} + F_{112}\tfrac{\partial^4 w_s(x,y,t)}{\partial x^4} + F_{122}\tfrac{\partial^4 w_s(x,y,t)}{\partial x^2 \partial y^2} + \\
&-F_{220}\tfrac{\partial^3 v(x,y,t)}{\partial y^3} + F_{221}\tfrac{\partial^4 w_b(x,y,t)}{\partial y^4} + F_{222}\tfrac{\partial^4 w_s(x,y,t)}{\partial y^4} - E_{232}\tfrac{\partial^2 \varphi(x,y,t)}{\partial y^2} + \\
&-G_{550}\tfrac{\partial^2 w_s(x,y,t)}{\partial x^2} - G_{550}\tfrac{\partial^2 \varphi(x,y,t)}{\partial x^2} + G_{551}\tfrac{\partial^2 w_s(x,y,t)}{\partial x^2} + G_{551}\tfrac{\partial^2 \varphi(x,y,t)}{\partial x^2} + \\
&+G_{661}\tfrac{\partial^2 w_s(x,y,t)}{\partial y^2} + G_{661}\tfrac{\partial^2 \varphi(x,y,t)}{\partial y^2} + K_1\tfrac{\partial^4 w_b(x,y,t)}{\partial x^2 \partial y^2} + \tfrac{1}{2}K_2\tfrac{\partial^4 \varphi(x,y,t)}{\partial x^2 \partial y^2} + \\
&+\tfrac{1}{2}K_3\tfrac{\partial^4 w_s(x,y,t)}{\partial x^2 \partial y^2} + \tfrac{1}{2}K_1\tfrac{\partial^4 w_s(x,y,t)}{\partial y^4} + \tfrac{1}{2}K\tfrac{\partial^4 w_b(x,y,t)}{\partial x^4} + \tfrac{1}{2}K_1\tfrac{\partial^4 w_s(x,y,t)}{\partial x^4} + \\
&+\tfrac{1}{4}K\tfrac{\partial^4 w_s(x,y,t)}{\partial x^4} + \tfrac{1}{4}K_0\tfrac{\partial^4 \varphi(x,y,t)}{\partial x^4} + K\tfrac{\partial^4 w_b(x,y,t)}{\partial x^2 \partial y^2} + \tfrac{1}{2}K\tfrac{\partial^4 w_s(x,y,t)}{\partial x^2 \partial y^2} + \\
&+K_1\tfrac{\partial^4 w_s(x,y,t)}{\partial x^2 \partial y^2} + \tfrac{1}{2}K_0\tfrac{\partial^4 \varphi(x,y,t)}{\partial x^2 \partial y^2} + \tfrac{1}{2}K\tfrac{\partial^4 w_b(x,y,t)}{\partial y^4} + \tfrac{1}{4}K\tfrac{\partial^4 w_s(x,y,t)}{\partial y^4} + \\
&\tfrac{1}{4}K_0\tfrac{\partial^4 \varphi(x,y,t)}{\partial y^4} - 2F_{440}\tfrac{\partial^3 v(x,y,t)}{\partial x^2 \partial y} + 4F_{441}\tfrac{\partial^4 w_b(x,y,t)}{\partial x^2 \partial y^2} + 4F_{442}\tfrac{\partial^4 w_s(x,y,t)}{\partial x^2 \partial y^2} + \\
&-2F_{440}\tfrac{\partial^3 u(x,y,t)}{\partial x \partial y^2} - C_d\tfrac{\partial w_b(x,y,t)}{\partial t} - C_d\tfrac{\partial w_s(x,y,t)}{\partial t} - C_d g\tfrac{\partial \varphi(x,y,t)}{\partial t} + \\
&+K_G\tfrac{\partial^2 w_s(x,y,t)}{\partial x^2} + K_G\tfrac{\partial^2 w_b(x,y,t)}{\partial x^2} + K_G g\tfrac{\partial^2 \varphi(x,y,t)}{\partial x^2} + K_G\tfrac{\partial^2 w_s(x,y,t)}{\partial y^2} + \\
&+K_G\tfrac{\partial^2 w_b(x,y,t)}{\partial y^2} + K_G g\tfrac{\partial^2 \varphi(x,y,t)}{\partial y^2} - K_W g\varphi(x,y,t) - K_W w_s(x,y,t) - K_W w_b(x,y,t) + \\
&-I_3\tfrac{\partial^3 u(x,y,t)}{\partial t^2 \partial x} + I_5\tfrac{\partial^4 w_b(x,y,t)}{\partial t^2 \partial x^2} + I_6\tfrac{\partial^4 w_s(x,y,t)}{\partial t^2 \partial x^2} - I_3\tfrac{\partial^3 v(x,y,t)}{\partial t^2 \partial y} + \\
&+I_5\tfrac{\partial^4 w_b(x,y,t)}{\partial t^2 \partial y^2} + I_6\tfrac{\partial^4 w_s(x,y,t)}{\partial t^2 \partial y^2} - I_0\tfrac{\partial^2 w_b(x,y,t)}{\partial t^2} - I_0\tfrac{\partial^2 w_s(x,y,t)}{\partial t^2} + \\
&-I_4\tfrac{\partial^2 \varphi(x,y,t)}{\partial t^2} = 0
\end{aligned}$$ (40)

$\delta\varphi:$

$$\begin{aligned}
&-\tfrac{1}{2}K_1\tfrac{\partial^4 w_b(x,y,t)}{\partial y^4} - \tfrac{1}{4}K_2\tfrac{\partial^4 \varphi(x,y,t)}{\partial y^4} - \tfrac{1}{4}K_3\tfrac{\partial^4 w_s(x,y,t)}{\partial y^4} - \tfrac{1}{2}K_1\tfrac{\partial^4 w_b(x,y,t)}{\partial x^4} + \\
&+\tfrac{1}{4}K_3\tfrac{\partial^4 w_s(x,y,t)}{\partial x^4} + \tfrac{1}{4}K_2\tfrac{\partial^4 \varphi(x,y,t)}{\partial x^4} + \tfrac{1}{4}K_4\tfrac{\partial^2 \varphi(x,y,t)}{\partial y^2} + \tfrac{1}{4}K_5\tfrac{\partial^2 w_s(x,y,t)}{\partial y^2} + \\
&+\tfrac{1}{4}K_5\tfrac{\partial^2 w_s(x,y,t)}{\partial x^2} + \tfrac{1}{4}K_4\tfrac{\partial^2 \varphi(x,y,t)}{\partial x^2} - G_{550}\tfrac{\partial^2 w_s(x,y,t)}{\partial x^2} - G_{550}\tfrac{\partial^2 \varphi(x,y,t)}{\partial x^2} + \\
&+G_{551}\tfrac{\partial^2 w_s(x,y,t)}{\partial x^2} + G_{551}\tfrac{\partial^2 \varphi(x,y,t)}{\partial x^2} + G_{661}\tfrac{\partial^2 w_s(x,y,t)}{\partial y^2} + G_{661}\tfrac{\partial^2 \varphi(x,y,t)}{\partial y^2} + \\
&-K_1\tfrac{\partial^4 w_b(x,y,t)}{\partial x^2 \partial y^2} - \tfrac{1}{2}K_2\tfrac{\partial^4 \varphi(x,y,t)}{\partial y^2 \partial x^2} - \tfrac{1}{2}K_3\tfrac{\partial^4 w_s(x,y,t)}{\partial y^2 \partial x^2} + \\
&+\tfrac{1}{4}K_0\tfrac{\partial^4 \varphi(x,y,t)}{\partial x^4} - \tfrac{1}{4}K\tfrac{\partial^4 w_s(x,y,t)}{\partial x^4} - \tfrac{1}{2}K\tfrac{\partial^4 w_b(x,y,t)}{\partial x^4} + \tfrac{1}{4}K_0\tfrac{\partial^4 \varphi(x,y,t)}{\partial y^4} + \\
&+K\tfrac{\partial^4 w_b(x,y,t)}{\partial y^2 \partial x^2} + \tfrac{1}{2}K\tfrac{\partial^4 w_s(x,y,t)}{\partial y^2 \partial x^2} + \tfrac{1}{2}K_0\tfrac{\partial^4 \varphi(x,y,t)}{\partial y^2 \partial x^2} + \tfrac{1}{4}K\tfrac{\partial^4 w_b(x,y,t)}{\partial y^4} + \\
&+\tfrac{1}{4}K\tfrac{\partial^4 w_s(x,y,t)}{\partial y^4} + E_{330}\varphi(x,y,t) + E_{230}\tfrac{\partial v(x,y,t)}{\partial y} - E_{131}\tfrac{\partial^2 w_b(x,y,t)}{\partial x^2} - E_{132}\tfrac{\partial^2 w_s(x,y,t)}{\partial x^2} + \\
&-E_{232}\tfrac{\partial^2 w_s(x,y,t)}{\partial y^2} - E_{231}\tfrac{\partial^2 w_b(x,y,t)}{\partial y^2} - G_{660}\tfrac{\partial^2 \varphi(x,y,t)}{\partial y^2} - G_{660}\tfrac{\partial^2 w_s(x,y,t)}{\partial y^2} + \\
&-C_d\tfrac{\partial w_b(x,y,t)}{\partial t} - C_d\tfrac{\partial w_s(x,y,t)}{\partial t} - C_d g^2\tfrac{\partial \varphi(x,y,t)}{\partial t} - K_W g w_b(x,y,t) - K_W g w_s(x,y,t) + \\
&-K_W g^2\varphi(x,y,t) + K_G g\tfrac{\partial^2 w_s(x,y,t)}{\partial x^2} + K_G g\tfrac{\partial^2 w_b(x,y,t)}{\partial x^2} + K_G g^2\tfrac{\partial^2 \varphi(x,y,t)}{\partial x^2} + \\
&+K_G g\tfrac{\partial^2 w_s(x,y,t)}{\partial y^2} + K_G g\tfrac{\partial^2 w_b(x,y,t)}{\partial y^2} + K_G g^2\tfrac{\partial^2 \varphi(x,y,t)}{\partial y^2} + \\
&-I_4\tfrac{\partial^2 w_b(x,y,t)}{\partial t^2} + I_4\tfrac{\partial^2 w_s(x,y,t)}{\partial t^2} + I_7\tfrac{\partial^2 \varphi(x,y,t)}{\partial t^2} = 0
\end{aligned}$$ (41)

More details about the coefficients in Equations (37)–(41), are reported in the Appendix A.

4. Analytical Solution Procedure

The differential equations of the Equations (37)–(41) are solved analytically according to the Navier's procedure in this section. Therefore, for a simply supported structure, we consider the following theoretical expressions for the displacement components [42]

$$\begin{aligned} u(x,y,t) &= U\cos(\alpha x)\sin(\beta y)e^{i\omega t}, \\ v(x,y,t) &= V\sin(\alpha x)\cos(\beta y)e^{i\omega t}, \\ w_b(x,y,t) &= W_b\sin(\alpha x)\sin(\beta y)e^{i\omega t}, \\ w_s(x,y,t) &= W_s\sin(\alpha x)\sin(\beta y)e^{i\omega t}, \\ \varphi(x,y,t) &= \Phi\sin(\alpha x)\sin(\beta y)e^{i\omega t} \end{aligned} \qquad (42)$$

in which U, V, W_s, W_b and Φ are the unknown coefficients. In addition, α and β are defined as $m\pi/a$ and $n\pi/b$, respectively, where m and n are the mode numbers along the length and width direction, respectively. After substituting Equation (42) into Equations (37)–(41), the equations of motion gain the following compact form

$$\left([K]_{5\times 5} + i\omega[C]_{5\times 5} - \omega^2[M]_{5\times 5}\right)\{d\} = 0 \qquad (43)$$

where $[K]$, $[C]$, and $[M]$ refer to the stiffness matrix, damping matrix, and mass matrix, respectively, whereas $\{d\}$ is the displacement vector. The natural frequencies of the structure are then obtained by solving the classical eigenvalue problem (43).

5. Numerical Results

In this section we illustrate the numerical results, in terms of vibration response, for a microsandwich plate with a honeycomb core made of Nomex or Glass phenolic, and Epoxy-reinforced GPLs as face sheets. The Nomex has the following properties: E_s = 3.2 GPa, ρ = 48 kg/m^3, and ν = 0.4. For the Glass phenolic, the same properties of Ref. [43] are assumed herein. The Epoxy matrix and GPLs reinforcement phase for the face sheets have the following properties [44]

$$\begin{aligned} E_{GPL} &= 1.01\,TPa, & \rho_{GPL} &= 1062.5\,kg/m^3, & \nu_{GPL} &= 0.186, \\ L_{GPL} &= 2.5\,\mu m, & w_{GPL} &= 1.5\,\mu m, & h_{GPL} &= 1.5\,nm, \\ E_M &= 130\,GPa, & \rho_M &= 8960\,kg/m^3, & \nu_M &= 0.34 \end{aligned}$$

The microplate has a total height equal to 150 μm, 80% of whose total height corresponds to the core, and the rest is equally divided between the two face sheets. The length of the square plate is ten-fold of its thickness. The internal cell angle, aspect ratio, and dimensionless cells thickness are assumed to be $\pi/6$, 1, and 0.1, respectively. Moreover, the material length-scale parameter is kept as 15 μm according to Ref. [36].

To check for the reliability of our formulation, we compare the results for a single-layer FG-GPL-reinforced square microplate with predictions by Thai et al. [45]. The comparative results are summarized in Table 2, in terms of dimensionless natural frequencies defined as $\Omega = \left(\omega a^2/h\right)\sqrt{\rho_M/E_M}$, for various mode numbers, while considering the effect of the aspect ratio (a/h) and length-scale parameter-to-total thickness (l_m/h). Based on Table 2, a very good agreement is observable between the results from our formulation and those ones from Ref. [45], where some negligible differences are related to the different kinematic assumptions, and/or different solution techniques.

In Table 3, we also summarize the natural frequencies for different mode numbers, as computed according to a MCST or a classical elasticity theory (CET), for a varying internal cell angle from 30° up to 60°. Based on results in Table 3, note that the mode number and internal cell angle yield a reverse effect on the natural frequency of the sandwich microplate, whereby an increasing mode number and a decreasing internal cell angle get higher values of the natural frequencies. It seems also that CET-based predictions are always more conservative than those once based on a MCST, in agreement with findings in Refs. [46–50] from the literature.

Table 2. Comparative evaluation between our results and those ones of Ref. [45] for a square microplate and different mode numbers.

a/h			l_m/h					
			0.0	0.2	0.4	0.6	0.8	1.0
5	Epoxy	Present	0.2145	0.2322	0.2786	0.3319	0.4143	0.4815
		Ref. [45]	0.2148	0.2301	0.2708	0.3271	0.3920	0.4615
	Uniform	Present	0.4460	0.4820	0.5794	0.7114	0.8622	1.0220
		Ref. [45]	0.4468	0.4789	0.5639	0.6813	0.8164	0.9613
10	Epoxy	Present	0.0586	0.0632	0.0752	0.0918	0.1109	0.1315
		Ref. [45]	0.0586	0.0629	0.0745	0.0905	0.1091	0.1290
	Uniform	Present	0.1219	0.1314	0.1564	0.1910	0.2308	0.2736
		Ref. [45]	0.1219	0.1310	0.1551	0.1885	0.2271	0.2686

Table 3. Effect of the internal cells angle of the honeycomb core on the structural response, as predicted by modified couple stress theory (MCST) and classical elasticity theory (CET).

		ω (MHz)		
	(m, n)	$\theta=30°$	$\theta=45°$	$\theta=60°$
MCST	(1, 1)	0.1789	0.1762	0.1728
	(2, 1)	0.3821	0.3691	0.3542
	(2, 2)	0.5113	0.4940	0.4758
CET	(1, 1)	0.1678	0.1633	0.1577
	(2, 1)	0.3555	0.3376	0.3166
	(2, 2)	0.4545	0.4310	0.4064

Another key aspect of the problem can be the sensitivity of the response to various GPLs dispersions in the Epoxy matrix over a wide range of mode numbers, as listed in Table 4. It is worth noticing that the sandwich structure becomes stiffer for an increased quantity of GPLs as reinforcing phase, and the natural frequency enhances dramatically in each mode number.

Table 4. Effect of the GPLs dispersion patterns on the natural frequencies of the micro structure, for different mode numbers.

	ω (MHz)			
(m, n)	Uniform ($\lambda_U = 1$)	Parabolic ($\lambda_P = 1$)	Linear ($\lambda_L = 2$)	Epoxy
(1, 1)	0.1745	0.1637	0.2115	0.1044
(2, 1)	0.3584	0.3368	0.4246	0.2192
(2, 2)	0.4832	0.4521	0.5670	0.2950
(3, 1)	0.5849	0.5505	0.6799	0.3656
(3, 2)	0.6794	0.6378	0.7860	0.4287
(3, 3)	0.8131	0.7606	0.9387	0.5016

It seems also that a linear dispersion of GPLs in the Epoxy matrix with $\lambda_L = 2$ is more effective than other types of distribution with $\lambda_P = \lambda_U = 1$ for an overall increase in the structural stiffness. This shows that the GPLs dispersion coefficient plays a crucial role, more than the type of GPLs dispersion, for an increase in the natural frequency.

Figure 2 shows the variation of the natural frequency for the sandwich microplate against the l_m/h ratio, for different dispersions of GPLs. By increasing l_m/h rational value, and keeping constant the total thickness of the sandwich model, the natural frequency increases monotonically, for each fixed value of λ_L, λ_P, λ_U. This behavior is due to a reduced flexibility of the sandwich microplate which corresponds to a stiffness and stability enhancement. For each type of GPLs dispersion, a higher distribution coefficient obtains higher natural frequencies.

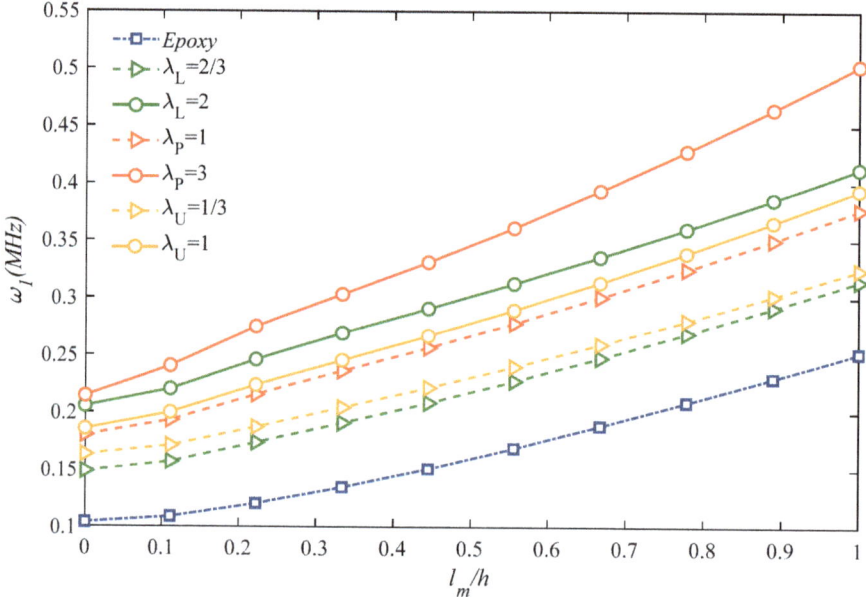

Figure 2. Size and GPLs amount effects on the fundamental natural frequency.

Figure 3 plots the effect of the aspect ratio, a/b, on the natural frequency of the microstructure for different GPLs dispersion coefficients. For each fixed GPLs dispersion coefficient and type, an increased aspect ratio up to one clearly reduces the natural frequency reaching the minimum value for a cubic sandwich structure. Once this minimum value is passed, the aspect ratio rolling up causes a monotonic increase in the natural frequency for each selected GPLs dispersion coefficient and type. A further systematic analysis is also performed to check for the sensitivity of the natural frequency alternation with the l_m/h ratio, under different assumptions for the honeycomb core material in Figure 4. Based the plots in this figure, it is worth observing that the most rigid sandwich microstructure is obtained for a uniform Nomex honeycomb core material, where the most flexible one is reached for an Epoxy/Glass Phenolic core material. All the other results based on an Epoxy/Nomex honeycomb or Uniform/Glass Phenolic core material assumption are very close to each other, and fall always within the previous two cases. As also plotted in Figure 5, the natural frequency decreases monotonically for an increasing geometrical ratio h_{GPL}/L_{GPL} of the reinforcing phase, as predicted by a CET or a MCST, respectively, while assuming three different rational values for L_{GPL}/W_{GPL}, namely, L_{GPL}/W_{GPL} = 1; 5/3; 2. This means that the GPLs length variations (reduction or enhancement) have a direct relationship with the natural frequency, stiffness and rigidity. For each selected theory, an increased value of L_{GPL}/W_{GPL} reduces gradually the natural frequency for each fixed value of h_{GPL}/L_{GPL}. Based on a comparative evaluation of the curves in Figure 5, it can be noted that MCST provides always a higher natural frequency compared to the CET.

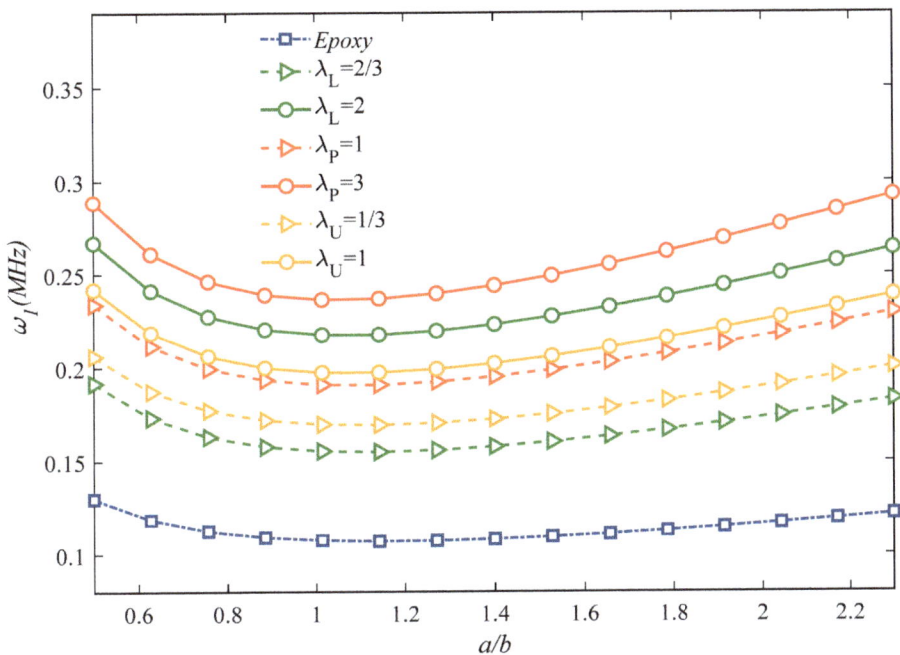

Figure 3. Effect of the aspect ratio and GPLs dispersion pattern on the structural response. ($S = 225 \times 10^{-12}$).

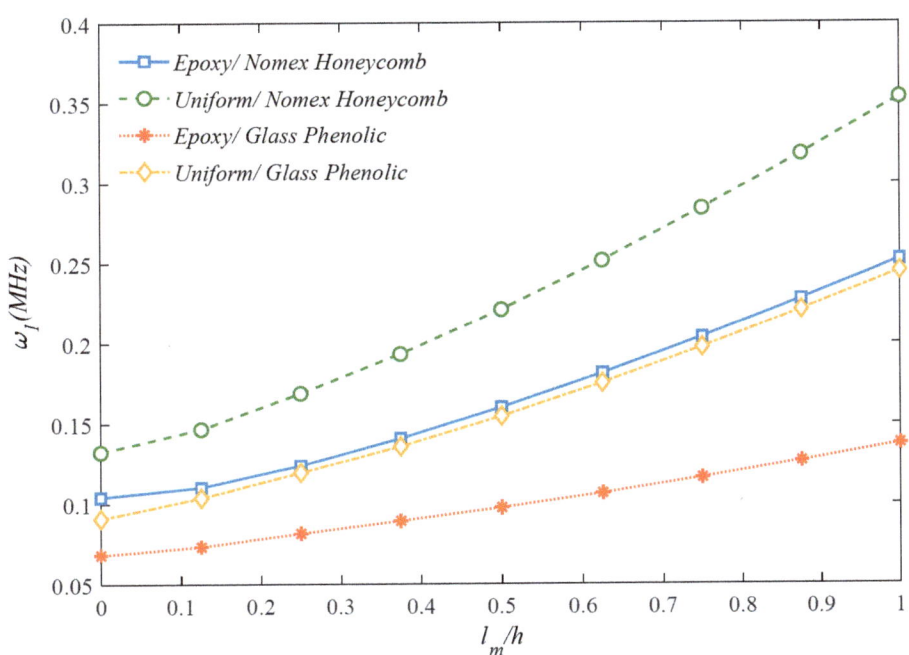

Figure 4. Effect of the core materials on the structural response.

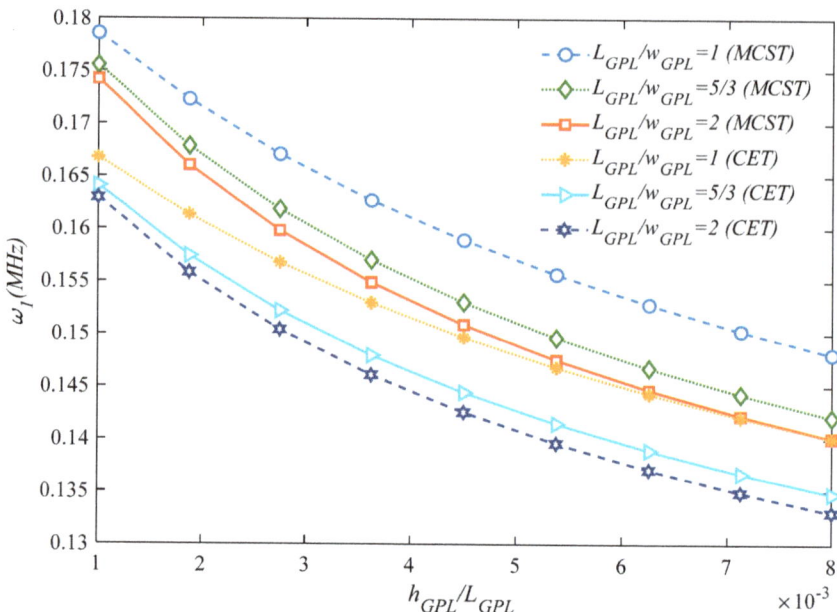

Figure 5. Effect of the GPLs geometry on the structural response, according to the MCST and CET.

In Figure 6 we analyze the effect of the viscoelastic foundation on the vibration response, while providing the 3D plot of the natural frequency for different combinations of K_W, K_G under three different assumptions for the damping parameter $C_d = 500; 1000; 1500$ (N·s/m) is provided. By increasing the Winkler and Pasternak parameters (K_W, K_G) the structural stiffness increases together with the natural frequency for each fixed value of C_d. Based on the three plots, it is worth mentioning the great damping effect on the frequency response, where a decreased value of C_d obtains higher frequencies for each fixed combination of K_W, K_G.

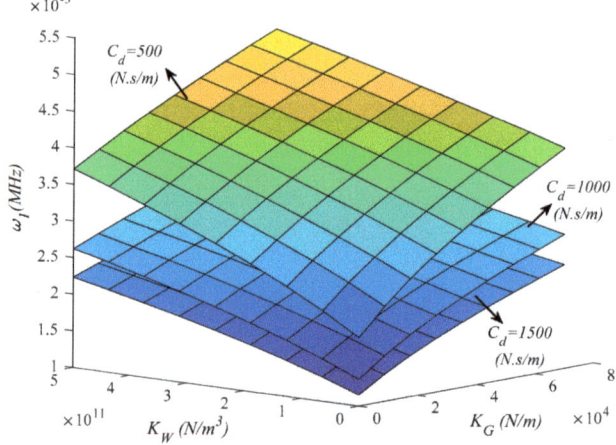

Figure 6. Effect of the viscoelastic foundation parameters on the first natural frequency of the structure ($\lambda_L = 2$).

The effect of the internal aspect ratio ϕ_0 and dimensionless cell thickness γ_0 on the first natural frequency of the sandwich microplate is plotted in Figure 7. Based on the results in this figure, larger magnitudes of γ_0 lead to an increased system stability. On the other hand, a clear reduction in the structural stiffness and frequency is gained by internal aspect ratio enhancement and honeycomb core thickness reduction in the case of fixed internal cells angle equal to 30°. This means that, for a constant value of total thickness, a lower face sheet thickness to core thickness ratio results in a higher stiffness and weaker flexibility.

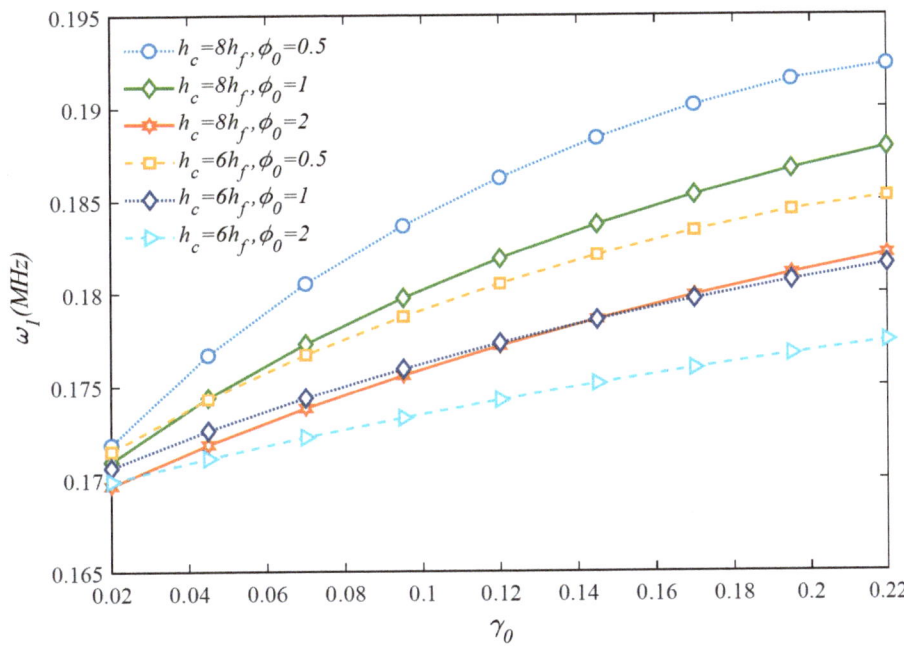

Figure 7. Effect of the honeycomb cells' geometrical parameters on the first natural frequency of the structure.

Moreover, based on the curvatures plotted in Figures 8 and 9 which represent the first natural frequency versus the honeycomb core internal cell angle ϕ_0 for different internal aspect ratios φ_0, it seems that an enhancement of both parameters gets a natural frequency reduction. In addition, Figure 9 illustrates that the thicker honeycomb core provides higher structural stiffness and natural frequency. As a final parametric investigation, we check for the variation of the first natural frequency with the l_m/h, based on the MCST or CET, under the assumption of three different core thicknesses.

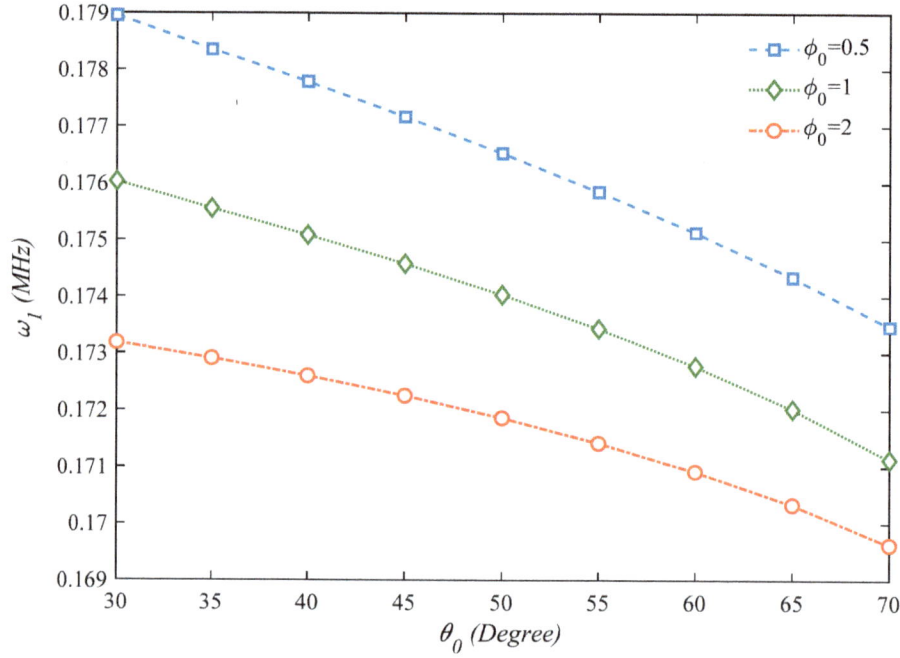

Figure 8. Effect of the internal cells angle of the honeycomb core on the first natural frequency of the structure ($\gamma_0 = 0.1$).

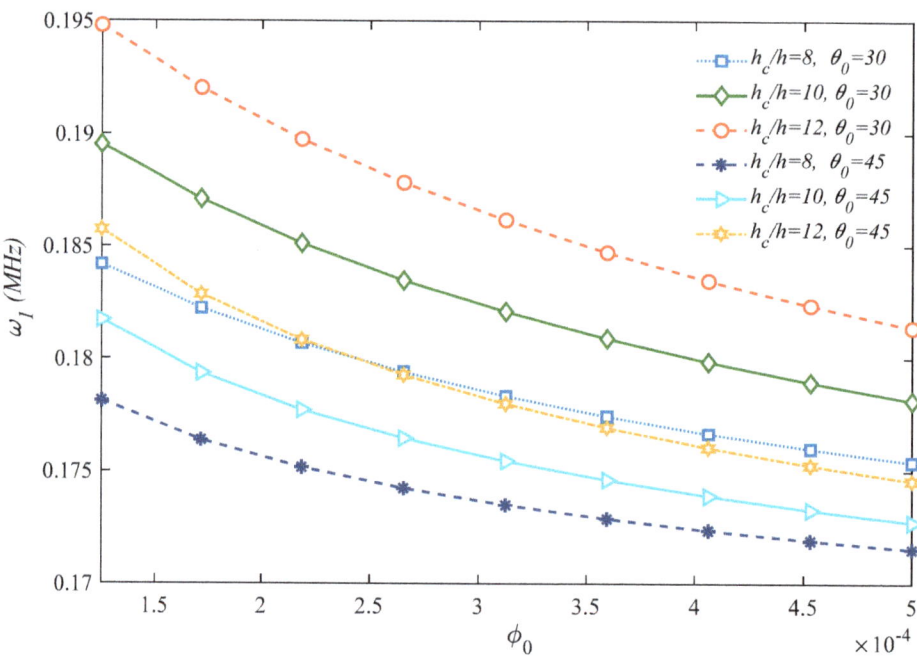

Figure 9. Effect of the thickness ratio on the first natural frequency of the structure.

Based on the plots in Figure 10, it should be noted that the natural frequency increases significantly for higher values of l_m/h ratio, when the problem is tackled by a MCST, whereas it remains almost unaffected by l_m/h according to a CET. This confirms, once again, the great importance of adopting a size-dependent approach instead of classical formulations.

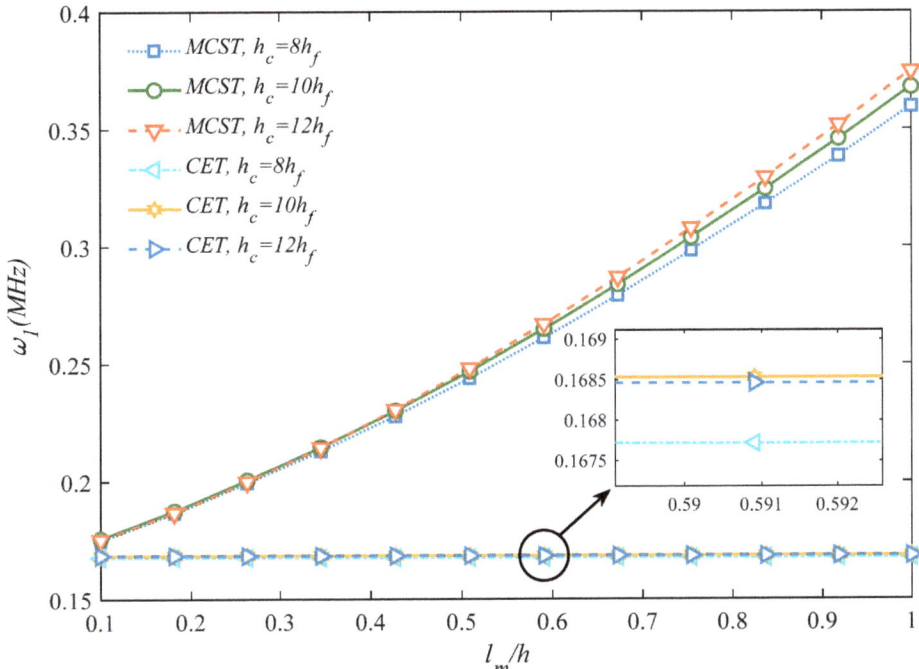

Figure 10. Comparison between results from a MCST and CET for different dimensionless length-scale parameters.

6. Conclusions

In this work, a QHSDT is employed to investigate the vibrational behavior of sandwich honeycomb microplates with two GPLs' composite face sheets, resting on elastic foundations. The equations of motion are obtained by applying the Hamilton's principle, where the Navier-type solutions are determined in analytical form. Based on a large systematic investigation, it is noted that an Epoxy/Nomex honeycomb core makes the sandwich structure less flexible than Epoxy/Glass phenolic and uniform glass phenolic core materials, whereby a uniform Nomex honeycomb core provides the highest structural stiffness. Moreover, a larger dimensionless cell thickness (γ_0) yields an increased stability in the system, whereas internal aspect ratio elevation provides structural stability reduction along with the system's stiffness and natural frequency. The results based on a MCST are compared to predictions from CET to provide a clear understanding about vibrational responses' sensitivity to size-dependent parameters. In agreement with findings from the literature, a CET always produces more conservative results compared to an MCST, which justifies the necessity of adopting non classical approaches instead of the classical ones. The proposed model together with our numerical results could be very useful for the design and manufacturing of many aerospace, automotive or shipbuilding engineering applications, where honeycomb structures are recommended for their great capability to tolerate high pressures and stresses despite their light structure.

Author Contributions: Conceptualization, H.A., M.K., R.D. and F.T.; Formal analysis, H.A., M.K., R.D. and F.T.; Investigation, H.A., Z.S.-J. and F.T.; Validation, H.A., M.-K., Z.S.-J., R.D. and F.T.; Writing—Original Draft, H.A.,

M.K., Z.S.-J., R.D. and F.T.; Writing—Review & Editing, R.D. and F.T.; Supervision, R.D. and F.T. All authors have read and agreed to the published version of the manuscript.

Funding: This research received no external funding.

Acknowledgments: The authors would like to thank Ehsan Arshid for his tireless efforts and guidance during this research.

Conflicts of Interest: The authors declare no conflict of interest.

Appendix A Appendix

The coefficients in Equations (37)–(41) are defined as in the following

$$C_{110}, C_{111}, C_{112} = \int_{\frac{h_c}{2}}^{\frac{h_c}{2}+h_t} C_{11f}(z)(1,z,z^2)dz + \int_{-\frac{h_c}{2}}^{\frac{h_c}{2}} C_{11c}(z)(1,z,z^2)dz + \int_{-\frac{h_c}{2}-h_b}^{-\frac{h_c}{2}} C_{11f}(z)(1,z,z^2)dz,$$

$$C_{120}, C_{121}, C_{122} = \int_{\frac{h_c}{2}}^{\frac{h_c}{2}+h_t} C_{12f}(z)(1,z,z^2)dz + \int_{-\frac{h_c}{2}}^{\frac{h_c}{2}} C_{12c}(z)(1,z,z^2)dz + \int_{-\frac{h_c}{2}-h_b}^{-\frac{h_c}{2}} C_{12f}(z)(1,z,z^2)dz,$$

$$C_{220}, C_{221}, C_{222} = \int_{\frac{h_c}{2}}^{\frac{h_c}{2}+h_t} C_{22f}(z)(1,z,z^2)dz + \int_{-\frac{h_c}{2}}^{\frac{h_c}{2}} C_{22c}(z)(1,z,z^2)dz + \int_{-\frac{h_c}{2}-h_b}^{-\frac{h_c}{2}} C_{22f}(z)(1,z,z^2)dz,$$

$$C_{440}, C_{441}, C_{442} = \int_{\frac{h_c}{2}}^{\frac{h_c}{2}+h_t} C_{44f}(z)(1,z,z^2)dz + \int_{-\frac{h_c}{2}}^{\frac{h_c}{2}} C_{44c}(z)(1,z,z^2)dz + \int_{-\frac{h_c}{2}-h_b}^{-\frac{h_c}{2}} C_{44f}(z)(1,z,z^2)dz,$$

$$F_{110}, F_{111}, F_{112} = \int_{\frac{h_c}{2}}^{\frac{h_c}{2}+h_t} C_{11f}(z)f(z)(1,z,f(z))dz + \int_{-\frac{h_c}{2}}^{\frac{h_c}{2}} C_{11c}(z)f(z)(1,z,f(z))dz + \int_{-\frac{h_c}{2}-h_b}^{-\frac{h_c}{2}} C_{11f}(z)f(z)(1,z,f(z))dz,$$

$$F_{120}, F_{121}, F_{122} = \int_{\frac{h_c}{2}}^{\frac{h_c}{2}+h_t} C_{12f}(z)f(z)(1,z,f(z))dz + \int_{-\frac{h_c}{2}}^{\frac{h_c}{2}} C_{12c}(z)f(z)(1,z,f(z))dz + \int_{-\frac{h_c}{2}-h_b}^{-\frac{h_c}{2}} C_{12f}(z)f(z)(1,z,f(z))dz,$$

$$F_{220}, F_{221}, F_{222} = \int_{\frac{h_c}{2}}^{\frac{h_c}{2}+h_t} C_{22f}(z)f(z)(1,z,f(z))dz + \int_{-\frac{h_c}{2}}^{\frac{h_c}{2}} C_{22c}(z)f(z)(1,z,f(z))dz + \int_{-\frac{h_c}{2}-h_b}^{-\frac{h_c}{2}} C_{22f}(z)f(z)(1,z,f(z))dz,$$

$$F_{440}, F_{441}, F_{442} = \int_{\frac{h_c}{2}}^{\frac{h_c}{2}+h_t} C_{44f}(z)f(z)(1,z,z^2)dz + \int_{-\frac{h_c}{2}}^{\frac{h_c}{2}} C_{44c}(z)f(z)(1,z,z^2)dz + \int_{-\frac{h_c}{2}-h_b}^{-\frac{h_c}{2}} C_{44f}(z)f(z)(1,z,z^2)dz,$$

$$E_{130}, E_{131}, E_{132} = \int_{\frac{h_c}{2}}^{\frac{h_c}{2}+h_t} C_{13f}(z)g'(z)(1,z,f(z))dz + \int_{-\frac{h_c}{2}}^{\frac{h_c}{2}} C_{13c}(z)g'(z)(1,z,f(z))dz + \int_{-\frac{h_c}{2}-h_b}^{-\frac{h_c}{2}} C_{13f}(z)g'(z)(1,z,f(z))dz,$$

$$E_{320}, E_{321}, E_{322} = \int_{\frac{h_c}{2}}^{\frac{h_c}{2}+h_t} C_{32f}(z)g'(z)(1,z,f(z))dz + \int_{-\frac{h_c}{2}}^{\frac{h_c}{2}} C_{32c}(z)g'(z)(1,z,f(z))dz + \int_{-\frac{h_c}{2}-h_b}^{-\frac{h_c}{2}} C_{32f}(z)g'(z)(1,z,f(z))dz,$$

$$E_{330} = \int_{\frac{h_c}{2}}^{\frac{h_c}{2}+h_t} C_{33f}(z)g'(z)^2 dz + \int_{-\frac{h_c}{2}}^{\frac{h_c}{2}} C_{33c}(z)g'(z)^2 dz + \int_{-\frac{h_c}{2}-h_b}^{-\frac{h_c}{2}} C_{33f}(z)g'(z)^2 dz,$$

$$G_{550}, G_{551} = \int_{\frac{h_c}{2}}^{\frac{h_c}{2}+h_t} C_{55f}(z)g(z)(1,f'(z))dz + \int_{-\frac{h_c}{2}}^{\frac{h_c}{2}} C_{55c}(z)g(z)(1,f'(z))dz$$
$$+ \int_{-\frac{h_c}{2}-h_b}^{-\frac{h_c}{2}} C_{55f}(z)g(z)(1,f'(z))dz,$$

$$G_{660}, G_{661} = \int_{\frac{h_c}{2}}^{\frac{h_c}{2}+h_t} C_{66f}(z)g(z)(1,f'(z))dz + \int_{-\frac{h_c}{2}}^{\frac{h_c}{2}} C_{66c}(z)g(z)(1,f'(z))dz$$
$$+ \int_{-\frac{h_c}{2}-h_b}^{-\frac{h_c}{2}} C_{66f}(z)g(z)(1,f'(z))dz,$$

$$K, K_0, K_1, K_2, K_3, K_4, K_5 = \int_{\frac{h_c}{2}}^{\frac{h_c}{2}+h_t} lm^2\mu_f(z)(1,g(z),f'(z),g(z)f'(z),f'(z)^2,g'(z)f''(z),f''(z)^2)dz+$$
$$+ \int_{-\frac{h_c}{2}}^{\frac{h_c}{2}} lm^2\mu_c(z)(1,g(z),f'(z),g(z)f'(z),f'(z)^2,g'(z)f''(z),f''(z)^2)dz+$$
$$+ \int_{-\frac{h_c}{2}-h_b}^{-\frac{h_c}{2}} lm^2\mu_f(z)(1,g(z),f'(z),g(z)f'(z),f'(z)^2,g'(z)f''(z),f''(z)^2)dz,$$

$$I_0, I_1, I_2, I_3, I_4, I_5, I_6, I_7 = \int_{\frac{h_c}{2}}^{\frac{h_c}{2}+h_t} \rho_f(z)(1,z,z^2,f(z),g(z),zf(z),f(z)^2,g(z)^2)dz+$$
$$+ \int_{-\frac{h_c}{2}}^{\frac{h_c}{2}} \rho_c(1,z,z^2,f(z),g(z),zf(z),f(z)^2,g(z)^2)dz+$$
$$+ \int_{-\frac{h_c}{2}-h_b}^{-\frac{h_c}{2}} \rho_f(z)(1,z,z^2,f(z),g(z),zf(z),f(z)^2,g(z)^2)dz$$

References

1. Maheri, M.R.; Adams, R.D. Steady-state flexural vibration damping of honeycomb sandwich beams. *Compos. Sci. Technol.* **1994**, *52*, 333–347. [CrossRef]
2. Guo, Y.; Zhang, J. Shock Absorbing Characteristics and Vibration Transmissibility of Honeycomb Paperboard. *Shock Vib.* **2004**, *11*, 521–531. [CrossRef]
3. Li, Y.; Jin, Z. Free flexural vibration analysis of symmetric rectangular honeycomb panels with SCSC edge supports. *Compos. Struct.* **2008**, *83*, 154–158. [CrossRef]
4. Liu, J.; Cheng, Y.S.; Li, R.F.; Au, F.T.K. A semi-analytical method for bending, buckling, and free vibration analyses of sandwich panels with square-hoeycomb cores. *Int. J. Struct. Stab. Dyn.* **2010**, *10*, 127–151. [CrossRef]
5. Burlayenko, V.N.; Sadowski, T. Influence of skin/core debonding on free vibration behavior of foam and honeycomb cored sandwich plates. *Int. J. Non Linear Mech.* **2010**, *45*, 959–968. [CrossRef]
6. Katunin, A. Vibration-based spatial damage identification in honeycomb-core sandwich composite structures using wavelet analysis. *Compos. Struct.* **2014**, *118*, 385–391. [CrossRef]
7. Mukhopadhyay, T.; Adhikari, S. Free-Vibration Analysis of Sandwich Panels with Randomly Irregular Honeycomb Core. *J. Eng. Mech.* **2016**, *142*, 06016008. [CrossRef]
8. Duc, N.D.; Seung-Eock, K.; Tuan, N.D.; Tran, P.; Khoa, N.D. New approach to study nonlinear dynamic response and vibration of sandwich composite cylindrical panels with auxetic honeycomb core layer. *Aerosp. Sci. Technol.* **2017**, *70*, 396–404. [CrossRef]
9. Piollet, E.; Fotsing, E.R.; Ross, A.; Michon, G. High damping and nonlinear vibration of sandwich beams with entangled cross-linked fibres as core material. *Compos. Part B Eng.* **2019**, *168*, 353–366. [CrossRef]
10. Kumar, S.; Renji, K. Estimation of strains in composite honeycomb sandwich panels subjected to low frequency diffused acoustic field. *J. Sound Vib.* **2019**, *449*, 84–97. [CrossRef]
11. Sobhy, M. Differential quadrature method for magneto-hygrothermal bending of functionally graded graphene/Al sandwich-curved beams with honeycomb core via a new higher-order theory. *J. Sandw. Struct. Mater.* **2020**. [CrossRef]
12. Li, Y.; Chen, Z.; Xiao, D.; Wu, W.; Fang, D. The Dynamic response of shallow sandwich arch with auxetic metallic honeycomb core under localized impulsive loading. *Int. J. Impact Eng.* **2020**, *137*, 103442. [CrossRef]

13. Chen, D.; Yang, J.; Kitipornchai, S. Nonlinear vibration and postbuckling of functionally graded graphene reinforced porous nanocomposite beams. *Compos. Sci. Technol.* **2017**, *142*, 235–245. [CrossRef]
14. Karimiasl, M.; Ebrahimi, F.; Mahesh, V. On Nonlinear Vibration of Sandwiched Polymer- CNT/GPL-Fiber Nanocomposite Nanoshells. *Thin-Walled Struct.* **2020**, *146*, 106431. [CrossRef]
15. Eyvazian, A.; Hamouda, A.M.; Tarlochan, F.; Mohsenizadeh, S.; Dastjerdi, A.A. Damping and vibration response of viscoelastic smart sandwich plate reinforced with non-uniform Graphene platelet with magnetorheological fluid core. *Steel Compos. Struct.* **2019**, *33*, 891–906.
16. Torabi, J.; Ansari, R. Numerical Phase-Field Vibration Analysis of Cracked Functionally Graded GPL-RC Plates. *Mech. Based Des. Struct. Mach.* **2020**, 1–20. [CrossRef]
17. Khoa, N.D.; Anh, V.M.; Duc, N.D. Nonlinear dynamic response and vibration of functionally graded nanocomposite cylindrical panel reinforced by carbon nanotubes in thermal environment. *J. Sandw. Struct. Mater.* **2019**. [CrossRef]
18. Ibrahim, H.H.; Tawfik, M.; Al-Ajmi, M. Thermal buckling and nonlinear flutter behavior of functionally graded material panels. *J. Aircr.* **2007**, *44*, 1610–1618. [CrossRef]
19. Li, Y.; Li, F.; He, Y. Geometrically nonlinear forced vibrations of the symmetric rectangular honeycomb sandwich panels with completed clamped supported boundaries. *Compos. Struct.* **2011**, *93*, 360–368. [CrossRef]
20. Mehar, K.; Panda, S.K. Thermal Free Vibration Behavior of FG-CNT Reinforced Sandwich Curved Panel Using Finite Element Method. *Polym. Compos.* **2017**. [CrossRef]
21. Nguyen, D.D.; Pham, C.H. Nonlinear dynamic response and vibration of sandwich composite plates with negative Poisson's ratio in auxetic honeycombs. *J. Sandw. Struct. Mater.* **2017**, *20*, 692–717. [CrossRef]
22. Tornabene, F.; Fantuzzi, N.; Bacciocchi, M.; Reddy, J.N. An equivalent layer-wise approach for the free vibration analysis of thick and thin laminated and sandwich shells. *Appl. Sci.* **2017**, *7*, 17. [CrossRef]
23. Tornabene, F.; Fantuzzi, N.; Bacciocchi, M.; Viola, E.; Reddy, J.N. A numerical investigation on the natural frequencies of FGM sandwich shells with variable thickness by the local generalized differential quadrature method. *Appl. Sci.* **2017**, *7*, 131. [CrossRef]
24. Jouneghani, F.Z.; Dimitri, R.; Tornabene, F. Structural response of porous FG nanobeams under hygro-thermo-mechanical loadings. *Compos. Part B Eng.* **2018**, *152*, 71–78. [CrossRef]
25. Tornabene, F.; Brischetto, S. 3D capability of refined GDQ models for the bending analysis of composite and sandwich plates, spherical and doubly-curved shells. *Thin Walled Struct.* **2018**, *129*, 94–124. [CrossRef]
26. Karimiasl, M.; Ebrahimi, F.; Mahesh, V. Nonlinear forced vibration of smart multiscale sandwich composite doubly curved porous shell. *Thin Walled Struct.* **2019**, *143*, 106152. [CrossRef]
27. Nejati, M.; Ghasemi-Ghalebahman, A.; Soltanimaleki, A.; Dimitri, R.; Tornabene, F. Thermal vibration analysis of SMA hybrid composite double curved sandwich panels. *Compos. Struct.* **2019**, *224*, 111035. [CrossRef]
28. Tornabene, F.; Fantuzzi, N.; Bacciocchi, M. Foam core composite sandwich plates and shells with variable stiffness: Effect of the curvilinear fiber path on the modal response. *J. Sandw. Struct. Mater.* **2019**, *21*, 320–365. [CrossRef]
29. Hebali, H.; Tounsi, A.; Houari, A.M.S.; Bessaim, A.; Abbes, E. New Quasi-3D Hyperbolic Shear Deformation Theory for the Static and Free Vibration Analysis of Functionally Graded Plates. *J. Eng. Mech.* **2014**, *140*, 374–383. [CrossRef]
30. Guerroudj, H.Z.; Yeghnem, R.; Kaci, A.; Zaoui, F.Z.; Benyoucef, S.; Tounsi, A. Eigenfrequencies of advanced composite plates using an efficient hybrid quasi-3D shear deformation theory. *Smart Struct. Syst.* **2018**, *22*, 121.
31. Amir, S.; Arshid, E.; Rasti-alhosseini, S.M.A.; Loghman, A. Quasi-3D tangential shear deformation theory for size-dependent free vibration analysis of three-layered FG porous micro rectangular plate integrated by nano-composite faces in hygrothermal environment. *J. Therm. Stresses* **2019**, *43*, 133–156. [CrossRef]
32. Benahmed, A.; Houari, A.M.S.; Benyoucef, S.; Belakhdar, K.; Tounsi, A. A novel quasi-3D hyperbolic shear deformation theory for functionally graded thick rectangular plates on elastic foundation. *Geomech. Eng.* **2017**, *12*, 9–34. [CrossRef]
33. Ebrahimi, F.; Karimiasl, M.; Mahesh, V. Vibration analysis of magneto-flexo-electrically actuated porous rotary nanobeams considering thermal effects via nonlocal strain gradient elasticity theory. *Adv. Nano Res.* **2019**, *7*, 223–231.
34. Torabi, K.; Afshari, H.; Aboutalebi, F.H. Vibration and flutter analyses of cantilever trapezoidal honeycomb sandwich plates. *J. Sandw. Struct. Mater.* **2019**, *21*. [CrossRef]

35. Arshid, E.; Amir, S.; Loghman, A. Static and Dynamic Analyses of FG-GNPs Reinforced Porous Nanocomposite Annular Micro-Plates Based on MSGT. *Int. J. Mech. Sci.* **2020**, *180*, 105656. [CrossRef]
36. Mohammad-Rezaei Bidgoli, E.; Arefi, M. Free vibration analysis of micro plate reinforced with functionally graded graphene nanoplatelets based on modified strain-gradient formulation. *J. Sandw. Struct. Mater.* **2019**. [CrossRef]
37. Amir, S.; Arshid, E.; Ghorbanpour Arani, M.R. Size-Dependent Magneto-Electro-Elastic Vibration Analysis of FG Saturated Porous Annular/Circular Micro Sandwich Plates Embedded with Nano-Composite Face sheets Subjected to Multi-Physical Pre Loads. *Smart Struct. Syst.* **2019**, *23*, 429–447.
38. Arshid, E.; Khorshidvand, A.R. Thin-Walled Structures Free vibration analysis of saturated porous FG circular plates integrated with piezoelectric actuators via differential quadrature method. *Thin Walled Struct.* **2018**, *125*, 220–233. [CrossRef]
39. Amir, S.; Soleymani-Javid, Z.; Arshid, E. Size-dependent free vibration of sandwich micro beam with porous core subjected to thermal load based on SSDBT. *Appl. Math. Mech.* **2019**, *99*, e201800334. [CrossRef]
40. Arshid, E.; Khorshidvand, A.R.; Khorsandijou, S.M. The Effect of Porosity on Free Vibration of SPFG Circular Plates Resting on visco-Pasternak Elastic Foundation Based on CPT, FSDT and TSDT. *Struct. Eng. Mech.* **2019**, *70*, 97–112.
41. Kiani, Y.; Dimitri, R.; Tornabene, F. Free vibration of FG-CNT reinforced composite skew cylindrical shells using the Chebyshev-Ritz formulation. *Compos. Part B Eng.* **2018**, *147*, 169–177. [CrossRef]
42. Amir, S.; BabaAkbar-Zarei, H.; Khorasani, M. Flexoelectric vibration analysis of nanocomposite sandwich plates. *Mech. Based Des. Struct. Mach.* **2020**, *48*, 146–163. [CrossRef]
43. Liu, Y.; Liu, W.; Gao, W.; Zhang, L.; Zhang, E. Mechanical responses of a composite sandwich structure with Nomex honeycomb core. *J. Reinf. Plast. Compos.* **2019**, *38*, 601–615. [CrossRef]
44. Lin, H.G.; Cao, D.Q.; Xu, Y.Q. Vibration, Buckling and Aeroelastic Analyses of Functionally Graded Multilayer Graphene-Nanoplatelets-Reinforced Composite Plates Embedded in Piezoelectric Layers. *Int. J. Appl. Mech.* **2018**, *10*, 1850023. [CrossRef]
45. Thai, C.H.; Ferreira, A.J.M.; Tran, T.D.; Phung-Van, P. A size-dependent quasi-3D isogeometric model for functionally graded graphene platelet-reinforced composite microplates based on the modified couple stress theory. *Compos. Struct.* **2020**, *234*, 111695. [CrossRef]
46. Dindarloo, M.H.; Li, L.; Dimitri, R.; Tornabene, F. Nonlocal Elasticity response of Doubly-Curved Nanoshells. *Symmetry* **2020**, *12*, 466. [CrossRef]
47. Karimi, M.; Khorshidi, K.; Dimitri, R.; Tornabene, F. Size-dependent hydroelastic vibration of FG microplates partially in contact with a fluid. *Compos. Struct.* **2020**, *244*, 112320. [CrossRef]
48. Khorasani, M.; Eyvazian, A.; Karbon, M.; Tounsi, A.; Lampani, L.; Sebaey, T.A. Magneto-electro-elastic vibration analysis of modified couple stress-based three-layered micro rectangular plates exposed to multi-physical fields considering the flexoelectricity effects. *Smart Struct. Syst.* **2020**, *26*, 331–343.
49. Arshid, E.; Kiani, A.; Amir, S. Magneto-electro-elastic vibration of moderately thick FG annular plates subjected to multi physical loads in thermal environment using GDQ method by considering neutral surface. *Mater. Des. Appl.* **2019**. [CrossRef]
50. Amir, S.; Khorasani, M.; BabaAkbar-Zarei, H. Buckling analysis of nanocomposite sandwich plates with piezoelectric face sheets based on flexoelectricity and first-order shear deformation theory. *J. Sandw. Struct. Mater.* **2020**, *22*, 2186–2209. [CrossRef]

Sample Availability: Samples of the compounds are not available from the authors.

Publisher's Note: MDPI stays neutral with regard to jurisdictional claims in published maps and institutional affiliations.

© 2020 by the authors. Licensee MDPI, Basel, Switzerland. This article is an open access article distributed under the terms and conditions of the Creative Commons Attribution (CC BY) license (http://creativecommons.org/licenses/by/4.0/).

Review

Kissinger Method in Kinetics of Materials: Things to Beware and Be Aware of

Sergey Vyazovkin

Department of Chemistry, University of Alabama at Birmingham, 901 S. 14th Street, Birmingham, AL 35294, USA; vyazovkin@uab.edu

Academic Editor: Giuseppe Cirillo
Received: 25 May 2020; Accepted: 16 June 2020; Published: 18 June 2020

Abstract: The Kissinger method is an overwhelmingly popular way of estimating the activation energy of thermally stimulated processes studied by differential scanning calorimetry (DSC), differential thermal analysis (DTA), and derivative thermogravimetry (DTG). The simplicity of its use is offset considerably by the number of problems that result from underlying assumptions. The assumption of a first-order reaction introduces a certain evaluation error that may become very large when applying temperature programs other than linear heating. The assumption of heating is embedded in the final equation that makes the method inapplicable to any data obtained on cooling. The method yields a single activation energy in agreement with the assumption of single-step kinetics that creates a problem with the majority of applications. This is illustrated by applying the Kissinger method to some chemical reactions, crystallization, glass transition, and melting. In the cases when the isoconversional activation energy varies significantly, the Kissinger plots tend to be almost perfectly linear that means the method fails to detect the inherent complexity of the processes. It is stressed that the Kissinger method is never the best choice when one is looking for insights into the processes kinetics. Comparably simple isoconversional methods offer an insightful alternative.

Keywords: crosslinking polymerization (curing); decomposition; degradation; liquid and solid state; phase transitions; thermal analysis

1. Introduction

The thermal behavior of materials is explored broadly by the techniques of differential scanning calorimetry (DSC), differential thermal analysis (DTA), and thermogravimetry (TGA). Kinetic analysis of the data obtained by these techniques provides important insights into the fundamental issues of the reactivity and stability of materials. When it comes to kinetics, the Kissinger method is by far the most popular way of evaluating the activation energy of thermally stimulated processes. The method was introduced in two successive publications [1,2] that, according to the Scopus database, have been cited over 12,000 times. Yet, science is not a popularity contest, so that a routine followed by the majority is not guaranteed to yield the best or even simply a correct result. As a matter of fact, the method is not among the techniques recommended [3] for advanced kinetic studies. Most of the time, it is employed in materials characterization work that among other quantities reports a number that presumably characterizes an energy barrier to a thermal process under study. In this situation, the Kissinger method provides an unbeatably simple way of estimating the activation energy.

While a desired trait, simplicity may carry the risk of trivializing the problem. This applies fully to the Kissinger method as its formalism largely oversimplifies the kinetics of the processes it treats. Nowadays, as never before, the kinetics community concerned with thermally stimulated processes has come to realize that these processes are commonly multi-step [3]. As such, they have more than one energy barrier that controls them, so that the temperature dependence of their rates cannot be described by a single activation energy. In contrast, the Kissinger method yields a single value of the activation energy regardless of the process complexity. In other words, this method is destined to generally miss the actual kinetic complexity. Clearly, this is an essential limitation of the method. Nonetheless, it is just as clear that exposing this limitation is unlikely to stop the enormous usage of the Kissinger method. Respectively, this is not an objective of the present work. Rather, it aims at helping those who use the method to do it in a conscientious manner. It means to use the method with the clear realization that its application does not usually provide adequate insights into the processes kinetics and typically yields the results of a very limited value.

To accomplish the task, we consider a number of the applications of the Kissinger method that include the most popular ones such as chemical reactions, crystallization and glass transition. For each of the processes considered we give a brief theoretical discussion to explain the origins of the complex temperature dependence of the respective rate. Then we apply the Kissinger method to experimental data to see whether the said complexity manifests itself in the application. In addition to that, we briefly discuss some general limitations that are associated with the underlying assumptions of the method.

2. Basics of the Method

The simplest form of the Kissinger equation for estimating the activation energy, E, is as follows:

$$E = -R \frac{d\ln\left(\frac{\beta}{T_p^2}\right)}{dT_p^{-1}} \tag{1}$$

where R is the gas constant, β is the heating rate and T_p is the temperature that corresponds to the position of the rate peak maximum. Most commonly T_p is determined as the temperature of the peak signal (maximum or minimum) measured by DSC, DTA or derivative thermogravimetry (DTG).

A curious fact about Equation (1) is that it had been proposed in an obscure paper by Bohun [4] a few years before the famous publications [1,2] by Kissinger. A more instructive form of the Kissinger equation is the integral one:

$$\ln\left(\frac{\beta}{T_p^2}\right) = \ln\left(-\frac{AR}{E}f'(\alpha_p)\right) - \frac{E}{RT_p} \tag{2}$$

where $f'(\alpha) = \frac{df(\alpha)}{d\alpha}$. Equation (2) originates from the basic rate equation of a single-step process:

$$\frac{d\alpha}{dt} = A\exp\left(\frac{-E}{RT}\right)f(\alpha) \tag{3}$$

where α is the extent of conversion of the reactant to products, t is the time, A is the preexponential factor and $f(\alpha)$ is the reaction model. A list of the models is available elsewhere [5].

Both Equations (1) and (2) suggest that the activation energy can be evaluated as the slope of the Kissinger plot of $\ln\left(\frac{\beta}{T_p^2}\right)$ vs. T_p^{-1}. However, Equation (2) indicates that for this plot to be linear the intercept should be a constant independent of the heating rate. This condition is not generally satisfied because α_p is known [6] to depend on β. To satisfy it, $f'(\alpha)$ should be independent of α. This is the case of a first order reaction model, $f(\alpha) = 1 - \alpha$, for which $f'(\alpha) = -1$. However, for the majority of the reaction models $f'(\alpha)$ depends on α and, thus, on β. This introduces some inaccuracy in estimating the value of E. As shown by Criado and Ortega [7] for a large variety of models the respective error does not exceed 5% as long as $E/RT > 10$.

It needs to be stressed that checking for potential systematic variation of α_p with β should be taken as a general prerequisite for using the Kissinger method. The existence of such variation can be an indication of the process complexity as discussed by Muravyev et al. [8]. They have also demonstrated that even a moderate change of 0.06 in α_p, caused by a change of β from 1 to 10 K min^{-1}, can result in a systematic error in E as large as 15%.

The error caused by the dependence of α_p on β is eliminated in isoconversional methods [5,9,10]. One of them is the Kissinger-Akahira-Sunose method [11]. It employs the same equations as the Kissinger method (i.e., Equations (1) and (2)) but replaces T_p with T_α. The latter is the temperature related to a given conversion at different heating rates. This is a more accurate way of estimating the activation energy. A critical advantage of an isoconversional method over the Kissinger one is that it affords determining the activation energy, E_α, as a function of conversion. A significant systematic variation of E_α with α reveals that the process under study involves more than one step. As shown later, this is an essential piece of kinetic information that the Kissinger method tends to miss.

Obviously, the method is not highly accurate but its numerical accuracy is rather tolerable for many practical purposes. Holba and Sestak [12] have additionally questioned the accuracy of the Kissinger method due to not accounting for the thermal inertia component of the heat flow as measured by heat flux DSC (or DTA). Indeed, it is typically assumed that the process rate is directly proportional to the heat flow, dQ/dt:

$$\frac{d\alpha}{dt} = \frac{1}{Q_0}\frac{dQ}{dt} \qquad (4)$$

where Q_0 is the total heat released or absorbed during the process. This assumption is due to Borchardt and Daniels [13], whose analysis suggests that the thermal inertia term can be neglected. Not correcting for this term should unavoidably cause some systematic error in the value of the activation energy. This is because the raw (i.e., uncorrected) DSC peaks appear at somewhat higher temperature than they should, and the magnitude of this temperature shift increases with increasing the heating rate. In regard to the Kissinger method, it means that the T_p values determined from uncorrected DSC peaks are shifted to a higher temperature.

The corrected heat flow is obtained via a relatively simple adjustment:

$$\frac{dQ}{dt} = RHF + \tau\frac{d(RHF)}{dt} \qquad (5)$$

where RHF is the raw heat flow as measured by DSC and the second addend is the thermal inertia term. This adjustment, however, requires estimating the time constant τ. This is done by analyzing the back tail of a DSC peak measured for melting of a pure metal [14]. The value of τ is proportional to the total heat capacity of the sample. On the other hand, the temperature correction is proportional to $\beta\tau$. It means that the effect of thermal inertia decreases with using smaller sample masses and slower heating rates. As a rule of thumb, the International Confederation of Thermal Anlaysis and Calrimetry (ICTAC) recommendations [15] suggest that for kinetic studies the product of the mass and heating rate should be kept under 100 mg K min^{-1}.

Our previous study has demonstrated [16] that ignoring thermal inertia in decomposition of 3 mg samples of polystyrene at the heating rates 2–20 °C min^{-1} causes statistically insignificant error when estimating the activation energy by an advanced isoconversional method [17]. Here we use the same data set [18] to compare the effect of thermal inertia on the activation energy estimated by the Kissinger method. The results are presented in Figure 1. As expected, the Kissinger plot for the corrected data is shifted to lower temperatures. The magnitude of the shift is barely detectable at slower heating rates but becomes larger at the faster ones. Upon accounting for thermal inertia, the E value has increased from 183 ± 6 to 191 ± 7 kJ mol^{-1}, that is, by only 4%. However, with account of the respective uncertainties, the t-test suggests that the difference is not statistically significant. This example does not mean that the effect of thermal inertia is negligible in general. Ultimately, the effect is determined by the magnitude of the temperature shift. As long as the latter does not rise above 2–3 °C, the effect should be negligible [16].

Another rarely considered issue related to the accuracy of the Kissinger method is its applicability to the temperature programs other than the one of linear heating. It is noteworthy that the linear heating rate is introduced in the Kissinger derivations by replacing dT/dt with β. However, the related equation is derived for the condition of the rate maximum, that is, the respective values have the meaning of the instantaneous heating rate, β_p. The latter can be defined for a nonlinear heating program that, in principle, affords extending the Kissinger method beyond the standard linear heating [19]. This is important in connection with the sample controlled thermal analysis [20], in which the temperature program is controlled by the response of the reaction rate to heating. This idea is implemented commercially in the techniques of high or maximum resolution TGA. The application of the Kissinger method in the case of nonlinear heating programs has been scrutinized by Sanchez-Jimenez et al. [21]. They have demonstrated that in the case of nonlinear heating the value of α_p can vary significantly with β_p, so that the respective Kissinger plot yields a completely erroneous value of the activation energy. Therefore, the major conclusion here is that before applying the method to the data obtained under a program other than simple linear heating, one needs to make sure that there is no significant variation of α_p with β_p.

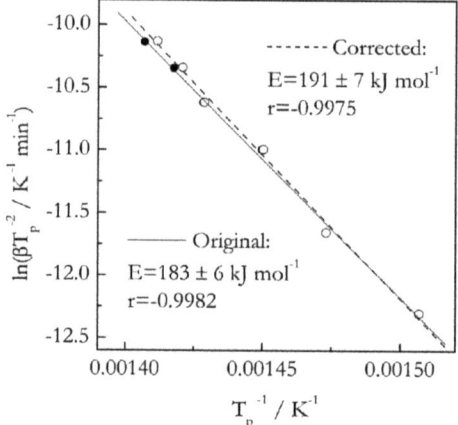

Figure 1. Kissinger plots for thermal degradation of isotactic polystyrene. Data from Liavitskaya and Vyazovkin [18] and Vyazovkin [16]. Solid and open circles represent T_p values obtained respectively from original (raw) data and data corrected for thermal inertia. The difference is obvious only for faster heating rates, 16 and 20 °C min^{-1}. Solid and dash lines are the least square fits to original and corrected data.

In addition, many types of runs are carried out under cooling temperature programs. Whether cooling is linear or not, the data obtained cannot be treated by the Kissinger method. A simple indication is that Equation (2) contains a logarithm of β. As already stated, the value of β replaces the value of dT/dt, so that on cooling β is necessarily negative. Obviously one cannot take a logarithm of a negative value, which means that Equation (2) cannot be used for cooling data. Forcing Equation (2) to treat cooling data by dropping the negative sign of β results in obtaining completely invalid values of the activation energy [22,23]. That is, the Kissinger method should never be applied to the data obtained on cooling.

3. Chemical Reactions

As already mentioned, the Kissinger method is based on Equation (3), which is the rate equation of a single step reaction. The temperature dependence of the latter is determined by a single activation energy that is readily evaluated by the Kissinger method. The problem, though, is that the reactions in the condensed (i.e., solid or liquid) phase typically involve multiple steps and, therefore, face more than a single energy barrier. This has important implications for the temperature dependence of the reaction rate.

Let us consider a very simple reaction that involves two competing steps, each of which follows the same model. The rate of this reaction is:

$$\frac{d\alpha}{dt} = k_1(T)f(\alpha) + k_2(T)f(\alpha) \equiv [k_1(T) + k_2(T)]f(\alpha) = k_{ef}(T)f(\alpha) \qquad (6)$$

where the subscripts 1 and 2 designate the rate constants, $k(T)$, respectively related to the individual steps. In its turn, the temperature dependence of the rate constant is defined via the Arrhenius equation:

$$k(T) = A \exp\left(\frac{-E}{RT}\right) \qquad (7)$$

Experimentally, the activation energy is determined from the slope of the plot of a logarithm of the rate constant vs. reciprocal temperature, that is, as the following derivative:

$$E = -R\frac{d \ln k(T)}{dT^{-1}} \qquad (8)$$

Plugging $k_{ef}(T)$ from Equation (6) into Equation (8) gives:

$$E = \frac{E_1 k_1(T) + E_2 k_2(T)}{k_1(T) + k_2(T)} \qquad (9)$$

Equation (9) suggests that the experimentally determined activation energy for the above reaction will be temperature dependent, which also means that the respective $\ln k(T)$ vs. T^{-1} plot will be nonlinear as illustrated elsewhere [24].

Comparing Equation (8) with Equation (1) suggests that the Kissinger plot for the above reaction would also be nonlinear and the respective activation energy temperature dependent. If this nonlinearity were easy to detect, the application area of the Kissinger method could be extended to multi-step kinetics. In reality, detecting such nonlinearity is not easy and contingent on several conditions. One of them is the number of heating rates used, that is, the number of points on the Kissinger plot. It is nearly impossible to detect the nonlinearity with less than 5 points. However, the typical application of the Kissinger method is limited to 3–4 heating rates. Another important factor, is the width of the temperature range of the T_p values. The wider the range, the better chances to detect the nonlinearity. In experimental terms, a wider range of T_p means a wider range of β. The ratio of the maximum to minimum heating rate should be no less than 5. Even if all these conditions are met, the nonlinearity may still escape detecting because the difference in the activation energies of the individual steps is not large enough.

A much more sensitive way of detecting the reaction complexity is to use an isoconversional method [5,9,10] that allows one to determine the activation energy as a function of conversion. As an example, Figure 2 provides a comparison of the Kissinger plot with a dependence of the isoconversional activation energy (E_α) on conversion for the thermal decomposition (dehydration) of calcium oxalate monohydrate as measured by DSC [25]. As one can see, the activation energy estimated by an isoconversional method varies from about 105 to 75 kJ mol^{-1}. This means that the respective Kissinger plot should be nonlinear. In particular, its lower temperature part should have a steeper slope than the high temperature one. This actually is the case, if one looks very closely. Yet, the change in the angle of the slope is a little over 3° and, thus, is easy to miss. Furthermore, treating this plot as linear, that is, ignoring the small nonlinearity, yields a high value of the correlation coefficient (r), which means that statistically this nonlinearity insignificant. All this illustrates how insensitive the Kissinger method is in detecting the reaction complexity. This is especially alarming considering that this Kissinger plot includes 10 heating rates ranging from 0.75 to 20 °C min^{-1} and covering a 53 °C interval.

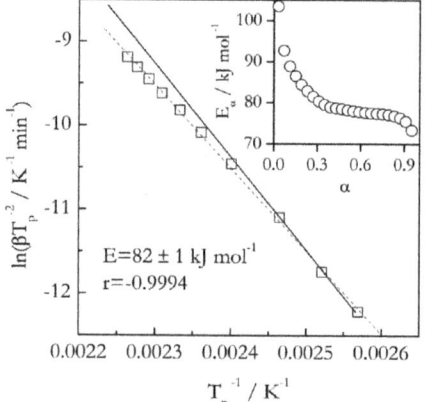

Figure 2. Kissinger plot for thermal dehydration of $CaC_2O_4 \cdot H_2O$. Dash line is least square fit to experimental data (squares). Solid line is a fit to the three lower temperature data points. It demonstrates that the angle of the slope changes for higher temperature data. Inset shows variation in isoconversional activation energy. Data from Liavitskaya and Vyazovkin [25].

Naturally, the question can be raised whether the statistically insignificant nonlinearity is important. An answer depends on the purpose of the kinetic study. If one simply needs to obtain a ballpark estimate for the activation energy, such nonlinearity can be ignored. However, it is critically important when one strives for a mechanistic understanding of the estimate. For instance, the decreasing dependence of E_α shown in Figure 2 is not just something encountered in a particular instance of dehydration of calcium oxalate monohydrate. It is a general phenomenon observed for a wide variety of reversible decompositions [26] that include dehydration of diverse crystal hydrates. For this type of processes, the activation energy depends on the equilibrium pressure, P_0, of the gas product as [18]:

$$E = E_1 + \Delta H^0 \frac{P}{P_0 - P} \quad (10)$$

where E_1 is the activation energy of the forward reaction, ΔH^0 is the reaction enthalpy and P is the partial pressure of the gas product. According to Equation (10), E should decrease with increasing temperature because P_0 increases, making the second addend increasingly smaller. Clearly, none of that information can be gained from the single value 82 ± 1 kJ mol^{-1} estimated by the Kissinger method (Figure 2).

It is worth noting that Agresti [27] has proposed a modification to the Kissinger method that accounts for the pressure dependence. As expected, the resulting Kissinger plots are nonlinear and their curvature increases dramatically in the vicinity of equilibrium temperatures.

Concerning the reaction complexity, there is a common belief that it has to manifest itself via the rate peaks that reveal shoulders or other aberrations of the regular bell-shaped form. Definitely, discovering such features is a sign of the reaction complexity. However, the opposite is not true, meaning that the absence of such features in DSC, DTA or DTG peaks does not mean that the reaction is simple, that is, single step. Figure 3 illustrates such situation in the case of epoxy-anhydride crosslinking polymerization (sometimes termed as curing). It is seen that the respective DSC peaks are of regular bell-shaped form without any obvious aberrations. Yet, the isoconversional activation energy demonstrates a significant increase from about 20 to 70 kJ mol^{-1}, which again results from a multi-step reaction mechanism [28] At the same time, the Kissinger plot is almost perfectly linear ($r = -0.9999$) and yields a single value of the activation energy, 71 kJ mol^{-1} [29]. The latter obviously gives no hints regarding the reaction complexity.

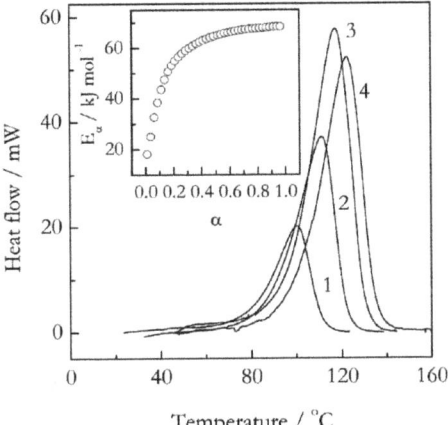

Figure 3. DSC curves for crosslinking polymerization of an epoxy-anhydride system. The numbers by the curves are heating rates in °C min^{-1}. The inset shows the conversion dependence of the isoconversional activation energy. Adapted with permission from Vyazovkin and Sbirrazzuoli [29]. Copyright 1999 Wiley-VCH.

Of course, there are cases of single-step reactions. These ordinarily are multi-step reactions whose overall kinetics is limited or dominated by one step. In these cases, the Kissinger method could serve as a basis for an adequate kinetic analysis. The problem, though, is that the method does not typically possess the sufficient sensitivity to differentiate reliably between the single and multi-step kinetics. The occurrence of single-step kinetics can be easily detected by an isoconversional method as the absence of any significant dependence of E_α on α. However, employing an isoconversional method for such purpose immediately makes the use of the Kissinger method redundant.

Last but not least, the Kissinger method is a part of the ASTM E698 technique for Arrhenius kinetic constants for thermally unstable materials [30]. This technique relies on using a single value of the activation energy estimated by the Kissinger method to make predictions of a material behavior under isothermal conditions. Unfortunately, in the case of the reaction complexity this technique produces rather poor predictions, that is, significantly less accurate than the ones produced via isoconversional methods that use variable activation energy [31,32].

4. Crystallization

There are about as many publications on the application of the Kissinger method to crystallization as to chemical reactions. The applications are so common that sometimes one can see the claims that the method was proposed by Kissinger for estimating the activation energy of crystallization. This, of course, is absolutely false. In reality, neither of his seminal papers [1,2] even contains the word "crystallization."

To understand the problems with this particular application, one needs to recognize that the temperature dependence of the crystallization rate differs dramatically from that of the reaction rate. Decreasing temperature makes chemical reactions to proceed slower. Crystallization rate depends on supercooling, $\Delta T = T_m - T$ with respect to the equilibrium melting temperature, T_m. At small supercoolings, crystallization accelerates with decreasing temperature until reaching the maximum rate at some temperature T_{max}. At large supercoolings, that is, at temperatures below T_{max}, the crystallization rate decreases with decreasing temperature.

This complex temperature dependence cannot be described by a single Arrhenius equation, which is the basis of the Kissinger method. It is described well by the models of nucleation or nuclei growth that combine the Arrhenius kinetics with the underlying thermodynamics of crystallization. The rate of crystallization can be limited by the formation of nuclei or by the growth of existing nuclei. The temperature dependence of the nucleation rate is adequately represented by the Turnbull and Fisher model [33]:

$$n = n_0 \exp\left(\frac{-E_D}{RT}\right) \exp\left(\frac{-\Delta G^*}{RT}\right) \tag{11}$$

where n is the nucleation rate constant, n_0 the preexponential factor, E_D is the activation energy of diffusion and ΔG^* is the free energy barrier to nucleation. The size of this barrier for a spherical nucleus is as follows:

$$\Delta G^* = \frac{16\pi\sigma^3 T_m^2}{3(\Delta H_m)^2 (\Delta T)^2} = \frac{\Omega}{(\Delta T)^2} \tag{12}$$

where σ is the surface energy (surface tension), ΔH_m is the enthalpy of melting per unit volume and Ω is a constant that collects all parameters that do not practically depend on temperature. The derivations can be found in various sources, for example, References [9,34–38].

If the crystallization rate, u, is limited by the growth of existing nuclei, it depends on temperature as follows [38]:

$$u = u_0 \exp\left(\frac{-E_D}{RT}\right)\left[1 - \exp\left(\frac{\Delta G}{RT}\right)\right] \tag{13}$$

where u_0 the preexponential factor and ΔG is the difference in the free energy of the final (crystalline) and initial (liquid) phase. Surprisingly, the monographic literature does not consider Equation (13) as commonly as Equation (11). Thus, it needs a few comments here. First, it is readily derived as the difference between the rates of the forward and reverse transition. For the forward rate, the energy barrier is E_D, whereas for the reverse rate it is $E_D - \Delta G$ ($\Delta G < 0$). The frequency (preexponential) factor is assumed the same for both rates.

Second, the free energy terms in Equations (11) and (13) differ entirely in their meaning. The ΔG^* term in Equation (11) is the energy barrier that makes it a positive value. The ΔG term in Equation (13) is the free energy change for a spontaneous process and, thus, negative. Note that Equation (13) is sometimes written with a negative sign in front of ΔG when it is referred to as the driving force. The latter in the strict thermodynamics sense [39] should be a positive quantity. This, however, may create extra confusion of dealing with positive ΔG for a spontaneous process. One way or another, the argument of the exponential function in the bracketed term of Equation (13) must be negative.

Despite their differences, Equations (11) and (13) suggest the existence of the rate maximum. In both equations, the acceleration at small supercooling is due to the thermodynamic terms. In Equation (11), it occurs because the nucleation barrier ΔG^* decreases with increasing the supercooling (Equation (12)), that is, with decreasing temperature. In Equation (13), the dependence on supercooling is introduced via an approximate equality [35,36,38]:

$$\Delta G = \Delta H_c\left(\frac{T_m - T}{T_m}\right) \equiv \Delta H_m\left(\frac{T - T_m}{T_m}\right) \tag{14}$$

where ΔH_c is the enthalpy of crystallization. Then, as temperature lowers and supercooling increases the value of ΔG becomes increasingly more negative. As a result, the exponential function in the bracketed term (Equation (13) decreases toward zero, whereas the term itself increases toward unity. Thus, the acceleration is associated with the bracketed term that becomes larger at larger supercoolings.

In both cases (Equations (11) and (13)), the thermodynamic acceleration is counteracted by the diffusional retardation that originates from continuously growing viscosity of the melt. This behavior is represented by the exponential term containing E_D. At some point the diffusional retardation starts to outweigh the thermodynamic acceleration, so that the rate begins to drop with decreasing temperature. As a result, the rate passes through a maximum.

Figure 4 (inset) displays the temperature dependence of the rate derived by combining Equations (13) and (14). Such dependence for the nucleation rate (n in Equation (11)) can be seen elsewhere [40]. Either of these dependencies passes through a maximum. However, in general the growth process tends to demonstrate the maximum at T_{max}, which is larger (closer to T_m) than that for the nucleation process.

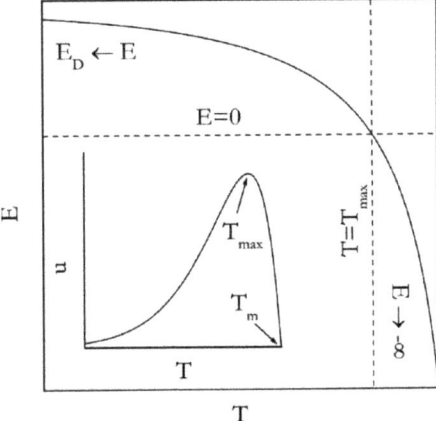

Figure 4. Temperature dependence of the activation energy according to Equation (15). Inset shows temperature dependence of the growth rate according to Equation (13).

Most relevant to the Kissinger analysis is that the existence of the rate maximum on the temperature dependence entails the inversion of the sign of the experimentally determined activation energy. This is unavoidable because the activation energy is determined by the sign of the temperature derivative of the rate. In the temperature region above T_{max} the rate decreases with increasing temperature; the sign of the experimental activation energy is negative. However, below T_{max} the rate increases as temperature rises so that the sign is positive. An analytic expression for the temperature dependence of the activation energy is arrived at by plugging u for $k(T)$ in Equation (8) that yields the following:

$$E = E_D + \frac{\Delta H_m \exp\left[\frac{\Delta H_m(T-T_m)}{RTT_m}\right]}{\exp\left[\frac{\Delta H_m(T-T_m)}{RTT_m}\right] - 1} \quad (15)$$

The resulting dependence is depicted in Figure 4. Indeed, it is seen that E takes on large negative values in the vicinity of T_m but increases toward 0 as temperature drops toward T_{max}. Yet, in the limit of large supercoolings ($T < T_{max}$) E is positive and decreases from E_D toward 0 as temperature rises. The nucleation model (Equation (11)) gives rise to a different analytic expression for the temperature dependence of the activation energy [9], viz.:

$$E = E_D - \Omega\left[\frac{2T}{(T_m - T)^3} - \frac{1}{(T_m - T)^2}\right] \quad (16)$$

Nevertheless, the dependence predicted by Equation (16) is very similar to the one predicted by Equation (15) and has exactly the same asymptotes, that is, E_D and $-\infty$ for the infinitely large and small supercooling respectively.

The above has direct relevance to the experimental studies of the crystallization kinetics. It needs be recognized that experimentally the temperature region above T_{max} is accessed by cooling melts, whereas the one below T_{max} by heating glasses. The applications of the Kissinger method to crystallization of melts are usually encountered in the field of polymers. As explained earlier, the Kissinger method cannot be applied to the data obtained on cooling and when forced to do that yields entirely erroneous values of the activation energy [22,23].

When the Kissinger method is applied to crystallization of glasses, it yields positive activation energies that are commonly reported as constant temperature independent values. On the other hand, the theory predicts the activation energy of the glass crystallization to decrease with increasing temperatures. It means that the corresponding Kissinger plots should be nonlinear, or, more precisely, concave down. As already noted, detecting the curvature requires using multiple heating rates spread over a broad range. For example, the effect is detectable (Figure 5) in the data obtained for crystallization of $Ga_{7.5}Se_{92.5}$ and $Si_{12.5}Te_{87.5}$ glasses [41,42]. Note that in both cases the range of the heating rates used is unusually broad, 5–90 °C min^{-1}. However, if one looks at these plots within a more typical range 5–25 °C min^{-1} (first five points corresponding to the lower temperatures) the curvature is practically unnoticeable. Expectedly, it is much easier to detect a variation in the activation energy by using an isoconversional method. For these two glasses the isoconversional activation energy decreases 1.5 times (from 85 to 55 kJ mol^{-1} for $Ga_{7.5}Se_{92.5}$ and from 200 to 130 kJ mol^{-1} for $Si_{12.5}Te_{87.5}$) in the temperature range of crystallization [41,42]. It should be emphasized that the curvature of the Kissinger plots for crystallization increases dramatically when using ultrafast scanning calorimetry [43]. This technique permits employing both much faster heating rates and much broader range of them.

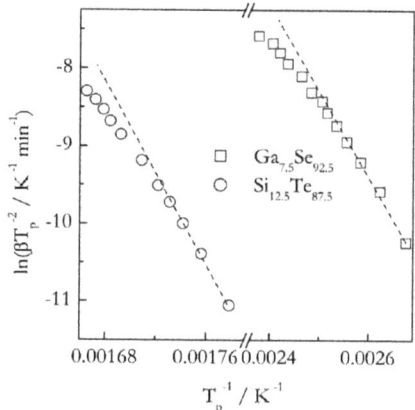

Figure 5. Kissinger plots for crystallization of $Ga_{7.5}Se_{92.5}$ (squares) and $Si_{12.5}Te_{87.5}$ (circles) glasses. Dash lines connecting three lowest temperature points are a guide for the eye to better visualize the nonlinearity. The heating rates range from 5 to 90 °C min^{-1}. Data from Abu El-Oyoun [41,42].

5. Glass Transition

The glass transition appears to be the third most important application area of the Kissinger method. One of popular models used for describing the glass transition kinetics is the Tool-Narayanaswami-Moynihan model (TNM) [44–46]. It can be presented as follows:

$$\ln \tau = \ln \tau_0 + \frac{xE}{RT} + \frac{(1-x)E}{RT_f} \qquad (17)$$

where τ is the relaxation time, τ_0 is the preexponential factor, x is the nonlinearity parameter and T_f is the fictive temperature. The relaxation time in Equation (17) is an analog of the reciprocal rate constant in the Arrhenius equation. The model indicates that the whole process of the glass transition is driven by a single constant activation energy. This is a simplification also known as thermorheological simplicity, which, by no means, is the general rule for the relaxation behavior [47]. An important limitation of the model is that it, in particular, predicts the Arrhenius temperature dependence for viscosity, whereas this dependence is generally of the Williams-Lander-Ferry (WLF) [48] or Vogel-Tammann-Fulcher (VTF) [37,49,50] type.

Based on Equation (17), Moynihan et al. [51] have proposed methods of estimating the activation energy of the glass transition from either cooling or heating data. For heating, E is evaluated from the heating rate dependence of the glass transition temperature:

$$E = -R\frac{d\ln \beta}{dT_g^{-1}}. \tag{18}$$

In their paper, Moynihan et al. point out that T_g can be defined from calorimetric measurements as temperature of the extrapolated onset or inflection point or the heat capacity maximum. The latter corresponds to the endothermic peak that appears on heating in DSC at the end of the glass transition event.

The appearance of this peak, sometimes referred to as the glass transition peak, has inspired numerous applications of the Kissinger method for determination of the activation energy of the glass transition. Unfortunately, most of these application have been wrong. Apparently, many believe that simply observing a DSC peak that shifts with the heating rate justifies the application of the Kissinger method. This is certainly not true in the case of the glass transition. The problem is not specific to the Kissinger method itself, although it may appear as such [52]. Rather, it arises from heating the glasses not having a proper thermal history. As stressed by Moynihan et al. [51,53,54], obtaining a correct value of E from Equation (18) requires creating a glass of a specific thermal history. Namely, immediately before heating the glass has to be cooled at the rate, which is equal (or proportional) to the rate of heating. Also, cooling must occur from the equilibrium state, that is, from well above T_g down to well below T_g. Using other thermal histories gives rise to the E values that can deviate dramatically from the correct one [54].

While demonstrated [54] in the case of Equation (18), the importance of using the proper thermal history applies fully to the application of the Kissinger method. As a matter of fact, both the Moynihan (18) and Kissinger (1) equations yield nearly identical values of E when T_g is estimated as the peak temperature of the glass transition and they both produce equally wrong values in the case of not using the proper thermal history [55]. Nevertheless, the application of Equation (18) to $T_g = T_p$ when using the proper thermal history does give rise to correct values of E [56]. Putting all these results together, we can conclude that one can use the Kissinger method to obtain the correct values of E as long as the glass transition measurements are performed on a sample exposed to the proper thermal history, for example, when using heating at β immediately preceded by cooling at $-\beta$, as mentioned before. More importantly, it makes no sense trying to determine the E values by the Kissinger or Moynihan method by heating the as-is glass samples. The resulting values would be largely the fortuitous ones that are impossible to interpret in a meaningful way.

Assuming that the glass transition measurements are performed under a proper thermal history, we can now return to aforementioned limitation of the TNM model associated with the oversimplified (i.e., Arrhenius) treatment of the temperature dependence of the relaxation time or viscosity. As stated, the more general is the WLF or the VTF dependence. The VTF equation for the relaxation time is:

$$\ln \tau = \ln \tau_0 + \frac{B}{T - T_0} \tag{19}$$

where B is a constant and T_0 is a reference temperature. The respective Arrhenius plot, $\ln \tau$ vs. T^{-1} is nonlinear and gives rise to the activation energy that decreases with increasing temperature as follows:

$$E = R\frac{BT^2}{(T - T_0)^2} \tag{20}$$

A similar dependence derives [48] from the WLF equation.

This brings about an important question as to whether the activation energy of the glass transition should be a constant or temperature dependent (i.e., variable) value. Based on the Adam-Gibbs theory [57], the activation energy in the Arrhenius equation is proportional to the size of the region that rearranges cooperatively during the transition. This size is inversely proportional to the configurational entropy that increases with T so that the experimental activation energy is expected to decrease with T. In practice, one may or may not detect this variation depending on the dynamic fragility [58] of the systems studied. According to Angell [58], there are strong and fragile glassformers that demonstrate distinctly different $\ln\tau$ vs. T^{-1} plots. For the strong glass formers, these plots are nearly linear, that is, Arrhenian. For the fragile ones, they are nonlinear, that is, of the WLF/VTF type. Typical examples of the strong and fragile glassformers respectively are inorganics and polymers. On the other hand, organics and metals tend to fall between those two limits.

In any event, when it comes to the Kissinger analysis of the glass transition one may obtain either linear or nonlinear Kissinger plots depending on the fragility of the systems studied. Examples of such plots are shown in Figure 6 for two glasses: boron oxide (B_2O_3) and polystyrene (PS). The T_p values have been extracted from the previously published DSC data [59,60] As seen from the figure, the Kissinger plot is practically linear for the strong glassformer, boron oxide, whereas it is nonlinear for the fragile one, polystyrene. That is, the activation energy of the glass transition is expected to be practically constant in the former case but temperature dependent in the later one. Note that detecting nonlinearity of the Kissinger plots requires using multiple heating rates in relatively broad range. Much more sensitive way of detecting a variation in the activation energy of the glass transition is to employ an isoconversional method that demonstrates clearly that the variability of the activation energy is proportional to the fragility [61]. Remarkably, the method has been capable of detecting a minor variation in the activation energy even in the case of boron oxide [60].

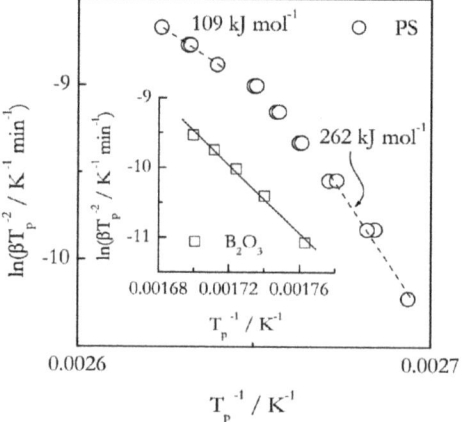

Figure 6. Kissinger plots for the glass transition in polystyrene (PS, circles) and boron oxide (B_2O_3, squares). For PS, the plot is visibly nonlinear, the activation energy decreases with temperature, the heating rate range 2.5–25 °C min^{-1}. For B_2O_3, the plot is practically linear, the activation energy is nearly constant, 203 ± 11 kJ mol^{-1}, the heating rate range 5–25 °C min^{-1}. Data from Vyazovkin et al. [59,60].

6. Melting and Other Processes

In addition to the three major application areas discussed above, the Kissinger method has been used for treating some other processes. Although these applications are relatively scarce, they are still of research interest. The most common and, perhaps, most confusing is melting. A detailed discussion regarding the theory and practice of the melting kinetics is given elsewhere [40]. Here, we only reiterate a few important points directly relevant to the Kissinger method.

The basic thermodynamics suggests that on heating of a substance its temperature remains constant throughout the melting process. This temperature is the equilibrium melting temperature, which is independent of the heating rate. The confusion may arise from the fact that the position of the DSC peak (i.e., T_p) measured for melting usually demonstrates a noticeable increase with the heating rate. This effect is observed because the DSC peaks are typically presented not as a function of the sample temperature but as a function of the reference (furnace) temperature. The latter obviously increases during continuous heating at the rate β. In this situation, the value of T_p represents the reference temperature, at which the substance has finished melting. This temperature shifts with the heating rate according to the following equation [62]:

$$T_p - T_m = \sqrt{2R_{sf}\Delta H_m m \beta} \qquad (21)$$

where R_{sf} is the thermal resistance and m is the mass.

As seen from Equation (21), for melting the shift of T_p with β is determined by physical parameters other than the activation energy. It means that, as a rule, the application of the Kissinger method to the T_p vs. β data would yield a number that does not represent the activation energy of melting. An exception to this rule are the compounds that undergo real superheating, that is, the compounds, whose temperature does rise during melting. For certain reasons [40], melting with superheating cannot be identified by plotting DSC signal against the sample temperature. A simple yet informative test [63] is checking whether the melting peak width increases with increasing either the heating rate or the sample mass. If it does, the melting occurs without superheating and the Kissinger method cannot be applied.

The absence of the aforementioned DSC peak broadening is indicative of melting with superheating. A more definitive (quantitative) criterion is based on determining the value of the exponent z in Equation (22):

$$T_p = T_m + B\beta^z \qquad (22)$$

where B and z are fit parameters. According to Toda [63], the z values markedly smaller than 0.5 indicate that melting occurs with superheating. In this circumstance, the Kissinger method can be applied to obtain an estimate for the activation energy of melting.

As an example of melting with superheating we consider the case of glucose. Superheating of this substance was reported by Tammann [64] and then reconfirmed by Hellmuth and Wunderlich [65]. For this compound, the z value (Equation (22)) has been found [66] to be 0.2 that confirms the occurrence of superheating. Applying the Kissinger method to the DSC data on glucose melting [67] gives rise to a plot (Figure 7) that is practically linear with a significant correlation coefficient ($r = -0.9986$). Correspondingly, the method yields a single constant activation energy 294 kJ mol^{-1}. Nonetheless, an isoconversional method yields activation energy that decreases from about 350 to 230 kJ mol^{-1}.

Important is that a decrease of the experimental activation energy of melting with increasing temperature is justified theoretically [40]. Theoretically, the kinetics of melting is treated by the nucleation and nuclei growth models (Equations (11) and (13)), that is, just as the kinetics of crystallization. Both models can be used to derive the temperature dependence of the activation energy. The growth model (Equation (13)) gives rise [67] to:

$$E = E_D - \frac{\Delta H_m \exp\left[\frac{\Delta H_m (T_m - T)}{RTT_m}\right]}{\exp\left[\frac{\Delta H_m (T_m - T)}{RTT_m}\right] - 1} \qquad (23)$$

and the nucleation model (Equation (11)) to:

$$E = E_D + \Omega\left[\frac{2T}{(T - T_m)^3} + \frac{1}{(T - T_m)^2}\right] \qquad (24)$$

Note that Equation (24) is mathematically identical with Equation (16) as can be found by replacing $(T - T_m)$ with $-(T - T_m)$. Either Equation (23) or Equation (24) predicts E to decrease with increasing T and they both have the same asymptotes. When T is close to T_m (i.e., at small superheatings) E tends to $+\infty$, whereas at large superheatings E tends to E_D.

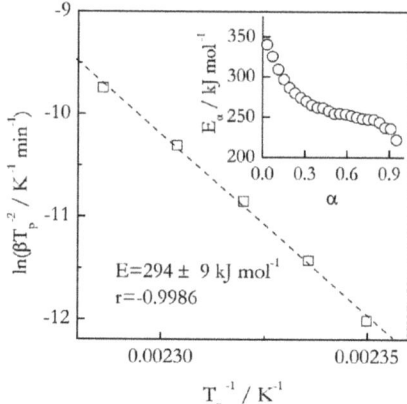

Figure 7. Kissinger plot for melting of glucose. Dash line is least square fit to experimental data (squares). Inset shows variation in isoconversional activation energy (circles). The heating rate range 1–11 °C min^{-1}. Data from Liavitskaya et al. [67]. Reproduced from Ref. [67] with permission from the PCCP Owner Societies.

It should be emphasized that the curvature of the Kissinger plots can be more prominent than the one seen in Figure 7. This has been found in the melting kinetic studies on poly(ethylene terephthalate) [68] and poly(ε-caprolactone) [69]. The curvature is larger when melting occurs closer to T_m. In such a case, it can be sufficient to evaluate the actual dependence of E vs. T that can further be used for estimating the parameters of the nucleation (Equation (24)) and nuclei growth (Equation (23)) models via fitting [67–69].

Lastly, one should keep in mind that the models of nucleation and nuclei growth are potentially applicable to a wide variety of phase transitions taking place on heating and cooling. Since it has been already explained that the Kissinger method cannot be applied to the processes taking place on cooling, we only mention examples of phase transitions occurring on heating. These include gelation of aqueous solutions of methylcellulose [70], the coil-to-globule transition in aqueous solutions of poly(N-isopropylacrylamide) [71], and solid-solid transition in some salts [72]. All these transitions demonstrate the E values that decrease with temperature. In addition, the curvature of the Kissinger plots has been large enough to determine the E vs. T dependence [71,72]. Fitting Equation (24) to this dependence has afforded estimating the magnitude of the free energy barrier (Equation (12)) and its dependence on temperature.

7. Conclusions

An overview of the problems associated with the Kissinger method has been presented. The method is not very accurate. There are different kinds of inaccuracies associated with the assumptions made to derive its basic equations. The assumption of the first-order reaction model normally does not give rise to a large error in the activation energy when the process follows another reaction model. However, this error may become exceedingly large when the process is studied under a temperature program other than the linear heating. The assumption that the kinetics is measured on heating is embedded in the final equation of the method. This makes the Kissinger method unsuitable for any data obtained on cooling.

The method yields a single activation energy that is consistent with the assumption of single-step kinetics. This feature of the method is seen as its most essential problem because the majority of the processes it is applied to are not single-step kinetics. The problem has been illustrated by using common applications of the method. The Kissinger plots tend to be almost perfectly linear in the case of the multi-step kinetics, readily revealed by isoconversional methods. It means that the method tends to yield a single activation energy when a process is controlled by more than one energy barrier. For this reason, the Kissinger method is usually incapable of providing adequate insights into the processes kinetics. As shown by comparison, isoconversional methods provide a significantly more insightful alternative.

To finalize we should note that over the years there have been some developments of the Kissinger method. In particular, efforts have been made to obtain an analytical [73] and exact [74] solution to the Kissinger equation, to adjust it to melt crystallization [12] and reversible decompositions [27], and to exploit the nonlinearity of the Kissinger plots [10]. So far, these developments have had little impact on the mainstream usage of the method that boils down to the straightforward application of the original equation and reporting a single value of the activation energy.

Funding: This research received no external funding.

Conflicts of Interest: The author declares no conflict of interest.

References

1. Kissinger, H.E. Variation of peak temperature with heating rate in differential thermal analysis. *J. Res. Natl. Bur. Stand.* **1956**, *57*, 217–221. [CrossRef]
2. Kissinger, H.E. Reaction kinetics in differential thermal analysis. *Anal. Chem.* **1957**, *29*, 1702–1706. [CrossRef]
3. Vyazovkin, S.; Burnham, A.K.; Favergeon, L.; Koga, N.; Moukhina, E.; Pérez-Maqueda, L.A.; Sbirrazzuoli, N. ICTAC Kinetics Committee recommendations for analysis of multi-step kinetics. *Thermochim. Acta* **2020**, *689*, 178597. [CrossRef]
4. Bohun, A. Thermoemission und Photoemission von Natriumchlorid. *Czechosl. J. Phys.* **1954**, *4*, 91–93. [CrossRef]
5. Vyazovkin, S.; Burnham, A.K.; Criado, J.M.; Pérez-Maqueda, L.A.; Popescu, C.; Sbirrazzuoli, N. ICTAC Kinetics Committee recommendations for performing kinetic computations on thermal analysis data. *Thermochim. Acta* **2011**, *520*, 1–19. [CrossRef]
6. Tang, T.B.; Chaudhri, M.M. Analysis of dynamic kinetic data from solid-state reactions. *J. Therm. Anal.* **1980**, *18*, 247–261. [CrossRef]
7. Criado, J.M.; Ortega, A. Non-Isothermal Transformation Kinetics: Remarks on the Kissinger Method. *J. Non-Cryst. Solids* **1986**, *87*, 302–311. [CrossRef]
8. Muravyev, N.V.; Pivkina, A.N.; Koga, N. Critical Appraisal of Kinetic Calculation Methods Applied to Overlapping Multistep Reactions. *Molecules* **2019**, *24*, 2298. [CrossRef]
9. Vyazovkin, S. *Isoconversional Kinetics of Thermally Stimulated Processes*; Springer: Heidelberg, Germany, 2015.
10. Vyazovkin, S. Modern isoconversional kinetics: From Misconceptions to Advances. In *The Handbook of Thermal Analysis & Calorimetry, Vol.6: Recent Advances, Techniques and Applications*, 2nd ed.; Vyazovkin, S., Koga, N., Schick, C., Eds.; Elsevier: Amsterdam, The Netherlands, 2018; pp. 131–172.
11. Akahira, T.; Sunose, T. Method of determining activation deterioration constant of electrical insulating materials. *Res. Rep. Chiba Inst. Technol.* **1971**, *16*, 22–31.
12. Holba, P.; Šesták, J. Imperfections of Kissinger Evaluation Method and Crystallization Kinetics. *Glass Phys. Chem.* **2014**, *40*, 486–495. [CrossRef]
13. Borchardt, H.J.; Daniels, F. The application of differential thermal analysis to the study of reaction kinetics. *J. Am. Chem. Soc.* **1957**, *79*, 41–46. [CrossRef]
14. Hohne, G.W.H.; Hemminger, W.F.; Flammersheim, H.J. *Differential Scanning Calorimetry*, 2nd ed.; Springer: Berlin, Germany, 2003.

15. Vyazovkin, S.; Chrissafis, K.; Di Lorenzo, M.L.; Koga, N.; Pijolat, M.; Roduit, B.; Sbirrazzuoli, N.; Suñol, J.J. ICTAC Kinetics Committee recommendations for collecting experimental thermal analysis data for kinetic computations. *Thermochim. Acta* **2014**, *590*, 1–23. [CrossRef]
16. Vyazovkin, S. How much is the accuracy of activation energy affected by ignoring thermal inertia? *Int. J. Chem. Kin.* **2020**, *52*, 23–28. [CrossRef]
17. Vyazovkin, S. Modification of the integral isoconversional method to account for variation in the activation energy. *J. Comput. Chem.* **2001**, *22*, 178–183. [CrossRef]
18. Liavitskaya, T.; Vyazovkin, S. Discovering the kinetics of thermal decomposition during continuous cooling. *Phys. Chem. Chem. Phys.* **2016**, *18*, 32021–32030. [CrossRef] [PubMed]
19. Chen, R.; Winer, S.A.A. Effects of Various Heating Rates on Glow Curves. *Appl. Phys.* **1970**, *41*, 5227–5232. [CrossRef]
20. Perez-Maqueda, L.A.; Criado, J.M.; Sanchez-Jimenez, P.E.; Dianez, M.J. Applications of sample-controlled thermal analysis (SCTA) to kinetic analysis and synthesis of materials. *J. Therm. Anal. Calorim.* **2015**, *120*, 45–51. [CrossRef]
21. Sanchez-Jimenez, P.E.; Criado, J.M.; Perez-Maqueda, L.A. Kissinger Kinetic Analysis of Data Obtained under Different Heating Schedules. *J. Therm. Anal. Calorim.* **2008**, *94*, 427–432. [CrossRef]
22. Vyazovkin, S. Is the Kissinger Equation Applicable to the Processes that Occur on Cooling? *Macromol. Rapid Commun.* **2002**, *23*, 771–775. [CrossRef]
23. Zhang, Z.; Chen, J.; Liu, H.; Xiao, C. Applicability of Kissinger model in nonisothermal crystallization assessed using a computer simulation method. *J. Therm. Anal. Calorim.* **2014**, *117*, 783–787. [CrossRef]
24. Vyazovkin, S. A time to search: Finding the meaning of variable activation energy. *Phys. Chem. Chem. Phys.* **2016**, *18*, 18643–18656. [CrossRef] [PubMed]
25. Liavitskaya, T.; Vyazovkin, S. Delving into the kinetics of reversible thermal decomposition of solids measured on heating and cooling. *J. Phys. Chem. C* **2017**, *121*, 15392–15401. [CrossRef]
26. Vyazovkin, S. Kinetic effects of pressure on decomposition of solids. *Int. Rev. Phys. Chem.* **2020**, *39*, 35–66. [CrossRef]
27. Agresti, F. An extended Kissinger equation for near equilibrium solid-gas heterogeneous transformations. *Thermochim. Acta* **2013**, *566*, 214–217. [CrossRef]
28. Vyazovkin, S.; Sbirrazzuoli, N. Isoconversional method to explore the mechanism and kinetics of multi-step epoxy cures. *Macromol. Rapid Commun.* **1999**, *20*, 387–389. [CrossRef]
29. Vyazovkin, S.; Sbirrazzuoli, N. Kinetic methods to study isothermal and nonisothermal epoxy-anhydride cure. *Macromol. Chem. Phys.* **1999**, *200*, 2294–2303. [CrossRef]
30. ASTM. *Standard Test Method for Arrhenius Kinetic Constants for Thermally Unstable Materials (ANSI/ASTM E698-79)*; ASTM: Philadelphia, PA, USA, 1979.
31. Vyazovkin, S.; Sbirrazzuoli, N. Mechanism and kinetics of epoxy-amine cure studied by differential scanning calorimetry. *Macromolecules* **1996**, *29*, 1867–1873. [CrossRef]
32. Vyazovkin, S.; Wight, C.A. Model-free and model-fitting approaches to kinetic analysis of isothermal and nonisothermal data. *Thermochim. Acta* **1999**, *340–341*, 53–68. [CrossRef]
33. Turnbull, D.; Fisher, J.C. Rate of nucleation in condensed systems. *J. Chem. Phys.* **1949**, *17*, 71–73. [CrossRef]
34. Mullin, J.W. *Crystallization*, 4th ed; Butterworth: Oxford, UK, 2004.
35. Mandelkern, L. *Crystallization of Polymers, v. 2*; Cambridge University Press: Cambridge, UK, 2004.
36. Papon, P.; Leblond, J.; Meijer, P.H.E. *The Physics of Phase Transitions*; Springer: Berlin, Germany, 1999.
37. Debenedetti, P.G. *Metastable Liquids: Concepts and Principles*; Princeton University Press: Princeton, NJ, USA, 1996.
38. Christian, J.W. *The Theory of Transformations in Metals and Alloys*; Pergamon Press: Amsterdam, The Netherlands, 2002.
39. Kondepudi, D.; Prigogine, I. *Modern Thermodynamics*; Wiley: Chichester, UK, 1998.
40. Vyazovkin, S. Activation energies and temperature dependencies of the rates of crystallization and melting of polymers. *Polymers* **2020**, *12*, 1070. [CrossRef]
41. Abu El-Oyoun, M. Evaluation of the transformation kinetics of $Ga_{7.5}Se_{92.5}$ chalcogenide glass using the theoretical method developed and isoconversional analyses. *J. Alloys Compd.* **2010**, *507*, 6–15. [CrossRef]

42. Abu El-Oyoun, M. DSC studies on the transformation kinetics of two separated crystallization peaks of Si12.5Te87.5 chalcogenide glass: An application of the theoretical method developed and isoconversional method. *Mater. Chem. Phys.* **2011**, *131*, 495–506. [CrossRef]
43. Orava, J.; Greer, A.L.; Gholipour, B.; Hewak, D.W.; Smith, C.E. Characterization of supercooled liquid $Ge_2Sb_2Te_5$, and its crystallization by ultrafast-heating calorimetry. *Nat. Mater.* **2012**, *11*, 279–283. [CrossRef] [PubMed]
44. Tool, A.Q. Relation between inelastic deformability and thermal expansion of glass in its annealing range. *J. Am. Ceram. Soc.* **1946**, *29*, 240–253. [CrossRef]
45. Narayanaswamy, O.S. A model of structural relaxation in glass. *J. Am. Ceram. Soc.* **1971**, *54*, 491–498. [CrossRef]
46. Moynihan, C.T.; Easteal, A.J.; De Bolt, M.A.; Tucker, J. Dependence of the fictive temperature of glass on cooling rate. *J. Am. Ceram. Soc.* **1976**, *59*, 12–16. [CrossRef]
47. Plazek, D.J. Bingham Medal Address: Oh, thermorheological simplicity, wherefore art thou? *J. Rheol.* **1996**, *40*, 987–1014. [CrossRef]
48. Williams, M.L.; Landel, R.F.; Ferry, J.D. The temperature dependence of relaxation mechanisms in amorphous polymers and other glass-forming liquids. *J. Am. Chem. Soc.* **1955**, *77*, 3701–3707. [CrossRef]
49. Matsuoka, S. *Relaxation Phenomena in Polymers*; Hanser: Munich, Germany, 1992.
50. Donth, E. *The Glass Transition*; Springer: Berlin, Germany, 2001.
51. Moynihan, C.T.; Easteal, A.J.; Wilder, J.; Tucker, J. Dependence of the glass transition temperature on heating and cooling rate. *J. Phys. Chem.* **1974**, *78*, 2673–2677. [CrossRef]
52. Svoboda, R.; Cicmanec, P.; Malek, J. Kissinger equation versus glass transition phenomenology. *J. Therm. Anal. Calorim.* **2013**, *114*, 285–293. [CrossRef]
53. Moynihan, C.T.; Lee, S.-K.; Tatsumisago, M.; Minami, T. Estimation of activation energies for structural relaxation and viscous flow from DTA and DSC experiments. *Thermochim. Acta* **1996**, *280*, 153–162. [CrossRef]
54. Crichton, S.N.; Moynihan, C.T. Dependence of the glass transition temperature on heating rate. *J. Non-Cryst. Solids* **1988**, *99*, 413–417. [CrossRef]
55. Svoboda, R.; Malek, J. Glass transition in polymers: (In)correct determination of activation energy. *Polymer* **2013**, *54*, 1504–1511. [CrossRef]
56. Svoboda, R. Novel equation to determine activation energy of enthalpy relaxation. *J. Therm. Anal. Calorim.* **2015**, *121*, 895–899. [CrossRef]
57. Adam, G.; Gibbs., J.H. On the temperature dependence of cooperative relaxation properties in glass-forming liquids. *J. Chem. Phys.* **1965**, *43*, 139–146. [CrossRef]
58. Angell, C.A. Relaxation in liquids, polymers and plastic crystals—Strong/fragile patterns and problems. *J. Non-Cryst. Solids* **1991**, *131*, 13–31. [CrossRef]
59. Vyazovkin, S.; Dranca, I. A DSC Study of α- and β-relaxations in a PS-clay system. *J. Phys. Chem. B* **2004**, *108*, 11981–11987. [CrossRef]
60. Vyazovkin, S.; Sbirrazzuoli, N.; Dranca, I. Variation of the effective activation energy throughout the glass transition. *Macromol. Rapid Commun.* **2004**, *25*, 1708–1713. [CrossRef]
61. Vyazovkin, S.; Sbirrazzuoli, N.; Dranca, I. Variation in activation energy of the glass transition for polymers of different dynamic fragility. *Macromol. Chem. Phys.* **2006**, *207*, 1126–1130. [CrossRef]
62. Illers, K.-H. Die Ermittlung des Schmelzpunktes von Kristallen Polymeren Mittels Warmeflusskalorimetrie (DSC). *Eur. Pol. J.* **1974**, *10*, 911–916. [CrossRef]
63. Toda, A. Heating rate dependence of melting peak temperature examined by DSC of heat flux type. *J. Therm. Anal. Calorim.* **2016**, *123*, 1795–1808. [CrossRef]
64. Tammann, A. Zur Uberhitzung von Kristallen. *Z. Phys. Chem.* **1910**, *68*, 257–269. [CrossRef]
65. Hellmuth, E.; Wunderlich, B. Superheating of linear high-polymer polyethylene crystals. *J. Appl. Phys.* **1965**, *36*, 3039–3044. [CrossRef]
66. Vyazovkin, S. Power law and Arrhenius approaches to the melting kinetics of superheated crystals: Are they compatible? *Cryst. Growth Des.* **2018**, *18*, 6389–6392. [CrossRef]
67. Liavitskaya, T.; Birx, L.; Vyazovkin, S. Melting Kinetics of Superheated Crystals of Glucose and Fructose. *Phys. Chem. Chem. Phys.* **2017**, *19*, 26056–26064. [CrossRef]
68. Vyazovkin, S.; Yancey, B.; Walker, K. Nucleation driven kinetics of poly(ethylene terephthalate) melting. *Macromol. Chem. Phys.* **2013**, *214*, 2562–2566. [CrossRef]

69. Vyazovkin, S.; Yancey, B.; Walker, K. Polymer melting kinetics appears to be driven by heterogeneous nucleation. *Macromol. Chem. Phys.* **2014**, *215*, 205–209. [CrossRef]
70. Chen, K.; Baker, A.N.; Vyazovkin, S. Concentration effect on temperature dependence of gelation rate in aqueous solutions of methylcellulose. *Macromol. Chem. Phys.* **2009**, *210*, 211–216. [CrossRef]
71. Farasat, R.; Vyazovkin, S. Coil-to-globule transition of poly(N-isopropylacrylamide) in aqueous solution: Kinetics in bulk and nanopores. *Macromol. Chem. Phys.* **2014**, *215*, 2112–2118. [CrossRef]
72. Farasat, R.; Vyazovkin, S. Nanoconfined solid-solid transitions: Attempt to separate the size and surface effects. *J. Phys. Chem. C* **2015**, *119*, 9627–9636. [CrossRef]
73. Roura, P.; Farjas, J. Analytical solution for the Kissinger equation. *J. Mater. Res.* **2009**, *24*, 3095–3098. [CrossRef]
74. Farjas, J.; Roura, P. Exact analytical solution for the Kissinger equation: Determination of the peak temperature and general properties of thermally activated transformations. *Thermochim. Acta* **2014**, *598*, 51–58. [CrossRef]

 © 2020 by the author. Licensee MDPI, Basel, Switzerland. This article is an open access article distributed under the terms and conditions of the Creative Commons Attribution (CC BY) license (http://creativecommons.org/licenses/by/4.0/).

Review

Nanomaterials in Electrochemical Sensing Area: Applications and Challenges in Food Analysis

Antonella Curulli

Istituto per lo Studio dei Materiali Nanostrutturati (ISMN) CNR, Via del Castro Laurenziano 7, 00161 Roma, Italy; antonella.curulli@cnr.it; Tel.: +39-06-4976-7643

Academic Editor: Giuseppe Cirillo
Received: 11 November 2020; Accepted: 4 December 2020; Published: 7 December 2020

Abstract: Recently, nanomaterials have received increasing attention due to their unique physical and chemical properties, which make them of considerable interest for applications in many fields, such as biotechnology, optics, electronics, and catalysis. The development of nanomaterials has proven fundamental for the development of smart electrochemical sensors to be used in different application fields such, as biomedical, environmental, and food analysis. In fact, they showed high performances in terms of sensitivity and selectivity. In this report, we present a survey of the application of different nanomaterials and nanocomposites with tailored morphological properties as sensing platforms for food analysis. Particular attention has been devoted to the sensors developed with nanomaterials such as carbon-based nanomaterials, metallic nanomaterials, and related nanocomposites. Finally, several examples of sensors for the detection of some analytes present in food and beverages, such as some hydroxycinnamic acids (caffeic acid, chlorogenic acid, and rosmarinic acid), caffeine (CAF), ascorbic acid (AA), and nitrite are reported and evidenced.

Keywords: nanomaterials; electrochemical sensors; hydroxycinnamic acids; caffeine; nitrite

1. Introduction

The introduction of novel functional nanomaterials and analytical technologies indicate the possibility for advanced electrochemical (bio)sensor platforms/devices for a wide number of applications, including biological, biotechnological, clinical and medical diagnostics, environmental and health monitoring, and food industries.

Nanoscale materials and nanomaterials are known as materials where any measurement is not as much as 100 nm. Nanomaterials reveal exciting properties that make them appeal to be exploited in electrochemistry and in the improvement of the (bio)sensors. Recent advances in nanotechnology have created a growing demand for their possible commercial application [1].

Nanotechnology involves the synthesis and characterization of nanomaterials, whereby nanomaterial can be defined as a natural or synthesized material containing particles, in an unbound state or as an aggregate or as an agglomerate, and where, for 50% or more of the particles in the number size distribution, one or more external dimensions is in the size range of 1–100 nm [2,3].

By reducing the material dimensions at the nanometre level, the chemical and physical properties of such a material can be modified and they are totally different with respect to the same corresponding bulk material [4–6].

Carbon nanotubes (CNTs) and gold nanoparticles (AuNPs) are among the most broadly explored nanomaterials because of their exceptional properties, which can be connected in different applications, e.g., detecting, and imaging. Yet, to date, the exploration field of advancement for the synthesis of new functionalized AuNPs and CNTs for sensing applications is a dynamic research territory. The combination of these nanomaterials has been developed, promoting improvements in controlling their size and shape [7,8].

Electroanalytical methods and electrochemical sensors have improved the analytical approach in different application fields, ranging from the biomedical to the environmental ones [9].

Particularly the modification and/or functionalization of the electrodic surface with nanomaterials involves an amplification of the corresponding electrochemical signal and it has proven very attractive for developing sensors with high sensitivity and selectivity [9].

In this review, we present a survey of the applications of different nanomaterials and nanocomposites as electrochemical sensing platforms for food analysis. The sensors analytical parameters, such as linearity range, detection limit, selectivity, and their possible applications to real samples are described and highlighted.

2. Electrochemical Techniques

Electrochemistry offers a wide range of electroanalytical techniques. A typical electrochemical experiment includes a working electrode made of a solid conductive material, such as platinum, gold, or carbon, a reference electrode, and a counter electrode, all the electrodes are generally immersed in a solution with a supporting electrolyte to guarantee the conductivity in the solution [10].

Electrochemical sensors belong to the largest family of chemical sensors. A chemical sensor can be defined as "a small device that, as the result of a chemical interaction or process between the analyte and the sensor device, transforms chemical or biochemical information of a quantitative or qualitative type into an analytically useful signal" [11]. This definition can be extended to the electrochemical ones modifying it in this way: a small device that, as the result of an electrochemical interaction or process between the analyte and the sensing device, transforms electrochemical information of a quantitative or qualitative type into an analytically useful signal [11]. The use of a nanomaterial together with the analyte kind and nature have proven crucial for the sensor sensitivity, selectivity, and stability [12]. As for all the chemical sensors, the critical parameters of electrochemical sensors are sensitivity, detection limit, dynamic range, selectivity, linearity, response time, and stability [13].

Several electrochemical methods have been employed for the detection of food additives, biological contaminants, and heavy metals [9].

In general, an electrochemical reaction can generate different measurable data, depending on the electrochemical technique adopted. In fact, a measurable current can be generated, and in this case, the corresponding electrochemical techniques are the amperometric ones. Alternatively, a potential can be measured and/or controlled, and in this case, the corresponding electrochemical techniques are the potentiometric ones. Finally, the electrochemical techniques, involving measurements of impedance at the electrode/solution interface are included in the electrochemical impedance spectroscopy (EIS) method [14].

Starting from the presentation of EIS, we propose a brief presentation and overview of the best known and used electrochemical techniques.

EIS is an electroanalytical method used for the evaluation of electron-transfer properties of the modified surfaces and in understanding of surface chemical transformations. EIS analysis provides mechanistic and kinetic information on a wide range of materials, such as batteries, fuel cells, corrosion inhibitors, etc. [14].

The overall electrochemical behaviour of an electrode can be represented by an equivalent circuit comprising resistance, inductance, and capacitance. The equivalent circuit elements, useful for analyte detection are resistance to the charge transfer R_{ct} and the double layer capacitance C_{dl}. The measured capacitance usually arises from the series combination of several elements, such as analyte binding (C_{anal}) to a sensing layer (C_{sens}) on the electrode (C_{el}). The sensitivity is then determined by the relative capacitance of the analyte layer and the sensing layer. One difficulty with capacitive sensors is that their sensitivity depends on obtaining the proper thickness of the original sensing layer.

Voltammetry belongs to the class of the amperometric techniques because the current produced from an electrochemical reaction is measured whilst varying the potential window. Since there are many ways to vary the potential, we can consider many voltammetric techniques. Among others, the most common and employed are the following: cyclic voltammetry (CV), linear sweep voltammetry (LSV), differential pulse voltammetry (DPV), and square wave voltammetry (SWV) [15–18].

CV and LSV are widely employed voltammetric techniques to study the electrochemical behaviour of an electroactive molecule.

DPV and SWV can be classified as pulse voltammetric techniques.

DPV and SWV in comparison with CV can be used to study the redox properties of extremely small amounts of electroactive compounds for several reasons, but principally: (1) in these measurements, the effect of the charging current can be minimized, so higher sensitivity is achieved and (2) only faradaic current is extracted, so electrode reactions can be analyzed more precisely.

Chronoamperometry (CA) is a potentiostatic technique, where the current is recorded as a time function. In Figure 1, an overview of the electrochemical methods of analysis, namely voltammetry, amperometry, electrochemical impedance spectroscopy (EIS), and potentiometry, is reported.

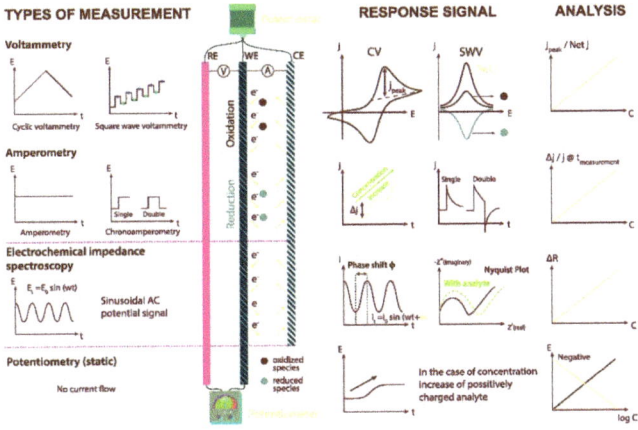

Figure 1. Overview of electrochemical methods of analysis: voltammetry, amperometry, electrochemical impedance spectroscopy (EIS), and potentiometry [19].

All the above-mentioned techniques have been widely employed in the development of electrochemical sensors for different application fields.

3. Nanomaterials, Nanotubes, Nanoparticles, and Nanocomposites

Recent developments in nanomaterial synthesis have allowed the development of advanced sensing systems [9,19].

Generally, we have considered modifications of the working electrode with different nanomaterials, ranging from the classical nanotubes to nanocomposites, among different nanostructures, such as graphene, metal nanoparticles, and/or nanostructured polymers. In addition, non-conventional sensing platforms, such as paper and/or screen-printed electrodes (SPE), also modified with different nanomaterials and/or nanostructures, have been considered. Hence, in this review, we report some example of these non-conventional sensing platforms, present in the literature and employed for different application fields [20,21].

Nanomaterials play a crucial role in the development of electrochemical sensors, improving the sensor stability, sensitivity, and selectivity in the presence of the common interferences.

In the following subparagraphs we introduce and describe nanomaterials and examples of the related electrochemical sensors just to show their applicability in the electrochemical sensing area.

3.1. Carbon-Based Nanomaterials

Carbon-based nanomaterials (single-walled carbon nanotubes (SWNTs), multi-walled carbon nanotubes (MWNTs), single-walled carbon nanohorns (SWCNHs), buckypaper, graphene, fullerenes (e.g., C_{60}), etc. present very interesting properties, such as high surface-to-volume ratio, high electrical conductivity, chemical stability/durability, and strong mechanical strength, and for these reasons they have found a large applicability in the sensing area [22–28].

Carbon nanotubes (CNTs) present several properties associated to their structure, functionality, morphology, and flexibility to be employed in synthesis of hybrid or composite materials due to their hollow cylindrical structure.

Carbon nanotubes can be classified as single-walled nanotubes (SWNTs), double-walled nanotubes (DWNTs), and multi-walled nanotubes (MWNTs) depending on the number of graphite layers. Functionalized CNTs have been used in several application fields. The chemical functionalities can easily be designed and tuned through the tubular structure modification.

Some interesting examples of carbon nanomaterials based electrochemical sensors, related to different application fields not only to food analysis, are illustrated below.

Venton and co-workers have used metal microelectrodes modified with CNTs for assembling an electrochemical sensor for detecting dopamine in vitro and in vivo. [29]. It has been found that CNTs-coated niobium (CNTs-Nb) microelectrode showed a low detection limit of 11 nM for dopamine. The CNTs-Nb sensor was also employed to detect stimulated dopamine release in anesthetized rats and showed high sensitivity for in vivo measurements.

The design and synthesis of functionalized CNTs for biological and biomedical applications are highly attractive because in vivo sensing requires high selectivity, accuracy, and long-term stability. Zhang et al. have prepared an electrochemical ascorbic acid sensor for measuring ascorbic acid in brain using aligned carbon nanotube fibers (CNF) as a microsensor [30], obtaining very interesting results. The sensor measured ascorbic acid concentration of 259.0 µM in the cortex, 264.0 µM in the striatum, and 261.0 µM in the hippocampus, respectively, under normal conditions.

Graphene is one of most applied nanomaterial in the sensing area. Different graphene-based materials have been produced (e.g., electrochemically and chemically modified graphene) using many procedures [31]. Graphene shows properties such as high conductivity, accelerating electron transfer, and a large surface area, very similar indeed to the corresponding properties of CNTS, so it is considered a good candidate for assembling sensors to determine several target molecules [9,31].

Graphene oxide (GO) is hydrophilic and can be dispersed in water solution because of hydrophilic functional groups (OH, COOH and epoxides) at the edge of the sheet and on the basal plane.

On the other hand, GO has a low conductivity in comparison to graphene, so reduced GO (rGO) is more employed as electrode modifier in electrochemical sensing/biosensing area [31].

Fluorine doped graphene oxide was used to prepare an electrochemical sensor for the detection of heavy metal ions such as Cd^{2+}, Pb^{2+}, Cu^{2+}, and Hg^{2+}. Square wave anodic stripping voltammetry was employed for the detection of the heavy metal ions. The authors have evidenced that the presence of fluorine builds a more appropriate platform for the stripping process, from the comparison between the sensor based on GO and the sensor based on F-GO [9].

Li and co-workers developed a based electrochemical sensor for the detection of metal ions, Pb^{2+} and Cd^{2+}, employing a Nafion–graphene composite film. The synergistic effect of graphene nanosheets and Nafion gave rise to a better sensitivity for detecting metal ion and enhanced the electrochemical sensor selectivity [32].

Li and co-workers have reported a graphene potassium doped modified glassy carbon electrode, for the determination of sulphite in water solution. A linear response in the concentration range of 2.5 µM–10.3 mM with a detection limit of 1.0 µM for SO_3^{2-} has been obtained. The graphene electronic properties resulted modified by the K doping [33].

A glassy carbon electrode was modified by means of hexadecyl trimethyl ammonium bromide (CTAB) functionalized GO/multiwalled carbon nanotubes (MWCNTs) for the detection of ascorbic acid (AA) and nitrite. The combination of GO and MWCNTs provided a sensing platform with high electrocatalytic activity, increased surface area, as well as good stability and sensitivity, allowing for the determination of nitrite and AA at the same time [34].

3.2. Gold Based Nanomaterials

Since the first examples of gold nanoparticles (AuNPs) synthesis, AuNPs have been employed for the assembly of different sensors. From an electroanalytical point of view, Au nanoparticles and in general gold nanomaterials were employed in the electrochemical sensing area because of their high conductivity, their compatibility, and a high surface to volume ratio [35,36]. Gold nanomaterials have been used for the selective oxidation processes, or rarely, for the reduction ones.

Important improvements have been performed in the Au nanoparticles and nanomaterials synthesis for electrochemical sensing. However, researchers are dealt with several challenges/criticalities, such as size control, morphology, and suitable dispersion and/or stabilizing agents and/or media. The recent development of nanoporous Au materials seems to be fundamental in overcoming these challenges and hierarchical gold nanoporous materials have been recently reported for biomedical applications [37]. Some interesting examples of gold nanomaterials based electrochemical sensors, related to different application fields not only to food analysis, are illustrated below.

An electrochemical sensor using a gold microwire electrode for the detection of heavy metals ions, such as copper and mercury, in seawater was described by Salaün and coworkers. The sensor using gold microwires was able to detect the two metals ions at the same time by means of anodic stripping voltammetry [38].

Concerning the same analytical issue, i.e., the detection of different heavy metals ions at the same time, Soares and co-workers developed an electrochemical sensor using a vibrating gold microwire, and determined arsenic, copper, mercury, and lead ions in freshwater by means of stripping voltammetry [39].

Recently, gold nanopores were synthesized by the alloying/dealloying method for increasing the electrochemical performance of an analyte under investigation and several examples are reported below.

Ding and co-workers prepared a nanoporous gold leaf by dealloying the Au/Ag alloy in nitric acid. The resulting 3D gold nanostructure allowed the small molecules movement inside. For example, it was employed to modify a GCE for detecting nitrite. The increased surface area and the high conductivity of the 3D nanostructured gold network justify the good performance of the sensor [40].

Other examples concerning the use of nanoporous gold have involved more properly the electrochemical investigation of different analytes with this new nanomaterial as proof of concept to apply it in the sensing area.

A green synthesis of nanoporous gold materials was proposed by Jia and co-workers by means of a cyclic alloying/de-alloying procedure. The resulting nanoporous gold film modified electrodes showed very interesting electrochemical performances in terms of a high surface area and good selectivity [41].

Lin et al. modified a GCE via the electrodeposition of Au nanoparticles on polypyrrole (PPy) nanowires. AuNPs enhanced the conductivity of the polymer nanowires and consequently the electron transfer rate resulted higher than that at bare GCE and/or at GCE just modified with the polymer nanowires [42].

Finally, a nanoporous Au 3D nanostructure was synthesized as a proof of concept to be applied as a sensing platform for detecting hydrazine, sulphite, and nitrite, present in the same sample. The nanostructured sensor showed good performances in terms of selectivity and sensitivity [43].

3.3. Hybrid Nanocomposites

To improve and amplify the performances of a sensor and/or a sensing platform, nanomaterials such as carbon and/or metal nanomaterials were incorporated in different polymers both natural (e.g., chitosan) or (electro)synthesized (e.g., PEDOT, polypyrrole). Some interesting examples of hybrid nanocomposite based electrochemical sensors, related to different application fields not only to food analysis, are illustrated below [44].

As a first example, we can introduce an electrochemical sensor using a polypyrrole–chitosan–titanium dioxide (PPy–CS–TiO$_2$) nanocomposite for glucose detection. Interactions between the TiO$_2$ nanoparticles and PPY enhanced the sensor properties in terms of sensitivity and selectivity [45].

Bimetallic Au–Pt nanoparticles have been incorporated in rGO and the electrochemical behaviour of resulting nanocomposite was investigated and compared with the electrochemical behaviour of the bimetallic nanoparticles and of the rGO. An enhanced electrochemical activity is observed, probably due to the increase of the surface area and to the increase of the nanocomposite conductivity respect to the bimetallic nanoparticles and to the rGO [46].

Feng et al. assembled a sensor for caffeic acid detection through a nanocomposite obtained by the combination of worm-like Au–Pd nanostructures and rGO [47]. From the spectroscopic characterization, the nanotubular worm-like Au–Pd nanostructures were uniformly distributed on rGO.

A carbon paste electrode was modified with a nanocomposite obtained by combining AuNPs and MWCNTs. The oxidation of nitrite was investigated at such a modified electrode and showed better performances in terms of electrocatalytic behaviour respect to those at the bare carbon paste electrode and/or to those at CPE modified with only AuNPs or with only MWCNTs [48].

A hybrid nanocomposite was prepared by assembling Pt nanoparticles on a graphene surface. The modified electrode was used for the detection of ascorbic acid (AA), uric acid (UA), and dopamine (DA), obtaining interesting results in terms of selectivity [49].

Chen and co-workers proposed a Pt nanocomposite combining Pt nanoparticles and single-walled carbon nanotubes (SWCNTs), instead of graphene or MWCNTs.

A GCE modified with this nanocomposite was used for the electrochemical detection of α-methylglyoxal. A good linearity in the concentration range of 0.1–100.0 μM, and a detection limit of 2.80 nM were obtained. The sensor was applied to detect α-methylglyoxal in real samples of wine and beer [50].

Yegnaraman and co-workers reported an Au based nanocomposite for the detection of AA, UA, and DA to test the selectivity for detecting analytes present in the same solution. The nanocomposite film was synthesized by introducing Au nanoparticles into the PEDOT polymer matrix. The modified GCE determined AA, UA, and DA simultaneously, with improved sensitivity and selectivity [51].

A glucose impedimetric biosensor [52] was assembled using a metal composite composed by a gold microtubes (AuμTs) architecture and polypyrrole overoxidized by Curulli and co-workers. A platinum (Pt) electrode was coated by gold microtubes, synthesized via electroless deposition within the pores of polycarbonate particle track-etched membranes (PTM). This platform was successfully used to deposit polypyrrole overoxidized film (OPPy) and to verify the possibility of developing a biosensor using OPPy, the characteristics of the H$_2$O$_2$ charge transfer reaction were studied before the enzyme immobilization. This composite material seems to be suitable in devices as biosensors based on oxidase enzymes, just because hydrogen peroxide is a side-product of the catalysis and could be directly related to the concentration of the analyte. Finally, a biosensor consisting of a Pt electrode modified with AuμTs, OPPy, and glucose oxidase was assembled to determine the glucose. The most important result of this biosensor was the wide linear range of concentration, ranging from 1.0 to 100 mM (18–1800 mg·dL^{-1}), covering the hypo- and hyperglycaemia range, useful in diabetes diagnosis, with limit of detection of 0.1 mM (1.8 mg·dL^{-1}) and limit of quantification of 1.0 mM (18 mg·dL^{-1}).

A wide range of glucose biosensor prototypes have been developed during these years, but the challenge is to obtain a biosensor capable of measuring the glucose concentration in the normal range (i.e., representative of healthy patients) and in the range of hyperglycaemia values (especially for diabetes patients, for clinical diagnosis of this widespread pathology). Very few examples propose biosensors with improved analytical performances, especially in terms of an extended linearity (still 50 mM ≈ 901 mg·dL^{-1}, useful for the hyperglycaemia pathology values), high selectivity toward the most common interferents and an improved stability [53–55].

4. Electrochemical Sensors for Food Analysis: Some Examples

In this section, several examples of the electrochemical sensors employing different nanomaterials and applied to detect different analytical targets such as some hydroxycinnamic acids (caffeic acid, chlorogenic acid, and rosmarinic acid), ascorbic acid, and nitrite were illustrated and discussed.

4.1. Phenolic Antioxidants

Natural antioxidants are species of great interest in many areas, as food chemistry, health care and clinical applications. They have beneficial effects on human health and could play an important role in the prevention and treatment of many pathologies (such as cardiovascular disorders, cancer, etc.) and to protect from oxidative stress [56–61]. The classification of antioxidants is commonly carried out based on the chemical structure, determining their reactivity. However, their antioxidant action is also strictly related to the redox properties and consequently their knowledge is very crucial for a better understanding of antioxidant mechanisms.

Among the natural antioxidants found in fruits and plants, hydroxycinnamic acids (HAs) are very important and present in all parts of the fruit and/or plant [62–64]. Undoubtedly, these compounds in food provide added value for their well-known health benefits, for their technological role, and marketing. The electrochemical methods have been extensively used to investigate the redox properties of various species and as analytical tool for the determination of redox target molecules. At present, as for other classes of antioxidants, the analysis of HAs and phenolic antioxidants is usually carried out using chromatographic techniques, which require sophisticated equipment and laborious analytical procedures [57]. The use of electrochemical methods for analytical purposes is receiving increasing interest [57], since they are fast, accurate, sensitive and can be used for the analysis of different and complex matrices with a low cost.

Both electrochemical sensors and biosensors are widely used for the determination of HAs and phenolic antioxidants. However, the electrochemical responses have been studied only from an analytical point of view, whereas the relationship between the antioxidant chemical structure and electrochemical behaviour has been neglected [65]. The understanding of key factors that affect the electrochemical response of analytes could promote the design of highly efficient sensors. To this aim, the role of the chemical structural features of antioxidants and nature of the electrode surface must be evaluated.

Recently, electrodes modified with nanocomposite films were successfully used for the analysis of antioxidants, such as caffeic acid, in complex matrices [66,67]. These films consist of gold nanoparticles (AuNPs) embedded into chitosan, a biodegradable and biocompatible polymer containing many hydroxyl and amino groups that can interact with the analytes. Later, the electrochemical response of structurally related antioxidants on electrodes modified with different gold–chitosan nanocomposite films was investigated and discussed [65,66].

To evaluate how the chemical structural features of analytes affect the electrochemical behaviour, different types of antioxidants, such as catechols, hydroxycinnamic acids, and flavonoids, structurally correlated and bearing different functional groups and steric hindrances, have been studied. This investigation [66] has demonstrated that the electrochemical response for structurally related antioxidants at AuNPs–chitosan modified electrodes depends on several parameters. The chemical structural features of the analytes affect the interaction with the electrode surface. However, their electrochemical behaviour cannot be explained only on these bases.

The nanostructure and surface functional groups of AuNPs-chitosan modified electrodes have also a key role. In particular, the formation of a collaborative network with interconnected metal nanoparticles in chitosan film significantly affects the electron transfer properties, whereas the surface functional groups can promote the interaction with the antioxidants. An overview of the behaviour of catechols, hydroxy cinnamic acids, and flavonoids derivatives at different AuNPs–chitosan modified electrodes has been illustrated by Curulli and co-workers [66].

A better response was observed for molecules with two hydroxyl groups in ortho position of the catechol ring, with a peculiar molecular symmetry (i.e., rosmarinic acid) and with a low steric hindrance, in particular for caffeic acid, chlorogenic acid, and rosmarinic acid. Moreover, the interaction with the antioxidants is also affected by the functional groups at modified electrode surface and the size and distribution of AuNPs into the polymeric matrix. The understanding of the parameters affecting the electrochemical behaviour of analytes at modified electrodes is a key issue and could significantly promote the design of highly efficient sensors.

4.1.1. Caffeic Acid

Caffeic acid (CA, 3,4-dihydroxycinnamic acid, chemical structure in Figure 2) is well known as a phenolic antioxidant. CA is present in wines, coffee, olive oil, as well as in some vegetables and fruits. It has several pharmacological functions and properties [62,68].

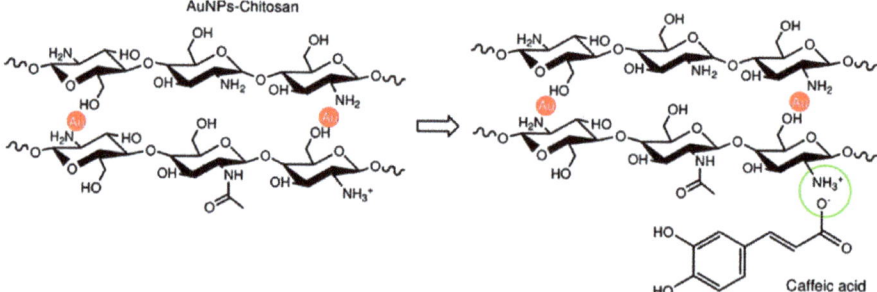

Figure 2. Schematic representation of the interaction between AuNPs/chitosan nanocomposite and caffeic acid. Reprinted with permission from [66] Copyright 2012, American Chemical Society.

According to the literature [66], several analytical methods have been employed for the detection of caffeic acid, from liquid chromatography to the electrochemical ones. Considering the electrochemical approach, involving nanomaterials, several examples are reported in Table 1.

Gold–chitosan nanocomposite is successfully proposed for assembling a sensitive and selective electrochemical sensor for the determination of an antioxidant such as caffeic acid by Curulli's group [66,67]. Taking advantage of the peculiar sensing performance of the nanocomposite, an analytical method based on differential pulse voltammetry for the determination of the polyphenol index in wines was proposed.

In this approach, colloidal gold nanoparticles (AuNPs) stabilized into a chitosan matrix were prepared using a green route. The synthesis was carried out by reducing AuIII to Au0 in an aqueous solution of chitosan and different organic acids. It has been demonstrated that by varying the nature of the acid it is possible to tune the reduction rate of the gold precursor ($HAuCl_4$) and to modify the morphology of the resulting metal nanoparticles. The use of chitosan enables the simultaneous synthesis and surface modification of AuNPs in one pot. Because of the excellent film-forming capability of this polymer, AuNPs–chitosan solutions were used to obtain hybrid nanocomposite films that combine highly conductive AuNPs with many organic functional groups. Figure 2, the proposed scheme for the interaction between the AuNPs-chitosan nanocomposite and caffeic acid is illustrated.

Table 1. An overview of recent electrochemical sensors for CA determination, using nanomaterials and/or nanocomposites.

Electrochemical Methods	Electrode Material	Linearity Range (mol·L^{-1})	LOD (mol·L^{-1})	Application	Reference
DPV	AuNps/Chitosan/AuE	5.00×10^{-8}–2.00×10^{-3}	2.50×10^{-8}	Red and white wines	[66,67]
SWS	Nafion/ER-GO/GCE	1.0×10^{-7}–1.0×10^{-6}	9.1×10^{-8}	White wines	[69]
DPV	MIS/AuE	5.00×10^{-7}–6.00×10^{-5}	1.50×10^{-7}	White wines	[70]
DPV	RGO@PDA/GCE	5.0×10^{-9}–4.55×10^{-4}	1.20×10^{-9}	Wines	[71]
DPV	Au–PEDOT/rGO/GCE	1.00×10^{-8}–4.60×10^{-5}	4.00×10^{-9}	Red wines	[72]
DPV	PdAu/PEDOT/rGO/GCE	1.90×10^{-9}–5.50×10^{-5}	3.70×10^{-10}	Red wines	[73]
Amperometry	SrV$_2$O$_6$/GCE	1.00×10^{-8}–2.07×10^{-4}	4.00×10^{-9}	No real samples	[74]
DPV	Au/PdNPs/GRF/GCE	3.00×10^{-8}–9.40×10^{-4}	6.00×10^{-9}	Fortified wines	[75]
DPV	Au@α-Fe$_2$O$_3$/RGO/GCE	1.90×10^{-5}–1.87×10^{-3}	9.80×10^{-8}	Coffee samples	[76]
DPV	PEDOT/rGO/PtE	5.0×10^{-9}–5.0×10^{-5}	2.0×10^{-9}	Teas	[77]
DPV	PtCu trifurcate nanocrystal/GCE	1.20×10^{-6}–1.90×10^{-3}	3.50×10^{-7}	Red wines	[78]
Amperometry	Cu$_2$S NDs@GOS NC/SPCE	5.50×10^{-8}–2.50×10^{-3}	2.20×10^{-10}	Soft drinks and red wines	[79]
DPV	PEDOT/GCE	thin film 1.50×10^{-7}–4.00×10^{-6} thick film 1.50×10^{-6}–4.75×10^{-5}		No real samples	[80]
DPV	MWCNTs-Bi/CTABCPE	6.0×10^{-8}–5.0×10^{-4}	1.91×10^{-9}	Coconut water, teas, and fruit juices	[81]
DPV	F-GO/GCE	5.00×10^{-7}–1.00×10^{-4}	1.80×10^{-8}	Red wines	[82]

Au-chitosan nanocomposites are successfully proposed as sensitive and selective electrochemical sensors for the determination of caffeic acid, by means of Differential Pulse Voltammetry. A linear response was obtained over a wide range of concentration from 5.00×10^{-8} M to 2.00×10^{-3} M, and the limit of detection was 2.50×10^{-8} M. Moreover, further analyses have demonstrated that a high selectivity toward caffeic acid can be achieved without interference from catechin or ascorbic acid (flavonoid and nonphenolic antioxidants, respectively). The quantification of the caffeic in red and white wines was also accomplished by standard addition method. The only sample treatment required in all cases consisted of an appropriate dilution with the supporting electrolyte solution, and the obtained results are in accordance with amounts reported in the literature for commercial samples [66,67].

An electrochemical sensor composed of Nafion–graphene nanocomposite film for the voltametric determination of caffeic acid (CA) has been proposed by Filik and co-workers [69]. A Nafion graphene oxide-modified glassy carbon electrode was fabricated by a simple drop-casting method and then graphene oxide was electrochemically reduced over the glassy carbon electrode. The electrochemical analysis method was based on the adsorption of caffeic acid on Nafion/ERGO/GCE and then the oxidation of CA during the stripping step. The resulting electrode showed an electrocatalytical response to the oxidation of caffeic acid (CA). At optimized test conditions, the calibration curve for CA showed two linear segments: The first linear segment increased from 0.1 to 1.5 µM and second linear segment increased up to10 µM. The detection limit was determined as 9.1×10^{-8} M using SWV. Finally, the proposed method was successfully used to determine CA in white wine samples.

An electrochemical sensor based on molecularly imprinted siloxanes (MIS) film was developed for the selective determination of CA by Kubota group [70]. The MIS film was prepared by sol-gel process, using the acid catalysed hydrolysis and condensation of tetraethoxysilane (TEOS), phenyltriethoxysilane (PTEOS), and 3-aminopropyltrimethoxysilane(3-APTMS) in the presence of CA as a template molecule. The MIS film was immobilized onto Au electrode surface pre-modified with 3-mercaptopropyltrimethoxysilane (3-MPTS). Under the optimized conditions, by using differential pulse voltammetry (DPV), the sensor showed a linear current response to the target CA concentration in the range from 0.500 to 60.0 µM, with a detection limit of 0.15 µM. The film exhibited high selectivity toward the template CA, as well as good stability and repeatability for CA determinations. Furthermore, the proposed sensor was applied to determine CA in wines samples and the results agreed with those obtained by a chromatographic method.

Curulli's group developed a PEDOT (poly(3,4-ethylenedioxy) thiophene) modified Pt sensor for the determination of caffeic acid (CA) in wine [83]. Cyclic voltammetry (CV) with the additions standard method was used to quantify the analyte at PEDOT modified electrodes. PEDOT films were electrodeposited on platinum electrode (Pt) in aqueous medium by galvanostatic method using sodium poly(styrene-4-sulfonate) (PSS) as electrolyte and surfactant. CV allows for detecting the analyte over a wide concentration range (10.0 nM–6.5 mM) with a detection limit of 3.0 nM. The electrochemical method proposed was applied to determine CA in white and red wines and the obtained results are in accordance with amounts reported in literature for commercial samples [65,66].

Thangavelu [71] reported a caffeic acid (CA) electrochemical sensor using reduced graphene oxide and polydopamine composite modified glassy carbon electrode (RGO@PDA/GCE). Cyclic voltammetry (CV) was used to investigate the electrochemical behaviour of different modified electrodes (bare GCE, RGO/GCE and RGO@PDA/GCE) toward oxidation of CA and the CV results showed that RGO@PDA composite has higher electrocatalytic activity to related CA oxidation than other modified electrodes. Differential pulse voltammetry was used for determination of CA and the response of CA was linear over the concentration ranging from 5.0 nM to 450.55 µM with the low detection limit of 1.2 nM. The composite modified electrode has acceptable selectivity in the presence of interfering species. The practical applicability of the composite was evaluated in wine samples and the obtained recovery of CA in wine samples confirms its potential for practical applications.

An Au-PEDOT/reduced graphene oxide nanocomposite (Au-PEDOT/rGO) modified glassy carbon electrode (GCE) showed significantly high electrocatalytic activity toward caffeic acid (CA) oxidation as compared to bare GCE and Au-PEDOT modified electrode [72]. Under optimized conditions, the Au–PEDOT/rGO constructed sensors exhibited a wide linear range of 0.01–46 µM with a detection limit as low as 0.004 µM for the detection of CA. To validate its possible application, the present sensor showed a satisfactory anti-interference performance, high reproducibility, and sensitivity for the determination of CA in the red wine sample.

Liu et al. proposed Pd-Au/PEDOT/graphene (Pd-Au/PEDOT/rGO) nanocomposites, easily prepared in an aqueous medium by a one-pot method [73]. In addition to the Pd-Au nanoparticles deposited on the PEDOT polymer structure, dendritic bimetallic Pd–Au nanoclusters were also observed in the composites. The morphology of the nanocomposite was greatly influenced by the molar ratio of the added metal salt precursors. X-ray photoelectron spectroscopy and X-ray diffraction analyses indicated the presence of synergism and electron exchange between the Pd–Au bimetallic nanoclusters. The resultant nanocomposite improved the electronalytical parameters of the obtained sensor for CA detection. Under optimized conditions, the Pd-Au/PEDOT/rGO modified glassy carbon electrode showed a linear response range of 0.001–55 µM and a detection limit of 0.37 nM. The practical application of the Pd-Au/PEDOT/rGO/GCE was tested to determine CA in red wine samples and a good recovery was obtained. The proposed method proved sufficient for practical applications without any sample pretreatment.

A well-defined one-dimensional (1D) rod-like strontium vanadate (SrV_2O_6) was prepared by simple hydrothermal method without using any other surfactants/templates [74]. The formation of rod-like SrV_2O_6 was confirmed by various analytical and spectroscopic techniques. The as-prepared rod-like SrV_2O_6 was employed as material for assembling an electrochemical sensor for the detection of caffeic acid (CA) as well as visible light active photocatalyst for the degradation of metronidazole (MNZ) antibiotic drug. As an electrochemical sensor, the SrV_2O_6 modified glassy carbon electrode (GCE) showed an electrocatalytic activity for the detection of CA by chronoamperometry (CA) and cyclic voltammetry (CV). In addition, the electrochemical sensor showed good selectivity, a linear response range of 0.01–207 µM, and a detection limit of 4 nM.

Shen-Ming Chen reported the simultaneous electrochemical deposition of gold and palladium nanoparticles on graphene flakes (Au/PdNPs-GRF) for the sensitive electrochemical determination of caffeic acid (CA) [75]. The electrochemical determination of CA at Au/Pd NPs deposited on GRF were studied by using cyclic voltammetry and differential pulse voltammetry. Au/PdNPs-GRF electrode exhibited electrocatalytic activity towards CA with wide linear range of 0.03–938.97 µM and detection limit of 6 nM, respectively. Moreover, the Au/PdNPs-GRF was found to be a selective and stable active material for the sensing of CA. In addition, the proposed sensor was employed successfully in real sample analysis of fortified wines.

As a proof-of-concept, an eco-friendly Au@α-Fe_2O_3@RGO ternary nanocomposite modified glass carbon electrode (GCE) was investigated for electrochemical detection of caffeic acid [76]. Characterization studies confirmed Au and α-Fe_2O_3 nanoparticles are uniformly distributed on the surfaces of reduced graphene oxide (RGO) nanosheets. The electrochemical mechanism involves the synergistic electrocatalytic activity of Au and α-Fe_2O_3 towards caffeic acid oxidation, with the RGO serving as an efficient electron shuttling mediator enhancing the sensor performance. The Au@α-Fe_2O_3@RGO modified GCE caffeic acid sensor showed a linear response range of 19–1869 µM and a detection limit of 0.098 µM. This ternary nanocomposite displays catalytic performance as well as selectivity toward caffeic acid. To demonstrate the potential application of the Fe_2O_3@RGO modified GCE caffeic acid sensor, caffeic acid in a coffee sample was measured.

Pt-PEDOT/reduced graphene oxide (Pt-PEDOT/rGO) nanocomposites were easily prepared by a simple and cost-effective one-pot method and used for modifying GCE for CA detection. The Pt-PEDOT/rGO nanocomposites modified glassy carbon electrode (Pt-PEDOT/rGO/GCE) displayed higher charge transfer efficiency and electrocatalytic activity to the oxidation of caffeic acid

(CA) in comparison with the bare GCE, and rGO modified electrodes (rGO/GCE). Furthermore, the Pt-PEDOT/rGO/GCE showed a linear range of 5.0×10^{-9} to 5.0×10^{-5} M with a low detection limit of 2.0×10^{-9} M for the CA determination. Finally, the modified electrode was also applied to detect CA in green tea and black tea samples with promising recovery results [77].

Pt/Cu nanocrystal with uniform trifurcate structure has been synthesized by using KI as structure-directing agent during the reduction process for developing an electrochemical sensor for CA detection [78]. Due to the large surface area of this synthesized dendritic trifurcate nanocrystal and the synergistic effect between Pt and Cu, this nanocatalyst showed interesting performances when applied to the electrochemical detection of caffeic acid. Specifically, the detection limit is 0.35 µM and the linear range is from 1.2–1.9 µM. Finally, the modified electrode was also applied to detect CA in red wine samples with good recovery data.

A nanocatalyst based on copper sulphide nanodots grown on graphene oxide sheets nanocomposite ($Cu_2SNDs@GOS$ NC) was synthesized by a simple sonochemical technique and applied in electrocatalytic sensing of caffeic acid (CA) [79]. The nanocomposite modified screen printed carbon electrode (SPCE) showed electrocatalytic performance towards CA oxidation. Under optimized conditions, Cu_2S NDs@GOS NC modified electrode showed fast and sensitive amperometric responses towards CA. The linear range was 0.055–2455 µM and the detection limit is 0.22 nM. The Cu_2S NDs@GOS NC/SPCE can quantify the amount of CA present in carbonated soft drinks and red wine without sample pretreatment.

Electrosynthesized PEDOT layers with different thickness and different doping counterions are investigated for their application in electrochemical CA sensing [80]. In terms of electroanalytical characteristics, adsorption control provides higher electroanalytical sensitivity in a narrow concentration range of linear response. Diffusion control results in markedly lower sensitivity values but extended range of linearity in the concentration dependence. The use of a non-linear calibration curve provides an extended concentration range for electroanalytical purpose under the established adsorption-controlled conditions. Doping counterions (polystyrene sulfonate and dodecyl sulphate) have no effect on the electroanalytical performance of PEDOT for CA oxidation.

An environmentally friendly sensor for the quantification of caffeic acid (CA) is reported [81]. Bismuth decorated multi-walled carbon nanotubes drop casted with cetyltrimethylammonium bromide demonstrated synergistic catalytic properties on enhancing the surface area of the carbon paste electrode. The proposed modified sensor was used to determine CA by differential pulse voltammetry (DPV) technique. The best results were obtained at physiological pH where the response was linear over a range of 6.0×10^{-8} to 5.0×10^{-4} M and with a limit of detection of 0.157 nM. The proposed sensor exhibited good performances vs. the most common interferents. The detection of CA in samples, such as coconut water, teas, and fruit juices without pretreatments was also reported.

Aicheng Chen reported a novel electrochemical sensor developed using fluorine-doped graphene oxide (F-GO) for the detection of caffeic acid (CA) [82]. The electrochemical behaviour of bare glassy carbon electrode (GCE), F-GO/GCE, and GO/GCE toward the oxidation of CA were studied using cyclic voltammetry (CV), and the results obtained from the CV investigation revealed that F-GO/GCE exhibited a high electrochemically active surface area and electrocatalytic activity. Differential pulse voltammetry (DPV) was employed for the CA detection, obtaining a linear concentration range from 0.5 to 100.0 µM with a limit of detection of 0.018 µM. Furthermore, the sensor selectivity was tested for the presence of other hydroxycinnamic acids and ascorbic acid. Moreover, the F-GO/GCE offered a good sensitivity, long-term stability, and a good reproducibility. The practical application of the electrochemical F-GO sensor was verified using several samples of commercially available wine. The developed electrochemical sensor can directly detect CA in wine samples without pretreatment, making it a promising candidate for food and beverage quality control.

4.1.2. Chlorogenic Acid

Chlorogenic acid (CGA, 5-O-caffeoylquinic acid, chemical structure in Figure 3) is a naturally occurring phenolic compound metabolized by several plants such as *Calendula officinalis* and *Echinacea purpurea* [84,85].

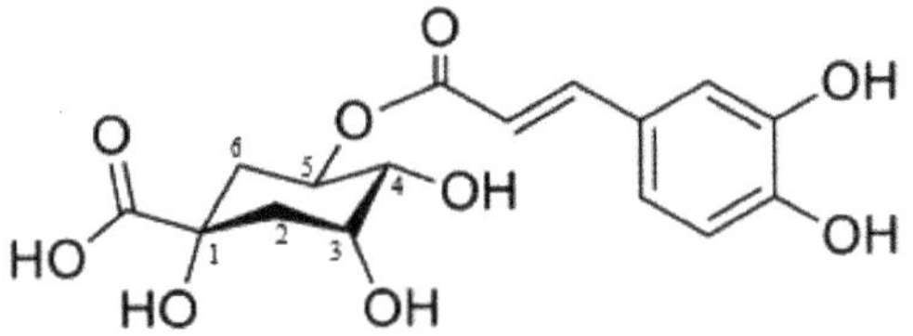

Figure 3. Chemical structure of chlorogenic acid [84].

Clinical investigations demonstrated that CGA can be classified as an anti-hypertension, anti-inflammatory, anti-carcinogenic, anti-bacterium, and antitumor agent [84]. Furthermore, there are also reports that CGA has a potential role in the regulation of glycemia levels in type 2 diabetes [84]. Therefore, the quantitative measurement of CGA in plant materials has attracted great interest, and different analytical methods have been employed for the determination of CGA, including high performance liquid chromatography, capillary electrophoresis, and electrochemical methods [85]. Nevertheless, electrochemical methods provide a rapid and simple procedure, relatively low-cost equipment, and possible in situ analysis. However, there are only few reports on the electrochemical behaviour/detection of CGA. Considering the electrochemical approach involving nanomaterials, examples are reported in Table 2.

Gao reported and described a multi-walled carbon nanotube modified screen-printed electrode (MWCNTs/SPE), employed for the electrochemical determination of CGA [86]. Cyclic voltammetry (CV) and differential pulse voltammetry (DPV) methods for the determination of CGA were proposed. Under the optimal conditions, the sensor exhibited linear ranges from 4.8×10^{-4}–4.4×10^{-2} M, and the detection limit for CGA of 3.38×10^{-4} M. According to the proposed analytical method, the MWCNTs/SPE was applied to the determination of CGA in coffee beans and the recovery was 94.74–106.65%. The result of CGA determination was in good agreement with those obtained by HPLC.

The preparation of a modified carbon paste electrode using highly defective mesoporous carbon (DMC) and room temperature ionic liquid 1-Butyl-3-methylimidazolium hexafluorophosphate (BMIM.PF$_6$) was described [85] and the obtained sensor was employed for the electrochemical detection of chlorogenic acid (CGA) in herbal extracts. DMC with defective structure was synthesized via a facile method using nanosilica as a hard template, sucrose as a carbon source, and KNO$_3$ as defect inducing agent. The proposed sensor exhibited good performances toward the electrochemical reaction of CGA in aqueous solution. The electrocatalytic behaviour was further considered as a detection procedure for the CGA determination by SWV. Under optimized conditions, the linear response range and detection limit were 2.00×10^{-8}–2.50×10^{-6} M and 1.00×10^{-8} M, respectively. The method was successfully applied for determination of CGA in extracts of *Calendula officinalis* and *Echinacea purpurea*.

Table 2. An overview of recent electrochemical sensors for CGA determination, using nanomaterials and/or nanocomposites.

Electrochemical Methods	Electrode Material	Linearity Range (mol·L^{-1})	LOD (mol·L^{-1})	Application	Reference
DPV	MWCNTs/SPE	4.8×10^{-4}–4.4×10^{-2}	3.38×10^{-4}	Coffee beans	[86]
DPV	DMC/BMIM.PF$_6$/CPE	2.00×10^{-8}–2.50×10^{-6}	1.00×10^{-8}	Herbal extracts of Calendula officinalis and Echinacea purpurea	[85]
DPV	AuNps@TAPB-DMTP-COFs/GCE	1.00×10^{-8}–4.00×10^{-5}	9.50×10^{-9}	Coffee, fruit juice and herbal extracts	[87]
Differential Pulse Voltammetry (DPV) WE	MWCNTs/CuONPs/LGN/GCE	5.00×10^{-3}–5.00×10^{-2}	1.25×10^{-5}	Coffee	[88]
Differential Pulse Voltammetry (DPV) WE	ZnO@PEDOT:PSS/GCE	3.00×10^{-8}–4.76×10^{-4}	2.00×10^{-8}	Coffee powder, soft drink	[89]

A novel gold nanoparticles-doped TAPB-DMTP-COFs (TAPB, 1,3,5-tris(4-aminophenyl)benzene; DMTP, 2,5-dimethoxyterephaldehyde; COFs, covalent organic frameworks) composite was prepared via COFs as the host matrix to support the growth of gold nanoparticles. Then, this composite was used to assemble an electrochemical sensor and presented a good electrocatalytic activity toward the oxidation of chlorogenic acid (CGA) in the phosphate buffer solution (pH 7.0) [87]. This electrochemical sensor displays a linear range of 1.00×10^{-8}–4.00×10^{-5} M, with a detection limit of 9.50×10^{-9} M. According to the proposed method, the sensor was applied to the determination of CGA in coffee, fruit juice, and in herbal extracts, and the recovery was 99.20–102.50%. The results of CGA determination were in good agreement with those obtained by HPLC.

An innovative nanocomposite of multiwalled carbon nanotubes (MWCNTs), copper oxide nanoparticles (CuONPs) and lignin (LGN) polymer were successfully synthesized and used to modify the glassy carbon electrode for the determination of chlorogenic acid (CGA). Cyclic voltammetry (CV) showed a quasi-reversible, adsorption-controlled and pH dependent behaviour. Differential pulse voltammetry (DPV) was applied and used for the quantitative detection of CGA. Under optimal conditions, the proposed sensor showed linear responses from 5 mM to 50 mM, whilst the limit of detection was found to be 0.0125 mM. The LGN-MWCNTs-CuONPs-GCE were applied to detect the CGA in real coffee samples with the recovery ranging from 97 to 106% [88].

The as synthesized ZnO@PEDOT:PSS modified glassy carbon electrode (GCE) was employed for the electrochemical detection of chlorogenic acid (CGA) [89]. The composite structure is formed by zinc oxide (ZnO) having the flower-like structure and by poly(3,4-ethylenedioxythiophene) polymer doped with poly(4-styrenesulfonate) (PEDOT: PSS). The nanocomposite was synthesized successfully through the one-pot chemical synthesis. The physicochemical properties of the synthesized composite were studied by using various analytical techniques. Specifically, ZnO@PEDOT:PSS-GCE towards CGA oxidation shows a linear range of 0.03–476.2 µM and limit of detection 0.02 µM, respectively. Furthermore, the ZnO@PEDOT:PSS-GCE effectiveness for the determination of CGA is tested by using coffee powder and soft drink as real samples.

The detection of chlorogenic acid in clinic samples for metabolic kinetics studies is a challenging task due to the low concentration and lack of sensitive analytical methods. Concerning this analytical issue, a porous pencil lead electrode (PLE) has been used to detect chlorogenic acid by SWV technique. The sensitivity was significantly improved due to the accumulation of CGA at the porous structure of the electrode. Under the optimized conditions, the concentration of chlorogenic acid was linear in the range of 7.7×10^{-8}–7.7×10^{-6} M with detection limits of 4.5×10^{-9} M. The electrode was used for the determination of chlorogenic acid in human urine with near 100% recovery [90].

4.1.3. Rosmarinic Acid

Rosmarinic acid (RA, chemical structure in Figure 4) is the ester of caffeic acid, predisposed to biological activities. [91]. It is also one of the natural antioxidants acting as an antiseptic, antiviral, antibacterial and anti-inflammatory agent. RA is an important and major secondary metabolite present in some plants and applied also in various fields such as medicine, cosmetics, and food industry. Due to the known impacts of RA on Alzheimer's disease, it was used to improve cognitive function as well as to reduce the severity of the renal disease [91].

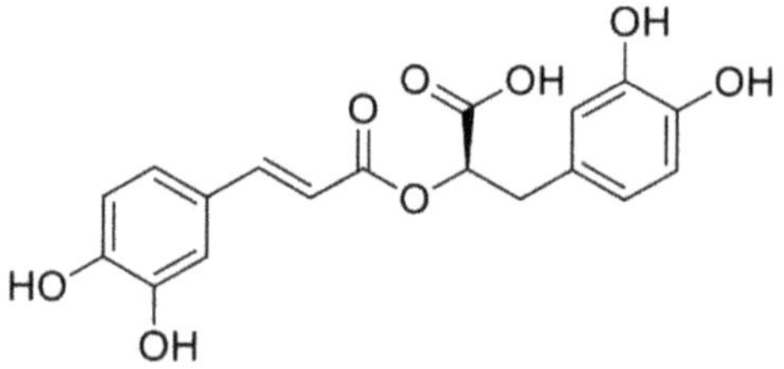

Figure 4. Chemical structure of rosmarinic acid [91].

Several studies have been reported concerning the determination of RA in different matrices using different analytical approaches such as chromatography, fluorescence, electrochemical sensors (very few papers), capillary electrophoresis [92].

Two different electrochemical sensing approaches are reported below the electrochemical behaviour of rosmarinic acid at the surface of a DNA-coated electrode was investigated using square-wave stripping voltammetry [93]. The voltammetric studies showed that rosmarinic acid is oxidized in two successive pH-dependent steps, each involving the transfer of two electrons and two protons. These oxidations correspond to two electroactive catechol groups. Moreover, strong interaction between the immobilized DNA and rosmarinic acid accumulates RA on the electrode surface resulting in an efficient preconcentration leading to high sensitivity of the sensor for rosmarinic acid determination. Several experimental parameters affecting the sensor response were optimized. Under optimized conditions, a linear concentration range of 0.040–1.5 µM with a detection limit of 0.014 µM was obtained. The proposed method was applied to the analysis of a rosemary extract. The obtained data was in good agreement with those obtained from HPLC analysis.

A modified carbon paste electrode (CPE) was designed and fabricated used magnetic functionalized molecularly imprinted polymer (MMIP) nanostructure for selective determination of RA in some plant extracts [93]. The MMIP nanostructure functionalized with –NH_2 group ($Fe_3O_4@SiO_2@NH_2$) was prepared by surface imprinting approach and then used as modifier for the sensor. The nano sized functionalized MMIP particles (i.e., $Fe_3O_4@SiO_2@NH_2$) superparamagnetic behaviour was analysed and verified. In addition, cyclic voltammetry (CV) and differential pulse voltammetry (DPV) techniques were used for study the electrochemical behaviour and determination of RA on the modified CPE. Based on the results, the modified sensor has good sensitivity and selectivity for the detection of RA in the presence of other species compared to unmodified electrodes. Under optimum conditions, two linear concentration ranges (0.1–100 µM and 100–500 µM) with a detection limit of 0.085 µM were obtained. Finally, the applicability of the designed sensor was examined for determination of RA in four plant extracts, including *Salvia officinalis*, *Zataria multiflora*, *Mentha longifolia*, and *Rosmarinus officinalis*.

4.2. Caffeine

Caffeine (CAF, chemical structure in Figure 5) a natural alkaloid, distributed in seeds, nuts, or leaves of a number of plants, mainly coffee, cocoa, tea, with the natural function as insecticide, is the most widely consumed psychoactive substance in human dietary. Many physiological effects of CAF are well known, from stimulation of the central nervous system, diuresis, and gastric acid secretion [94] to nausea, seizures, trembling, and nervousness. Mutation effects on DNA have been also reported. Moreover, it is considered a risk molecule for cardiovascular diseases. Recently, also an antioxidant activity has been suggested for CAF, showing protective effects against oxidative stress. The presence of CAF in many beverages and drug formulations of worldwide economic importance [94] makes it an analyte of great interest and although many different analytical methods are currently applied, novel analytical methods for fast, sensitive and reliable determination of CAF are always necessary, especially for particular purposes, as the determination in specific matrix in the presence of interfering agents, or in a specific concentration range, besides under beneficial conditions in terms of time consumption, material cost, and procedure ease [94].

Figure 5. Scheme of the approach and method used for the caffeine detection reprinted with the permission from [94]. Copyright 2017 Elsevier.

Electrochemical determination of CAF on different electrode materials, usually bare and miscellaneously modified carbon-based electrodes, has been also reported in literature [94]. The electrochemical methods offer a series of practical advantages as procedure ease, not too expensive instruments, the possibility of miniaturization, besides good sensitivity, wide linear concentration range, suitability for real-time detection, and reduced sensitivity to matrix effects. Several examples, concerning CAF detection via nanomaterials based electrochemical sensors are reported in Table 3.

As a first example of nanomaterials application to detect caffeine, we would like to mention a voltammetric sensor for caffeine, based on a glassy carbon electrode modified with Nafion and graphene oxide (GO) [95]. It exhibits a good affinity for caffeine (resulting from the presence of Nafion), and good electrochemical response (resulting from the presence of GO) for the oxidation of caffeine. The electrode enables the determination of caffeine in the range from 4.0×10^{-7} to 8.0×10^{-5} M, with a detection limit of 2.0×10^{-7} M. The sensor displays good stability, reproducibility, and high sensitivity. It was successfully applied to the quantitative determination of caffeine in beverages. Kubota group described the development of a novel sensitive molecularly imprinted electrochemical sensor for the detection of caffeine [96]. The sensor was prepared on a glassy carbon electrode modified with multiwall carbon nanotubes (MWCNTs)/vinyltrimethoxysilane (VTMS) recovered by a molecularly imprinted siloxane (MIS) film prepared by sol–gel process. MWCNTs/VTMS was produced by a simple grafting of VTMS on MWCNTs surface by in situ free radical polymerization. The siloxane layer was obtained from the acid-catalyzed hydrolysis/condensation of a solution constituted by tetraethoxysilane (TEOS), methyltrimethoxysilane (MTMS), 3-(aminopropyl)trimethoxysilane (APTMS), and caffeine as a template molecule. The MIS/MWCNTs-VTMS/GCE sensor was tested in a solution of the caffeine and other similar molecules. After optimization of the experimental conditions, the sensor showed a linear response range from 0.75 to 40 µM with a detection limit (LOD) of 0.22 µM. The imprinted sensor was successfully tested to detect caffeine in real samples.

Pumera group reported and discussed the electrochemical detection of caffeine at different chemically modified graphene (CMG) surfaces with different defects and oxygen functionalities [97]. The analytical performances of graphite oxide (GPO), graphene oxide (GO), and electrochemically reduced graphene oxide (ERGO) were compared for the detection of caffeine. It was found that ERGO showed the best analytical parameters, such as lower oxidation potential, sensitivity, linearity range (5.00×10^{-5}–3.00×10^{-4} M) and reproducibility of the response. ERGO was then used for the analysis of real samples. Caffeine levels of soluble coffee, teas, and energy drinks were measured without the need of any sample pre-treatment.

Table 3. An overview of electrochemical sensors for CAF determination, using nanomaterials and/or nanocomposites.

Electrochemical Methods	Electrode Material	Linearity Range (mol·L^{-1})	LOD (mol·L^{-1})	Application	Reference
DPV	Nafion/GO/GCE	4.00×10^{-7}–8.00×10^{-5}	2.00×10^{-7}	Soft and energy drinks, cola beverage	[95]
DPV	MIS/MWCNTs/VTMS/GCE	7.50×10^{-7}–4.00×10^{-5}	2.20×10^{-7}	Coffees, energy drinks,	[96]
DPV	ERGO/GCE	5.00×10^{-5}–3.00×10^{-4}	Not declared	Cola beverage, tea, and soluble coffee	[97]
DPV	AuNps/chitosan–ionic liquid/Gr/GCE	2.50×10^{-8}–2.49×10^{-6}	4.42×10^{-9}	Energy drink, teas, drugs	[98]
DPV	AuNps@PPY/PGE	2.00×10^{-9}–5.00×10^{-8} 5.00×10^{-8}–1.00×10^{-6}	9.00×10^{-10}	Soft and energy drinks, green tea, human plasma, drugs and urine	[99]
DPV	AuNps/chitosan/AuE	2.00×10^{-6}–5.00×10^{-2}	1.00×10^{-6}	Cola beverages, energy drink, teas	[94]
DPV	GO/RG/CPE	8.00×10^{-6}–8.00×10^{-4}	1.53×10^{-7}	Cola beverages, energy drink, teas, and drugs	[100]
DPV	CoON/CPE	5.00×10^{-6}–6.00×10^{-4}	1.60×10^{-8}	Coffees	[101]

A solution containing anisotropic gold nanoparticles (AuNPs), chitosan (CHIT), and ionic liquid (IL, i.e., 1-butyl-3-methylimidazolium tetrafluoroborate, (BMIM) (BF_4)) was cast on a graphene(r-GO) modified glassy carbon electrode to assemble an electrochemical sensor for the determination of theophylline (TP) and caffeine (CAF) [98]. The factors influencing this analytical approach are investigated, including the ingredients of the hybrid film, the concentrations of r-GO, $HAuCl_4$, and IL, and the buffer solution pH. Under the optimized conditions, the linear response ranges are 2.50×10^{-8}–2.10×10^{-6} M and 2.50×10^{-8}–2.49×10^{-6} M for TP and CAF, respectively. The detection limits are 1.32×10^{-9} M and 4.42×10^{-9} M, respectively. The electrochemical sensor shows good reproducibility, stability, and selectivity, and it has been applied to the determination of TP and CAF in real samples. A sensitive and selective nanocomposite imprinted electrochemical sensor for the determination of caffeine has been proposed by Rezaei [99]. The imprinted sensor was fabricated on the surface of pencil graphite electrode (PGE) via one-step electropolymerization of the imprinted conductive polymer, including gold nanoparticles (AuNPs), and caffeine. Because of the presence of specific binding sites on the molecularly imprinted polymer (MIP), the sensor responded quickly to caffeine. AuNPs were introduced for the enhancement of electrical response. The linear ranges of the MIP sensor were from 2.0 to 50.0 and 50.0 to 1000.0 nM, with the limit of detection of 0.9 nM. Furthermore, the proposed method was applied for the determination of caffeine in real samples (urine, plasma, tablet, green tea, energy, and soda drink).

Curulli group developed a simple and selective method for the determination of caffeine also in complex matrix at a gold electrode modified with gold nanoparticles (AuNPs) synthetized in a chitosan matrix in the presence of oxalic acid [94]. A scheme of the approach and method used for the caffeine detection is illustrated in Figure 5.

The electrochemical behaviour of caffeine at both gold bare and gold electrode modified with AuNPs was carried out in acidic medium by cyclic voltammetry (CV), differential pulse voltammetry (DPV) and electrochemical impedance spectroscopy (EIS). Electrochemical parameters were optimized to improve the electrochemical response to caffeine. The most satisfactory result was obtained using a gold electrode modified with AuNPs synthetized in a chitosan matrix in the presence of oxalic acid, in aqueous solution containing $HClO_4$ 0.4 M as supporting electrolyte. The performance of the sensor was then evaluated in terms of linearity range (2.0×10^{-6}–5.0×10^{-2} M), operational and storage stability, reproducibility, limit of detection (1.0×10^{-6} M) and response to a series of interfering compounds as ascorbic acid, citric acid, gallic acid, caffeic acid, ferulic acid, chlorogenic acid, glucose, catechin and epicatechin. The sensor was then successfully applied to determine the caffeine content in commercial beverages and the results were compared with those obtained with HPLC-PDA as an independent method and with those declared from manufacturers.

A novel graphene oxide-reduced glutathione modified carbon paste (GORGCP) sensor has been assembled for sensitive determination of CAF in 0.01 M phosphoric acid (H_3PO_4) solution of pH range (1.0–5.0) in both aqueous and surfactant media (0.5 mM sodium dodecylsulfate (SDS) [100]. The interaction of CAF over the surface of sensor was investigated using different electrochemical techniques. The linear detection range of CAF was between 8–800 μM with a limit of detection of 0.153 μM. The sensor showed a good selectivity and was applied for detecting CAF in real samples with recoveries, ranging from 97.76% to 101.36%.

Walcarius and co-workers [101] reported the more recent sensor for CAF detection involving nanomaterials. A nano-cobalt (II, III) oxide modified carbon paste (NCOMCP) sensor was prepared and applied to the determination of CAF in both aqueous (0.01 M H_2SO_4) and micellar media (0.5 mM sodium dodecylsulfate, SDS). The electrochemical behaviour of CAF was studied by means the most common electrochemical techniques. Using DPV, CAF was detected in a concentration range between 5.0 and 600 μM, with a limit of detection of 0.016 μM. The sensor showed good selectivity vs. the most common interferences and was applicable for CAF sensing in several commercial coffee samples with recoveries ranging from 98.9% to 101.9%.

4.3. Ascorbic Acid

L-Ascorbic acid (AA) or Vitamin C is a water-soluble vitamin and antioxidant, present in many biological systems and food. The biochemically and physiologically active form is the L-enantiomer, with a γ-lactone structure [102]. AA is an ideal scavenger free-radical and singlet oxygen or act as a chelating agent. As a strong antioxidant, it acts as a two-electron donor involving hydrogen atom transfer, giving rise to the ascorbate radical ion first, and finally to dehydroascorbic acid. The AA action prevents the oxidation of several compounds present in food and/or beverages. The deficiency of AA can cause several diseases, such as rheumatoid arthritis, Parkinson's and Alzheimer's diseases, and even cancer [102]. The excess of AA can result to several other diseases such as gastric irritation. Moreover, in the presence of heavy metal cations, the excess of AA has other drawbacks, because it can act as a pro-oxidant, in other words it limits its own antioxidant action tills to produce the reactive oxygen species, causing oxidative stress. Therefore, the determination of AA in biological fluids is very important for the diagnosis of such diseases. The quantitative determination of AA is also necessary for different application fields, including among others, cosmetics, drugs, and food [103].

Conventional bare electrodes, like Pt, Au, and glassy carbon, were used for the ascorbic acid detection but because of the vitamin C overpotential, fouling of the electrode surface was observed, involving low sensitivity and selectivity. Nanomaterials were used for the modification of conventional electrode to reduce the overpotential and to enhance the electrode sensitivity, selectivity, and reproducibility.

In Table 4, some interesting examples whereby the electrochemical sensors employing nanomaterials were designed and successfully applied for the AA detection in food analysis are reported.

Table 4. An overview of electrochemical sensors for AA determination, using nanomaterials and/or nanocomposites.

Electrochemical Methods	Electrode Material	Linearity Range (mol·L^{-1})	LOD (mol·L^{-1})	Application	Reference
DPV	NiCoO$_2$/C/GCE	1.00×10^{-5}–2.63×10^{-3}	5.00×10^{-7}	Fetal bovine serum, Vitamin C tableys, Vitamin C drinks	[104]
SWV	AgNPs@onion extracts/CPE	4.00×10^{-7}–4.50×10^{-4}	1.00×10^{-7}	Orange, kiwi and apple juices	[105]
DPV	ZnO-CuO NLs/GCE	1.00×10^{-7}–1.00×10^{-1}	1.20×10^{-8}	Human, mouse, and rabbit serum, orange juice, and urine	[106]
DPV	CNO-NiMoO$_4$-MnWO$_4$/GCE	1.00×10^{-6}–1.00×10^{-4}	3.30×10^{-7}	Orange, strawberry, tomato, pineapple juices	[107]
Amperometry	Mesoporous CuCo$_2$O$_4$/GCE	1.00×10^{-4}–1.05×10^{-3}	2.10×10^{-7}	Vitamin C tablets, Vitamin C effeverscent tablets and urine	[108]
LSV	Au-gr/CVE	1.00×10^{-6}–5.75×10^{-3}	5.00×10^{-8}	Cherry-apple juice, apple juice for children, apple juice and apple nectar clarified	[109]
FIA-Amperometry	rGO/GCE	Linearity range not declared	4.70×10^{-6}	Milk, fermented milk, chocolate milk and multivitamin supplement	[110]
DPV	HKUST-1/ITO	1.00×10^{-5}–2.65×10^{-3}	3.00×10^{-6}	Vitamin C pills, Vitamin C tablets, Vitamin C effervescent tablets	[111]
Amperometry	Ni$_6$ NCs/CB/GCE	1.00×10^{-6}–3.21×10^{-3}	1.00×10^{-7}	Vitamin C tablets	[112]

A smartphone-based integrated voltammetry system using modified electrode was developed for simultaneous detection of biomolecules and it is a fine example to combine nanomaterials and innovative technological approach. The system included a disposable sensor, a coin-size detector, and a smartphone equipped with application program [113]. Screen-printed electrodes were used, where reduced graphene oxide and gold nanoparticles were electrochemically deposited. The smartphone is the core device to communicate with the detector, elaborate the data, and record voltammograms in real-time. Then, the system was applied to detect standard solutions of the biomolecules and their mixtures as examples. The results showed that the method allowed the selective detection of the target molecules. The system was then tested to detect the different molecules, including AA, in artificial urine and in beverages. with linear and high sensitivity.

Arduini and co-workers reported an interesting development of a carbon black nanomodified inkjet-printed sensor for ascorbic acid detection for AA detection [114]. The possibility to use the paper employed in printed electronics as a substrate to develop a paper-based sensor is analysed and evaluated. In Figure 6, the procedure to manufacture paper-based inkjet printed electrodes and the dimensions of the printed electrodes are shown.

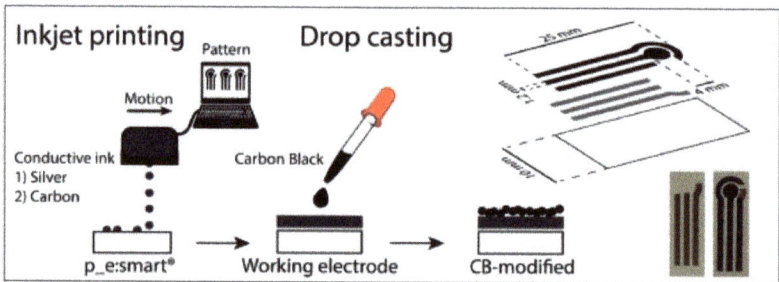

Figure 6. Procedure to manufacture paper-based inkjet printed electrodes and dimensions of the printed electrodes Reprinted with permission from [114]. Copyright 2018, Elsevier.

To improve the electrochemical performances of the inkjet-printed sensor, a dispersion of carbon black nanoparticles was used to modify the working electrode, for assembling a nanomodified electrochemical sensor platform. This disposable sensor was characterized both electrochemically and morphologically, and it has been used to detect ascorbic acid as model analyte. It has been evidenced that the presence of carbon black as nano modifier decreased the overpotential for ascorbic acid oxidation in comparison to the unmodified sensor. Under optimized experimental conditions, this printed electrochemical sensor was successfully employed for the detection of ascorbic acid in a dietary supplement, and the results were in agreement with those declared from the manufacturers.

Rahman and co-coworkers modified with ZnO·CuO nanoleaves a flat GCE, fabricated through a conducting nafion polymer matrix in order to detect selective acetylcholine and ascorbic acid simultaneously [106].

Concerning ascorbic acid, a linearity range from 100.0 pM to 100.0 mM was obtained with a detection limit of 12.0 pM. ZnO·CuO nanoleaves were synthesized by using a wet-chemical technique. This sensor applied for the determination of acetylcholine and ascorbic acid in real samples, i.e., human, mouse, and rabbit serum, orange juice, and urine, with satisfactory results.

Decorated carbon nanoonions (CNOs) with nickel molybdate ($NiMoO_4$) and manganese tungstate ($MnWO_4$) nanocomposite were used to modify a glassy carbon electrode (GCE) and to detect AA in real samples [107].

The high conductivity and large surface area of the CNOs and the electrocatalytic effect of $NiMoO_4$ and $MnWO_4$ nanocomposite provide a synergistic effect on the electrooxidation processes of AA. The modified GCE surface with a synthesized nanocomposite of $CNO-NiMoO_4-MnWO_4$ was used to determine ascorbic acid using differential pulse voltammetry (DPV) method. The developed ascorbic acid sensor displayed a wide dynamic calibration range (1–100 µM) with a detection limit of 0.33 µM. The sensor was applied to determine AA in real samples of orange, tomato, strawberry, and pineapple juices with satisfactory recovery.

Vidrevich and co-workers developed a voltammetric sensor based on a carbon veil (CV) and phytosynthesized gold nanoparticles (Au-gr) for ascorbic acid (AA) determination [108].

Extract from strawberry leaves was used as source of antioxidants (reducers) for Au-gr phytosynthesis. The sensor exhibits a linear response to AA in a concentration range from 1 µM to 5.75 mM and a limit of detection of 0.05 µM. The sensor was successfully applied for the determination of AA content in fruit juices without samples preparation. A correlation (r = 0.9867) between the results of AA determination obtained on the developed sensor and integral antioxidant activity of fruit juices was observed.

A nonenzymatic amperometric sensor for determination ascorbic acid has been developed using the glassy carbon electrode (GCE) modified with mesoporous $CuCo_2O_4$ rods [109].

The mesoporous $CuCo_2O_4$ rods were prepared by hydrothermal method. The sample is pure spinel $CuCo_2O_4$ rods with mesoporous structure. The mesoporous $CuCo_2O_4$ rods showed a linearity range of AA concentration (1–1000 mM) with a linearity range from 1.00 to 1000 µM. The detection limit is down to 0.21 µM. The $CuCo_2O_4$/GCE AA sensor exhibited good performance in terms of stability, repeatability and selectivity and was appled to tetect AA in vitamin C tablets.

A simple, sensitive and precise electroanalytical method was developed using flow injection analysis (FIA) with amperometric detection and reduced graphene oxide sensor for ascorbic acid determination in samples of multivitamin beverages, milk, fermented milk, and milk chocolate [110].

No interference of sample matrix was observed, avoiding solvent extraction procedures (samples were only diluted). The FIA allowed a detection limit of 4.7 µM. Good precision (RSD < 7%) and accuracy (recoveries between 91 and 108%) evidenced the robustness of the method. The method was compared with ultra-fast liquid chromatography (UFLC), obtaining statistically similar results (95% confidence level).

A Cu-based nanosheet metal–organic framework (MOF), HKUST-1, was synthesised using a solvent method at room temperature. An indium tin oxide (ITO) electrode) was modified with this material to catalyse the electrochemical oxidation of ascorbic acid [111].

Under optimal conditions, the oxidation peak current at +0.02 V displayed a linear relationship with the concentration of AA within the ranges of 0.01–25 and 25–265 µM, respectively. The limit of detection was 3 µM. The porous nanosheet structure of HKUST-1 could explain the significative enhancement of the effective surface area and the electron transfer capaability. Moreover, the novel AA sensor demonstrated good reproducibility, stability, and high selectivity towards glucose, uric acid (UA), dopamine (DA), and several amino acids. It was also successfully applied to the real sample testing of various AA containing tablets.

Metal nanoclusters (NCs) are highly desirable as active catalysts due to their highly active surface atoms. Among the reported metal clusters, nickel nanoclusters (Ni NCs) have been less well developed than others, such as gold, silver and copper. A simple method was developed by Chen to synthesize atomically precise Ni clusters with the molecular formula of $Ni_6(C_{12}H_{25}S)_{12}$ [112].

The Ni_6 NCs can be self-assembled into nanosheets due to their uniform size. It was found that the $Ni_6(C_{12}H_{25}S)_{12}$ clusters were loaded on carbon black and the resulting nanocomposite (Ni_6 NCs/CB) was used to modify a GCE. The modified GCE was used to determine AA. The Ni_6 NCs/CB based sensor showed high sensing performance for AA with a wide linear range (1–3212 µM) and a detection limit of 0.1 µM. The high catalytic activity could be due to the high fraction of surface Ni atoms with low coordination in the sub-nanometer clusters. Finally, Ni clusters can be used as highly efficient catalysts for the electrochemical detection of AA in tablets.

4.4. Nitrite

Nitrite is an additive used to extend the shelf-life of beverages and foods such as ham, salami, and other cured meats [9]. On the other hand, it proves harmful for the human body if it is added to food and beverages at levels higher than those indicated by the safety standards. The World Health Organization (WHO) has established a concentration limit of 3.0 mg·L^{-1} (65.22 µM) for nitrite in drinking water [115]. Therefore, the detection and quantification of nitrite is one of the critical issues of food analysis. Novel nanomaterials have been synthesized and used for the design of advanced sensors for the nitrite detection. It has been shown that nanomaterial-based electrocatalysts significantly improve analytical performances for the nitrite determination and several examples are reported in Table 5. We must evidence that the majority of them are concerned with the detection in water samples, but very significant examples are concerned with cured meats and milk.

As first example, we present a sensor based on silver nanoparticles stabilized by polyamidoamine (PAMAM) dendrimer, synthesized by a simple chemical reduction method [116]. The Ag-PAMAM modified electrode exhibited good electrocatalytic performance for the oxidation of nitrite. Under optimized conditions, in a pH 6.0 phosphate buffer solution, the sensor displayed a linear range from 4.0 µM to 1.44 mM with a detection limit of 0.4 µM. Furthermore, the proposed sensor was used for the determination of nitrite in tap water and milk samples.

Du and co-workers have prepared a promising electrochemical sensor based on Pt nanoparticles synthesized by a one-pot hydrothermal method and distributed on the surface of reduced graphene oxide (RGO) sheets [117]. The morphology and composition of as-prepared PtNPs-RGO composites have been characterized by different spectroscopic methods. The Pt-RGO modified electrode shows a good linearity between the peak current and concentration of nitrite (the range is not declared) with a detection limit of 0.1 µM. Compared to the Pt nanoparticles or the RGO modified glassy carbon electrode, the nanocomposite-based sensor shows a good reproducibility, stability, and anti-interference electrocatalytic performance toward nitrite sensing. Finally, it has been applied to detect nitrite in beverages.

Cellulose is the most abundant, renewable, biodegradable, natural polymer resource on earth and can be a good substrate for catalysis. For this reason, Zhang and co-workers used straw cellulose, oxidized by 2,2,6,6 tetra-methylpiperidine-1-oxylradical (TEMPO), and then they synthesized via a hydrothermal method a TEMPO oxidized straw cellulose/molybdenum sulphide (TOSC-MoS$_2$) composite for modifying a GCE [118]. The TOSC-MoS$_2$ modified glassy carbon electrode is used as electrochemical sensor for detecting nitrite. Cyclic Voltammetry (CV) results showed TOSC-MoS$_2$ has electrocatalytic activity for the oxidation of nitrite. The amperometric response results indicate the TOSC-MoS$_2$ modified GCE can be used to determine nitrite with linear ranges of 6.0–3140 and 3140–4200 µM and a detection limit of 2.0 µM. The proposed sensor has good anti-interference performance, and it has been applied to real sample analysis (river and drinking water).

Shen-Ming Chen group has prepared and evaluated a palladium nanoparticles (PdNPs) decorated functionalized multiwalled carbon nanotubes (f-MWCNT) modified glassy carbon electrode for the amperometric determination of nitrite in different water samples [119]. The f-MWCNT/PdNPs composite modified electrode was prepared by electrodeposition of PdNPs on the surface of f-MWCNT. The resultant sensor exhibits excellent electrocatalytic activity towards the oxidation of nitrite compared to MWCNT, f-MWCNT, and PdNPs modified electrodes. The amperometric method was used to determine nitrite and the response of the nitrite on modified electrode was linear in the concentration range from 0.05 to 2887.6 µM with a detection limit of 22 nM. Sensor was further applied to detect nitrite in different water samples (river, pond, and drinking water).

Table 5. An overview of electrochemical sensors for nitrite determination, using nanomaterials and/or nanocomposites.

Electrochemical Methods	Electrode Material	Linearity Range (mol·L^{-1})	LOD (mol·L^{-1})	Application	Reference
DPV	AgNPs@PAMAM/GCE	4.00×10^{-6}–1.44×10^{-3}	4.00×10^{-7}	Milk and tap water	[116]
DPV	PtNPs/rGO/GCE	Linearity range not declared	1.00×10^{-7}	Beverages	[117]
DPV	TOSC-MoS$_2$/GCE	6.00×10^{-6}–4.20×10^{-3}	2.00×10^{-6}	River and drinking water	[118]
Amperometry	f-MWCNT/PdNPs/GCE	5.00×10^{-8}–3.00×10^{-6}	2.20×10^{-8}	River, pond, and drinking water	[119]
DPV	GNs/GCE	1.00×10^{-6}–1.05×10^{-4}	2.20×10^{-7}	Tap water	[120]
DPV	Pd/Fe$_3$O$_4$/polyDOPA/RGO/GCE	2.50×10^{-6}–6.47×10^{-3}	5.00×10^{-7}	River water and sausage	[121]
Amperometry	rGO/Acr/GCE	4.00×10^{-7}–3.60×10^{-3}	1.20×10^{-7}	Milk, mineral and tap water	[122]
Amperometry	MnO$_2$/GO-SPE	1.00×10^{-7}–1.00×10^{-3}	9.00×10^{-8}	Tap and mineral water	[123]
DPV	AuNPs/carbosilane-dendrimer/GCE	1.00×10^{-5}–5.00×10^{-3}	2.00×10^{-7}	Natural water	[124]
SWV	Cu/MWCNT/RGO/GCE	1.00×10^{-7}–7.50×10^{-5}	3.00×10^{-8}	Tap and mineral waters, sausages, salami, and cheese	[125]
DPV	MOFs-derived α-Fe$_2$O$_3$/CNTs/GCE	1.00×10^{-7}–7.50×10^{-5}	3.00×10^{-8}	Tap and mineral waters, sausages, salami, and cheese	[126]
DPV	CoCNM/GCE	5.00×10^{-6}–7.05×10^{-6}	1.80×10^{-7}	Tap water	[127]
DPV	Ag/CuNCs/MWCNTs/GCE	1.00×10^{-6}–1.00×10^{36}	2.00×10^{-7}	Lake water, drinking water and seawater.	[128]
DPV	Co$_3$O$_4$@rGO/CNTs/GCE	8.00×10^{-6}–5.60×10^{-2}	1.60×10^{-8}	Tap water	[129]
DPV	Ni@Pt/Gr/GCE	1.00×10^{-5}–1.50×10^{-2}	Not declared	Tap water	[130]
DPV	AuNPs@Cu-MOF/GCE	1.00×10^{-7}–1.00×10^{-2}	8.20×10^{-8}	River water	[131]

A highly sensitive electrochemical method was developed by Kalcher and co-workers for the determination of nitrites in tap water using a glassy carbon electrode modified with graphene nanoribbons (GNs/GCE) [120]. Graphene nanoribbons (GNs) have been synthetized and aligned to the surface of glassy carbon electrode (GCE) and showed good electrocatalytic activity for nitrite oxidation. Studies about electrochemical behaviour and optimization of the most important experimental conditions were performed using cyclic voltammetry (CV) in Britton–Robinson buffer solution (BRBS), while quantitative studies were undertaken with amperometry. The influence of most common interferents was found to be negligible. Under optimized experimental conditions linear calibration curves were obtained in the range from 0.5 to 105 µM with the detection limit of 0.22 µM. The proposed method and sensor are successfully applied for the determination of nitrite present tap water samples without any pretreatment.

A green route was used to synthesize $Pd/Fe_3O_4/polyDOPA/RGO$ composite. The in-situ nucleation and growth of Fe_3O_4 and Pd nanostructures was performed on reduced graphene oxide (RGO), based on poly DOPA (3,4-Dihydroxy-l-phenylalanine, DOPA) [121]. This composite showed an electrocatalytic activity toward nitrite oxidation. The amperometric results indicated that $Pd/Fe_3O_4/poly$ DOPA/RGO modified glassy carbon electrode was employed to determine nitrite with a linear range of 2.5–6470 µM and a detection limit of 0.5 µM. With good anti-interference behaviour and good stability, the proposed sensor was successfully applied for the determination of nitrite in Yellow River water and sausage extract with satisfactory results.

Flexible and free-standing reduced graphene oxide (rGO) papers, doped with dye molecules such as phenazine, phenothiazine, phenoxazine, xanthene, acridine, and thiazole, were used to assemble an electrochemical sensor for nitrite detection by Aanyalıoglu [122]. The assembling procedure included two basic steps: vacuum filtration of a dispersion containing GO sheets and dye molecule through a membrane and reduction of free-standing GO/dye paper to rGO/dye paper by treatment with hydriodic acid. Electrical conductivity and electrochemical performance studies indicated that acriflavine (Acr) doped rGO paper electrode showed the best performance for the oxidation of nitrite. The resulted amperometric sensor was very stable, flexible, and reproducible, with a detection limit of 0.12 µM and linear range of 0.40–3600 µM. Under optimized conditions, it was applied to detect nitrite in milk as well as mineral and tap water samples.

A screen-printed amperometric sensor based on a carbon ink bulk-modified with graphene oxide decorated with MnO_2 nanoparticles (MnO_2/GO-SPE) was prepared as a sensor for nitrite detection by Banks and co-workers [123]. The MnO_2/GO-SPE showed an electrocatalytic activity for the electrochemical determination of nitrite in 0.1 M phosphate buffer solution (pH 7.4), with a limit of detection of 0.09 µM and with two linear ranges of 0.1–1 µM and 1–1000 µM, respectively. Additionally, the nitrite sensor presented a good selectivity, reproducibility, and stability and was applied to detect nitrite in tap and mineral water samples.

A paper-based, low-cost, disposable electrochemical sensing platform was developed for nitrite analysis based on graphene nanosheets and gold nanoparticles by Miao [132]. In comparison with the electrochemical responses at bare gold and glassy carbon electrodes, a considerably higher oxidation current was recorded. At the paper-based electrode, the fouling effect due to the oxidation products adsorption was negligible. Moreover, this paper-based sensing platform was applied to determine nitrite in environmental and food samples.

Glassy carbon and platinum electrodes were modified with composite films including gold nanoparticles (AuNPs) and a carbosilane-dendrimer with peripheral electronically communicated ferrocenyl units (Dend), by Alonso and co-workers [124]. The modified electrodes exhibited interesting electrocatalytic activity toward the oxidation of nitrite, giving higher peak currents at lower oxidation potentials than those at the bare electrode and at the dendrimer-modified electrode. Under the optimized conditions the sensor has a linear response in the 10 µM to 5 mM concentration range, and a detection limit of 2.0×10^{-7} M. The sensor was successfully applied to the determination of nitrite in natural water samples.

An electrochemical sensor based on Cu metal nanoparticles distributed on multiwall carbon nanotubes-reduced graphene oxide nanosheets (Cu/MWCNT/RGO) for determination of nitrite and nitrate ions was assembled by Rezaei group [125]. The morphology of the nanocomposite was investigated using different spectroscopic methods. Under optimized experimental conditions, the modified GCE showed catalytic activity toward the oxidation of nitrite with a significant increase in oxidation peak currents in comparison with those at bare GCE. Using SWV the sensor showed a linear concentration range from 0.1 to 75 µM with detection limit of 30 nM for nitrite. Furthermore, the sensor was applied to the detection of nitrite in the tap and mineral waters, sausages, salami, and cheese samples.

Highly ordered MOFs-derived α-Fe_2O_3/CNTs hybrids through in-situ insertion of carbon nanotubes and their subsequent calcination were prepared for the electrochemical detection of nitrite. Metal organic frameworks (MOFs), composed of metal ions and organic ligands connected each other through strong coordination bonds, have been widely applied in gas storage and separation [0]. The in-situ insertion of CNTs improves the electron-transfer between the α-Fe_2O_3 and CNTs, making α-Fe_2O_3 highly active for oxidation of nitrite. The sensor demonstrated good performances of nitrite detection with the limit of detection of 0.15 µM, and the linear range from 0.5 µM to 4000 µM. Finally, it was applied for the detection of nitrite in tap and pond water.

Another MOFs was used to prepare a sensor for detecting nitrite by Zhou and co-workers [127]. The MOFs were calcined in argon atmosphere to prepare magnetic Co nanocages (CoCN). The synthesized CoCN were deposited on the surface of glassy carbon electrode obtaining a magnetic sensing platfor(CoCN/MGCE). The prepared CoCN/MGCE displayed interesting performance for electrocatalytic oxidation of nitrite, with a linear concentration range of 5–705 µM with a 0.18 µM detection limit. The proposed sensor was applied for detecting nitrite in real samples of tap water. A glassy carbon electrode modified with Ag/Cu nanoclusters and multiwalled carbon nanotubes to detect nitrite was prepared using the electrodeposition method [128]. This electrode showed electrocatalytic activity for the oxidation of nitrite. Upon an increase of the nitrite concentration from 1.0 µM to 1.0 mM, the current response increased linearly, with a detection limit of 2×10^{-7} M. The electrode exhibited good stability, and it was applied in the detection of nitrite in lake water, drinking water and seawater.

An electrochemical sensor based on cobalt oxide decorated reduced graphene oxide and carbon nanotubes (Co_3O_4-rGO/CNTs) has been prepared for the nitrite detection. The sensor showed a linear concentration range from 8 µM to 56 mM and a detection limit of 0.016 µM. It was also applied to determine the nitrite level in tap water real samples with satisfactory recovery [129].

Core-shell Ni@Pt nanoparticles were synthesised on graphene [Ni@Pt/Gr] by Medhany [130]. This synthesized nanocomposite was characterized by different spectroscopic methods. Ni@Pt/Gr electrode showed electrocatalytic activity towards nitrite oxidation. Linear calibration curves were recorded at Ni@Pt/Gr electrodes obtaining a linear concentration range of 10–15 mM. The prepared nanocomposite could successfully estimate nitrite concentration in drinking water samples. Finally, another example of a sensor using a MOF system has been added. An electrochemical sensing platform based on a Cu-based metal-organic framework (Cu-MOF) decorated with gold nanoparticles (AuNPs) was assembled by Chen for the detection of nitrite [131]. AuNPs were electrodeposited on Cu-MOF modified glassy carbon electrode (Cu-MOF/GCE) using the potentiostatic method. The AuNPs decorated Cu-MOF (Cu-MOF/Au) displayed a catalytic effect for the oxidation of nitrite due to the high surface area and porosity of Cu-MOF, preventing the aggregation of AuNPs. The amperometry was adopted for quantitative determination of nitrite. The prepared electrochemical sensing platform demonstrates high sensitivity, selectivity, and good stability for the detection of nitrite. It shows two wide linear ranges of 0.1–4000 and 4000–10,000 µM, and a low detection limit of 82 nM. Moreover, the sensing platform can also be used for the nitrite detection in river water samples.

5. Conclusions

An overview of the most recent applications of nanomaterials and electrochemical sensors in food analysis has been presented.

It is evident that a plethora of nanomaterials including hybrid nanocomposites has been prepared, characterized, and used for different sensing applications, in many cases, the design has involved an accurate and targeted control of shape and dimensions of the nanostructure.

In many cases smart nanomaterials/nanocomposites, based on carbon nanostructures, metal nanoparticles/clusters and polymer both natural and/or synthetic, have been prepared and coupled with the most common electroanalytical techniques with very interesting results in terms of sensitivity, linearity range amplitude and selectivity. A challenge for the future is the application of all these promising results for defining widespread analytical protocols for addressing the food security issue, employing low cost, biocompatible materials and polymers as well as greener synthetic approaches and devices, e.g., screen printed electrodes, where the criticalities due to fouling are eliminated.

Although the majority the described sensors demonstrated a possible usefulness for food analysis area, they have been validated only in the laboratory. Generally, there are many electrochemical sensors for the food analysis of additives and contaminants. However, the number of the corresponding commercially available sensors is very limited. Precise and accurate validation studies are strongly suggested for a real employment of these electrochemical sensors for food and beverage safety testing. In addition, the impacts of environmental constraints, storage, selectivity, the matrix effects due to the complexity of food samples, and stability under real operative conditions must be verified. The validation and testing of statistically relevant numbers of samples, comparability, and interlaboratory studies to validate the robustness of such sensing platforms are the next critical steps for the achievement of industry level acceptance and regulatory approvals. It would be highly beneficial if these sensors were to be introduced by the food industry to monitor the safety and the quality of processed food and beverages.

Moreover, studies of the toxicity and degradation of these nanomaterials are required. All these issues should be further addressed before the introduction to the sensors market.

Finally, the development of smart sensors is linked to the development of portable devices. Improved portability may be achieved through the connectivity and integration of electrochemical sensors with devices such as smartphones and tablets, but very few examples are available [113]. Such an integration of two distinct areas of research (sensors and ICT), addressing the day-to-day needs of people, could facilitate the introduction of the next generation of smart sensors into the food processing industries to increase the quality and safety of food and beverages.

Funding: This research received no external funding.

Acknowledgments: The author would thank Alessandro Trani for the technical support.

Conflicts of Interest: The author declares no conflict of interest.

References

1. Logothetidis, S. *Nanostructured Materials and Their Applications*; Springer: Berlin/Heidelberg, Germany, 2012; pp. 1–22.
2. Roco, M.C.; Mirkin, C.A.; Dincer Hersam, M.C. *Nanotechnology Research Directions for Societal Needs in 2020: Retrospective and Outlook*; Springer Science Business Media: Berlin/Heidelberg, Germany, 2011.
3. Sattler, K.D. *Handbook of Nanophysics: Principles and Methods*; CRC Press: Boca Raton, FL, USA, 2010.
4. Chen, A.; Chatterjee, S. Nanomaterials based electrochemical sensors for biomedical applications. *Chem. Soc. Rev.* **2013**, *42*, 5425–5438. [CrossRef]
5. Umasankar, Y.; Adhikari, B.-R.; Chen, A. Effective immobilization of alcohol dehydrogenase on carbon nanoscaffolds for ethanol biofuel cell. *Bioelectrochemistry* **2017**, *118*, 83–90. [CrossRef]

6. Martínez-Periñán, E.; Foster, W.C.; Down, P.M.; Zhang, Y.; Ji, X.; Lorenzo, E.; Kononovs, D.; Saprykin, I.A.; Yakovlev, N.V.; Pozdnyakov, A.G.; et al. Graphene encapsulated silicon carbide nanocomposites for high and low power energy storage applications. *J. Carbon Res.* **2017**, *3*, 20. [CrossRef]
7. Duran, N.; Marcato, P.D. Nanobiotechnology perspectives. Role of nanotechnology in the food industry: A review. *Int. J. Food Sci. Technol.* **2013**, *48*, 1127–1134. [CrossRef]
8. Srilatha, B. Nanotechnology in Agriculture. *J. Nanomed. Nanotechnol.* **2011**, *2*, 5.
9. Manikandan, V.S.; Adhikarib, B.R.; Chen, A. Nanomaterial based electrochemical sensors for the safety and quality control of food and beverages. *Analyst* **2018**, *143*, 4537–4554. [CrossRef] [PubMed]
10. Bartlett, P.N. *Bioelectrochemistry 45: Fundamentals, Experimental Techniques, and Applications*; John Wiley & Sons: Hoboken, NJ, USA, 2008.
11. Durst, R.A. Chemically modified electrodes: Recommended terminology and definitions. *Pure Appl. Chem.* **1997**, *69*, 1317–1324. [CrossRef]
12. Hierlemann, A.; Gutierrez-Osuna, R. Higher-Order Chemical Sensing. *Chem. Rev.* **2008**, *108*, 563–613. [CrossRef]
13. Bonizzoni, M.; Anslyn, E.V. Combinatorial Methods for Chemical and Biological Sensors. *J. Am. Chem. Soc.* **2009**, *131*, 14597–14598. [CrossRef]
14. Suni, I.I. Impedance methods for electrochemical sensors using nanomaterials. *Trends Anal. Chem.* **2008**, *27*, 604–611. [CrossRef]
15. Katz, E.; Willner, I. Probing Biomolecular Interactions at Conductive and Semiconductive Surfaces by Impedance Spectroscopy: Routes to Impedimetric Immunosensors, DNA-Sensors, and Enzyme Biosensors. *Electroanalysis* **2003**, *15*, 913–947. [CrossRef]
16. Wang, J. *Analytical Electrochemistry*, 2nd ed.; Wiley/VCH: New York, NY, USA, 2000.
17. Bockris, J.O.M.; Reddy, A.K.N.; Gamboa-Aldeco, M. *Modern Electrochemistry 2A: Fundamentals of Electrodics*, 2nd ed.; Kluwer Academic/Plenum Publishers: New York, NY, USA, 2000; Volume 2.
18. Bard, A.J.; Faulkner, L.R. *Electrochemical Methods: Fundamentals and Applications*, 2nd ed.; John Wiley & Sons: New York, NY, USA, 2001.
19. Dincer, C.; Bruch, R.; Costa-Rama, E.; Fernández-Abedul, M.T.; Merkoçi, A.; Manz, A.; Urban, G.A.; Güder, F. Disposable Sensors in Diagnostics, Food, and Environmental Monitoring. *Adv. Mater.* **2019**, *31*, 1806739. [CrossRef] [PubMed]
20. Sharma, S.; Singha, N.; Tomar, V.; Chandra, R. A review on electrochemical detection of serotonin based on surface modified electrodes. *Biosens. Bioelectron.* **2018**, *107*, 76–93. [CrossRef] [PubMed]
21. Brownson, D.A.; Banks, C.E. Graphene electrochemistry: An overview of potential applications. *Analyst* **2010**, *135*, 2768–2778. [CrossRef] [PubMed]
22. Teradal, N.L.; Jelinek, R. Carbon Nanomaterials: Carbon Nanomaterials in Biological Studies and Biomedicine. *Adv. Healthc. Mater.* **2017**, *6*, 1700574. [CrossRef]
23. Porto, L.S.; Silva, N.; de Oliveira, A.E.F.; Pereira, A.C.; Borges, K.B. Carbon nanomaterials: Synthesis and applications to development of electrochemical sensors in determination of drugs and compounds of clinical interest. *Rev. Anal. Chem.* **2019**, *38*, 20190017. [CrossRef]
24. Bobrinetskiy, I.I.; Knezevic, N.Z. Graphene-based biosensors for on-site detection of contaminants in food. *Anal. Methods* **2018**, *10*, 5061–5071. [CrossRef]
25. Bounegru, A.V.; Apetrei, C. Carbonaceous Nanomaterials Employed in the Development of Electrochemical Sensors Based on Screen-Printing Technique-A Review. *Catalysts* **2020**, *10*, 680. [CrossRef]
26. Kirchner, E.-M.; Hirsch, T. Recent developments in carbon-based two-dimensional materials: Synthesis and modification aspects for electrochemical sensors. *Microchim. Acta* **2020**, *187*, 441–462. [CrossRef]
27. Kour, R.; Arya, S.; Young, S.-J.; Gupta, V.; Bandhoria, P.; Khosla, A. Review-Recent Advances in Carbon Nanomaterials as Electrochemical Biosensors. *J. Electrochem. Soc.* **2020**, *16*, 037555. [CrossRef]
28. Pandey, H.; Khare, P.; Singh, S.; Pratap Singh, S. Carbon nanomaterials integrated molecularly imprinted polymers for biological sample analysis: A critical review. *Mater. Chem. Phys.* **2020**, *239*, 121966. [CrossRef]
29. Yang, C.; Jacobs, C.B.; Nguyen, M.D.; Ganesana, M.; Zestos, A.G.; Ivanov, I.N.; Puretzky, A.A.; Rouleau, C.M.; Geohegan, D.B.; Venton, B.J. Carbon Nanotubes Grown on Metal Microelectrodes for the Detection of Dopamine. *Anal. Chem.* **2016**, *88*, 645–652. [CrossRef]

30. Zhang, L.; Liu, F.; Sun, X.; Wei, G.F.; Tian, Y.; Liu, Z.P.; Huang, R.; Yu, Y.; Peng, H. Engineering Carbon Nanotube Fiber for Real-Time Quantification of Ascorbic Acid Levels in a Live Rat Model of Alzheimer's Disease. *Anal. Chem.* **2017**, *89*, 1831–1837. [CrossRef]
31. Pumera, M. Graphene-based nanomaterials and their electrochemistry. *Chem. Soc. Rev.* **2010**, *39*, 4146–4157. [CrossRef] [PubMed]
32. Li, J.; Guo, S.; Zhai, Y.; Wang, E. Nafion–graphene nanocomposite film as enhanced sensing platform for ultrasensitive determination of cadmium. *Electrochem. Commun.* **2009**, *11*, 1085–1088. [CrossRef]
33. Li, X.-R.; Liu, J.; Kong, F.-Y.; Liu, X.-C.; Xu, J.-J.; Chen, H.-Y. Potassium-doped graphene for simultaneous determination of nitrite and sulfite in polluted water. *Electrochem. Commun.* **2012**, *20*, 109–112. [CrossRef]
34. Yang, Y.J.; Li, W. CTAB functionalized graphene oxide/multiwalled carbon nanotube composite modified electrode for the simultaneous determination of ascorbic acid, dopamine, uric acid and nitrite. *Biosens. Bioelectron.* **2014**, *56*, 300–306. [CrossRef]
35. Mustafa, F.; Andreescu, S. Nanotechnology-based approaches for food sensing and packaging applications. *RSC Adv.* **2020**, *10*, 19309. [CrossRef]
36. Xiaoa, T.; Huang, J.; Wang, D.; Meng, T.; Yang, X. Au and Au-Based nanomaterials: Synthesis and recent progress in electrochemical sensor applications. *Talanta* **2020**, *206*, 120210. [CrossRef]
37. Liu, Z.; Nemec-Bakk, A.; Khaper, N.; Chen, A. Sensitive Electrochemical Detection of Nitric Oxide Release from Cardiac and Cancer Cells via a Hierarchical Nanoporous Gold Microelectrode. *Anal. Chem.* **2017**, *89*, 8036–8043. [CrossRef]
38. Salaün, P.; van den Berg, C.M. Voltammetric Detection of Mercury, and Copper in Seawater Using a Gold Microwire Electrode. *Anal. Chem.* **2006**, *78*, 5052–5060. [CrossRef] [PubMed]
39. Alves, G.M.; Magalhães, J.M.; Salaün, P.; Van den Berg, C.M.; Soares, H.M. Simultaneous electrochemical determination of arsenic, copper, lead, and mercury in unpolluted fresh waters using a vibrating gold microwire electrode. *Anal. Chim. Acta* **2011**, *703*, 1–7. [CrossRef] [PubMed]
40. Xing, G.; Liqin, W.; Zhaon, L.; Ding, Y. Nanoporous Gold Leaf for Amperometric Determination of Nitrite. *Electroanalysis* **2011**, *23*, 381–386.
41. Jia, F.; Yu, C.; Ai, Z.; Zhang, L. Fabrication of Nanoporous Gold Film Electrodes with Ultrahigh Surface Area and Electrochemical Activity. *Chem. Mater.* **2007**, *19*, 3648–3653. [CrossRef]
42. Li, J.; Lin, X. Electrocatalytic oxidation of hydrazine and hydroxylamine at gold nanoparticle-polypyrrole nanowire modified glassy carbon electrode. *Sens. Actuators B* **2007**, *126*, 527–535. [CrossRef]
43. Manikandan, V.S.; Liu, Z.; Chen, A. Simultaneous detection of hydrazine, sulfite, and nitrite based on a nanoporous gold microelectrode. *J. Electroanal. Chem.* **2018**, *819*, 524–532. [CrossRef]
44. Yoon, H. Current Trends in Sensors Based on Conducting Polymer Nanomaterials. *Nanomaterials* **2013**, *3*, 524–549. [CrossRef]
45. Al-Mokaram, A.; Amir, M.A.; Yahya, R.; Abdi, M.M.; Mahmud, N.M.E. The Development of Non-Enzymatic Glucose Biosensors Based on Electrochemically Prepared Polypyrrole–Chitosan–Titanium Dioxide Nanocomposite Films. *Nanomaterials* **2017**, *7*, 129. [CrossRef]
46. Govindhan, M.; Amiri, M.; Chen, A. Au nanoparticle/graphene nanocomposite as a platform for the sensitive detection of NADH in human urine. *Biosens. Bioelectron.* **2015**, *66*, 474–480. [CrossRef]
47. Li, S.-S.; Hu, Y.-Y.; Wang, A.-J.; Weng, X.; Chen, J.-R.; Feng, J.-J. Simple synthesis of worm-like Au–Pd nanostructures supported on reduced graphene oxide for highly sensitive detection of nitrite. *Sens. Actuators B* **2015**, *208*, 468–474. [CrossRef]
48. Afkhami, A.; Soltani-Felehgari, F.; Madrakian, T.; Ghaedi, H. Surface decoration of multi-walled carbon nanotubes modified carbon paste electrode with gold nanoparticles for electro-oxidation and sensitive determination of nitrite. *Biosens. Bioelectron.* **2014**, *51*, 379–385. [CrossRef] [PubMed]
49. Sun, C.-L.; Lee, H.-H.; Yang, J.-M.; Wu, C.-C. The simultaneous electrochemical detection of ascorbic acid, dopamine, and uric acid using graphene/size-selected Pt nanocomposites. *Biosens. Bioelectron.* **2011**, *26*, 3450–3455. [CrossRef] [PubMed]
50. Chatterjee, S.; Chen, A. Voltammetric detection of the α-dicarbonyl compound: Methylglyoxal as a flavoring agent in wine and beer. *Anal. Chim. Acta* **2012**, *751*, 66–70. [CrossRef] [PubMed]
51. Mathiyarasu, J.; Senthilkumar, S.; Phani, K.L.N.; Yegnaraman, V. PEDOT-Au nanocomposite film for electrochemical sensing. *Mater. Lett.* **2008**, *62*, 571–573. [CrossRef]

52. Bianchini, C.; Zane, D.; Curulli, A. Gold microtubes assembling architecture for an impedimetric glucose biosensing system. *Sens. Actuators B* **2015**, *220*, 734–742. [CrossRef]
53. Wang, J.; Musameh, M. Carbon-nanotubes doped polypyrrole glucose biosensor. *Anal. Chim. Acta* **2005**, *539*, 209–2013. [CrossRef]
54. Valentini, F.; Galache Fernandez, L.; Tamburri, E.; Palleschi, G. Single Walled Carbon Nanotubes/polypyrrole–GOx composite films to modify gold microelectrodes for glucose biosensors: Study of the extended linearity. *Biosens. Bioelectron.* **2013**, *43*, 75–78. [CrossRef]
55. Adamson, T.L.; Eusebio, F.A.; Cook, C.B.; La Belle, J.T. The promise of electrochemical impedance spectroscopy as novel technology for the management of patients with diabetes mellitus. *Analyst* **2012**, *137*, 4179–4187. [CrossRef]
56. Denisov, E.T.; Afanas'ev, I.B. *Oxidation and Antioxidants in Organic Chemistry and Biochemistry*; CRC Press: Andover, MA, USA, 2005.
57. Quideau, S.; Deffieux, D.; Douat-Casassus, C.; Pouysegu, L. Plant Polyphenols: Chemical Properties, Biological Activities, and Synthesis. *Angew. Chem. Int. Ed.* **2011**, *50*, 586–621. [CrossRef]
58. Gordon, M.H. Significance of Dietary Antioxidants for Health. *Int. J. Mol. Sci.* **2012**, *13*, 173–179. [CrossRef]
59. Rice-Evans, C.A.; Miller, N.J.; Paganga, G. Structure-antioxidant activity relationships of flavonoids and phenolic acids. *Free Radic. Biol. Med.* **1996**, *20*, 933–956. [CrossRef]
60. Bors, W.; Heller, W.; Michel, C.; Stettmaier, K. Flavonoids, and polyphenols: Chemistry and biology. In *Handbook of Antioxidants*; Cadenas, E., Packer, L., Eds.; Marcel Dekker: New York, NY, USA, 1996; pp. 409–466.
61. Guo, Q.; Yue, Q.; Zhao, J.; Wang, L.; Wang, H.; Wei, X.; Liu, J.; Jia, J. How far can hydroxyl radicals travel? An electrochemical study based on a DNA mediated electron transfer process. *Chem. Commun.* **2011**, *47*, 11906–11908. [CrossRef] [PubMed]
62. Pandey, K.B.; Rizvi, S.I. Plant Polyphenols as Dietary Antioxidants in Human Health and Disease. *Oxid. Med. Cell. Longev.* **2009**, *2*, 270–278. [CrossRef]
63. Apak, R.; Demirci Çekiç, S.; Üzer, A.; Çelik, S.E.; Bener, M.; Bekdeşer, B.; Can, Z.; Sağlam, Ş.; Önem, A.N.; Erçağ, E. Novel Spectroscopic and Electrochemical Sensors and Nanoprobes for the Characterization of Food and Biological Antioxidants. *Sensors* **2018**, *18*, 186. [CrossRef]
64. Bounegru, A.V.; Apetrei, C. Voltammetric Sensors Based on Nanomaterials for Detection of Caffeic Acid in Food Supplements. *Chemosensors* **2020**, *8*, 41. [CrossRef]
65. Di Carlo, G.; Curulli, A.; Trani, A.; Zane, D.; Ingo, G.M. Enhanced electrochemical response of structurally related antioxidant at nanostructured hybrid films. *Sens. Actuators B* **2014**, *191*, 703–710. [CrossRef]
66. Di Carlo, G.; Curulli, A.; Toro, R.G.; Bianchini, C.; De Caro, T.; Padeletti, G.; Zane, D.; Ingo, G.M. Green Synthesis of Gold–Chitosan Nanocomposites for Caffeic Acid Sensing. *Langmuir* **2012**, *28*, 5471–5479. [CrossRef]
67. Curulli, A.; Di Carlo, G.; Ingo, G.M.; Riccucci, C.; Zane, D.; Bianchini, C. Chitosan Stabilized Gold Nanoparticle-Modified Au Electrodes for the Determination of Polyphenol Index in Wines: A Preliminary Study. *Electroanalysis* **2012**, *24*, 897–904. [CrossRef]
68. Kang, N.J.; Lee, K.W.; Shin, B.J.; Jung, S.K.; Hwang, M.K.; Bode, A.M.; Heo, Y.-S.; Lee, H.J.; Dong, Z. Caffeic acid, a phenolic phytochemical in coffee, directly inhibits Fyn kinase activity and UVB-induced COX-2 expression. *Carcinogenesis* **2008**, *30*, 321–330. [CrossRef]
69. Filik, H.; Çetintaş, G.; Aslıhan Avan, A.; Aydar, S.; Naci Koç, S.; Boz, İ. Square-wave stripping voltammetric determination of caffeic acid on electrochemically reduced graphene oxide–Nafion composite film. *Talanta* **2013**, *116*, 245–250. [CrossRef]
70. Leite, F.; de Jesus Rodrigues Santos, F.R.W.; Tatsuo Kubota, L. Selective determination of caffeic acid in wines with electrochemical sensor based on molecularly imprinted siloxanes. *Sens. Actuators B* **2014**, *193*, 238–246. [CrossRef]
71. Thangavelu, K.; Palanisamy, S.; Chen, S.M.; Velusamy, V.; Chen, T.W.; Kannan Ramarajc, S. Electrochemical Determination of Caffeic Acid in Wine Samples Using Reduced Graphene Oxide/Polydopamine Composite. *J. Electrochem. Soc.* **2016**, *163*, B726–B731. [CrossRef]
72. Liu, Z.; Xu, J.; Yue, R.; Yang, T.; Gao, L. Facile one-pot synthesis of Au–PEDOT/rGO nanocomposite for highly sensitive detection of caffeic acid in red wine sample. *Electrochim. Acta* **2016**, *196*, 1–12. [CrossRef]

73. Liu, Z.; Lu, B.; Gao, Y.; Yang, T.; Yue, R.; Xu, J.; Gao, L. Facile one-pot preparation of Pd–Au/PEDOT/graphene nanocomposites and their high electrochemical sensing performance for caffeic acid detection. *Rsc Adv.* **2016**, *6*, 89157–89166. [CrossRef]
74. Karthik, R.; Vinoth Kumar, J.; Chen, S.-M.; Senthil Kumar, P.; Selvam, V.; Muthuraj, V. A selective electrochemical sensor for caffeic acid and photocatalyst for metronidazole drug pollutant-A dual role by rod-like SrV$_2$O$_6$. *Sci. Rep.* **2017**, *7*, 7254. [CrossRef] [PubMed]
75. Thangavelu, K.; Raja, N.; Chen, S.-M.; Liao, W.-C. Nanomolar electrochemical detection of caffeic acid in fortified wine samples based on gold/palladium nanoparticles decorated graphene flakes. *J. Colloids Interface Sci.* **2017**, *501*, 77–85. [CrossRef]
76. Bharath, G.; Alhseinat, E.; Madhu, R.; SM Mugo Alwasel, S.; Harrath, A.H. Facile synthesis of Au@α-Fe2O3@RGO ternary nanocomposites for enhanced electrochemical sensing of caffeic acid toward biomedical applications. *J. Alloys Compd.* **2018**, *750*, 819–827. [CrossRef]
77. Gao, L.; Yue, R.; Xu, J.; Liu, Z.; Chai, J. Pt-PEDOT/rGO nanocomposites: One-pot preparation and superior electrochemical sensing performance for caffeic acid in tea. *J. Electroanal. Chem.* **2018**, *816*, 14–20. [CrossRef]
78. Shi, Y.; Xu, H.; Gu, Z.; Wang, C.; Du, Y. Sensitive detection of caffeic acid with trifurcate PtCu nanocrystals modified glassy carbon electrode. *Colloids Surf. A* **2019**, *567*, 27–31. [CrossRef]
79. Chen, T.-W.; Rajaji, U.; Chen, S.-M.; Govindasamy, M.; Selvin, S.S.P.; Manavalan, S.; Arumugam, R. Sonochemical synthesis of graphene oxide sheets supported Cu2S nanodots for high sensitive electrochemical determination of caffeic acid in red wine and soft drinks. *Compos. Part B* **2019**, *158*, 419–427. [CrossRef]
80. Karabozhikova, V.; Tsa, V. Electroanalytical determination of caffeic acid–Factors controlling the oxidation reaction in the case of PEDOT-modified electrodes. *Electrochim. Acta* **2019**, *293*, 439–446. [CrossRef]
81. Erady, V.; Mascarenhas, R.J.; Satpati, A.K.; Bhakta, A.K.; Mekhalif, Z.; Delhalle, J.; Dhason, A. Carbon paste modified with Bi decorated multi-walled carbon nanotubes and CTAB as a sensitive voltammetric sensor for the detection of Caffeic acid. *Microchem. J.* **2019**, *146*, 73–82. [CrossRef]
82. Manikandantr, V.S.; Sidhureddy, B.; Thiruppathi, A.R.; Chen, A. Sensitive Electrochemical Detection of Caffeic Acid in Wine Based on Fluorine-Doped Graphene Oxide. *Sensors* **2019**, *19*, 1604. [CrossRef] [PubMed]
83. Bianchini, C.; Curulli, A.; Pasquali, M.; Zane, D. Determination of caffeic acid in wine using PEDOT film modified electrode. *Food Chem.* **2014**, *156*, 81–86. [CrossRef]
84. Santana-Gálvez, J.; Cisneros-Zevallos, L.; Jacobo-Velázquez, D.A. Chlorogenic Acid: Recent Advances on Its Dual Role as a Food Additive and a Nutraceutical against Metabolic Syndrome. *Molecules* **2017**, *22*, 358. [CrossRef]
85. Mohammadi, N.; Najafi, M.; Adeh, N.B. Highly defective mesoporous carbon–Ionic liquid paste electrode as sensitive voltammetric sensor for determination of chlorogenic acid in herbal extracts. *Sens. Actuators B* **2017**, *243*, 838–846. [CrossRef]
86. Ma, X.; Yang, H.; Xiong, H.; Li, X.; Gao, J.; Gao, Y. Electrochemical Behavior and Determination of Chlorogenic Acid Based on Multi-Walled Carbon Nanotubes Modified Screen-Printed Electrode. *Sensors* **2016**, *16*, 1797. [CrossRef]
87. Zhang, T.; Chen, Y.; Huang, W.; Wang, Y.; Hu, X. A novel AuNPs-doped COFs composite as electrochemical probe for chlorogenic acid detection with enhanced sensitivity and stability. *Sens. Actuators B* **2018**, *276*, 362–369. [CrossRef]
88. Chokkareddy, R.; Redhi, G.G.; Karthick, T. A lignin polymer nanocomposite based electrochemical sensor for the sensitive detection of chlorogenic acid in coffee samples. *Heliyon* **2019**, *5*, e01457. [CrossRef]
89. Manivannan, K.; Sivakumar, M.; Cheng, C.-C.; Luc, C.-H.; Chen, J.-K. An effective electrochemical detection of chlorogenic acid in real samples: Flower-like ZnO surface covered on PEDOT: PSS composites modified glassy carbon electrode. *Sens. Actuators B* **2019**, *301*, 127002. [CrossRef]
90. Huang, Z.; Zhang, Y.; Sun, J.; Chen, S.; Chen, Y.; Fang, Y. Nanomolar detection of chlorogenic acid at the cross-section surface of the pencil lead electrode. *Sens. Actuators B* **2020**, *321*, 128550. [CrossRef]
91. Nadeem, M.; Imran, M.; Aslam Gondal, T.; Shahbaz, A.I.M.; Amir, R.M.; Wasim Sajid, M.; Batool Qaisrani, T.; Atif, M.; Hussain, G.; Salehi, B.; et al. Therapeutic Potential of Rosmarinic Acid: A Comprehensive Review. *Appl. Sci.* **2019**, *9*, 3139. [CrossRef]
92. Mohamadi, M.; Mostafavi, A.; Torkzadeh-Mahanic, M. Voltammetric Determination of Rosmarinic Acid on Chitosan/Carbon Nanotube Composite-Modified Carbon Paste Electrode Covered with DNA. *J. Electrochem. Soc.* **2015**, *162*, B344–B349. [CrossRef]

93. Alipour, S.; Azar, A.; Husain, S.W.; Rajabi, H.R. Determination of Rosmarinic acid in plant extracts using a modified sensor based on magnetic imprinted polymeric nanostructures. *Sens. Actuators B* **2020**, *323*, 128668. [CrossRef]
94. Trani, A.; Petrucci, R.; Marrosu, G.; Zane, D.; Curulli, A. Selective electrochemical determination of caffeine at a gold-chitosan nanocomposite sensor: May little change on nanocomposites synthesis affect selectivity? *J. Electroanal. Chem.* **2017**, *788*, 99–106. [CrossRef]
95. Zhao, F.; Wang, F.; Zhao, W.; Zhou, J.; Liu, Y.; Zou, L.; Ye, B. Voltammetric sensor for caffeine based on a glassy carbon electrode modified with Nafion and graphene oxide. *Microchim. Acta* **2011**, *174*, 383–390. [CrossRef]
96. de Jesus Rodrigues Santos, W.; Santhiago, M.; Pagotto Yoshida, I.V.; Kubota, L.T. Electrochemical sensor based on imprinted sol–gel and nanomaterial for determination of caffeine. *Sens. Actuators B* **2012**, *166–167*, 739–745. [CrossRef]
97. Yu Heng Khoo, W.; Pumera, M.; Bonanni, A. Graphene platforms for the detection of caffeine in real samples. *Anal. Chim. Acta* **2013**, *804*, 92–97. [CrossRef]
98. Yang, G.; Zhao, F.; Zeng, B. Facile fabrication of a novel anisotropic gold nanoparticle–chitosan–ionic liquid/graphene modified electrode for the determination of theophylline and caffeine. *Talanta* **2014**, *127*, 116–122. [CrossRef] [PubMed]
99. Rezaei, B.; Khalili Boroujeni, M.; Ensafi, A.A. Caffeine electrochemical sensor using imprinted film as recognition element based on polypyrrole, sol-gel, and gold nanoparticles hybrid nanocomposite modified pencil graphite electrode. *Biosens. Bioelectron.* **2014**, *60*, 77–83. [CrossRef] [PubMed]
100. Shehata, M.; Azab, S.M.; Fekry, A.M. May glutathione and graphene oxide enhance the electrochemical detection of caffeine on carbon paste sensor in aqueous and surfactant media for beverages analysis? *Synth. Met.* **2019**, *256*, 116122. [CrossRef]
101. Fekry, A.M.; Shehata, M.; Azab, S.M.; Walcarius, A. Voltammetric detection of caffeine in pharmacological and beverages samples based on simple nano- Co (II, III) oxide modified carbon paste electrode in aqueous and micellar media. *Sens. Actuators B* **2020**, *302*, 127172. [CrossRef]
102. Dhara, K.; Mahapatra Debiprosad, R. Review on nanomaterials-enabled electrochemical sensors for ascorbic acid detection. *Anal. Biochem.* **2019**, *586*, 113415. [CrossRef] [PubMed]
103. Siddeeg, S.M.; Salem Alsaiari, N.; Tahoon, M.A.; Ben Rebah, F. The Application of Nanomaterials as Electrode Modifiers for the Electrochemical Detection of Ascorbic Acid: Review. *Int. J. Electrochem. Sci.* **2020**, *15*, 3327–3346. [CrossRef]
104. Zhang, X.; Yu, S.; He, W.; Uyama, H.; Xie, Q.; Zhang, L.; Yang, F. Electrochemical sensor based on carbon supported $NiCoO_2$ nanoparticles for selective detection of ascorbic acid. *Biosens. Bioelectron.* **2014**, *55*, 446–451. [CrossRef]
105. Khalilzadeh, M.A.; Borzoo, M. Green synthesis of silver nanoparticles using onion extract and their application for the preparation of a modified electrode for determination of ascorbic acid. *J. Food Drug Anal.* **2016**, *24*, 796–803. [CrossRef] [PubMed]
106. Hussain, M.M.; Asiri AM Rahman, M.M. Non-enzymatic simultaneous detection of acetylcholine and ascorbic acid using ZnO·CuO nanoleaves: Real sample analysis. *Microchem. J.* **2020**, *159*, 105534. [CrossRef]
107. Mohammadnia, M.S.; Khosrowshahi, E.M.; Naghian, E.; Keihan, A.H.; Sohouli, E.; Plonska-Brzezinska, M.E.; Sobhani-Nasab, A.; Rahimi-Nasrabadi, M.; Ahmadi, F. Application of carbon nanoonion $NiMoO_4$-$MnWO_4$ nanocomposite for modification of glassy carbon electrode: Electrochemical determination of ascorbic acid. *Microchem. J.* **2020**, *159*, 105470. [CrossRef]
108. Brainina Khiena, Z.; Bukharinova, M.A.; Stozhko, N.Y.; Sokolkov, S.V.; Tarasov, A.V.; Vidrevich, M.B. Electrochemical Sensor Based on a Carbon Veil Modified by Phytosynthesized Gold Nanoparticlesmfor Determination of Ascorbic Acid. *Sensors* **2020**, *20*, 1800. [CrossRef]
109. Xiao, X.; Zhang, Z.; Nan, F.; Zhao, Y.; Wang, P.; He, F.; Wang, Y. Mesoporous $CuCo_2O_4$ rods modified glassy carbon electrode as a novel non-enzymatic amperometric electrochemical sensors with high sensitive ascorbic acid recognition. *J. Alloys Compd.* **2021**, *852*, 157045. [CrossRef]
110. de Faria, L.V.; Pedrosa Lisboa, T.; Marques de Farias, D.; Moreira Araujo, F.; Moura Machado, M.; Arromba de Sousa, R.; Costa Matos, M.A.; Abarza Muñoz, R.A.; Camargo Matos, R. Direct analysis of ascorbic acid in food beverage samples by flow injection analysis using reduced graphene oxide sensor. *Food Chem.* **2020**, *319*, 126509. [CrossRef] [PubMed]

111. Shen, T.; Liu, T.; Mo, H.; Yuan, Z.; Cui, F.; Jin, Y.; Chen, X. Cu-based metal–organic framework HKUST-1 as effective catalyst for highly sensitive determination of ascorbic acid. *Rsc Adv.* **2020**, *10*, 22881. [CrossRef]
112. Zhuanga, Z.; Chen, W. One-step rapid synthesis of $Ni_6(C_{12}H_{25}S)_{12}$ nanoclusters for electrochemical sensing of ascorbic acid. *Analyst* **2020**, *145*, 2621–2631. [CrossRef] [PubMed]
113. Jia, D.; Liu, Z.; Liu, L.; Shin Low, S.; Lua, Y.; Yu, X.; Zhu, L.; Li, C.; Liu, Q. Smartphone-based integrated voltammetry system for simultaneous detection of ascorbic acid, dopamine, and uric acid with graphene and gold nanoparticles modified screen-printed electrodes. *Biosens. Bioelectron.* **2018**, *119*, 55–62. [CrossRef]
114. Cinti, S.; Colozza, N.; Cacciotti, I.; Moscone, D.; Polomoshnov, M.; Sowade, E.; Baumann, R.R.; Arduini, F. Electroanalysis moves towards paper-based printed electronics: Carbon black nanomodified inkjet-printed sensor for ascorbic acid detection as a case study. *Sens. Actuators B* **2018**, *265*, 155–160. [CrossRef]
115. WHO. *Guidelines for Drinking-Water Quality*, 3rd ed.; World Health Organization: Geneva, Switzerland, 2004; p. 417.
116. Ning, D.; Zhang, H.; Zheng, J. Electrochemical sensor for sensitive determination of nitrite based on the PAMAM dendrimer-stabilized silver nanoparticles. *J. Electroanal. Chem.* **2014**, *717–718*, 29–33. [CrossRef]
117. Yang, B.; Bin, D.; Wang, H.; Zhu, M.; Yang, P.; Du, Y. High quality Pt-graphene nanocomposites for efficient electrocatalytic nitrite sensing. *Colloids Surf. A Phys. Eng. Asp.* **2015**, *481*, 43–50. [CrossRef]
118. Wang, H.; Wen, F.; Chen, Y.; Sun, T.; Meng, Y.; Zhang, Y. Electrocatalytic determination of nitrite based on straw cellulose/Molybdenum sulphide nanocomposite. *Biosens. Bioelectron.* **2016**, *85*, 692–697. [CrossRef]
119. Thirumalraj, B.; Palanisamy, S.; Chen, S.-M.; Zhao, D.-H. Amperometric detection of nitrite in water samples by use of electrodes consisting of palladium-nanoparticle-functionalized multi-walled carbon nanotubes. *J. Colloid Interface Sci.* **2016**, *478*, 413–420. [CrossRef]
120. Mehmeti, E.; Stanković, D.M.; Hajrizi, A.; Kalcher, K. The use of graphene nanoribbons as efficient electrochemical sensing material for nitrite determination. *Talanta* **2016**, *159*, 34–39. [CrossRef]
121. Zhao, Z.; Xia, Z.; Liu, C.; Huang, H.; Ye, W. Green synthesis of Pd/Fe_3O_4 composite based on poly DOPA functionalized reduced graphene oxide for electrochemical detection of nitrite in cured food. *Electrochim. Acta* **2017**, *256*, 146–154. [CrossRef]
122. Aksu, Z.; Alanyalıoglu, M. Fabrication of free-standing reduced graphene oxide composite papers doped with different dyes and comparison of their electrochemical performance for electrocatalytical oxidation of nitrite. *Electrochim. Acta* **2017**, *258*, 1376–1386. [CrossRef]
123. Jaiswal, N.; Tiwari, I.; Foster, C.W.; Banks, C.E. Highly sensitive amperometric sensing of nitrite utilizing bulk modified MnO2 decorated Graphene oxide nanocomposite screen-printed electrodes. *Electrochim. Acta* **2017**, *227*, 255–266. [CrossRef]
124. Losada, J.; García Armada, M.P.; García, E.; Casado, C.M.; Alonso, B. Electrochemical preparation of gold nanoparticles on ferrocenyl-dendrimer film modified electrodes and their application for the electrocatalytic oxidation and amperometric detection of nitrite. *J. Electroanal. Chem.* **2017**, *788*, 14–22. [CrossRef]
125. Bagheri, H.; Hajian, A.; Rezaei, M.; Shirzadmehr, A. Composite of Cu metal nanoparticles-multiwall carbon nanotubes-reduced graphene oxide as a novel and high performance platform of the electrochemical sensor for simultaneous determination of nitrite and nitrate. *J. Hazard. Mater.* **2017**, *324*, 762–772. [CrossRef]
126. Wang, K.; Wu, C.; Wang, F.; Liu, C.; Yu, C.; Jiang, G. In-situ insertion of carbon nanotubes into metal-organic frameworks-derived-Fe2O3 polyhedrons for highly sensitive electrochemical detection of nitrite. *Electrochim. Acta* **2018**, *285*, 128–138. [CrossRef]
127. Zhou, X.; Zhou, Y.; Hong, Z.; Zheng, X.; Lv, R. Magnetic Co@carbon nanocages for facile and binder-free nitrite sensor. *J. Electroanal. Chem.* **2018**, *824*, 45–51. [CrossRef]
128. Zhang, Y.; Nie, J.; Wei, H.; Xu, H.; Wang, Q.; Cong, Y.; Tao, J.; Zhang, Y.; Chu, L.; Zhou, Y.; et al. Electrochemical detection of nitrite ions using Ag/Cu/MWNT nanoclusters electrodeposited on a glassy carbon electrode. *Sens. Actuators B* **2018**, *258*, 1107–1116. [CrossRef]
129. Zhao, Z.; Zhang, J.; Wang, W.; Sun, Y.; Li, P.; Hu, J.; Chen, L.; Gong, W. Synthesis and electrochemical properties of Co3O4-rGO/CNTs composites towards highly sensitive nitrite detection. *Appl. Surf. Sci.* **2019**, *485*, 274–282. [CrossRef]
130. Hameed, R.M.A.; Medany, S.S. Construction of core-shell structured nickel@platinum nanoparticles on graphene sheets for electrochemical determination of nitrite in drinking water samples. *Microchem. J.* **2019**, *145*, 354–366. [CrossRef]

131. Chen, H.; Yang, T.; Liu, F.; Li, W. Electrodeposition of gold nanoparticles on Cu-based metal-organic framework for the electrochemical detection of nitrite. *Sens. Actuators B* **2019**, *286*, 401–407. [CrossRef]
132. Wang, P.; Wang, M.; Zhou, F.; Yang, G.; Qu, L.; Miao, X. Development of a paper-based, inexpensive, and disposable electrochemical sensing platform for nitrite detection. *Electrochem. Commun.* **2017**, *81*, 74–78. [CrossRef]

Publisher's Note: MDPI stays neutral with regard to jurisdictional claims in published maps and institutional affiliations.

© 2020 by the author. Licensee MDPI, Basel, Switzerland. This article is an open access article distributed under the terms and conditions of the Creative Commons Attribution (CC BY) license (http://creativecommons.org/licenses/by/4.0/).

Communication

Mixed Amphiphilic Polymeric Nanoparticles of Chitosan, Poly(vinyl alcohol) and Poly(methyl methacrylate) for Intranasal Drug Delivery: A Preliminary In Vivo Study

Inbar Schlachet, Hen Moshe Halamish and Alejandro Sosnik *

Laboratory of Pharmaceutical Nanomaterials Science, Department of Materials Science and Engineering, Technion-Israel Institute of Technology, Technion City, Haifa 3200003, Israel; inbarschlachet@gmail.com (I.S.); chen.moshe.num1@gmail.com (H.M.H.)
* Correspondence: sosnik@technion.ac.il or alesosnik@gmail.com; Tel.: +972-77-887-1971

Academic Editor: Giuseppe Cirillo
Received: 5 August 2020; Accepted: 28 September 2020; Published: 30 September 2020

Abstract: Intranasal (i.n.) administration became an alternative strategy to bypass the blood–brain barrier and improve drug bioavailability in the brain. The main goal of this work was to preliminarily study the biodistribution of mixed amphiphilic mucoadhesive nanoparticles made of chitosan-*g*-poly(methyl methacrylate) and poly(vinyl alcohol)-*g*-poly(methyl methacrylate) and ionotropically crosslinked with sodium tripolyphosphate in the brain after intravenous (i.v.) and i.n. administration to Hsd:ICR mice. After i.v. administration, the highest nanoparticle accumulation was detected in the liver, among other peripheral organs. After i.n. administration of a 10-times smaller nanoparticle dose, the accumulation of the nanoparticles in off-target organs was much lower than after i.v. injection. In particular, the accumulation of the nanoparticles in the liver was 20 times lower than by i.v. When brains were analyzed separately, intravenously administered nanoparticles accumulated mainly in the "top" brain, reaching a maximum after 1 h. Conversely, in i.n. administration, nanoparticles were detected in the "bottom" brain and the head (maximum reached after 2 h) owing to their retention in the nasal mucosa and could serve as a reservoir from which the drug is released and transported to the brain over time. Overall, results indicate that i.n. nanoparticles reach similar brain bioavailability, though with a 10-fold smaller dose, and accumulate in off-target organs to a more limited extent and only after redistribution through the systemic circulation. At the same time, both administration routes seem to lead to differential accumulation in brain regions, and thus, they could be beneficial in the treatment of different medical conditions.

Keywords: central nervous system (CNS); blood–brain barrier (BBB); self-assembled polymeric nanoparticles; intranasal delivery; biodistribution

1. Introduction

The treatment of diseases of the central nervous system (CNS) by systemic drug administration is challenging, owing to the presence of the blood–brain barrier (BBB) and the blood–cerebrospinal fluid barrier [1]. The BBB excludes more than 95% of the small-molecule and biological drugs from crossing into the brain [2,3]. In addition, the BBB displays different efflux transporters that transport substrate molecules (e.g., drugs) out of the brain endothelium, against a concentration gradient [4,5]. Drugs that do not comply with fundamental physicochemical characteristics such as high lipid solubility, low molecular weight and less than 8–10 H bonds with water cannot cross the BBB and their bioavailability and pharmacological efficacy diminished [2].

New delivery approaches that increase drug delivery to the CNS are under intense investigation [6]. The transient disruption of the BBB by osmotic shrinkage of the endothelial cells together with the opening of BBB tight junctions by intracarotid arterial infusion of non-diffusive solutes such as mannitol is one of them [7]. A main drawback is that increased permeability might also enable the passage of plasma proteins and result in abnormal neuronal function [8]. Another strategy is the use of carrier-mediated transport systems that transport nutrients such as glucose and amino acids into the CNS [8]. Drugs with the proper molecular design that do not always comply with the structure–activity relationship can be recognized by these influx transporters and show high permeability across the BBB [2,8,9]. Over recent decades, a plethora of nanotechnology strategies have been investigated to overcome the limited ability to deliver active molecules from the systemic circulation into the CNS [10–15]. For this, drug-loaded nanoparticles of a different nature (e.g., lipid, polymeric) and size are surface-decorated with ligands that bind receptors overexpressed in the BBB and cross the BBB by transcytosis [16,17].

The existence of a nose-to-brain pathway that bypasses the BBB has been evidenced by the accumulation and harm caused by environmental nanoparticulate matter in the CNS [18–21]. With the emergence of nanomedicine, different types of pure drug nanocrystals and nanoparticles were designed and their CNS bioavailability following intranasal (i.n.) administration assessed [22–26]. Recent studies confirmed the advantage of i.n. administration of nanocarriers over the intravenous (i.v.) route to increase drug bioavailability in the olfactory bulb. Among the different nanotechnology delivery platforms, polymeric micelles are among the most promising [22]. For example, in an early study, we demonstrated that the i.n. administration of the antiretroviral efavirenz nanoencapsulated within core-corona polymeric micelles made of poly(ethylene oxide)-*b*-poly(propylene oxide) (PEO-PPO) block copolymers significantly increases its bioavailability in the brain of rats with respect to the i.v. counterpart [27]. In this context, fundamental nanoparticle features (e.g., size, shape, surface chemistry) that govern the nose-to-brain transport remain to be elucidated [28]. For instance, the size, the shape and/or the surface chemistry and charge could change the transport pathway.

PEO-PPO polymeric micelles show limited encapsulation capacity for many hydrophobic cargos and low physical stability upon dilution over time. In this context, we developed mucoadhesive amphiphilic polymeric nanoparticles produced by the aggregation of chitosan (CS) and poly(vinyl alcohol) (PVA) graft copolymers hydrophobized in the side-chain with different hydrophobic blocks such as poly(methyl methacrylate) (PMMA) and their physical stabilization by non-covalent crosslinking of CS and PVA domains with sodium tripolyphosphate (TPP) and boric acid, respectively [29–33]. These nanoparticles display a multimicellar nanostructure [32,33] and very high physical stability under extreme dilution [29–33]. Aiming to understand the cellular pathways involved in this transport, we recently investigated the interaction of these polymeric nanoparticles with primary olfactory sensory neurons, cortical neurons and microglia isolated from olfactory bulb, olfactory epithelium and cortex of newborn rats [34]. Our results strongly suggested the involvement of microglia (and not cortical or olfactory neurons) in the nose-to-brain transport of nanoparticulate matter.

CS has been extensively investigated as a mucoadhesive drug nanocarrier, and its cytotoxicity is a matter of debate [35,36], including for i.n. drug delivery [37–40]. Most works reported on the good cell compatibility of this polysaccharide that is classified as "generally recognized as safe" (GRAS) by the US Food and Drug Administration (FDA) [41]. However, CS nanoparticles have been also associated with cell toxicity because of the electrostatic interaction of the positively-charged surface with the negatively-charged cell membrane, and the toxicity level depends on the cell type [42].

Aiming to capitalize on the potential of our versatile amphiphilic nanocarriers in mucosal drug delivery in general and in nose-to-brain administration in particular, we preliminary investigated the cell compatibility of CS-*g*-PMMA nanoparticles in human primary nasal epithelial cells and showed their high toxicity [42]. In this context, we produced mixed CS-*g*-PMMA:PVA-*g*-PMMA (1:1 weight ratio) nanoparticles that display better human nasal cell compatibility than the CS-based counterparts owing to a decrease in the surface charge density, as expressed by a less positive zeta-potential (Z-potential). In addition, we demonstrated that they cross a model of nasal epithelium in vitro [43]. These nanoparticles encapsulated two experimental anticancer drugs [43]. In this work, we preliminarily investigated the biodistribution of these mixed amphiphilic nanoparticles after i.n. administration to Hsd:ICR mice and compared it to the i.v. route for the first time.

2. Results and Discussion

Mixed nanoparticles were produced by the solvent casting method that comprised co-dissolution of identical amounts of CS-PMMA30 (a CS-g-PMMA copolymer containing 30% *w/w* of PMMA) and PVA-PMMA16 (a PVA-g-PMMA copolymer containing 16% *w/w* of PMMA) in dimethyl sulfoxide (DMSO), drying under vacuum and redispersion in water [43,44]. Self-assembly takes place once the critical aggregation concentration (CAC) is surpassed. The CAC of CS-PMMA30 and PVA-PMMA16 is in the 0.04–0.05% *w/v* range [30–33]. Since the self-assembly process is random, by utilizing this method, we anticipated the formation of mixed nanoparticles with very similar qualitative and quantitative composition. To physically stabilize the nanoparticle, CS domains were crosslinked by the formation of a polyelectrolyte complex with TPP. The size, size distribution and Z-potential of 0.1% *w/v* non-crosslinked and TPP-crosslinked mixed CS-PMMA30:PVA-PMMA16 nanoparticles before the in vivo studies were analyzed by dynamic light scattering (DLS), at 25 °C [42]. Non-crosslinked and crosslinked nanoparticles showed monomodal size distribution (one size population), while the polydispersity index (PDI), which is a measure of the size distribution, slightly changed after the ionotropic crosslinking; e.g., non-crosslinked mixed CS-PMMA30:PVA-PMMA16 nanoparticles showed a hydrodynamic diameter (D_h) of 193 ± 62 nm and a PDI of 0.23 (Table S1) [42]. This size is similar to that shown by pure CS-PMM30 (D_h of 184 ± 4 nm; PDI of 0.20) nanoparticles and larger than that of pure PVA-PMMA16 counterparts of the same concentration (D_h of 92 ± 4 nm and PDI of 0.14) [33]. Crosslinking of a 0.1% *w/v* nanoparticle suspension with TPP solution in water (1% *w/v*; 2.5 µL per mL of nanoparticles) resulted in an increase in the size to 249 ± 26 nm and in the PDI to 0.26, and their full physical stabilization [42]. Pure crosslinked CS-PMM30 nanoparticles are larger—332 ± 54 nm (PDI of 0.33)—owing to nanoparticle bridging [32], a phenomenon that is less likely in mixed particles that contain 50% *w/w* of non-ionic PVA-PPMA16, a copolymer that does not interact with TPP.

The surface charge of nanoparticulate matter affects their cell compatibility, and usually, positively-charged particles are more cytotoxic than neutral and negatively-charged ones [45]. CS has been extensively reported as a biocompatible polysaccharide, and it is approved in the food industry [41]. However, it may elicit cell toxicity in vitro due to a highly positively-charged surface [36,37,46]. Crosslinking of self-assembled CS-based nanoparticles was implemented to physically stabilize them and to partly neutralize the net positive surface charge and increase their cell compatibility [30]. This modification was not enough to ensure their good compatibility with human primary nasal epithelium cells [42]. Thus, we produced mixed nanoparticles that reduce the effective CS concentration on the surface and thus, its charge density, while preserving the nanoencapsulation capacity of the nanoparticles and its mucoadhesiveness [42]. Further crosslinking reduced the Z-potential and improved the compatibility of the nanoparticles in primary nasal epithelial cells [42,43].

We visualized the morphology of non-crosslinked and crosslinked nanoparticles by high resolution-scanning electron microscopy (HR-SEM). The size was in line with DLS analysis, considering that in HR-SEM, the nanoparticles underwent drying as opposed to DLS where the D_h is measured (Figure 1). Some aggregation during sample preparation could not be prevented, though these aggregates are not present in the nanoparticle suspension.

Since in a previous work, we showed that these nanoparticles cross a model of the human nasal epithelium in vitro [42], we hypothesized that they could effectively reach the CNS upon i.n. administration. In this framework, the main goal of this work was to investigate for the first time the biodistribution and accumulation in the brain and other organs of mixed CS-PMMA30:PVA-PMMA16 nanoparticles after i.n. administration to Hsd:ICR mice and compare it to the i.v. route. Since crosslinked nanoparticles are physically stable, as opposed to the non-crosslinked counterparts, for this preliminary study, 0.1% w/v crosslinked mixed CS-PMMA30:PVA-PMMA16 nanoparticles were labeled with the near infrared (NIR) dye NIR-797 and 200 µL of the nanoparticle suspension was injected i.v. through the tail vein (total nanoparticle dose of 8 mg/kg), or 20 µL of the same formulation was administered i.n. (total nanoparticle dose of 0.8 mg/kg). It is important to highlight that in this preliminary study, the nanoparticle dose administered i.n. was 10-fold smaller than i.v. At predetermined time points, live animal screening was performed using IVIS Spectrum In Vivo Imaging System. After i.v. administration, nanoparticles reach the systemic circulation and interact with the reticuloendothelial system, a system of macrophages mostly in the liver that could sequester the nanoparticles due the recognition of opsonins (serum proteins), while nanoparticles with size of up to 5–10 nm could undergo renal filtration [47]. At different time points post-administration (0–24 h), mice were sacrificed, the different organs carefully dissected to prevent cross contamination and the average fluorescence radiance (AFR) of each organ was quantified by subtracting the basal signal of each organ in control (untreated) animals (Figure 2). After i.v. administration, the highest accumulation at the different time points was observed in the liver (Figure 2a), as described for other nanoparticles of similar size and composition upon i.v. administration [48–50]. Other organs showed lower AFR associated with a more limited nanoparticle off-target accumulation. According to the size (several hundreds of nanometers), these nanoparticles do not undergo renal filtration. Thus, their detection in the kidneys is most probably related to their accumulation in the renal tissue (e.g., proximal tube epithelium) [50].

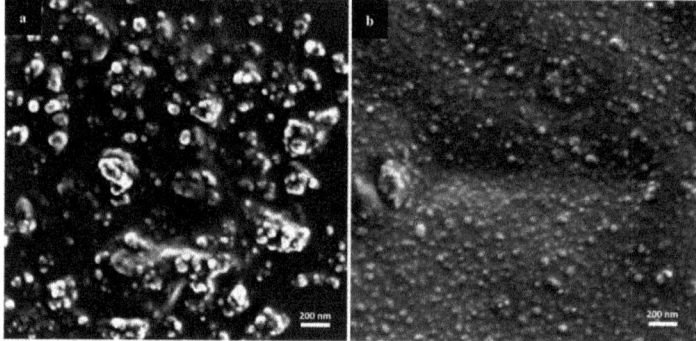

Figure 1. HR-SEM micrographs of mixed CS-g-PMMA:PVA-g-PMMA nanoparticles. (**a**) non-crosslinked and (**b**) TPP-crosslinked nanoparticles.

Intranasal is a local administration route that capitalizes on the nose-to-brain transport to surpass the BBB and target different parts of the brain. Thus, accumulation in peripheral organs such as the liver was expected to take place to a very limited extent [51]. After i.n. administration, the accumulation of the nanoparticles in off-target organs was much lower than after i.v. injection (Figure 2b). Moreover, a comparison of the AFR values in the different organs at different time points (0–4 h) after i.n. and i.v. administration revealed that some of the differences between both administration routes were statistically significant (Table S2). In particular, the accumulation of the nanoparticles in the liver, which is the main clearance organ for nanoparticulate matter in this size range, was up to 20 times lower after i.n. administration than by i.v. even though the dose was 10 times smaller; intranasally administered nanoparticles could reach peripheral organs after redistribution from the CNS to the systemic circulation [52,53]. These results highlight the benefit of i.n. administration to reduce off-target delivery and toxicity.

The imaging system used in this study normalizes the AFR to the organ that displays the maximum intensity, in this case the liver. Thus, we imaged the brains separately from the other organs (in triplicates) at different time points (0–24 h, depending on the administration route) and estimated the nanoparticle accumulation in the "top" brain (i.v. and i.n. administration), and "bottom" brain and head (i.n. administration) (Figure 3a). Upon i.v. administration (0–4 h), our nanoparticles accumulated mainly in the "top" brain. Later time points were not investigated in this preliminary study because we previously showed the relatively limited bioavailability of this type of nanoparticle in the CNS of mice after i.v. administration and the need for the surface modification with ligands that bind receptors expressed on the apical side of the BBB endothelium [49]. We were more interested in exploring the behavior of intranasally administered nanoparticles, and thus, we tracked them for 24 h. Different i.n. administration methods and formulations could affect the biodistribution of the drug-loaded nanoparticles in the CNS and the pharmacological outcome. For example, Martins et al. showed that the i.n. administration of oxytocin with a nebulizer leads to a different pharmacological outcome compared to a standard nasal spray [54]. These results highlight the complexity of this transport pathway and the difficulty of comparing among works that used different formulations, doses and administration regimens. After i.n. administration, particles are expected to enter the CNS through the olfactory region and accumulate in the nose, the nose-to-brain tract (e.g., olfactory bulb) and the "bottom" brain (Figure 3a), before they disseminate to all the brain [21,55].

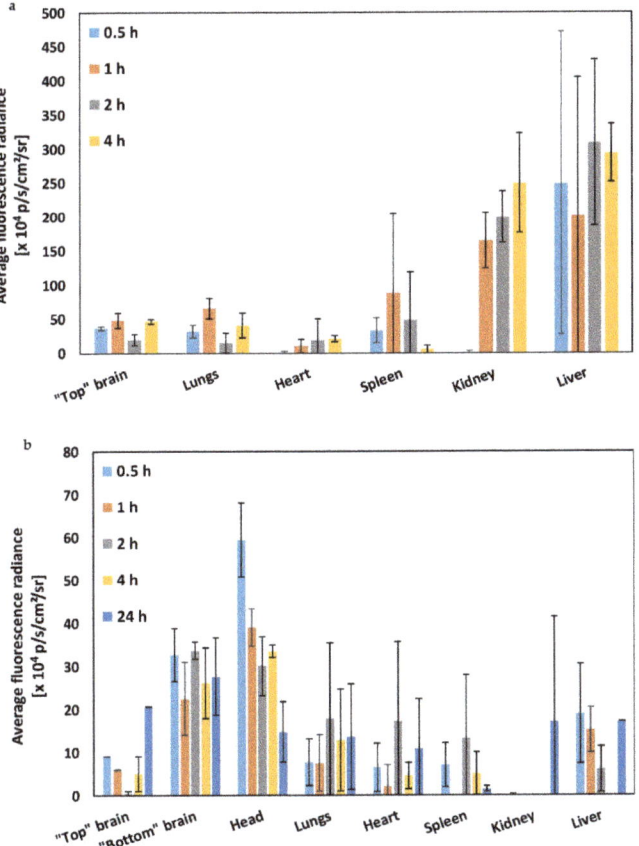

Figure 2. Biodistribution of NIR-797-labeled crosslinked 0.1% *w/v* mixed CS-PMMA30:PVA-PMMA16 nanoparticles after (**a**) i.v. administration and (**b**) i.n. administration to Hsd:ICR mice (n = 3). The measurement was performed after organ dissection at each time point. Average fluorescence radiance was measured using Living Imaging analysis software. Bars represent the average of mice at each time point. The error bars are S.D. from the mean. Statistical comparisons are summarized in Table S2.

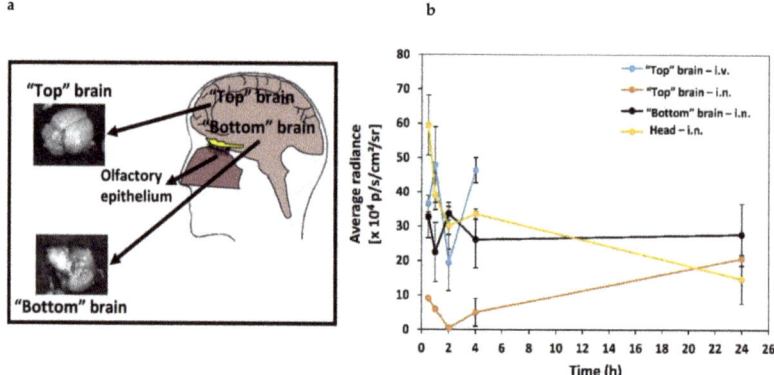

Figure 3. Ex vivo analysis of the distribution of NIR-797-labeled crosslinked 0.1% *w/v* mixed CS-PMMA30: PVA-PMMA16 nanoparticles in the brain following i.v. and i.n. administration to Hsd:ICR mice (n = 3). (**a**) Scheme of the top and bottom brain and (**b**) average radiance over time obtained after the subtraction of the control (untreated mice brain) radiance (n = 3).

The highest AFR value was measured in the "top" brain 1 h after i.v. injection (Figure 3b). At this point, we analyzed the bottom side of the brain after i.n. administration because after penetrating through the olfactory epithelium, the nanoparticles could be accumulated in this area of the CNS close to the pons and serve as a reservoir from which an encapsulated drug could be released. Two hours after i.n. administration, the accumulation in the "bottom" brain was significantly higher than upon i.v. injection. We further calculated the AFR in the brain (with subtraction of the control signal) at each time point and compared values of area under the curve (AUC) between 0 and 4 h (AUC_{0-4h}). Nanoparticle accumulation in the "bottom" brain after i.n. administration (AUC_{0-4h} = 110 ± 10 × 10^4 p/s/cm^2/sr) was similar and not significantly different from that of intravenously administered nanoparticles in the "top" brain (AUC_{0-4h} = 130 ± 20 × 10^4 p/s/cm^2/sr) (Figure 3b, Table 1). These results indicate that a similar brain bioavailability could be reached with a 10 times smaller dose. At the same time, it is important to point out that in the case of i.n. administration, these mucoadhesive nanoparticles could be initially retained in the nasal mucosa and accumulate in the nose-to-brain tract and, at a later stage, be released and diffuse across the brain tissue to reach more distant areas.

Table 1. Average fluorescence radiance obtained from different brain regions and the head after the subtraction of the control (untreated mice) radiance at different time points and calculated area under the curve $(AUC)_{0-4h}$ values.

Brain Region	AUC_{0-4h} (× 10^4 p/s/cm^2/sr) ± S.D.	AUC_{0-24h} (× 10^4 p/s/cm^2/sr) ± S.D.
"Top" brain—i.v.	130 ± 20 *	N.D.
"Top" brain—i.n.	15 ± 4	41 ± 3 *****
"Bottom" brain—i.n.	110 ± 18 **	164 ± 7 **,****,*****
Head—i.n.	138 ± 17 ***	186 ± 6 *****

N.D.: Not determined. * Statistically significant difference between AUC_{0-4h} in the "top" brains of mice after i.v. and i.n. administration ($p < 0.05$); ** statistically significant AUC_{0-4h} and AUC_{0-24h} difference between the "top" and "bottom" brains of mice after i.n. administration ($p < 0.05$); *** statistically significant AUC_{0-4h} and AUC_{0-24h} difference between the "top" brain and the head of mice after i.n. administration ($p < 0.05$); ****statistically significant AUC_{0-24h} difference between the "bottom" brain and the head of mice after i.n. administration ($p < 0.05$); *****statistically significant differences between AUC_{0-24h} and AUC_{0-4h} after i.n. administration ($p < 0.05$).

In general, two beneficial phenomena were observed after i.n. administration when compared to i.v.: (i) there was higher accumulation in the brain [51] and (ii) the accumulation was less spread, enabling targeting of the nanoparticles to more specific CNS regions, which is associated with the nanoparticle properties [56] and probably with the type of nasal nanoformulation [54]. After i.n.

administration, our mucoadhesive nanoparticles reached the brain quickly (less than 1 h) and could be detected mainly in the head due to their retention in the nose and accumulated mainly in the "bottom" brain, while the AFR in the "top" brain was relatively low. The fast delivery of nanoparticles to the brain upon i.n. administration has been reported in the literature [57,58]. It is also important to stress that nose tissues were not isolated, and thus, the whole head without the brain was imaged to estimate the retention of the nanoparticles in the nasal mucosa. The $AUC_{0-4\,h}$ in the "bottom" brain was significantly higher than that in the "top" brain, indicating that the nanoparticles initially accumulate in the nose-brain tract (Figure 3b, Table 1). At 24 h, the AFR in the "bottom" brain remained almost constant (AFR ratio between 24 and 4 h was 1.06) and it increased by 4-fold in the "top" brain with respect to 4 h, at the expense of the AFR detected in the head/nose. A similar trend was followed by the $AUC_{0-24\,h}$ that showed a moderate increase in the "bottom" brain from $110 \pm 18 \times 10^4$ to $164 \pm 7 \times 10^4$ p/s/cm²/sr and a more pronounced one in the "top" brain from $15 \pm 4 \times 10^4$ to $41 \pm 3 \times 10^4$ p/s/cm²/sr (Table 1). These findings suggest that the nanoparticles are transported from the nasal mucosa and the "bottom" brain to other brain areas at a slower rate, leading to these changes in the $AUC_{0-24\,h}$ values. Most previous research utilizing i.n. delivery of nanomedicines disregarded the possible differential biodistribution and assumed that all the brain areas are exposed to similar nanoparticle concentrations. Our results with mixed CS-PMMA30:PVA-PMMA16 nanoparticles strongly suggest that they are not homogeneously distributed in the brain soon after administration. In addition, they strongly suggest that with the proper nanoparticle design, specific structures of the CNS could be targeted to treat different medical conditions affecting them.

Having said this, more studies need to be conducted to realize this potential. Future studies will investigate the pharmacokinetics of encapsulated drugs in the CNS upon i.v. and i.n. administration and will also include later time points.

3. Methods

3.1. Synthesis of the Chitosan-g-Poly(methyl methacrylate) and Poly(vinyl alcohol)-g-Poly(methyl methacrylate) Copolymers

A CS-*g*-PMMA copolymer containing 30% *w/w* of PMMA (CS-PMMA30, as determined by proton-nuclear magnetic resonance [31,32]) was synthesized by the graft free radical polymerization of MMA (99% purity, Alfa Aesar, Heysham, UK) onto the CS backbone in water. For this, low molecular weight CS (0.4 g, degree of deacetylation of 94%; viscosity ≤100 mPa.s, Glentham Life Sciences, Corsham, UK) was dissolved in nitric acid 70% (0.05 M in water, 100 mL) that was degassed by sonication (30 min, Elmasonic S 30, Elma Schmidbauer GmbH, Singen, Germany). Then, a tetramethylethylenediamine (TEMED, Alfa Aesar) solution (0.18 mL in 50 mL degassed water) was poured into the CS solution and purged with nitrogen for 30 min at room temperature (RT). The purged CS solution was magnetically stirred and heated to 35 °C, and 142 μL MMA (distilled under vacuum to remove inhibitors before use) was added to the degassed water (48 mL) and then mixed with the CS solution. Finally, a cerium (IV) ammonium nitrate (CAN, Strem Chemicals, Inc., Newburyport, MA, USA) solution (0.66 g in 2 mL degassed water) was added to the polymerization reaction that was allowed to proceed for 3 h at 35 °C and under continuous nitrogen flow. After 3 h, the polymerization was quenched by adding 0.13 g of hydroquinone (Merck, Hohenbrunn, Germany). The product was purified by dialysis against distilled water using a regenerated cellulose dialysis membrane with a molecular weight cut-off (MWCO) of 12–14 kDa (Spectra/Por® 4 nominal flat width of 75 mm, diameter of 48 mm and volume/length ratio of 18 mL/cm; Spectrum Laboratories, Inc., Rancho Dominguez, CA, USA) for 48–72 h and freeze-dried. The same chemical pathway with small modifications was used to synthesize a PVA-*g*-PMMA copolymer containing 16% *w/w* of PMMA (PVA-PMMA16) [33]. First, PVA (0.4 g) was dissolved in distilled water (100 mL) at RT, and TEMED (0.18 mL in 50 mL degassed water) was dissolved in 70% nitric acid (0.45 mL). Then, TEMED and PVA solutions were degassed by sonication for 30 min, mixed and purged with nitrogen for 30 min at RT. The solution was heated to 35 °C and 142 μL MMA dispersed in degassed water (48 mL) and added to the reaction mixture. Finally, a CAN solution (0.66 g in 2 mL degassed water) was added and the reaction allowed to proceed for 2 h at 35 °C. The reaction product was purified by dialysis and freeze-dried. Products were stored at 4 °C until use.

For biodistribution studies (see below), CS-PMMA30 and PVA-PMMA16 copolymers were fluorescently-labeled with the near infrared tracer NIR-797 isothiocyanate (Sigma-Aldrich, St. Louis, MO, USA). For this, CS-PMMA30 (80 mg) was dissolved in 8 mL water supplemented with acetic acid (pH = 5.5), prepared by diluting 70 μL glacial acetic acid (Bio-Lab Ltd., Jerusalem, Israel) in 1 L water under magnetic stirring. Then, NIR-797 (0.4 mg) was dissolved in *N,N*-dimethylformamide (0.2 mL, DMF, Bio-Lab Ltd.), added to the copolymer solution and the mixture stirred for 16 h protected from light, at RT. Finally, the product was dialyzed (48 h, regenerated cellulose dialysis membrane, MWCO of 3500 Da, Membrane Filtration Products, Inc., Seguin, TX, USA), freeze-dried (72–96 h) and stored protected from light at 4 °C until use. In the case of PVA-PMMA16, the copolymer (100 mg) was dissolved in 3 mL DMF under magnetic stirring. Then, NIR-797 (0.8 mg) was dissolved in DMF (0.1 mL), added to the copolymer solution and the mixture stirred for 16 h protected from light, at RT. The reaction mixture was diluted with deionized water (1:2 *v/v*), dialyzed (48 h, regenerated cellulose dialysis membrane, MWCO of 3500 Da) to remove unreacted NIR-797, freeze-dried (72–96 h) and stored protected from light at 4 °C until use. The theoretical NIR-797 content was between 0.5 and 0.8% *w/w*.

3.2. Preparation and Characterization of Mixed Chitosan-g-Poly(methyl methacrylate):Poly(vinyl alcohol)-g-Poly(methyl methacrylate) Nanoparticles

Identical amounts of CS-PMMA30 and PVA-PMMA16 were dissolved in DMSO to reach a total copolymer concentration of 0.5% *w/v* under continuous stirring (24 h), at 37 °C. Subsequently, the solution was dried under vacuum utilizing a freeze-dryer, the copolymer mixture re-dispersed in water supplemented with acetic acid (pH = 5.5) to reach a final total copolymer concentration of 0.1% *w/v* and filtered (1.2 μm cellulose acetate syringe filter, Sartorius Stedim Biotech GmbH, Göttingen, Germany). For physical stabilization, 0.1% *w/v* nanoparticles were crosslinked by the addition of 1% *w/v* TPP (Sigma-Aldrich) aqueous solution (2.5 μL of crosslinking solution per mL of 0.1% *w/v* nanoparticle dispersion).

The size (expressed as hydrodynamic diameter, D_h), size distribution (estimated by the PDI) and the Z-potential (an estimation of the surface charge density) of 0.1% *w/v* systems were measured using the Zetasizer Nano-ZS in the same media detailed above for the different samples. Z-potential measurements of the same samples required the use of laser Doppler microelectrophoresis in the Zetasizer Nano-ZS. Each value obtained is expressed as the mean ± standard deviation (S.D.) of at least three independent samples, while each DLS or Z-potential measurement is an average of at least seven runs.

The morphology of mixed nanoparticles before and after crosslinking was visualized by HR-SEM (carbon coating, acceleration voltage of 2–4 kV, Ultraplus, Zeiss, Oberkochen, Germany). For this, mixed nanoparticle suspensions (0.5% *w/v* total copolymer concentration) were drop-casted on silicon wafer, dried at 37 °C in the oven and carbon-coated. Images were obtained using an in-lens detector at 3–4 mm working distance. The nanosuspensions were sprayed on top of a silicon wafer (cz polished silicon wafers <100> oriented, highly doped N/Arsenic, SHE Europe Ltd., Livingston, UK) by introducing high pressure nitrogen which allowed an even spread of the nanoparticles on the wafer. Next, the wafer was attached to the grid using carbon-tape, and additional tape was placed on its frame. At the corners of the frame, silver paint (SPI# 05002-AB—Silver, SPI supplies, West Chester, PA, USA) was applied and the samples were carbon coated.

3.3. Biodistribution of Mixed Chitosan-g-Poly(methyl methacrylate):poly(vinyl alcohol)-g-Poly(methyl methacrylate) Nanoparticles

For biodistribution studies, CS-PMMA30 and PVA-PMMA16 copolymers were fluorescently labeled with NIR-797, as described above.

Hsd:ICR mice (Envigo, Jerusalem, Israel) were maintained at the Gutwirth animal facility of the Technion-Israel Institute of Technology. All animal experiments were approved and performed according to the guidelines of the Institutional Animal Research Ethical Committee at the Technion (ethics approval number IL-052-05-18). Animal welfare was monitored daily by the staff veterinarians. Mice fasted for 12 h prior to experiments. Crosslinked mixed CS-PMMA30:PVA-PMMA16 nanoparticles (200 μL, 0.1% w/v) were injected i.v. into the tail vein. For i.n. administration, mice were lightly anesthetized with 2.5% isoflurane (USP Terrel™ Piramal Critical Care, Bethlehem, PA, USA), and fixed in a supine position for the administration of 10 μL of the nanoparticles in each nostril using a pipette (total volume of 20 μL, 0.1% w/v). After 0.5, 1, 2, 4 and 24 h post-i.v. injection or i.n. administration, animals were sacrificed by dislocation and organs (liver, spleen, kidney, lungs, heart, brain, and head/nose) were dissected. Organ screening was performed ex vivo using an Imaging System (IVIS, PerkinElmer, Waltham, MA, USA) with an excitation at 795 nm and an emission at 810 nm. Then, at the same conditions (see above), image analysis was performed using Living Imaging analysis software (PerkinElmer). The auto fluorescence of organs of the control (untreated) mice were subtracted. Mice were used in triplicates for each time point. Then, the average radiance in the brain (with subtraction of the control signal) at each time point was calculated, and the values of $AUC_{0-4\,h}$ determined according to Equation (1) [59]

$$AUC = \sum_i \frac{(t_{i+1} - t_i)}{2 \times (AR_i + AR_{i+1})} \quad (1)$$

where t_i is the starting time point, t_{i+1} is the finishing time point (0.5, 1, 2 and 4 h), AR_i is the starting value of average fluorescence measured and AR_{i+1} is the finishing value for each measurement over time.

3.4. Statistical Analysis

Statistical analysis of the different experiments was performed by a t-test on raw data (Excel, Microsoft Office 2013, Microsoft Corporation). P values smaller than 0.05 were regarded as statistically significant.

Supplementary Materials: The following are available online, Table S1: hydrodynamic diameter (D_h), size distribution (PDI) and Z-potential of CS-PMMA30, PVA-PMMA16 and CS-PMMA30:PVA-PMMA16 nanoparticles (total copolymer concentration of 0.1% w/v), as determined by DLS; Table S2: statistical analysis of AFR data in the different organs and time points after i.v. and i.n. administration of mixed CS-PMMA30:PVA-PMMA16 nanoparticles, as analyzed by the IVIS Spectrum In Vivo Imaging System ($p < 0.05$).

Author Contributions: I.S. designed the methodology, conducted the experiments, analyzed the results, and assisted in the writing and the edition of the original and the revised manuscript; H.M.H. conducted the experiments and A.S. achieved the funding, conceptualized and supervised the research, analyzed the results, and wrote and edited the original and the revised manuscript. All authors have read and agreed to the published version of the manuscript.

Funding: This work was supported by the Israel Science Foundation (ISF, Grant #269/15) and the Teva National Network of Excellence in Neuroscience Research Grant. Partial support of the Russell Berrie Nanotechnology Institute (Technion) is also acknowledged.

Conflicts of Interest: The authors declare no conflict of interest.

References

1. Blanco-Prieto, M. Drug Delivery to the Central Nervous System: A Review. *J. Pharm. Sci.* **2003**, *6*, 252–273.
2. Pardridge, W. Non-Invasive Drug Delivery to the Human Brain Using Endogenous Blood-Brain Barrier Transport Systems. *Pharm. Sci. Technol. Today* **1999**, *2*, 49–59. [CrossRef]
3. Oller-Salvia, B.; Sanchez-Navarro, M.; Giralt, E.; Teixido, M. Blood-Brain Barrier Shuttle Peptides: An Emerging Paradigm for Brain Delivery. *Chem. Soc. Rev.* **2016**, *45*, 4690–4707. [CrossRef] [PubMed]
4. Löscher, W.; Potschka, H. Blood-Brain Barrier Active Efflux Transporters: ATP-Binding Cassette Gene Family. *NeuroRx* **2005**, *2*, 86–98. [CrossRef]

5. Dallas, S.; Miller, D.S.; Bendayan, R. Multidrug Resistance-Associated Proteins: Expression and Function in the Central Nervous System. *Pharmacol. Rev.* **2006**, *58*, 140–161. [CrossRef]
6. Zeiadeh, I.; Najjar, A.; Karaman, R. Strategies for Enhancing the Permeation of CNS-Active Drugs through the Blood-Brain Barrier: A Review. *Molecules* **2018**, *23*, 1289. [CrossRef]
7. Rapoport, S.I. Osmotic Opening of the Blood–Brain Barrier: Principles, Mechanism, and Therapeutic Applications. *Cell Mol. Neurobiol.* **2000**, *20*, 217–230. [CrossRef]
8. Ohtsuki, S.; Terasaki, T. Contribution of Carrier-Mediated Transport Systems to the Blood–Brain Barrier as a Supporting and Protecting Interface for the Brain; Importance for CNS Drug Discovery and Development. *Pharm. Res.* **2007**, *24*, 1745–1758. [CrossRef]
9. Tsuji, A. Small Molecular Drug Transfer Across the Blood-Brain Barrier via Carrier-Mediated Transport Systems. *NeuroRx* **2005**, *2*, 54–62. [CrossRef]
10. Gomes, M.J.; Mendes, B.; Martins, S.; Sarmento, B. Nanoparticle Functionalization for Brain Targeting Drug Delivery and Diagnostic. In *Handbook of Nanoparticles*; Aliofkhazraei, M., Ed.; Springer: Cham, Switzerland, 2016.
11. Gabathuler, R. Approaches to Transport Therapeutic Drugs Across the Blood-Brain Barrier to Treat Brain Diseases. *Neurobiol. Dis.* **2010**, *37*, 48–57. [CrossRef]
12. Fang, F.; Zou, D.; Wang, W.; Yin, Y.; Yin, T.; Hao, S.; Wang, B.; Wang, G.; Wang, Y. Non-Invasive Approaches for Drug Delivery to the Brain Based on the Receptor Mediated Transport. *Mater. Sci. Eng. C* **2017**, *76*, 1316–1327. [CrossRef] [PubMed]
13. Tam, V.H.; Sosa, C.; Liu, R.; Yao, N.; Priestley, R.D. Nanomedicine as a Non-Invasive Strategy for Drug Delivery Across the Blood Brain Barrier. *Int. J. Pharm.* **2016**, *515*, 331–342. [CrossRef] [PubMed]
14. Gaillard, P.J.; Visser, C.C.; Appeldoorn, C.C.M.; Rip, J. Enhanced Brain Drug Delivery: Safely Crossing the Blood-Brain Barrier. *Drug Discov. Today Technol.* **2012**, *9*, 155–160. [CrossRef]
15. Wong, H.L.; Wu, X.Y.; Bendayan, R. Nanotechnological Advances for the Delivery of CNS Therapeutics. *Adv. Drug Deliv. Rev.* **2012**, *64*, 686–700. [CrossRef] [PubMed]
16. Saraiva, C.; Praca, C.; Ferreira, R.; Santos, T.; Ferreira, L.; Bernardino, L. Nanoparticle-Mediated Brain Drug Delivery: Overcoming Blood-Brain Barrier to Treat Neurodegenerative Diseases. *J. Control. Release* **2016**, *235*, 34–47. [CrossRef]
17. Georgieva, J.V.; Hoekstra, D.; Zuhorn, I.S. Smuggling Drugs into the Brain: An Overview of Ligands Targeting Transcytosis for Drug Delivery across the Blood Brain Barrier. *Pharmaceutics* **2014**, *6*, 557–583. [CrossRef]
18. Lucchini, R.G.; Dorman, D.C.; Elder, A.; Veronesi, B. Neurological Impacts from Inhalation of Pollutants and the Nose-Brain Connection. *Neurotoxicology* **2012**, *33*, 838–841. [CrossRef]
19. Oberdörster, G.; Sharp, Z.; Atudorei, V.; Elder, A.; Gelein, R.; Kreyling, W.; Cox, C. Translocation of Inhaled Ultrafine Particles to the Brain. *Inhal. Toxicol.* **2004**, *16*, 437–445. [CrossRef]
20. Elder, A.; Gelein, R.; Silva, V.; Feikert, T.; Opanashuk, L.; Carter, J.; Potter, R.; Maynard, A.; Ito, Y.; Finkelstein, J.; et al. Translocation of Inhaled Ultrafine Manganese Oxide Particles to the Central Nervous System. *Environ. Health. Perspect.* **2006**, *114*, 1172–1178. [CrossRef]
21. Babadjouni, R.; Patel, A.; Liu, Q.; Shkirkova, K.; Lamorie-Foote, K.; Connor, M.; Hodis, D.M.; Cheng, H.; Sioutas, C.; Morgan, T.E.; et al. Nanoparticulate Matter Exposure Results in Neuroinflammatory Changes in the Corpus Callosum. *PLoS ONE* **2018**, *13*, e0206934. [CrossRef]
22. Pires, P.C.; Santos, A.O. Nanosystems in Nose-To-Brain Drug Delivery: A Review of Non-Clinical Brain Targeting Studies. *J. Control. Release* **2018**, *270*, 89–100. [CrossRef]
23. Feng, Y.; He, H.; Li, F.; Lu, Y.; Qi, J.; Wu, W. An Update on the Role of Nanovehicles in Nose-to-Brain Drug Delivery. *Drug Discov. Today* **2018**, *23*, 1079–1088. [CrossRef] [PubMed]
24. Hanson, L.R.; Frey, W.H. Intranasal Delivery Bypasses the Blood-Brain Barrier to Target Therapeutic Agents to the Central Nervous System and Treat Neurodegenerative Disease. *BMC Neurosci.* **2008**, *9*, S5. [CrossRef] [PubMed]
25. Perez, A.P.; Mundiña-Weilenmann, C.; Romero, E.L.; Morilla, M.J. Increased Brain Radioactivity by Intranasal P-labeled siRNA Dendriplexes Within in Situ-Forming Mucoadhesive Gels. *Int. J. Nanomed.* **2012**, *7*, 1373–1385.
26. Kumar, A.; Pandey, A.N.; Jain, S.K. Nasal-Nanotechnology: Revolution for Efficient Therapeutics Delivery. *Drug Deliv.* **2016**, *23*, 671–683. [CrossRef]

27. Chiappetta, D.A.; Hocht, C.; Opezzo, J.A.W.; Sosnik, A. Intranasal Administration of Antiretroviral-Loaded Micelles for Anatomical Targeting to the Brain in HIV. *Nanomedicine* **2013**, *8*, 223–237. [CrossRef]
28. Kanazawa, T.; Taki, H.; Tanaka, K.; Takashima, Y.; Okada, H. Cell-Penetrating Peptide-Modified Block Copolymer Micelles Promote Direct Brain Delivery via Intranasal Administration. *Pharm. Res.* **2011**, *28*, 2130–2139. [CrossRef]
29. Raskin Menaker, M.; Schlachet, I.; Sosnik, A. Mucoadhesive Nanogels by Ionotropic Crosslinking of Chitosan-*g*-Oligo(NiPAam) Polymeric Micelles as Novel Drug Nanocarriers. *Nanomedicine* **2016**, *11*, 217–233. [CrossRef]
30. Moshe, H.; Davizon, Y.; Menaker Raskin, M.; Sosnik, A. Novel Poly(Vinyl Alcohol)-Based Amphiphilic Nanogels by Boric Acid Non-Covalent Crosslinking of Polymeric Micelles. *Biomater. Sci.* **2017**, *5*, 2295–2309. [CrossRef]
31. Noi, I.; Schlachet, I.; Kumarasamy, M.; Sosnik, A. Permeability of Chitosan-*g*-Poly(Methyl Methacrylate) Amphiphilic Nanoparticles in a Model of Small Intestine In Vitro. *Polymers* **2018**, *10*, 478. [CrossRef]
32. Schlachet, I.; Trousil, J.; Rak, D.; Knudsen, K.D.; Pavlova, E.; Nyström, B.; Sosnik, A. Chitosan-*graft*-Poly(Methyl Methacrylate) Amphiphilic Nanoparticles: Self-Association and Physicochemical Characterization. *Carbohydr. Polym.* **2019**, *212*, 412–420. [CrossRef] [PubMed]
33. Moshe Halamish, H.; Trousil, J.; Rak, D.; Knudsen, K.D.; Pavlova, E.; Nyström, B.; Štěpánek, P.; Sosnik, A. Self-Assembly and Nanostructure of Poly(Vinyl Alcohol)-*graft*-Poly(Methyl Methacrylate) Amphiphilic Nanoparticles. *J. Colloid Interface Sci.* **2019**, *553*, 512–523. [CrossRef] [PubMed]
34. Kumarasamy, M.; Sosnik, A. The Nose-to-Brain Transport of Polymeric Nanoparticles is Mediated by Immune Sentinels and Not by Olfactory Sensory Neurons. *Adv. Biosyst.* **2019**, *3*, 1900123. [CrossRef] [PubMed]
35. Mohammed, M.A.; Syeda, J.T.M.; Wasan, K.M.; Wasan, E.K. An Overview of Chitosan Nanoparticles and Its Application in Non-Parenteral Drug Delivery. *Pharmaceutics* **2017**, *9*, 53. [CrossRef]
36. Ways, M.; Lau, W.M.; Khutoryanskiy, V.V. Chitosan and Its Derivatives for Application in Mucoadhesive Drug Delivery Systems. *Polymers* **2018**, *10*, 267. [CrossRef]
37. Alam, S.; Khan, Z.I.; Mustafa, G.; Kumar, M.; Islam, F.; Bhatnagar, A.; Ahmad, F.J. Development and Evaluation of Thymoquinone-Encapsulated Chitosan Nanoparticles for Nose-to-Brain Targeting: A Pharmacoscintigraphic Study. *Int. J. Nanomed.* **2012**, *7*, 5705–5718. [CrossRef]
38. Casettari, L.; Illum, L. Chitosan in Nasal Delivery Systems for Therapeutic Drugs. *J. Control. Release* **2014**, *190*, 189–200. [CrossRef]
39. Rassu, G.; Soddu, E.; Cossu, M.; Gavini, E.; Giunchedi, P.; Dalpiaz, A. Particulate Formulations Based on Chitosan for Nose-to-Brain Delivery of Drugs. A Review. *J. Drug Deliv. Sci. Technol.* **2016**, *32*, 77–87. [CrossRef]
40. Marques, C.; Som, C.; Schmutz, M.; Borges, O.; Borchard, G. How the Lack of Chitosan Characterization Precludes Implementation of the Safe-by-Design Concept. *Front. Bioeng. Biotechnol.* **2020**, *8*, 165. [CrossRef]
41. Hu, Y.-L.; Qi, W.; Han, F.; Shao, J.-Z.; Gao, J.-Q. Toxicity Evaluation of Biodegradable Chitosan Nanoparticles Using a Zebrafish Embryo Model. *Int. J. Nanomed.* **2011**, *6*, 3351–3359.
42. Schlachet, I.; Sosnik, A. Mixed Mucoadhesive Amphiphilic Polymeric Nanoparticles Cross a Model of Nasal Septum Epithelium In Vitro. *ACS Appl. Mater. Interfaces* **2019**, *11*, 21360–21371. [CrossRef] [PubMed]
43. Schlachet, I. Innovative Nano-Biomaterials for the Improved delivery of Antitumorals to the Central Nervous System in the Therapy of Pediatric Brain Tumors. Ph.D. Thesis, Technion-Israel Institute of Technology, Haifa, Israel, 2019.
44. Gaucher, G.; Dufresne, M.-H.; Sant, V.P.; Kang, N.; Maysinger, D.; Leroux, J.-C. Block Copolymer Micelles: Preparation, Characterization and Application in Drug Delivery. *J. Control. Release* **2005**, *109*, 169–188. [CrossRef] [PubMed]
45. Fröhlich, E. The Role of Surface Charge in Cellular Uptake and Cytotoxicity of Medical Nanoparticles. *Int. J. Nanomed.* **2012**, *7*, 5577–5591. [CrossRef] [PubMed]
46. Rizeq, B.R.; Younes, N.N.; Rasool, K.; Nasrallah, G.K. Synthesis, Bioapplications, and Toxicity Evaluation of Chitosan-Based Nanoparticles. *Int. J. Mol. Sci.* **2019**, *20*, 5776. [CrossRef]
47. Ernsting, M.J.; Murakami, M.; Roy, A.; Li, S.D. Factors Controlling the Pharmacokinetics, Biodistribution and Intratumoral Penetration of Nanoparticles. *J. Control. Release* **2013**, *172*, 782–794. [CrossRef]
48. He, C.; Hu, Y.; Yin, L.; Tang, C.; Yin, C. Effects of Particle Size and Surface Charge on Cellular Uptake and Biodistribution of Polymeric Nanoparticles. *Biomaterials* **2010**, *31*, 3657–3666. [CrossRef]

49. Bukchin, A.; Sanchez-Navarro, M.; Carrera, A.; Teixidó, M.; Carcaboso, A.M.; Giralt, E.; Sosnik, A. Amphiphilic Polymeric Nanoparticles Modified with a Retro-Enantio Peptide Shuttle Target the Brain of Mice. *Chem. Mater.* **2020**, *32*, 7679–7693. [CrossRef]
50. Williams, R.M.; Shah, J.; Ng, B.D.; Minton, D.R.; Gudas, L.J.; Park, C.Y.; Heller, D.A. Mesoscale Nanoparticles Selectively Target the Renal Proximal Tubule Epithelium. *Nano Lett.* **2015**, *15*, 2358–2364. [CrossRef]
51. Yadav, S.; Gattacceca, F.; Panicucci, R.; Amiji, M.M. Comparative Biodistribution and Pharmacokinetic Analysis of Cyclosporine—A in the Brain upon Intranasal or Intravenous Administration in an Oil-in-Water Nanoemulsion Formulation. *Mol. Pharm.* **2015**, *12*, 1523–1533. [CrossRef]
52. Gao, X.; Chen, J.; Chen, J.; Wu, B.; Chen, H.; Jiang, X. Quantum Dots Bearing Lectin-Functionalized Nanoparticles as a Platform for In Vivo Brain Imaging. *Bioconjug. Chem.* **2008**, *19*, 2189–2195. [CrossRef]
53. Sekerdag, E.; Lüle, S.; Bozdağ Pehlivan, S.; Öztürk, N.; Kara, A.; Kaffashi, A.; Vural, I.; Işıkay, I.; Yavuz, B.; Oguz, K.K.; et al. A Potential Non-Invasive Glioblastoma Treatment: Nose-to-Brain Delivery of Farnesylthiosalicylic Acid Incorporated Hybrid Nanoparticles. *J. Control. Release* **2017**, *261*, 187–198. [CrossRef] [PubMed]
54. Martins, D.A.; Mazibuko, N.; Zelaya, F.; Vasilakopoulou, S.; Loveridge, J.; Oates, A.; Maltezos, S.; Mehta, M.; Wastling, S.; Howard, M.; et al. Effects of Route of Administration on Oxytocin-Induced Changes in Regional Cerebral Blood Flow in Humans. *Nat. Commun.* **2020**, *11*, 1160. [CrossRef] [PubMed]
55. Shiga, H.; Taki, J.; Yamada, M.; Washiyama, K.; Amano, R.; Matsuura, Y.; Matsui, O.; Tatsutomi, S.; Yagi, S.; Tsuchida, A.; et al. Evaluation of the Olfactory Nerve Transport Function by SPECT-MRI Fusion Image with Nasal Thallium-201 Administration. *Mol. Imaging Biol.* **2011**, *13*, 1262–1266. [CrossRef] [PubMed]
56. McGowan, J.W.D.; Shao, Q.; Vig, P.J.S.; Bidwell, G.L. Intranasal Administration of Elastin-like Polypeptide for Therapeutic Delivery to the Central Nervous System. *Drug Des. Devel. Ther.* **2016**, *10*, 2803–2813. [CrossRef]
57. Xia, H.; Gao, X.; Gu, G.; Liu, Z.; Zeng, N.; Hu, Q.; Song, Q.; Yao, L.; Pang, Z.; Jiang, X.; et al. Low molecular weight protamine-functionalized naoparticles for drug delivery to the brain after intranasal administration. *Biomaterials* **2011**, *32*, 9888–9898. [CrossRef]
58. Md, S.; Khan, R.A.; Mustafa, G.; Chuttani, K.; Baboota, S.; Sahni, J.K.; Ali, J. Bromocriptine loaded chitosan nanoparticles intended for direct nose to brain delivery: Pharmacodynamic, pharmacokinetic and scintigraphy study in mice model. *Eur. J. Pharm. Sci.* **2013**, *48*, 393–405. [CrossRef]
59. Pardi, N.; Hogan, M.J.; Naradikian, M.S.; Parkhouse, K.; Cain, D.W.; Jones, L.; Moody, M.A.; Verkerke, H.P.; Myles, A.; Willis, E.; et al. Nucleoside-Modified MRNA Vaccines Induce Potent T Follicular Helper and Germinal Center B Cell Responses. *J. Exp. Med.* **2018**, *215*, 1571–1588. [CrossRef]

© 2020 by the authors. Licensee MDPI, Basel, Switzerland. This article is an open access article distributed under the terms and conditions of the Creative Commons Attribution (CC BY) license (http://creativecommons.org/licenses/by/4.0/).

Alginate Bioconjugate and Graphene Oxide in Multifunctional Hydrogels for Versatile Biomedical Applications

Giuseppe Cirillo [1,*], Elvira Pantuso [2], Manuela Curcio [1], Orazio Vittorio [3,4,5], Antonella Leggio [1], Francesca Iemma [1], Giovanni De Filpo [6] and Fiore Pasquale Nicoletta [1]

1. Department of Pharmacy, Health and Nutritional Sciences, University of Calabria, 87036 Rende (CS), Italy; manuela.curcio@unical.it (M.C.); antonella.leggio@unical.it (A.L.); francesca.iemma@unical.it (F.I.); fiore.nicoletta@unical.it (F.P.N.)
2. National Research Council of Italy (CNR)—Institute on Membrane Technology (ITM), 87036 Rende (CS), Italy; e.pantuso@itm.cnr.it
3. Children's Cancer Institute, Lowy Cancer Research Centre, UNSW Sydney, Sydney, NSW 2031, Australia; OVittorio@ccia.unsw.edu.au
4. School of Women's and Children's Health, Faculty of Medicine, UNSW Sydney, Sydney, NSW 2052, Australia
5. ARC Centre of Excellence for Convergent BioNano Science and Technology, Australian Centre for NanoMedicine, UNSW Sydney, Sydney, NSW 2052, Australia
6. Department of Chemistry and Chemical Technologies, University of Calabria, 87036 Rende (CS), Italy; giovanni.defilpo@unical.it
* Correspondence: giuseppe.cirillo@unical.it; Tel.: +39-098-449-3208

Abstract: In this work, we combined electrically-conductive graphene oxide and a sodium alginate-caffeic acid conjugate, acting as a functional element, in an acrylate hydrogel network to obtain multifunctional materials designed to perform multiple tasks in biomedical research. The hybrid material was found to be well tolerated by human fibroblast lung cells (MRC-5) (viability higher than 94%) and able to modify its swelling properties upon application of an external electric field. Release experiments performed using lysozyme as the model drug, showed a pH and electro-responsive behavior, with higher release amounts and rated in physiological vs. acidic pH. Finally, the retainment of the antioxidant properties of caffeic acid upon conjugation and polymerization processes (Trolox equivalent antioxidant capacity values of 1.77 and 1.48, respectively) was used to quench the effect of hydrogen peroxide in a hydrogel-assisted lysozyme crystallization procedure.

Keywords: hybrid hydrogels; controlled drug delivery; protein crystallization; lysozyme

1. Introduction

Hydrogels are valuable materials in biomedicine, including tissue engineering, drug delivery, and discovery, by virtue of their toughness, softness, flexibility, and elasticity [1,2]. More importantly, the significant wetting tendency of their hydrophilic and porous surfaces are interesting features in the preparation of biomaterials suitable for different applications, including the fabrication of either release devices or templates for crystallization of biologically relevant proteins [3]. The possibility of designing effective therapeutic strategies via modulating the amount and rate of release, offers the possibility to match different therapeutic needs with a single device [4,5]. In addition, the determination of complex molecular structures through X-ray diffraction (XRD) analyses is a key finding for elucidating the molecular basis of human pathologies and discovering new therapeutic targets [6–8].

The functional features and chemical composition of hydrogels can be finely tuned using a combination of components with different physical or chemical properties at the nanometer or molecular level [9] and these so-called hybrid hydrogels show performances superior to those of individual components [10,11]. The high biocompatibility, low-immunogenicity, biodegradability, chemical versatility, and natural abundance of polysaccharides and proteins can be coupled with significant stability, high purity, and absence of variability between batches (e.g., molecular weight) of synthetic polymers, for the formation of versatile natural/synthetic hybrid materials [12,13]. Moreover, the incorporation of inorganic components (e.g., carbon nanostructures, CNs) via covalent or non-covalent interactions [14,15] results in organic/inorganic hybrid materials with improved properties (e.g., thermal and mechanical stability) for long-term applications [16,17]. In addition, the electrical conductivity of CNs is retained, and therefore the hydrogel properties (e.g., water uptake and affinity for a loaded therapeutic agent) can be modulated by applying an external electric field [18,19].

More recently, it has been proven that functional hydrogels with antioxidant properties can be synthesized by the conjugation of active molecules (e.g., polyphenols) to polymer chains [20,21], offering interesting solutions for mitigating oxidative stress. This is a significant challenge in both drug delivery and discovery, since the possibility to reduce oxidative damages is a key property for any material designed for interactions with living tissues [22,23], as well as in X-ray crystallography, where the oxidation of the protein sample in a crystallization droplet can lead to failure of the crystallization process [24,25].

In this work, we synthesized a multifunctional hybrid hydrogel film (HACG) obtained by co-polymerization of acrylate monomers in the presence of graphene oxide (GO), and a new sodium alginate-caffeic acid conjugate (AlgCF). The performance of HACG in biomedical applications was evaluated by exploring two main issues, namely the fabrication of a smart delivery vehicle able to modulate the release of bioactive macromolecules by electric stimulation, and the obtainment of a template able to confine and concentrate macromolecules in its porous structure, thus, facilitating crystallization under oxidation conditions.

2. Results and Discussion

In this study, the proposed multifunctional hybrid hydrogel (i.e., HACG) is composed of an acrylate network where AlgCF conjugate and graphene oxide (GO) are inserted as functional components, conferring peculiarities suitable for applications in biomedicine.

Acrylate polymers have been extensively studied for biomedical applications, due to their high biocompatibility and chemical versatility [26,27].

Sodium alginate (Alg) is a natural polysaccharide with β-D-mannuronic acid and α-L-guluronic acid repeating units [28], which shows the typical advantages of polysaccharides for biomedical applications, together with a high degradation rate allowing applications such as cell encapsulating and wound healing devices [29–31], as well as a coadjuvant in protein crystallization [32,33].

Caffeic acid (CF), which belongs to the hydroxycinnamic acids with a phenylpropenoic acid structure containing a 3,4-dihydroxylated aromatic ring attached to a carboxyl group through a trans-ethylene wire, is a powerful antioxidant that can both scavenge and inhibit the generation of free radical species [34,35]. The hexagonal lattice structure of hybridized sp^2 carbon atoms of GO is responsible for superior electrical conductivity and ability to interact with bioactive molecules of either low or high molecular weight via π–π, hydrophobic, electrostatic interactions, and hydrogen bonding [36,37]. Furthermore, due to the presence of oxygen-rich functionalities (e.g., epoxide, phenolic, hydroxyl, and carboxylic groups), GO can be covalently incorporated in hydrogel networks via condensation/polymerization methods, obtaining hybrid hydrogels for drug delivery, biosensors, and tissue scaffold applications [38–40].

The synthetic procedure involved the synthesis of the AlgCF conjugate, using immobilized laccase as a biocatalyst, and its subsequent co-polymerization in the presence of GO and acrylate monomer mixture, consisting of hydroxyethyl acrylate (HEA) and polyethylene glycol dimethacrylate 750 (PEGDMA) as plasticizing and crosslinking monomers, respectively (Figure 1).

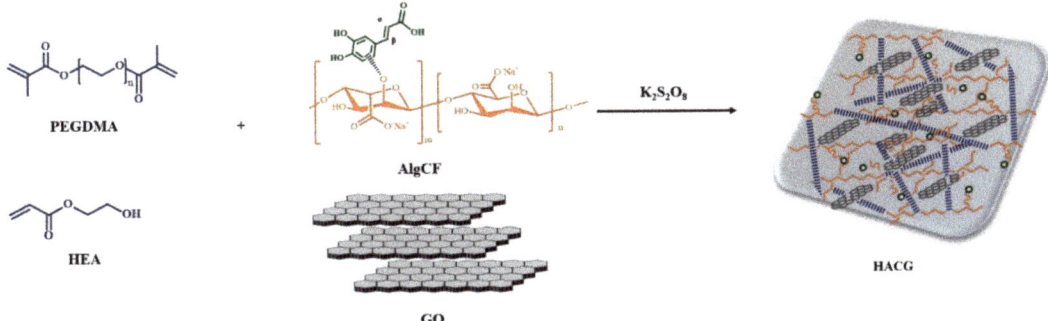

Figure 1. Schematic representation of the synthesis of multifunctional hybrid hydrogels.

2.1. Synthesis and Characterization of Sodium Alginate-Caffeic Acid Conjugate

AlgCF conjugate was synthesized by means of laccase chemistry via a heterogeneous catalysis approach previously developed for the conjugation of different polyphenolic compounds to polysaccharide and protein materials [41]. This methodology can be conducted in a totally green environment, ensuring the absence of any trace of toxicity and a high purity of the final product, which are key advantages for any material designed for biomedical applications. The experimental procedure involved immobilization of laccase into acrylate polymer networks. Laccase promotes a one-electron oxidation of caffeic acid (CF) [42,43] favoring the coupling by reactive groups in the Alg side chains, although the actual reaction mechanism is not well understood [44].

After purification by dialysis procedure to ensure removal of any trace of unconjugated CF, chemical characterization of AlgCF was performed by means of ^{1}H-NMR and calorimetric analyses to assess the effective conjugation and the effect of CF on the thermal properties of the conjugate, respectively, while static light scattering measurements were used to estimate the average molecular weight (Mw) of Alg and AlgCF.

The ^{1}H-NMR spectra of Alg and AlgCF are reported in Figure 2. Signals in the range of 4.2–5.1 ppm were assigned to the Alg and AlgCF anomeric protons of β-D-mannuronic acid and α-L-guluronic acid repeating units, and to the methine protons adjacent to the carboxyl groups, the resonance of the remaining protons of sugar rings fall in between 3.2 and 4.1 ppm [45,46]. In the AlgCF spectrum, new signals not recorded in the spectrum of native Alg and assigned to CF residues were detected and considered to be experimental evidence for the effective formation of the conjugate, namely the olefinic protons at around 6.2 (α) and 7.4 (β) ppm and the aromatic protons in the range of 6.6–7.3 ppm [47].

The differential scanning calorimetry (DSC) thermograms of Alg and AlgCF are shown in Figure 3.

Alg showed the typical DSC curve of polysaccharide materials with hydroxyl and carboxyl functionalities in the repeating units, with a first transition at 95 °C, assigned to the evaporation of moisture from the polymer [48], and a broad exothermic at 243 °C, expressing the formation of CO_2, CH_4, and H_2O from polysaccharide chains due to decomposition at a higher temperature [49]. The decomposition involves different kinds of chemical reactions, including depolymerization, elimination of oxygen-rich functionalities in the side chains, chain scissions, recombination, and cross-linking, which can be accelerated in the presence of radical species [50].

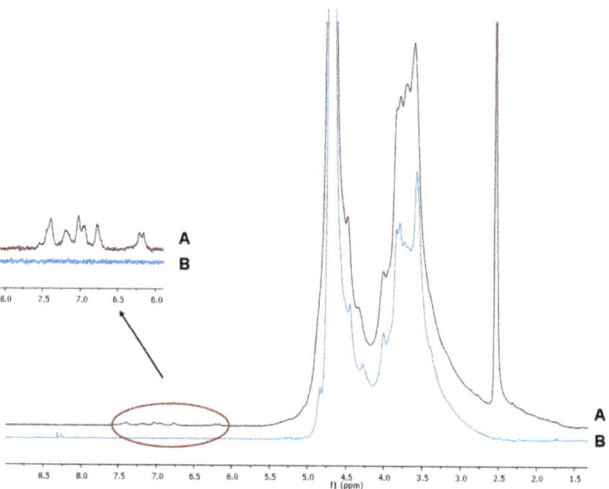

Figure 2. ^{1}H-NMR of purified (**A**) sodium alginate-caffeic acid conjugate (AlgCF) and (**B**) sodium alginate (Alg) in D$_2$O.

The covalent conjugation with CF moieties was expected to enhance the thermal stability of the AlgCF conjugate, as confirmed by the shift of the exothermal degradation peak to higher temperature values (251 °C).

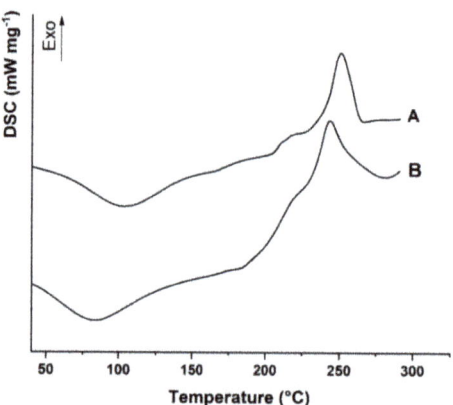

Figure 3. Differential scanning calorimetry (DSC) thermograms of (**A**) AlgCF and (**B**) Alg. Curves were vertically shifted for readability.

The results of the molecular weight determination (145 ± 15 kDa for both Alg and AlgCF) indicated that the molecular weight did not significantly change upon conjugation of CF, suggesting a functionalization degree below 10%.

2.2. Synthesis and Characterization of Hybrid Hydrogels

We previously reported on the possibility of covalently incorporating either carbon nanostructures (carbon nanotubes and graphene oxide) or polyphenol conjugates into acrylate hydrogels via free radical polymerization, obtaining functional materials suitable for drug delivery and healing applications [23,51].

Here, following the same synthetic approach, AlgCF and GO were simultaneously incorporated into a hydrogel network (HACG) based on HEA and PEGDMA, optimizing the reagent ratio (11.5% AlgCF, 1.15% GO, 42.30% HEA, and 45.05% PEGDMA, w/w) in order to maximize the amount of incorporated GO and AlgCF and the water affinity, avoiding, at the same time, hydrogel breakage upon drying. A lower amount of plasticizing and crosslinker, indeed, carried out to fragile hydrogels, while higher amounts of PEGDMA were responsible for less swellable materials.

Control samples, HAG and HAC, were synthesized by replacing AlgCF with native Alg and not inserting GO in the pre-polymerization feed, respectively, to determine the influence of either CF or GO moieties on the device performance. In detail, HAC was used to highlight the electro-responsivity in the release experiments, while HAG was useful in the crystallization experiments.

HACG was characterized by a rough and porous surface, as per morphological investigation (Figure 4a), while human fibroblast lung cells (MRC-5) viability values higher than 94.0% ± 2.5 upon incubation of HACG (0.1 to 1.0 mg mL^{-1} concentration range, Figure 4b), confirmed its biocompatibility. These cells, indeed, are widely recognized as an in vitro model to check the toxicity of different kinds of biologically oriented materials, due to their specific metabolic features and high sensibility to almost any types of chemicals [52].

Similar results were observed when HAG and HAC were used.

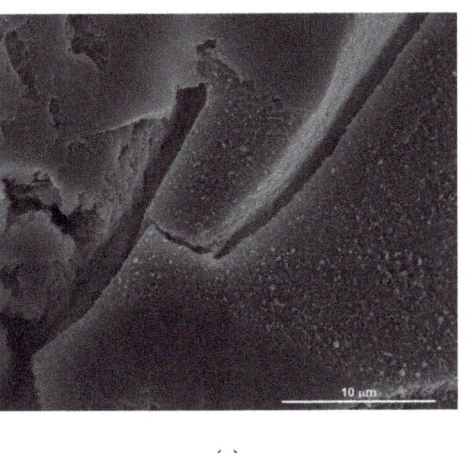
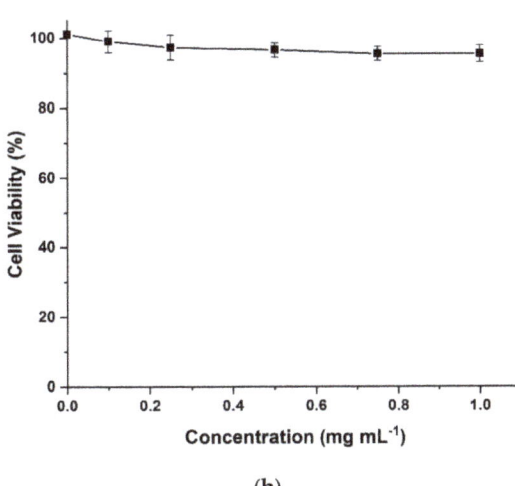

(a) (b)

Figure 4. (a) SEM image of multifunctional hybrid hydrogel film (HACG) sample; (b) Human fibroblast lung cells (MRC-5) viability after 72 h incubation with pulverized HACG sample suspended in Dulbecco's modified Eagle's medium (DMEM) + FCS 10%.

Swelling experiments were performed in order to evaluate the HACG water affinity, a required key property for materials proposed for interactions with living tissues. Due to the chemical features of Alg and GO, such determination was performed in two pH conditions, mimicking an acidic (pH 5.5) and a physiological environment (pH 7.4), respectively, and in the presence or absence of electric stimulation at 12, 24, and 48 V (Table 1).

From the data in Table 1, it is evident that the carboxylic functionalities in both Alg and GO were responsible for the higher water absorption properties at pH 7.4 vs. pH 5.5, because of their different ionization statuses, carrying out electrostatic repulsion between the carboxylate anions (pKa of COOH in the range 4–5) [53]. The effect of applying an external voltage on the swelling profile can be highlighted by comparing the swelling of blank and hybrid hydrogels and introducing the swelling ratio parameter S_r, according to Equation (1):

$$S_r = \frac{WR_v - WR_0}{WR_0} \times 100 \tag{1}$$

where WR_v and WR_0 are the swelling degrees at the selected voltage (12, 24, and 48 V) and 0 V, respectively.

Table 1. Swelling behavior at equilibrium of blank and hybrid hydrogels in different pH and voltage conditions.

Sample	pH	Voltage	WR (%)	$S_r = \frac{WR_v - WR_0}{WR_0} \times 100$
HAC	5.5	0	332 ± 3.1	---
		12	334 ± 3.3	0.6
		24	337 ± 2.9	1.5
		48	341 ± 3.4	2.7
HACG	5.5	0	351 ± 2.8	---
		12	401 ± 3.0	14.2
		24	578 ± 3.2	64.7
		48	554 ± 3.1	57.8
HAC	7.4	0	605 ± 2.7	---
		12	611 ± 2.9	1.0
		24	617 ± 2.7	2.0
		48	620 ± 2.8	2.5
HACG	7.4	0	696 ± 3.1	---
		12	805 ± 3.0	15.7
		24	849 ± 2.7	22.0
		48	835 ± 2.8	20.0

As reported in the literature, this effect can be explained as the result of two main phenomena, i.e., ionization of the COOH functionalities and generation of an osmotic pressure between the inner and outer portions of the hydrogel network, due to the rearrangement of mobile ions moving to the opposite electrodes [54].

The swelling degree of HAC was weakly affected by the applied voltages (Sr below 2.7 at both pH values), while the GO sp^2 carbon layer conferred significant electro-responsivity to HACG, being more evident at acidic than physiological conditions. At pH 5.5, the larger number of available undissociated COOH could be effectively ionized by the electric stimulation, resulting in an increased electrostatic repulsion, and thus higher hydrogel swelling. On the contrary, at pH 7.4, a more significant number of COO$^-$ groups was already formed even in the absence of the external voltage, and lower Sr values were recorded. Furthermore, although the ionization of COOH groups was the main driving force at 12 and 24 V, osmotic pressure started to be predominant at higher voltages (48 V), carrying out a more evident network deformation and reduced Sr values (Sr_{48} 57.8 and 20.0 vs. Sr_{24} 64.7 and 22.0 at pH 5.0 and 7.4, respectively).

2.3. Lysozyme Loading and In Vitro Release Studies

Electric stimulation is a valuable tool for fine tuning the delivery of therapeutic agents, in terms of both total release amount and rate, because of the possibility to modulate key parameters such as voltage intensity and duration [55]. Here, we explored the possibility of using HACG as a carrier for lysozime (LZM), a naturally occurring protein (14.4 kDa molecular weight) that shows high biocompatibility and antibacterial activities through degradation of cell walls of Gram-positive bacteria [56], and is widely proposed as a bioactive component of hydrogel systems for tissue engineering applications [57].

LZM was loaded into HAC and HACG through a soaking–drying procedure to reach the same LZM to carrier ratio in both cases (6.0%). The loading of LZM on HAC is the result of the formation of strong electrostatic interactions between the negatively charged carboxylic functionalities of AlgCF (pKa between 4 and 5) and positively charged groups on LZM (isoelectric point = 11) [58].

Release experiments were recorded as a function of pH variation, selecting the physiological (7.4) and the acidic (5.5) pH value of infection site where LZM is planned to elicit its antimicrobial activity.

For a more detailed comparison of the behavior of blank and hybrid hydrogels, experimental data were analyzed by applying four different mathematical models, considering the release kinetics as zero order (Equation (2)), first-order (Equation (3)) kinetics, or a combination of Fickian and anomalous diffusion (Equations (4) and (5)) [59].

The first model is the zero-order kinetic expressed by Equation (2) as follows:

$$\frac{M_t}{M_0} = K_0 t \quad (2)$$

where M_t is the amount of released LZM at time t, M_0 the total amount of loaded LZM, K_0 is the zero-order kinetic constant, and t the time of release.

Equation (3) describes a first-order kinetic as follows:

$$\frac{M_t}{M_0} = a\left(1 - e^{-K_1 t}\right) \quad (3)$$

K_1 is the first-order kinetic constant, t is the time of the release, and a is the release coefficient.

The third model is given by the Ritger–Peppas Equation (4):

$$\frac{M_t}{M_0} = K t^n \quad (4)$$

where K is the kinetic constant, t is the time of the release, and n is the coefficient indicating the mechanism of the release; $n \leq 0.43$ indicates a Fickian diffusion mechanism, n = 0.84 a Case II transport, and 0.43 < n < 0.85 anomalous transport mechanism.

The last model is described by the Peppas–Sahlin Equation (5):

$$\frac{M_t}{M_0} = K_F t^{1/2} + K_a t \quad (5)$$

where K_F and K_a are the kinetic constants of Fickian and anomalous diffusion, respectively.

The LZM release profiles from HAC at acidic pH are depicted in Figure 5.

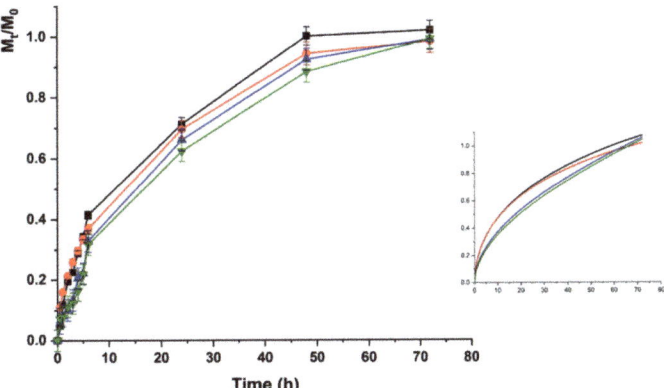

Figure 5. Lysozyme (LZM) release profiles from HAC at pH 5.5 and (■) 0; (●) 12; (▼) 24; and (▲) 48 V. Inset: Fitting curve by Equation (5).

Upon pH increase to 7.4 (Figure 6), the LZM release from HAC can be explained as the result of two opposite phenomena modulating the protein to hydrogel interactions.

The LZM ionization equilibrium moved to the undissociated form, thus, promoting the release, while a prevalence of dissociated COOH functionalities occurred on the polymer network, enhancing the LZM affinity to the hydrogel through a higher density of negative charges. The prevalent phenomenon was the modification of ionized LZM concentration at equilibrium, resulting in an increased release at pH 7.4 vs. 5.5. The lower interactions between LZM and hydrogels at pH 7.4 were responsible for release kinetics, better described by first-order kinetics (higher R^2 values in Table 2), while at pH 5.5 Ritger–Peppas and Peppas–Sahlin diffusion models (Equations (4) and (5)) better fitted the experimental data. Thus, the analysis of data at pH 5.5 was performed by using Equation (5).

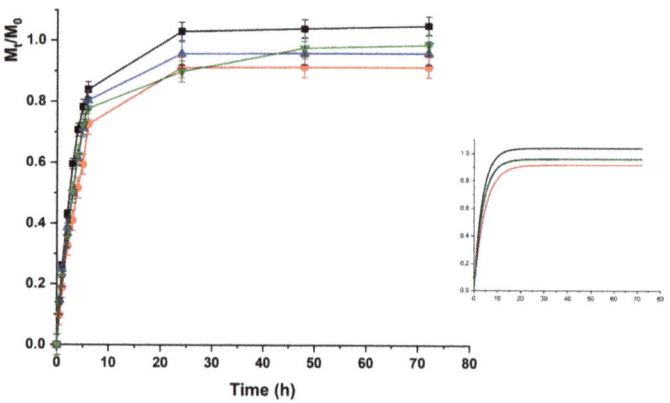

Figure 6. LZM release profiles from HAC at pH 7.4 and (■) 0; (●) 12; (▼) 24; and (▲) 48 V. Inset: Fitting curve by Equation (3).

Table 2. Kinetic parameters for LZM release obtained by applying the kinetic models.

Sample	pH	Voltage	Zero Order		First Order		Ritger–Peppas			Peppas–Sahlin			
			K_0	R^2	K_1	R^2	n	K	R^2	K_F	K_a (10^{-2})	$\frac{K_F}{K_a}$	R^2
HAC	5.5	0	0.0178	0.8384	0.0780	0.9778	0.45	0.1590	0.9894	0.1633	0.43	38	0.9831
		12	0.0170	0.8301	0.0719	0.9490	0.43	0.1678	0.9906	0.1629	0.51	32	0.9949
		24	0.0167	0.8517	0.0621	0.9480	0.47	0.1358	0.9647	0.1417	0.21	67	0.9683
		48	0.0165	0.8655	0.0579	0.9434	0.49	0.1292	0.9670	0.1323	0.14	94	0.9724
HACG	5.5	0	0.0072	0.9490	0.0098	0.9640	0.67	0.0277	0.9808	0.0302	0.31	10	0.9744
		12	0.0155	0.8831	0.0458	0.9783	0.50	0.1117	0.9839	0.1175	0.04	293	0.9859
		24	0.0106	0.8664	0.0187	0.8646	0.48	0.0854	0.9854	0.0871	0.16	54	0.9902
		48	0.0108	0.8587	0.0196	0.8481	0.46	0.0924	0.9858	0.0921	0.17	54	0.9896
HAC	7.4	0	0.0202	0.5454	0.3006	0.9952	0.23	0.4349	0.7824	0.3785	3.10	12	0.9478
		12	0.0176	0.5958	0.1895	0.9783	0.27	0.3251	0.8145	0.2977	2.27	13	0.9555
		24	0.0186	0.5493	0.2495	0.9928	0.23	0.3950	0.7848	0.3456	2.82	12	0.9522
		48	0.0187	0.5702	0.2427	0.9896	0.38	0.3795	0.7999	0.3330	2.64	13	0.9378
HACG	7.4	0	0.0104	0.5154	0.3316	0.9825	0.20	0.2508	0.8266	0.2063	1.75	11	0.9474
		12	0.0186	0.3959	0.5843	0.9915	0.14	0.5839	0.5334	0.4509	4.25	10	0.8118
		24	0.0166	0.5205	0.2992	0.9901	0.22	0.3776	0.7383	0.3249	2.74	11	0.9313
		48	0.0168	0.4862	0.3653	0.9905	0.19	0.4241	0.7335	0.3502	3.06	11	0.9248

The same trend was recorded for HACG, where a faster release was recorded at physiological vs. acidic pH (higher kinetic constants), although the maximum amount of release was around 50% in both cases (Figures 7 and 8).

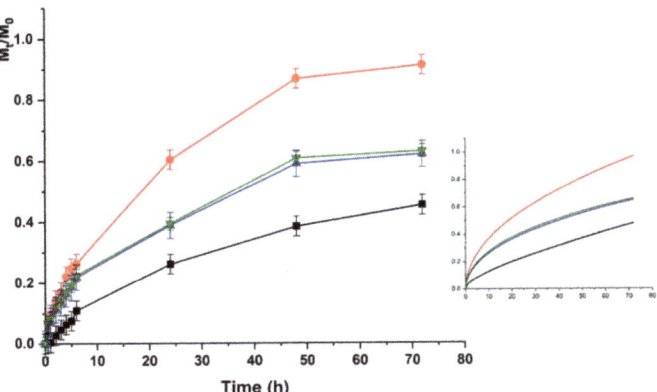

Figure 7. LZM release profiles from HACG at pH 5.5 and (■) 0; (●) 12; (▼) 24; and (▲) 48 V. Inset: Fitting curve by Equation (5).

Here, the presence of GO affected the LZM to hydrogel interaction since GO is able to interact with the bioactive protein via both electrostatic (COOH) and π–π stacking interactions, thus, resulting in higher affinity at both pH conditions.

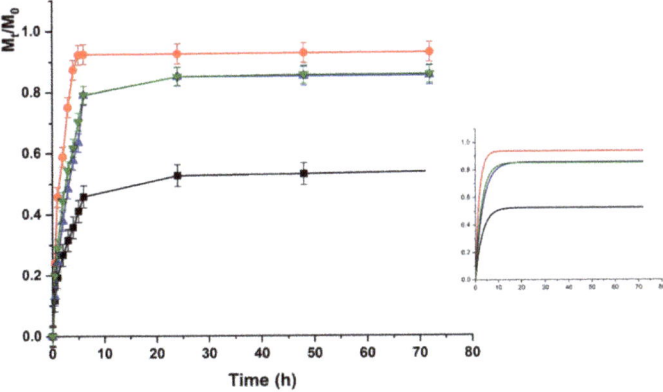

Figure 8. LZM release profiles from HACG at pH 7.4 and (■) 0; (●) 12; (▼) 24; and (▲) 48 V. Inset: Fitting curve by Equation (3).

Further considerations can be done by considering the release upon application of an external electric field. As expected, no significant modification in the release amount and rate was recorded when HAC was used, due to the absence of any electro-conductive component in the polymer network. A different behavior was observed for the HACG hybrid hydrogel. The application of an electric field with different voltages resulted in a fine tuning of the delivery profiles, with the release at physiological pH being faster than that at acidic pH in all voltage conditions, due to the above-mentioned modulation of LZM to hydrogel interactions upon pH variation.

The differences in the release profiles at each selected voltage could be attributed to the modulation of both the swelling degree and the ionization state of the entire system (LZM + HACG) by the electrical stimulation. In detail, a significant increase in the release was recorded when 12 V was applied, as a consequence of the higher degree of swelling of the polymer network, promoting LZM diffusion to the surrounding environment. A further increase in the applied voltage (24 V), carried out to an enhanced ionization of COOH residues, allowed the formation of a higher number of negative charges suitable for interaction with LZM, and thus reduced the release rate/amount. A further increase in the voltage did not result in a significative modification of the release kinetics at both pHs, probably because of the formation of an equilibrium state between the LZM–HACG interaction and the strong osmotic pressure formed across the network hindering the diffusion. Further confirmations of this hypothesis can be obtained by investigating the modification of K_F and K_a kinetic constants in Equation (5) (indicating the diffusional and anomalous contributions, respectively) upon application of an external voltage. In all cases, the Fickian diffusion is the predominant effect ($K_F/K_a > 10$), with the fast release at 12 vs. 0 V determining a significant increase in the K_F/K_a. When 24 and 48 V were applied, the insurgence of the osmotic stress contribution was highlighted by the simultaneous K_F reduction and K_a increase, with the K_F/K_a value reduced by a half.

2.4. Lysozyme Crystallization upon Oxidative Stress Condition

The large wetting tendency of hydrophilic and porous surfaces of hydrogels was successfully employed in protein crystallization by virtue of their ability to reduce the activation energy to nucleation, thus, reducing the induction time and increasing the crystal growth rate [60,61]. This is importance in biomedicine, where significant research efforts have been expended for analyzing tertiary structures of proteins in order collect key information about the molecular mechanisms underlying cellular biological and pathological processes [62]. Nevertheless, chemical reactions occurring during crystallization can lead to poor reproducibility or even failure of the crystallization. Among others, the oxidization process in crystallization droplets is a key phenomenon to be considered, because of the formation of irreversible intermolecular disulfide bridges, oxidation films, and protein precipitates [24]. To address this issue, we explored the possibility of using the CF functionalities of HACG as a scavenging agent against the oxidative stress induced by H_2O_2 during LZM crystallization, preserving the protein tertiary structure.

To prove this hypothesis, the retainment of CF antioxidant potency upon conjugation to Alg and the further co-polymerization in the acrylate network were assessed by determining the available phenolic groups by the Folin–Ciocalteu tests [63], the Trolox (6-hydroxy-2,5,7,8-tetramethylchroman-2-carboxylic acid) equivalent antioxidant capacity (TEAC) [64], and the concentration needed for a 50% decay (IC_{50}) of DPPH (1,1-diphenyl-2-picrylhydrazyl) radical [65].

The determination of available phenolic groups allowed estimating a functionalization degree of 7.7 mg CF per g of AlgCF. The TEAC value gives a clear indication about the number of radicals that can be quenched by a tested antioxidant compound; in our experimental conditions, TEAC values of 1.85 and 1.77 were obtained for free and conjugated CF (referred to the amount of conjugated CF as per Folin–Ciocalteu test), respectively, clearly demonstrating that the scavenging ability of CF was almost unchanged after insertion in the AlgCF side chains. The determination of IC_{50} values for DPPH radical (0.85 mg mL^{-1}, corresponding to 6.7 µg mL^{-1} CF equivalent concentration), close to that of free CF (6.1 µg mL^{-1}, $p > 0.05$) was used as a further confirmation of this statement.

HACG was found to possess an available phenolic content of 0.63 mg CF equivalent per g of dry hydrogel (suggesting an incorporation of 81.8 mg AlgCF per g), a TEAC value of 1.48 (referred to the CF content as per the Folin–Ciocalteu test), and an IC_{50} value of 27.7 mg mL^{-1} in the DPPH assay (17.45 µg mL^{-1} CF equivalent concentration).

Then, the functional hydrogel HACG was employed as substrate for LZM crystallization in standard conditions [66]. The conventional hanging drop crystallization technique was used to investigate the best conditions for obtaining good results in terms of the quality of the lysozyme crystals. In general, in a crystallization process the best results in terms of crystal quality are obtained when the crystals displayed on the drop at the end of the process are few and large. However, several parameters could influence the steps of crystal nucleation and growth. Nucleation starts when the protein solution reaches an optimal level of oversaturation. In an ideal condition, key parameters related to the solution such as pH, temperature, and precipitating agents, lead the protein solution to the narrow area of oversaturation, at which the protein can undergo a spontaneous nucleation. When the crystalline nuclei are formed, and therefore the level of over-saturation is reduced, the metastable zone is reached, where the growth of the crystal is favored [67].

As a general rule, an oversaturation condition is favored by adding suitable precipitating agents to the protein solution, such as PEG, organic solvents, or even inorganic salts such as sodium chloride. These contribute to the achievement of oversaturation in the solution by varying the chemical-physical characteristics (temperature, ionic strength, and pH). In detail, inorganic salts serve this purpose by influencing the ionic strength of the protein solution.

In the case of the crystallization process performed in vapor phase, the polymeric hydrogel film, used as support by virtue of its porosity, promotes the establishment of a balance between the protein drop and the stripping solution present in the well, through the solvent evaporation from the drop to the stripping solution. Under these conditions, the oversaturation needed for nucleation is reached more easily. The best conditions, among those analyzed in this study, were a lysozyme protein solution concentration of 10 mg mL^{-1} in sodium acetate buffer 0.1 M, pH 4.6, and 7% w/v precipitating agent (NaCl) diluted in the same buffer acetate (both in the drop and in the well).

The formation of LZM crystals on the surface of either HAG or HACG is shown in Figure 9, confirming that the CF residues did not interfere with the process.

In the presence of H_2O_2, the induced stress caused the crystallization process to fail when HAG was used as a support, obtaining small powder structures (Figure 10a–c).

Interestingly, when HACG was used as a support, the radical scavenging ability of CF residues counteracted the oxidizing activity of H_2O_2, resulting in the formation of well-defined LZM crystals (Figure 10d–f).

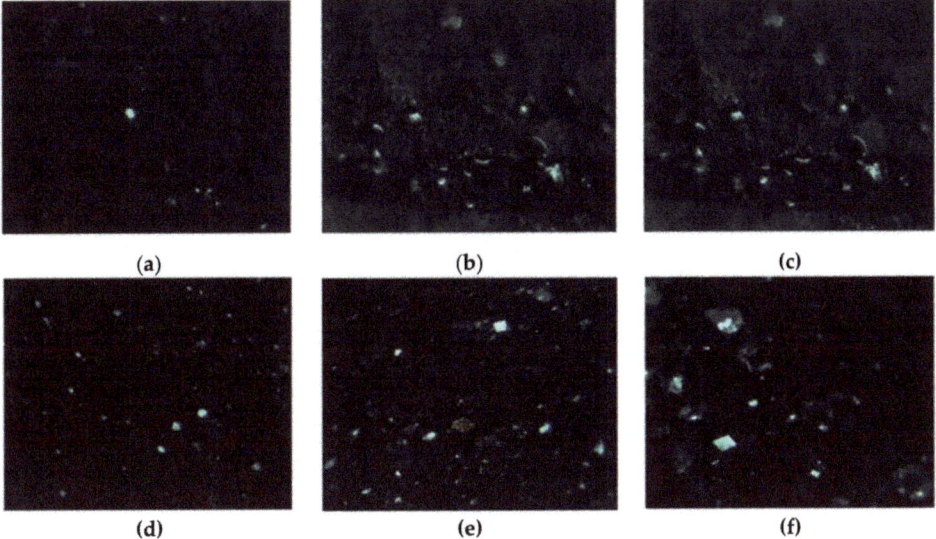

Figure 9. LZM crystals. (**a**–**c**) Observed on HAG; (**d**–**f**) Observed on HACG, after (**a**,**d**) 24, (**b**,**e**) 48, and (**c**,**f**) 72 h.

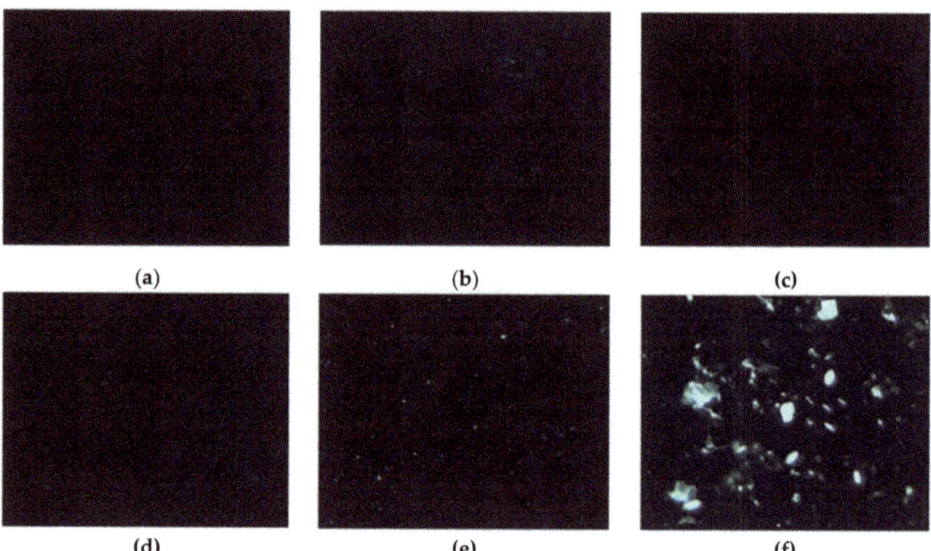

Figure 10. LZM crystals. (**a**–**c**) Observed on HAG; (**d**–**f**) Observed on HACG, after (**a**,**d**) 24, (**b**,**e**) 48, and (**c**,**f**) 72 h in the presence of H_2O_2.

Our preliminary results can be considered to be proof of the potential applicability of HACG as functional material for LZM crystallization, with further experiments aimed to evaluate the applicability for crystallization of more complex oxidizable proteins.

3. Materials and Methods

3.1. Synthesis of Alginate-Caffeic Acid Conjugate

AlgCF was synthesized by means of heterogeneous catalysis involving the use of a previously developed immobilized laccase as solid biocatalyst [43]. Briefly, 100 mg Alg and 10 mg CF were dissolved in 3.5 mL H_2O containing 5% DMSO, and, after the addition of 250 mg biocatalyst (11.5 U), reacted at 37 °C under 70 rpm. After 12 h, the conjugate was purified by dialysis in 6–27/32″ dialysis tubes, MWCO 12,000–14,000 Da (Medicell International LTD, Liverpool, UK) against DMSO/H_2O mixture solution (5%) until complete removal of unreacted CF in the washing media. Then, DMSO was removed by dialysis against water, and the solution was dried with a freeze drier (Micro Modulyo, Edwards Lifesciences, Irvine, CA, USA) to afford a vaporous solid. The presence of CF in the washing media was analyzed by high-pressure liquid chromatography (HPLC) in the following conditions: Jasco PU-2089 Plus liquid chromatography equipped with a Rheodyne 7725i injector (20 µL loop), a Tracer Excel 120 ODS-A column particle size 5 µm, 15 × 0.4 cm (Barcelona, Spain), a mobile phase consisting of acetonitrile-water containing 0.1% phosphoric acid (70:30) running at a flow rate of 1.5 mL min^{-1}, a Jasco UV-2075 HPLC detector operating at 330 nm, and a Jasco-Borwin integrator (Jasco Europe S.R.L., Milan, Italy) [68].

All chemicals were from Merck/Sigma Aldrich, Darmstadt, Germany.

3.2. Synthesis of Hybrid Hydrogel

Hybrid hydrogels were prepared via a previously developed polymerization procedure [51] consisting of preliminary dispersion of 5.0 mg GO in 3.5 mL water containing 50 mg AlgCF water solution by a cup-horn high intensity ultrasonic homogenizer (SONOPULS) with a cylindrical tip (amplitude 70%, time 30 min). Then, 184 mg HEA and 196 mg PEGDMA were added, and the solution was purged with gaseous nitrogen for 20 min. After adding ammonium persulfate (10% *w/w*), the polymerization mixture was allowed to react at 40 °C after placement between two 5.0 × 5.0 cm² glass plates, separated with a Teflon spacer (0.6 mm) and brought together by binder clips. The obtained hybrid hydrogels were extensively washed with water to remove unreacted species, and then dried overnight under vacuum at 40 °C. Control hydrogels were prepared with the same procedure without the insertion of GO (HAC) or replacing AlgCF conjugate with native Alg (HAG).

All chemicals were from Merck/Sigma Aldrich, Darmstadt, Germany.

3.3. Characterization Procedures

3.3.1. Instruments

^{1}H-NMR spectra (300 MHz, D_2O) were recorded using a Bruker Avance 300 (Bruker Italy, Milan, Italy).

Calorimetric analyses were carried out using a DSC200 PC differential scanning calorimeter (Netzsch, Selb, Germany). Following a standard procedure, about 5.0 mg of dried sample was placed in an aluminum pan, and then sealed tightly by an aluminum lid. The thermal analyses were performed from 60 to 300 °C under a dry nitrogen atmosphere with a flow rate of 25 mL min^{-1} and heating rate of 5 °C min^{-1}.

Freeze-dried grounded samples were deposited onto self-adhesive, conducting carbon tape (Plano GmbH, Wetzlar, Germany) and scanning electron microscope images were acquired using a NOVA NanoSEM 200 (0–30 kV) (Thermo Fisher Scientific, Hillsboro, OR, USA).

Static light scattering measurements (Zetasizer Nano ZS instrument, Malvern Panalytical, Malvern, UK) were performed on Alg and AlgCF water solutions with a concentration between 0.1 and 5.0 mg mL^{-1} in order to determine the average molecular weight [69]. All the solutions were filtered using a 0.22 µm filter (Millex®-GV syringe-driven filter unit, Merck/Sigma Aldrich, Darmstadt, Germany), and then placed in quartz cuvettes. The light intensity and its time autocorrelation function were measured at 173° scattering angle after 2 min of equilibration at 25 °C using automatic time settings.

The Debye plots were generated by using Debye's light scattering Equation (6):

$$\frac{KC}{R_\theta} = \frac{1}{M_W} + 2B_{22}C \qquad (6)$$

where R_θ is the excess Rayleigh ratio of the polymer in a solution with a polymer concentration C and M_W is the average molecular weight. K is the optical constant and is defined as reported in the following Equation (7):

$$K = \frac{4\pi^2 n^2}{N_A \lambda^4}\left(\frac{dn}{dc}\right)^2 \qquad (7)$$

where n is the solvent refractive index, dn/dc is the refractive index increment, λ is the wavelength of the incident light, and N_A is the Avogadro's number. The average molecular weight was obtained from the inverse of the intercept of the linear Debye plot of KC/R_θ versus the polymer concentration C.

3.3.2. Antioxidant Tests

The amount of total phenolic equivalents, expressed as CF equivalent per g of sample was determined using Folin–Ciocalteu assay, as reported in [63], using the calibration curve of the free antioxidant. The TEAC values were determined according to a previously reported protocol with slight modifications [64]. In separate experiments, AlgCF and swollen HACG at a CF equivalent concentration of 2.0 µg mL^{-1} were added to 2,2′-azinobis(3-ethylbenzothiazoline-6-sulphonic acid) (ABTS) solution in the 0–1.23 10^{-4} mol L^{-1} concentration range and incubated at 37 °C for 6 min in the dark. Then, the absorbance was measured at 734 nm and the following Equation (8) was used to calculate TEAC:

$$TEAC = \frac{1}{1.9\,[CF]} \qquad (8)$$

where 1.9 is the number of molecules that can be scavenged per mol trolox, [CF] is the CF equivalent concentration (mol L^{-1}) in the sample. The maximal amount of ABTS scavenged by the CF at the tested concentration, C, was calculated by plotting the reduction of ABTS concentration against its initial concentration according to the following Equation (9):

$$y = C\left(1 - e^{-bx}\right) \qquad (9)$$

where x and y are the initial ABTS concentration and the reduction in ABTS concentration, respectively.

Radical scavenging properties were evaluated measuring the inhibition (%) of the stable 2,2′-diphenyl-1-picrylhydrazyl radical (DPPH) radical by AlgCF (from 0.25 to 1.50 mg mL^{-1}) and swollen HACG (from 15.0 to 30.0 mg mL^{-1}), according to the literature protocol and the following Equation (10) [65]:

$$Inhibition(\%) = \frac{A_0 - A_1}{A_0} \times 100 \qquad (10)$$

where A_0 and A_1 are the absorbances of DPPH solution in the absence or presence of polymer samples, respectively.

The water affinity of HAC and HACG was investigated in PBS (pH 7.4) and sodium acetate buffer (pH 5.5) under different voltage conditions (0, 12, 24, and 48 V). Briefly, 1.0 cm^2 specimens were cut from each sample, weighed, and immersed into the swelling medium at 37 °C. Excess water was removed after 24 h, and samples blotted with a tissue to remove surface moisture and weighed. The water content percentage (WR) was expressed by the following Equation (11):

$$WR(\%) = \frac{W_s - W_d}{W_d} \times 100 \qquad (11)$$

where W_s and W_d are the sample weights in their swollen and dry state.

All chemicals were from Merck/Sigma Aldrich, Darmstadt, Germany.

3.3.3. Cytotoxicity Studies

The cytotoxic effects of HACG were assessed on human fibroblast lung cells MRC-5 (ATCC CCL-171). Briefly, cells were cultured in Dulbecco's modified Eagle's medium (DMEM) supplemented with 10% FBS and 1% L-glutamate, grown as a monolayer at 37 °C in 5% CO_2, and seeded into 96-well plates (100 mL per well) at a predetermined density (20,000 cells/well) to achieve 90% confluency by the endpoint of the assay.

For the cytotoxicity determination, HACG was pulverized, and incubated with MRC-5 cells after being suspended in DMEM + FCS 10% (0.1–1.0 mg mL^{-1} range of concentration). After 72 h of incubation, the media containing treatment was replaced with 10% AlamarBlue in fresh media. The metabolic activity was detected by spectrophotometric analysis by assessing the absorbance of AlamarBlue® (difference between 570 nm and 595 nm) using a Bio Rad multiplate reader. Cell viability was determined and expressed as the percentage of viability of untreated control cells.

All chemicals were from Merck/Sigma Aldrich, Darmstadt, Germany.

3.4. Lysozyme Loading and In Vitro Release Studies

LZM was loaded into hybrid hydrogels at 6.0% (by weight) by soaking 50 mg of dried samples (HAC or HACG) with 3.5 mL LZM solution, in water (1.0 mg mL^{-1}) for 3 days, and then samples were dried to a constant weight at reduced pressure in the presence of P_2O_5.

The in vitro LZM release was investigated by dissolution method with alternate shaking both in the absence and in the presence of an external electric voltage (12, 24, and 48 V). In separate experiments, specimens of ~1 cm^2 of loaded hydrogels were weighted and immersed in flasks containing 10 mL PBS (pH 7.4) and sodium acetate buffer (pH 5.5) solutions at 37.0 ± 0.1 °C in a water bath. At suitable time intervals, 1.0 mL release medium was withdrawn and replaced with fresh medium to ensure sink conditions during the experiment. After filtration (Iso-DiscTM Filters PTFE 25–4 25 mm × 0.45 µm, Supelco/Merck, Darmstadt, Germany), released LZM was measured by UV–Vis analysis on an Evolution 201 spectrophotometer (ThermoFisher Scientific, Hillsboro, OR, USA) operating with 1.0 cm quartz cells at 280 nm [70].

All chemicals were from Merck/Sigma Aldrich, Darmstadt, Germany.

3.5. Lysozyme Crystallization

The crystallization tests were performed by conventional hanging drop vapor diffusion method using the prepared thin films of hydrogel as crystallization supports. Lysozyme was dissolved in sodium acetate buffer (0.1 M, pH 4.6) at the initial concentration of 10 mg mL^{-1}. The precipitant and stripping solutions were composed of sodium chloride, NaCl (7.0 wt.%), dissolved in the same buffer. A drop of protein solution (5 µL), pipetted on the surface of the hydrogel membrane and added with an equal volume of precipitant solution, was left equilibrating with 6 mL of stripping solution in the well. The final crystallization solution, after mixing the protein and precipitant solutions, was 5 mg mL^{-1}.

To assure the result reproducibility, 5 replica experiments for each tested condition were carried out. Then, the crystallization system was incubated at 20 ± 0.1 °C before optical microscopy inspection of the droplets after time intervals of 24, 48, and 72 h. Protein crystals were observed under an optical microscope (Axiovert 25, Zeiss, Oberkochen, Germany) equipped with a video camera. The same experimental conditions were used when H_2O_2 (15% *w/v*) was added to the hanged drop and the reservoir solution.

4. Conclusions

We provided experimental evidence that a novel multifunctional hybrid hydrogel is an effective platform to either provide the electro-responsive release of biologically active molecules such as LZM or facilitate its crystallization under oxidative stress. This ability arises from the combination of the peculiar features of the network component; GO was responsible for the electro-conductivity and high affinity to LZM, while the AlgCF conjugate was the functional element with antioxidant properties.

The synthetic strategy consisted of two steps. First, AlgCF was synthesized by enzyme catalysis, and then inserted into the acrylate polymer network together with GO showing high biocompatibility and water affinity. The evaluation of the LSM release profile highlighted a pH- and electro-responsivity reliance, because of the variation of the ionization degree of carboxyl functionalities on AlgCF and GO, and the insurgence of an osmotic pressure within the swollen hydrogels upon application of an electric field.

Finally, the LZM crystallization experiments conducted in the presence of H_2O_2 proved the suitability of hybrid hydrogel to counteract protein denaturation, thus, facilitating the formation of well-defined crystals under oxidative conditions.

Overall, our results have shown the potential to perform subsequent studies for the development of further experimental protocols to evaluate the applicability of the proposed hydrogel system as a support for the delivery or the investigation of the three-dimensional (3D) structure of biologically relevant proteins.

Author Contributions: Conceptualization, G.C. and F.P.N.; data curation, A.L., F.I., G.D.F. and F.P.N.; formal analysis, G.C., A.L. and G.D.F.; investigation, E.P., M.C. and O.V.; methodology, G.C., O.V. and F.P.N.; resources, F.P.N.; supervision, F.I. and F.P.N.; validation, G.D.F. and F.P.N.; visualization,

M.C. and F.I.; writing—original draft, G.C.; writing—review and editing, O.V., A.L. and F.P.N. All authors have read and agreed to the published version of the manuscript.

Funding: This research received no external funding.

Institutional Review Board Statement: Not Applicable.

Informed Consent Statement: Not Applicable.

Data Availability Statement: Not Applicable.

Conflicts of Interest: The authors declare no conflict of interest.

Sample Availability: Samples of the compounds are available from the authors.

References

1. Yi, J.; Choe, G.; Park, J.; Lee, J.Y. Graphene oxide-incorporated hydrogels for biomedical applications. *Polym. J.* **2020**, *52*, 823–837. [CrossRef]
2. Buwalda, S.J.; Boere, K.W.M.; Dijkstra, P.J.; Feijen, J.; Vermonden, T.; Hennink, W.E. Hydrogels in a historical perspective: From simple networks to smart materials. *J. Control. Release* **2014**, *190*, 254–273. [CrossRef]
3. Di Profio, G.; Polino, M.; Nicoletta, F.P.; Belviso, B.D.; Caliandro, R.; Fontananova, E.; De Filpo, G.; Curcio, E.; Drioli, E. Tailored Hydrogel Membranes for Efficient Protein Crystallization. *Adv. Funct. Mater.* **2014**, *24*, 1582–1590. [CrossRef]
4. Tibbitt, M.W.; Dahlman, J.E.; Langer, R. Emerging Frontiers in Drug Delivery. *J. Am. Chem. Soc.* **2016**, *138*, 704–717. [CrossRef] [PubMed]
5. Kost, J.; Langer, R. Responsive polymeric delivery systems. *Adv. Drug. Deliver. Rev.* **2012**, *64*, 327–341. [CrossRef]
6. Wang, L.; He, G.H.; Ruan, X.H.; Zhang, D.S.; Xiao, W.; Li, X.C.; Wu, X.M.; Jiang, X.B. Tailored Robust Hydrogel Composite Membranes for Continuous Protein Crystallization with Ultrahigh Morphology Selectivity. *ACS Appl. Mater. Interfaces* **2018**, *10*, 26653–26661. [CrossRef] [PubMed]
7. Salehi, S.M.; Manju, A.C.; Belviso, B.D.; Portugal, C.A.M.; Coelhoso, I.M.; Mirabelli, V.; Fontananova, E.; Caliandro, R.; Crespo, J.G.; Curcio, E.; et al. Hydrogel Composite Membranes Incorporating Iron Oxide Nanoparticles as Topographical Designers for Controlled Heteronucleation of Proteins. *Cryst. Growth Des.* **2018**, *18*, 3317–3327. [CrossRef]
8. Fernandez-Leiro, R.; Scheres, S.H.W. Unravelling biological macromolecules with cryo-electron microscopy. *Nature* **2016**, *537*, 339–346. [CrossRef]
9. Gu, H.B.; Liu, C.T.; Zhu, J.H.; Gu, J.W.; Wujcik, E.K.; Shao, L.; Wang, N.; Wei, H.G.; Scaffaro, R.; Zhang, J.X.; et al. Introducing advanced composites and hybrid materials. *Adv. Compos. Hybrid Mater.* **2018**, *1*, 1–5. [CrossRef]
10. Wegst, U.G.K.; Bai, H.; Saiz, E.; Tomsia, A.P.; Ritchie, R.O. Bioinspired structural materials. *Nat. Mater.* **2015**, *14*, 23–36. [CrossRef]
11. Jia, X.Q.; Kiick, K.L. Hybrid Multicomponent Hydrogels for Tissue Engineering. *Macromol. Biosci.* **2009**, *9*, 140–156. [CrossRef] [PubMed]
12. Prusty, K.; Swain, S.K. Polypropylene oxide/polyethylene oxide-cellulose hybrid nanocomposite hydrogels as drug delivery vehicle. *J. Appl. Polym. Sci.* **2021**, *138*, 49921. [CrossRef]
13. Hinderer, S.; Layland, S.L.; Schenke-Layland, K. ECM and ECM-like materials—Biomaterials for applications in regenerative medicine and cancer therapy. *Adv. Drug Deliv. Rev.* **2016**, *97*, 260–269. [CrossRef]
14. Iglesias, D.; Bosi, S.; Melchionna, M.; Da Ros, T.; Marchesan, S. The Glitter of Carbon Nanostructures in Hybrid/Composite Hydrogels for Medicinal Use. *Curr. Top. Med. Chem.* **2016**, *16*, 1976–1989. [CrossRef] [PubMed]
15. Gooneh-Farahani, S.; Naimi-Jamal, M.R.; Naghib, S.M. Stimuli-responsive graphene-incorporated multifunctional chitosan for drug delivery applications: A review. *Expert Opin. Drug Deliv.* **2019**, *16*, 79–99. [CrossRef] [PubMed]
16. Zhang, Q.; Deng, H.; Li, H.J.; Song, K.Y.; Zeng, C.; Rong, L. Preparation of Graphene Oxide-Based Supramolecular Hybrid Nanohydrogel Through Host-Guest Interaction and Its Application in Drug Delivery. *J. Biomed. Nanotechnol.* **2018**, *14*, 2056–2065. [CrossRef] [PubMed]
17. Gonzalez-Dominguez, J.M.; Martin, C.; Dura, O.J.; Merino, S.; Vazquez, E. Smart Hybrid Graphene Hydrogels: A Study of the Different Responses to Mechanical Stretching Stimulus. *ACS Appl. Mater. Interfaces* **2018**, *10*, 1987–1995. [CrossRef]
18. Ganguly, S.; Ray, D.; Das, P.; Maity, P.P.; Mondal, S.; Aswal, V.K.; Dhara, S.; Das, N.C. Mechanically robust dual responsive water dispersible-graphene based conductive elastomeric hydrogel for tunable pulsatile drug release. *Ultrason. Sonochem.* **2018**, *42*, 212–227. [CrossRef]
19. Cirillo, G.; Curcio, M.; Spizzirri, U.G.; Vittorio, O.; Tucci, P.; Picci, N.; Iemma, F.; Hampel, S.; Nicoletta, F.P. Carbon nanotubes hybrid hydrogels for electrically tunable release of Curcumin. *Eur. Polym. J.* **2017**, *90*, 1–12. [CrossRef]
20. Kim, B.; Kang, B.; Vales, T.P.; Yang, S.K.; Lee, J.; Kim, H.J. Polyphenol-Functionalized Hydrogels Using an Interpenetrating Chitosan Network and Investigation of Their Antioxidant Activity. *Macromol. Res.* **2018**, *26*, 35–39. [CrossRef]
21. Feng, Y.; Xiao, K.; He, Y.; Du, B.; Hong, J.; Yin, H.; Lu, D.; Luo, F.; Li, Z.; Li, J.; et al. Tough and biodegradable polyurethane-curcumin composited hydrogel with antioxidant, antibacterial and antitumor properties. *Mater. Sci. Eng. C* **2021**, *121*, 111820. [CrossRef] [PubMed]

22. Witzler, M.; Alzagameem, A.; Bergs, M.; El Khaldi-Hansen, B.; Klein, S.E.; Hielscher, D.; Kamm, B.; Kreyenschmidt, J.; Tobiasch, E.; Schulze, M. Lignin-Derived Biomaterials for Drug Release and Tissue Engineering. *Molecules* **2018**, *23*, 1885. [CrossRef]
23. Di Luca, M.; Curcio, M.; Valli, E.; Cirillo, G.; Voli, F.; Butini, M.E.; Farfalla, A.; Pantuso, E.; Leggio, A.; Nicoletta, F.P.; et al. Combining antioxidant hydrogels with self-assembled microparticles for multifunctional wound dressings. *J. Mater. Chem. B* **2019**, *7*, 4361–4370. [CrossRef]
24. Senda, M.; Senda, T. Anaerobic crystallization of proteins. *Biophys. Rev.* **2018**, *10*, 183–189. [CrossRef] [PubMed]
25. Senda, M.; Kishigami, S.; Kimura, S.; Senda, T. Crystallization and preliminary X-ray analysis of the reduced Rieske-type [2Fe-2S] ferredoxin derived from *Pseudomonas* sp. strain KKS102. *Acta Cryst. Sect. F Struct. Biol. Cryst. Commun.* **2007**, *63*, 311–314. [CrossRef] [PubMed]
26. Liu, Y.; Li, Y.F.; Keskin, D.; Shi, L.Q. Poly(beta-Amino Esters): Synthesis, Formulations, and Their Biomedical Applications. *Adv. Health Mater.* **2019**, *8*, 1801359. [CrossRef]
27. Adlington, K.; Nguyen, N.T.; Eaves, E.; Yang, J.; Chang, C.Y.; Li, J.N.; Gower, A.L.; Stimpson, A.; Anderson, D.G.; Langer, R.; et al. Application of Targeted Molecular and Material Property Optimization to Bacterial Attachment-Resistant (Meth)acrylate Polymers. *Biomacromolecules* **2016**, *17*, 2830–2838. [CrossRef]
28. Das, D.; Pham, H.T.T.; Lee, S.; Noh, I. Fabrication of alginate-based stimuli-responsive, non-cytotoxic, terpolymric semi-IPN hydrogel as a carrier for controlled release of bovine albumin serum and 5-amino salicylic acid. *Mat. Sci. Eng. C Mater.* **2019**, *98*, 42–53. [CrossRef]
29. Chen, Y.W.; Lu, C.H.; Shen, M.H.; Lin, S.Y.; Chen, C.H.; Chuang, C.K.; Ho, C.C. In vitro evaluation of the hyaluronic acid/alginate composite powder for topical haemostasis and wound healing. *Int. Wound J.* **2020**, *17*, 394–404. [CrossRef]
30. Zhang, M.; Zhao, X. Alginate hydrogel dressings for advanced wound management. *Int. J. Biol. Macromol.* **2020**, *162*, 1414–1428. [CrossRef]
31. Puscaselu, R.G.; Lobiuc, A.; Dimian, M.; Covasa, M. Alginate: From Food Industry to Biomedical Applications and Management of Metabolic Disorders. *Polymers* **2020**, *12*, 2417. [CrossRef]
32. Sugahara, M. A Technique for High-Throughput Protein Crystallization in Ionically Cross-Linked Polysaccharide Gel Beads for X-Ray Diffraction Experiments. *PLoS ONE* **2014**, *9*, 95017. [CrossRef]
33. Willaert, R.; Zegers, I.; Wyns, L.; Sleutel, M. Protein crystallization in hydrogel beads. *Acta Crystallogr. Sect. D Struct. Biol.* **2005**, *61*, 1280–1288. [CrossRef] [PubMed]
34. Espindola, K.M.M.; Ferreira, R.G.; Narvaez, L.E.M.; Rosario, A.C.R.S.; da Silva, A.H.M.; Silva, A.G.B.; Vieira, A.P.O.; Monteiro, M.C. Chemical and Pharmacological Aspects of Caffeic Acid and Its Activity in Hepatocarcinoma. *Front. Oncol.* **2019**, *9*, 541. [CrossRef]
35. Raja, S.T.K.; Thiruselvi, T.; Aravindhan, R.; Mandal, A.B.; Gnanamani, A. In vitro and in vivo assessments of a 3-(3,4-dihydroxyphenyl)-2-propenoic acid bioconjugated gelatin-based injectable hydrogel for biomedical applications. *J. Mater. Chem. B* **2015**, *3*, 1230–1244. [CrossRef]
36. Goenka, S.; Sant, V.; Sant, S. Graphene-based nanomaterials for drug delivery and tissue engineering. *J. Control. Release* **2014**, *173*, 75–88. [CrossRef] [PubMed]
37. Liu, J.Q.; Cui, L.; Losic, D. Graphene and graphene oxide as new nanocarriers for drug delivery applications. *Acta Biomater.* **2013**, *9*, 9243–9257. [CrossRef] [PubMed]
38. Kuila, T.; Bose, S.; Mishra, A.K.; Khanra, P.; Kim, N.H.; Lee, J.H. Chemical functionalization of graphene and its applications. *Prog. Mater. Sci.* **2012**, *57*, 1061–1105. [CrossRef]
39. Yu, X.W.; Cheng, H.H.; Zhang, M.; Zhao, Y.; Qu, L.T.; Shi, G.Q. Graphene-based smart materials. *Nat. Rev. Mater.* **2017**, *2*, 17046. [CrossRef]
40. Curcio, M.; Farfalla, A.; Saletta, F.; Valli, E.; Pantuso, E.; Nicoletta, F.P.; Iemma, F.; Vittorio, O.; Cirillo, G. Functionalized Carbon Nanostructures Versus Drug Resistance: Promising Scenarios in Cancer Treatment. *Molecules* **2020**, *25*, 2102. [CrossRef]
41. Vittorio, O.; Curcio, M.; Cojoc, M.; Goya, G.F.; Hampel, S.; Iemma, F.; Dubrovska, A.; Cirillo, G. Polyphenols delivery by polymeric materials: Challenges in cancer treatment. *Drug Deliv.* **2017**, *24*, 162–180. [CrossRef]
42. Yang, J.; Sun, J.N.; An, X.J.; Zheng, M.X.; Lu, Z.X.; Lu, F.X.; Zhang, C. Preparation of ferulic acid-grafted chitosan using recombinant bacterial laccase and its application in mango preservation. *RSC Adv.* **2018**, *8*, 6759–6767. [CrossRef]
43. Vittorio, O.; Cojoc, M.; Curcio, M.; Spizzirri, U.G.; Hampel, S.; Nicoletta, F.P.; Iemma, F.; Dubrovska, A.; Kavallaris, M.; Cirillo, G. Polyphenol Conjugates by Immobilized Laccase: The Green Synthesis of Dextran-Catechin. *Macromol. Chem. Phys.* **2016**, *217*, 1488–1492. [CrossRef]
44. Sampaio, S.; Taddei, P.; Monti, P.; Buchert, J.; Freddi, G. Enzymatic grafting of chitosan onto Bombyx mori silk fibroin: Kinetic and IR vibrational studies. *J. Biotechnol.* **2005**, *116*, 21–33. [CrossRef] [PubMed]
45. Fertah, M.; Belfkira, A.; Dahmane, E.M.; Taourirte, M.; Brouillette, F. Extraction and characterization of sodium alginate from Moroccan *Laminaria digitata* brown seaweed. *Arab. J. Chem.* **2017**, *10*, S3707–S3714. [CrossRef]
46. Brus, J.; Urbanova, M.; Czernek, J.; Pavelkova, M.; Kubova, K.; Vyslouzil, J.; Abbrent, S.; Konefal, R.; Horsky, J.; Vetchy, D.; et al. Structure and Dynamics of Alginate Gels Cross-Linked by Polyvalent Ions Probed via Solid State NMR Spectroscopy. *Biomacromolecules* **2017**, *18*, 2478–2488. [CrossRef] [PubMed]

47. Lopez-Martinez, L.M.; Santacruz-Ortega, H.; Navarro, R.E.; Sotelo-Mundo, R.R.; Gonzalez-Aguilar, G.A. A H-1 NMR Investigation of the Interaction between Phenolic Acids Found in Mango (*Manguifera indica cv Ataulfo*) and Papaya (*Carica papaya cv Maradol*) and 1,1-diphenyl-2-picrylhydrazyl (DPPH) Free Radicals. *PLoS ONE* **2015**, *10*, 0140242. [CrossRef] [PubMed]
48. Siddaramaiah; Swamy, T.M.M.; Ramaraj, B.; Lee, J.H. Sodium alginate and its blends with starch: Thermal and morphological properties. *J. Appl. Polym. Sci.* **2008**, *109*, 4075–4081. [CrossRef]
49. Nair, R.M.; Bindhu, B.; Reena, V.L. A polymer blend from Gum Arabic and Sodium Alginate-preparation and characterization. *J. Polym. Res.* **2020**, *27*, 154. [CrossRef]
50. Soares, J.P.; Santos, J.E.; Chierice, G.O.; Cavalheiro, E.T.G. Thermal behavior of alginic acid and its sodium salt. *Eclet. Quim.* **2004**, *29*, 57–63. [CrossRef]
51. Di Luca, M.; Vittorio, O.; Cirillo, G.; Curcio, M.; Czuban, M.; Voli, F.; Farfalla, A.; Hampel, S.; Nicoletta, F.P.; Iemma, F. Electro-responsive graphene oxide hydrogels for skin bandages: The outcome of gelatin and trypsin immobilization. *Int. J. Pharm.* **2018**, *546*, 50–60. [CrossRef] [PubMed]
52. Pasquier, E.; Street, J.; Pouchy, C.; Carre, M.; Gifford, A.J.; Murray, J.; Norris, M.D.; Trahair, T.; Andre, N.; Kavallaris, M. Beta-blockers increase response to chemotherapy via direct antitumour and anti-angiogenic mechanisms in neuroblastoma. *Br. J. Cancer* **2013**, *108*, 2485–2494. [CrossRef] [PubMed]
53. Gogoi, N.; Chowdhury, D. Novel carbon dot coated alginate beads with superior stability, swelling and pH responsive drug delivery. *J. Mater. Chem. B* **2014**, *2*, 4089–4099. [CrossRef]
54. Saikia, A.K.; Aggarwal, S.; Mandal, U.K. Electrically induced swelling and methylene blue release behaviour of poly (N-isopropylacrylamide-co-acrylamido-2-methylpropyl sulphonic acid) hydrogels. *Colloid Polym. Sci.* **2015**, *293*, 3533–3544. [CrossRef]
55. Servant, A.; Bussy, C.; Al-Jamal, K.; Kostarelos, K. Design, engineering and structural integrity of electro-responsive carbon nanotube-based hydrogels for pulsatile drug release. *J. Mater. Chem. B* **2013**, *1*, 4593–4600. [CrossRef]
56. Tan, H.Q.; Jin, D.W.; Qu, X.; Liu, H.; Chen, X.; Yin, M.; Liu, C.S. A PEG-Lysozyme hydrogel harvests multiple functions as a fit-to-shape tissue sealant for internal-use of body. *Biomaterials* **2019**, *192*, 392–404. [CrossRef]
57. Yang, Y.W.; Zhang, C.N.; Cao, Y.J.; Qu, Y.X.; Li, T.Y.; Yang, T.G.; Geng, D.; Sun, Y.K. Bidirectional regulation of i-type lysozyme on cutaneous wound healing. *Biomed. Pharm.* **2020**, *131*, 110700. [CrossRef] [PubMed]
58. Ben Amara, C.; Degraeve, P.; Oulahal, N.; Gharsallaoui, A. pH-dependent complexation of lysozyme with low methoxyl (LM) pectin. *Food Chem.* **2017**, *236*, 127–133. [CrossRef] [PubMed]
59. Unagolla, J.M.; Jayasuriya, A.C. Drug transport mechanisms and in vitro release kinetics of vancomycin encapsulated chitosan-alginate polyelectrolyte microparticles as a controlled drug delivery system. *Eur. J. Pharm. Sci.* **2018**, *114*, 199–209. [CrossRef] [PubMed]
60. Fermani, S.; Falini, G.; Minnucci, M.; Ripamonti, A. Protein crystallization on polymeric film surfaces. *J. Cryst. Growth* **2001**, *224*, 327–334. [CrossRef]
61. Liu, Y.X.; Wang, X.J.; Lu, J.; Ching, C.B. Influence of the roughness, topography, and physicochemical properties of chemically modified surfaces on the heterogeneous nucleation of protein crystals. *J. Phys. Chem. B* **2007**, *111*, 13971–13978. [CrossRef] [PubMed]
62. Escolano-Casado, G.; Contreras-Montoya, R.; Conejero-Muriel, M.; Castellvi, A.; Juanhuix, J.; Lopez-Lopez, M.T.; de Cienfuegos, L.A.; Gavira, J.A. Extending the pool of compatible peptide hydrogels for protein crystallization. *Crystals* **2019**, *9*, 244. [CrossRef]
63. Zare, M.; Sarkati, M.N.; Tashakkorian, H.; Partovi, R.; Rahaiee, S. Dextran-immobilized curcumin: An efficient agent against food pathogens and cancer cells. *J. Bioact. Compat. Polym.* **2019**, *34*, 309–320. [CrossRef]
64. Arts, M.J.T.J.; Dallinga, J.S.; Voss, H.P.; Haenen, G.R.M.M.; Bast, A. A new approach to assess the total antioxidant capacity using the TEAC assay. *Food Chem.* **2004**, *88*, 567–570. [CrossRef]
65. Li, Q.; Li, Q.; Tan, W.Q.; Zhang, J.J.; Guo, Z.Y. Phenolic-containing chitosan quaternary ammonium derivatives and their significantly enhanced antioxidant and antitumor properties. *Carbohyd. Res.* **2020**, *498*, 108169. [CrossRef] [PubMed]
66. Di Profio, G.; Curcio, E.; Cassetta, A.; Lamba, D.; Drioli, E. Membrane crystallization of lysozyme: Kinetic aspects. *J. Cryst. Growth* **2003**, *257*, 359–369. [CrossRef]
67. Li, G.P.; Xiang, Y.; Zhang, Y.; Wang, D.C. A simple and efficient innovation of the vapor-diffusion method for controlling nucleation and growth of large protein crystals. *J. Appl. Cryst.* **2001**, *34*, 388–391. [CrossRef]
68. Iranshahi, M.; Amanzadeh, Y. Rapid isocratic HPLC analysis of caffeic acid derivatives from *Echinacea purpurea* cultivated in Iran. *Chem. Nat. Compd.* **2008**, *44*, 190–193. [CrossRef]
69. Pantuso, E.; Mastropietro, T.F.; Briuglia, M.L.; Gerard, C.J.J.; Curcio, E.; ter Horst, J.H.; Nicoletta, F.P.; Di Profio, G. On the Aggregation and Nucleation Mechanism of the Monoclonal Antibody Anti-CD20 Near Liquid-Liquid Phase Separation (LLPS). *Sci. Rep.* **2020**, *10*, 8902. [CrossRef] [PubMed]
70. Farinha, S.; Moura, C.; Afonso, M.D.; Henriques, J. Production of Lysozyme-PLGA-Loaded Microparticles for Controlled Release Using Hot-Melt Extrusion. *AAPS PharmSciTech* **2020**, *21*, 274. [CrossRef]

MDPI
St. Alban-Anlage 66
4052 Basel
Switzerland
Tel. +41 61 683 77 34
Fax +41 61 302 89 18
www.mdpi.com

Molecules Editorial Office
E-mail: molecules@mdpi.com
www.mdpi.com/journal/molecules

www.ingramcontent.com/pod-product-compliance
Lightning Source LLC
LaVergne TN
LVHW070148100526
838202LV00015B/1913